SAMER ADEEB
UNIVERSITY OF ALBERTA

Kendall Hunt
publishing company

Kendall Hunt
publishing company

www.kendallhunt.com
Send all inquiries to:
4050 Westmark Drive
Dubuque, IA 52004-1840

ISBN 978-0-7575-8879-2

Printed in the United States of America
10 9 8 7 6 5 4 3 2 1

Contents

Preface

Developed as an evolution from classical mechanics, solid mechanics applies Newton's equations of equilibrium to predict the behavior of deformable objects under various mechanical loads. However, because of the complexity of the required mathematics, its applications were limited. As a result, solid mechanics was a theoretical discipline with a limited set of applications. It was not until the second half of the twentieth century that the advancement of the computational power enabled a wide range of applications for solid mechanics. Consequently, the applications of solid mechanics evolved from modeling objects with uniform geometries, such as beams under small deformations, to applications as complex as modeling the large deformation of rubber.

Of the various numerical methods, Finite Element Analysis (FEA) emerged as a natural choice to handle the equations of solid mechanics. Today, FEA is used in almost every discipline that involves modeling the deformation of an object due to mechanical loading. The development of commercial FEA software packages has extended the application of the method to many fields of mechanical design and research, including aerospace, the auto industry, bridges, buildings, and even medical devices. As a result, the past two decades witnessed an explosion in available FEA software packages that expanded the reach of the tools through hiding the theory's complexity.

With the expanded reach comes a price. FEA enables its users to make decisions regarding the mechanical behavior of an object without possessing the adequate solid mechanics background. The results can be prone to misjudgements. One such example is the use of the "von–Mises stress"—a stress measure that governs the development of damage in metals—to describe the stress in bone and other materials. Von–Mises stress is an insufficient measure for this application because it does not differentiate between tension and compression—a necessary differentiation for modeling bones. As a result, many users are interested in understanding the theory underneath these tools. I hope this book provides the theoretical background needed for users who plan on developing or using FEA models.

This book in its current form evolved as a combination of the notes I have used to teach undergraduate and graduate courses in solid mechanics and finite element analysis. I have always called for trying to present the particular examples to the student before teaching the general or abstract principles. Students should go through the same thought process that was required by scientists and theoreticians to reach the abstract principles. I believe that by giving the students the abstract principles before the examples, we deprive them from the empowering feeling of the deep understanding of the abstract concepts they can reach by themselves. However, after reading what I have written, I realized that I have followed the same path that many others have followed, starting with the abstract and then giving the examples afterward. Perhaps this is most evident in starting the first chapter with the abstract definition of linear vector spaces. However, and mostly due to the constant requests of many of my students, I have tried to include as many examples as possible. The examples rely on the very powerful Mathematica software.

While I want students to understand the underlying theory for achieving results, I certainly do not want them to waste their time manipulating equations and formulas. The purpose of this book is not to teach the students how to calculate the eigenvalues or the eigenvectors, but rather to learn what they mean and understand their importance. As a result, I chose to use Mathematica to liberate the students from the cumbersome tasks of manipulating equations and formulas. Choosing Mathematica, naturally, comes with the burden of learning a new tool; however, I have tried to introduce the logic behind the software in a step–by–step approach in hopes that the students are able to quickly and easily adopt the software. The Mathematica code implemented within the text is formatted as follows:

```
(*Any boxed text formatted with this font is a piece of code that can be
directly input in a Mathematica notebook*)
```

The book starts with a chapter on introductory linear algebra, which I consider to be the most important chapter in the whole book. I recommend that students interested in a deep understanding of the concepts of stresses and strains to devote the time to understand the first chapter. Once the first chapter is mastered, the remaining chapters will become mere applications—I tried to the best of my ability to relate the concepts in the subsequent chapters to the topics in the first chapter. In the final chapter, the introduction to FEA, I present the FEA as just an example of the different methods capable of solving Newton's equations of equilibrium applied to solid objects. I was hoping that, in this presentation, the students can develop an understanding of the method in light of the theoretical equations behind it.

Finally, I would like to acknowledge the support and help I received from my wife and best friend, Lindsey Westover. She has spent almost as many hours proofreading and correcting the manuscript as I spent developing it. Without her love, encouragement, and dedication, this book would have not come to existence. I also would like to acknowledge the help and support of my brother and my other best friend, Ramy Adeeb. His critical opinion and advice along the way have been so important in the development of this manuscript.

Samer M. Adeeb
July 25, 2011. Edmonton, AB, Canada

Mathematical Symbols

The following table contains a list of the mathematical symbols that are most commonly used throughout the text along with some illustrative examples. The student is expected to carefully study these symbols before attempting to read the text.

Symbol	Description	Additional information/examples
$\mathbb{R}$	The set of Real numbers	
$\{\ldots\}$	A set of elements	$A = \{1,2,5\}$ $A = \{x_1, x_2, x_3\} = \{x_i\}^3_{i=1}$
$\in$	element of (belongs to)	$5 \in \mathbb{R}$
$\subset$	subset of	$\{5,3,2.5\} \subset \mathbb{R}$ $A = \{1,2,3,\ldots.\}, A \subset \mathbb{R}$
$f: A \to B$	f is a function (mapping) from set A to set B	$A = [0,1]\, f: \mathrm{A} \to \mathbb{R}$ is equivalent to $\forall x \in A, f(x) \in \mathbb{R}$
$f: x \in A \to y \in B$	f maps the element x in set A to the element y in set B	$A = [0,1], f: x \in \mathrm{A} \to y \in \mathbb{R}$ is equivalent to $A = [0,1], \forall x \in A: \exists y = f(x) \in \mathbb{R}$
$A \times B$	The set of ordered pairs made of elements of A and B	$A = \{1,2\}, B = \{4,5,6\}$ $A \times B = \{(1,4), (1,5), (1,6), (2,4), (2,5), (2,6)\}$
$\forall$	For every	$A \subset \mathbb{R}, \forall x \in A: x > 0$ Therefore, any element in A if exists has to be greater than 0
s.t.	such that	
$\exists$	There exists	$A \subset \mathbb{R}, \exists x \in A: x > 0$ Therefore, there is an element in A that is greater than 0
$\exists!$	There exists exactly one (unique)	$A \subset \mathbb{R}, \exists! x \in A \, s.t. \, x > 0$ Therefore, there is only one element in the set A that is greater than 0
$+: X \times X \to X$	The operation $+$ maps every ordered pair element in the set $X \times X$ to another element in X	$+: \mathbb{R} \times \mathbb{R} \to \mathbb{R}$ Meaning: $\forall x, y \in \mathbb{R}: z = x + y \in \mathbb{R}$
$\Leftrightarrow$	If an only if (iff). The two statements on both sides are equivalent	$\mathrm{A} \subset \mathbb{R} \Leftrightarrow \forall x \in A: x \in \mathbb{R}$
$\Rightarrow$	The statement on the left side leads to the statement on the right side	$f = 4 \Rightarrow f \in \mathbb{R}$
$\{x \mid p(x)\}$	The set of elements x satisfying the property $p(x)$	$\mathrm{V} = \{x \mid x \in \mathbb{R}, 0 \leq x \leq 1\} \Rightarrow V = [0,1]$

CHAPTER 1

Introductory Linear Algebra

Linear Algebra is one of the most important mathematical tools in engineering. The students of every engineering discipline use the fundamentals of linear algebra to solve for unknowns in systems of linear equations. In particular, continuum mechanics can be considered as a direct application of Linear Algebra combined with vector calculus. The term Linear Algebra entails two fundamental ingredients: linear spaces and linear maps between linear spaces. In this chapter, the definitions of each ingredient will be presented, along with examples to clarify the concepts. The basic definitions are usually presented for completeness. An engineering student is not required to prove the concepts, but rather, to understand them. The student will be required to solve different problems using hand calculations and the Mathematica software.

1.1. LINEAR VECTOR SPACES

1.1.1. Basic Definition

A SET X is called a **linear vector space over** the field of real numbers $\mathbb{R}$ if the two operations (addition) $+ : X \times X \to X$ and (multiplication by a scalar) $\cdot : \mathbb{R} \times X \to X$ are defined and satisfy the following axioms:

$$\forall x, y, z \in X, \forall \alpha, \beta, \gamma \in \mathbb{R}:$$

1. Commutativity: $x + y = y + x$
2. Associativity: $(x + y) + z = x + (y + z)$
3. Distributivity of vector addition over scalar multiplication: $\alpha \cdot (x + y) = (\alpha \cdot x) + (\alpha \cdot y)$
4. Distributivity of scalar addition over scalar multiplication: $(\alpha + \beta) \cdot x = (\alpha \cdot x) + (\beta \cdot x)$
5. Compatibility of scalar multiplication with field multiplication: $\alpha \cdot (\beta \cdot x) = (\alpha \times \beta) \cdot x$
6. Identity/Zero element: There exists $\hat{0} \in X$ such that $\forall x \in X: x + \hat{0} = x$

7. Inverse element of addition: $\forall x \in X{:}\exists \tilde{x} \in X$ such that $x + \tilde{x} = \hat{0}$ ($\tilde{x}$ is represented by $-x$)
8. Identity element of scalar multiplication: $\forall x \in X{:}1 \cdot x = x$

The elements of a linear vector space are called **vectors**.

Problem: Use the above axioms to prove the following statements:

1. If $\hat{0}$ exists, then it is unique.
2. If $\tilde{x}$ exists, then it is unique.
3. $\forall \alpha \in \mathbb{R}$ and $\forall x \in X$ one has $0 \cdot x = \hat{0}$ and $\alpha \cdot \hat{0} = \hat{0}$.
4. $\tilde{x} = -1 \cdot x$.

(The symbol $\hat{0}$ is used to describe the zero vector of X, and it should not be confused with the scalar value $0 \in \mathbb{R}$)

Solution:

Statement 1: To prove that $\hat{0}$ is unique, assume that there are two zero elements θ_1 and θ_2 satisfying axiom 6 above, then, $\forall x \in X$:

$$x = x + \theta_1 \quad \text{and} \quad x = x + \theta_2$$

Substituting θ_2 for x in the first equation and θ_1 for x in the second equation leads to:

$$\theta_1 = \theta_2$$

Statement 2: The uniqueness of inverse element of addition can be proven similarly. Let $x \in X$ and, assuming the existence of $\tilde{x}_1$ and $\tilde{x}_2$ satisfying axiom 7 above:

$$\hat{0} = x + \tilde{x}_1 \quad \text{and} \quad \hat{0} = x + \tilde{x}_2$$

Adding $\tilde{x}_2$ to the first equation and $\tilde{x}_1$ to the second equation leads to the uniqueness of the inverse element of addition:

$$\hat{0} + \tilde{x}_2 = \hat{0} + \tilde{x}_1 \Longrightarrow \tilde{x}_2 = \tilde{x}_1$$

Note that the uniqueness of $\hat{0}$ and $\tilde{x}$ leads to the general result:

$$x + y_1 = x + y_2 \Longleftrightarrow y_1 = y_2$$

Statement 3: Let $x \in X$. Axioms 4 and 8 above lead to the following:

$$x = 1 \cdot x = (1 + 0) \cdot x = x + (0 \cdot x)$$

The uniqueness of the zero vector can be used to show the required result:

$$x + (0 \cdot x) = x + \hat{0} \Longrightarrow 0 \cdot x = \hat{0}$$

Additionally, let $x \in X, \alpha \in R$, and then, using axiom 5:

$$\alpha \cdot \hat{0} = \alpha \cdot (0 \cdot x) = (\alpha \times 0) \cdot x = 0 \cdot x = \hat{0}$$

Statement 4: Let $x \in X$, and then, using the uniqueness of the inverse element of x:

$$x + \tilde{x} = \hat{0} = 0 \cdot x = (1 - 1) \cdot x = x + (-1) \cdot x \Longrightarrow \tilde{x} = (-1) \cdot x$$

The inverse element of x is denoted $(-1) \cdot x = -x$

1.1.2. Subspaces

A SET Y is called a subspace of a linear vector space X (over $\mathbb{R}$) if $Y \subset X$ and Y is closed under addition and multiplication by a scalar. i.e., $\forall x, y \in Y, \forall \alpha, \beta \in \mathbb{R}: \alpha x + \beta y \in Y$.

1.1.3. Dimension and Basis of a Linear Vector Space

Let V be a linear vector space over $\mathbb{R}$.

DEFINITION **Linear independence:** A set of vectors $\{v_1, v_2, \ldots, v_n\} \subset V$ is said to be linearly inde– pendent if none of the vectors of the set can be expressed in terms of the other elements in the set; i.e., the only combination that would produce the zero element is the trivial combination:

$$\alpha_1 v_1 + \alpha_2 v_2 + \cdots + \alpha_n v_n = \hat{0} \Longleftrightarrow \forall i: \alpha_i = 0$$

Otherwise, the set of vectors is termed **linearly dependent**.

DEFINITION **Basis:** A basis B of a linear vector space V is a set of **linearly independent nonzero vectors** such that $\forall x \in V: x$ can be expressed as a linear combination of the elements of the set.

Alternatively: $B = \{e_1, e_2, \ldots, e_n\}$ and $\forall i: e_i \neq \hat{0}$ is a basis of V **if and only if**:

- $\alpha_1 e_1 + \alpha_2 e_2 + \cdots + \alpha_n e_n = \hat{0} \Longleftrightarrow \forall i: \alpha_i = 0$
- $\forall x \in V: \exists x_i \in \mathbb{R}$ such that: $x = \sum_{i=1}^{n} x_i e_i$

DEFINITION **Dimension:** The dimension of a linear vector space V is the number of elements in its basis set.

Problem: Show that if $B = \{e_1, e_2, \ldots, e_n\}$ is a basis of V, then every element of V has a unique decomposition in terms of the elements of B. In other words, show that $\forall x \in V: \exists! x_i \in \mathbb{R}$ such that: $x = \sum_{i=1}^{n} x_i e_i$

Solution: Assume that the decomposition in terms of the elements of the basis set is not unique. Let $x \in V$. Assume that there are two decompositions for x in terms of the basis elements; i.e., assume $\exists x_i, y_i \in \mathbb{R}$ such that:

$$x = \sum_{i=1}^{n} x_i e_i = \sum_{i=1}^{n} y_i e_i$$

We also have that the zero vector is unique:

$$x - x = \hat{0} = \sum_{i=1}^{n} (x_i - y_i)\, e_i$$

However, since the basis set is linearly independent, the only linear combination producing the zero vector is the trivial combination:

$$\forall i{:}(x_i - y_i) = 0 \Longrightarrow x_i = y_i$$

This proves the uniqueness of the decomposition; i.e., $\forall x \in V{:}\exists! x_i \in \mathbb{R}$ such that: $x = \sum_{i=1}^{n} x_i e_i$

1.1.4. Examples of Linear Vector Spaces

1.1.4.a. $\mathbb{R}^n$ Vector Space

For the majority of engineering purposes, the important examples of linear vector spaces are $\mathbb{R}^2 = \mathbb{R} \times \mathbb{R}$, $\mathbb{R}^3 = \mathbb{R} \times \mathbb{R} \times \mathbb{R}$ and in general $\mathbb{R}^n = \mathbb{R} \times \mathbb{R} \times \mathbb{R} \ldots$ (n times), where $n \in \mathbb{N}$.

An element of $\mathbb{R}^2$ is a list of two numbers or a vector of two dimensions and can be written as:

$$x = \{1, 5\} = \begin{pmatrix} 1 \\ 5 \end{pmatrix}$$

Notice that the notation here does not differentiate between a row vector and a column vector, except when a matrix is multiplied by a vector. The "addition" and "multiplication by a scalar" operations in $\mathbb{R}^2$ are defined such that the corresponding numbers in the list are added together and each number in the list is multiplied by the scalar, respectively. Let $x = \{1, 5\}$, $y = \{-1, 3\}$, $\alpha = 2$, $\beta = -1$, and then:

$$x + y = \{0, 8\} = \begin{pmatrix} 0 \\ 8 \end{pmatrix}$$

$$\alpha x + \beta y = \{2, 10\} + \{1, -3\} = \{3, 7\} = \begin{pmatrix} 3 \\ 7 \end{pmatrix}$$

$$\tilde{y} = -y = \{1, -3\} = \begin{pmatrix} 1 \\ -3 \end{pmatrix}$$

$$y + \tilde{y} = \{0, 0\} = \begin{pmatrix} 0 \\ 0 \end{pmatrix}$$

Notice that the zero element $\hat{0}$ in $\mathbb{R}^2$ is the list $\{0, 0\} = \begin{pmatrix} 0 \\ 0 \end{pmatrix}$.

Basis:

There are infinitely many possible choices of basis in any linear vector space. The traditional choices for $\mathbb{R}^2$ (and similarly for $\mathbb{R}^3$) are $e_1 = \begin{pmatrix} 1 \\ 0 \end{pmatrix}$ and $e_2 = \begin{pmatrix} 0 \\ 1 \end{pmatrix}$. These two vectors indeed form a basis since the following is satisfied:

- Linear Independence: $\alpha_1 e_1 + \alpha_2 e_2 = \begin{pmatrix} \alpha_1 \\ \alpha_2 \end{pmatrix} = \hat{0} \Leftrightarrow \alpha_1$ and $\alpha_2 = 0$
- $\forall x \in \mathbb{R}^2 : x = \begin{pmatrix} x_1 \\ x_2 \end{pmatrix}$. i.e. x can be *uniquely* written as the linear combination:
  $x = x_1 e_1 + x_2 e_2$

Problem 1: Show that $e_1 = \begin{pmatrix} 1 \\ 1 \end{pmatrix}$ and $e_2 = \begin{pmatrix} 2 \\ 2 \end{pmatrix}$ do not form a basis for $\mathbb{R}^2$

Solution: The first requirement of a basis set is that the vectors have to be linearly independent. However, it is obvious that e_1 and e_2 are linearly dependent. To view this, we first write a general linear combination of the vectors e_1 and e_2:

$$\alpha_1 e_1 + \alpha_2 e_2 = \begin{pmatrix} \alpha_1 + 2\alpha_2 \\ \alpha_1 + 2\alpha_2 \end{pmatrix}$$

It is obvious that at least one nontrivial linear combination is producing the zero vector. For example, setting $\alpha_1 = 2$ and $\alpha_2 = -1$ produces the zero vector and, thus, e_1 and e_2 are linearly dependent, indicating that $B = \{e_1, e_2\}$ cannot from a basis. Notice that there are many vectors in $\mathbb{R}^2$ that cannot be expressed as a linear combination of e_1 and e_2 **(Can you find an example?)**.

Problem 2: Show that $e_1 = \begin{pmatrix} 1 \\ 1 \end{pmatrix}$ and $e_2 = \begin{pmatrix} -1 \\ 1 \end{pmatrix}$ form a basis for $\mathbb{R}^2$ and find the unique expansion of $x = \{10, 12\}$ in the basis set $B = \{e_1, e_2\}$.

Solution: $B = \{e_1, e_2\}$ forms a basis set since the two requirements for a basis set are satisfied:

- $\alpha_1 e_1 + \alpha_2 e_2 = \begin{pmatrix} \alpha_1 - \alpha_2 \\ \alpha_1 + \alpha_2 \end{pmatrix} = \hat{0}$ is possible **if and only if** $\alpha_1 = 0$ and $\alpha_2 = 0$ indicating that e_1 and e_2 are linearly independent.
- $\forall x \in \mathbb{R}^2 : x = \begin{pmatrix} x_1 \\ x_2 \end{pmatrix}$. x can be written with the unique linear combination

  $x = \alpha_1 e_1 + \alpha_2 e_2 = \begin{pmatrix} \alpha_1 - \alpha_2 \\ \alpha_1 + \alpha_2 \end{pmatrix}$. The components α_1 and α_2 can be obtained as follows:

  $$\begin{pmatrix} x_1 \\ x_2 \end{pmatrix} = \begin{pmatrix} \alpha_1 - \alpha_2 \\ \alpha_1 + \alpha_2 \end{pmatrix} \Longrightarrow \begin{pmatrix} \alpha_1 \\ \alpha_2 \end{pmatrix} = \frac{1}{2}\begin{pmatrix} x_1 + x_2 \\ -x_1 + x_2 \end{pmatrix}$$

The unique expansion of the vector $x = \begin{pmatrix} 10 \\ 12 \end{pmatrix}$ in the basis set B can be found by substituting the values of 10 and 12 for x_1 and x_2 in the above equations, which yields:

$$\begin{pmatrix} \alpha_1 \\ \alpha_2 \end{pmatrix} = \frac{1}{2}\begin{pmatrix} x_1 + x_2 \\ -x_1 + x_2 \end{pmatrix} = \frac{1}{2}\begin{pmatrix} 22 \\ -2 \end{pmatrix} = \begin{pmatrix} 11 \\ -1 \end{pmatrix}$$

Thus, $x = 11e_1 - e_2$

Problem 3: Show that the set $V = \{\alpha e_1 | e_1 = \{1, 0\}, \alpha \in \mathbb{R}\} = \text{span}\{e_1\}$ forms a subspace in $\mathbb{R}^2$.

Note that the set span $\{e_1\}$ is the set composed of all the possible linear combinations of the vectors in brackets.

Solution: The set V is the set of vectors that have the form $x = \{\alpha, 0\}$ where α is any real number. The set V is a subset of $\mathbb{R}^2$. To show that this set forms a subspace means we must show that this set is closed under addition and multiplication by a scalar. Indeed, if any two arbitrary vectors (x and y) in V are added together:

$$x = \begin{pmatrix} x_1 \\ 0 \end{pmatrix}, \quad y = \begin{pmatrix} y_1 \\ 0 \end{pmatrix}, \quad x + y = \begin{pmatrix} x_1 + y_1 \\ 0 \end{pmatrix}$$

the resulting vector $x + y$ is also an element of V and, thus, we say that the set V is closed under addition.

If a vector $x \in V$ is multiplied by a scalar α, the resulting vector is also an element of V:

$$x = \begin{pmatrix} x_1 \\ 0 \end{pmatrix}, \quad \alpha x = \begin{pmatrix} \alpha x_1 \\ 0 \end{pmatrix}$$

Thus, we say that the set V is closed under multiplication by a scalar. Therefore, V is a subspace of $\mathbb{R}^2$.

▶ **BONUS EXERCISE:** Does the set $V = \{\alpha e_1 \mid e_1 = \{1, 0\}, 0 \le \alpha \le 1\}$ form a subspace in $\mathbb{R}^2$? Explain your answer.

1.1.4.b. $\mathbb{M}^n$ Vector Space

Another example of linear vector spaces is the space of square matrices $\mathbb{M}^n$. For example, $\mathbb{M}^2$, the space of all square matrices that have the dimensions 2×2, is a linear vector space. The elements of this space are lists of two rows and two columns, and the "addition" and "multiplication by a scalar" operations are defined similarly. Let $x = \{\{1, 5\}, \{2, 3\}\} = \begin{pmatrix} 1 & 5 \\ 2 & 3 \end{pmatrix}$, $Y = \{\{2, 2\}, \{7, 8\}\} = \begin{pmatrix} 2 & 2 \\ 7 & 8 \end{pmatrix}$, $\alpha = 2, \beta = -1$, then:

$$X + Y = \begin{pmatrix} 3 & 7 \\ 9 & 11 \end{pmatrix}$$

$$\alpha X + \beta Y = \begin{pmatrix} 0 & 8 \\ -3 & -2 \end{pmatrix}$$

$$\tilde{Y} = -Y = \begin{pmatrix} -2 & -2 \\ -7 & -8 \end{pmatrix}$$

$$Y + \tilde{Y} = \begin{pmatrix} 0 & 0 \\ 0 & 0 \end{pmatrix}$$

Notice that the zero element $\hat{0}$ in $\mathbb{M}^2$ is the list $\begin{pmatrix} 0 & 0 \\ 0 & 0 \end{pmatrix}$.

Basis:

The traditional choice of basis in $\mathbb{M}^2$ (and similarly for $\mathbb{M}^n$) is $E_1 = \begin{pmatrix} 1 & 0 \\ 0 & 0 \end{pmatrix}$, $E_2 = \begin{pmatrix} 0 & 1 \\ 0 & 0 \end{pmatrix}$, $E_3 = \begin{pmatrix} 0 & 0 \\ 1 & 0 \end{pmatrix}$ and $E_4 = \begin{pmatrix} 0 & 0 \\ 0 & 1 \end{pmatrix}$

1.1.4.c. Function Spaces

The set of all real valued functions defined on the interval [0, 1] can be written as follows:

$$V = \left\{ f{:}[0,1] \to \mathbb{R} \,\middle|\, \int_0^1 f(x)dx < \infty \right\}$$

This set satisfies all the axioms of a linear vector space over $\mathbb{R}$ (**Why? What is the dimension of V?**).

1.1.4.d. Counter Examples

The following sets do not satisfy one or more of the axioms of a linear vector space (**Why?**):

- The set of all possible colors
- The set $B \subset \mathbb{R}{:}B = \{x | 1 \leq x \leq 2\}$ (i.e., the set of all the real numbers in the interval [1,2])

1.1.5. Notation in $\mathbb{R}^3$ and $\mathbb{M}^3$

In solid mechanics, we will mostly be dealing with elements of $\mathbb{R}^3$ and $\mathbb{M}^3$.

If $x \in \mathbb{R}^3$ then x has three different scalar components and will be normally represented as x_1, x_2 and x_3 such that $x = \begin{pmatrix} x_1 \\ x_2 \\ x_3 \end{pmatrix}$. We can also use x_i to represent a general component where i can take any value between 1 and 3.

If $M \in \mathbb{M}^3$ then M has nine different components and will be normally represented as $M_{11}, M_{12}, \ldots etc$ such that $M = \begin{pmatrix} M_{11} & M_{12} & M_{13} \\ M_{21} & M_{22} & M_{23} \\ M_{31} & M_{32} & M_{33} \end{pmatrix}$. We can also use M_{ij} to represent a general component where i and j can take any values between 1 and 3.

1.1.6. Additional Operations in $\mathbb{R}^n$ (Norm, Distances, Dot Product)

1.1.6.a. Norm

"Norm" is a function that assigns a strictly positive number to every nonzero vector. The norm is used to describe the size of elements in a linear vector space.

Let V be a linear vector space over $\mathbb{R}$. A norm is defined as a function $\|\ \|:V \rightarrow [0, \infty[$ that satisfies the following properties. $\forall x, y \in V, \forall \alpha \in \mathbb{R}$:

1. Positive homogeneity or positive scalability: $\|\alpha x\| = |\alpha|\|x\|$
2. Triangle inequality: $\|x + y\| \leq \|x\| + \|y\|$
3. Positive definiteness: $\|x\| = 0 \Longleftrightarrow x = \hat{0}$

The "**Euclidian norm**" is by far the most commonly used norm in $\mathbb{R}^n$. The Euclidian norm is defined as:

$$\forall x \in \mathbb{R}^n \|x\| := \sqrt{x_1^2 + x_2^2 + \cdots + x_n^2} = \sqrt{\sum_{i=1}^{n} x_i^2}$$

It should be noted, however, that there are many other possible norms to define the size of a vector. An interested reader should consult a book on linear spaces.

▶ **BONUS EXERCISES:**

- Show that the triangle inequality can also be expressed as: $|\|x\| - \|y\|| \leq \|x - y\|$.
- Show that the Euclidian norm satisfies the properties of a norm function.

1.1.6.b. Distances between Vectors

The distances between elements of a linear vector space can be represented with a metric or a distance function.

Let V be a linear vector space. A metric or a distance function is defined as a function $\rho_m:V \times V \longrightarrow [0, \infty[$ that satisfies the following properties. $\forall x, y, z \in V$:

1. $\rho_m(x, y) = 0 \Longleftrightarrow x = y$
2. $\rho_m(x, y) = \rho_m(y, x)$
3. $\rho_m(x, y) \leq \rho_m(x, z) + \rho_m(z, y)$

In the last two sections, we defined the size of, and the distance between, vectors. In their abstract definition, they are not related. However, the distance function can be chosen as the size of the difference between two vectors:

$$\rho_m(x, y) := \|x - y\|$$

The "**Euclidian metric**" is the metric function generated by the "**Euclidian norm**":

$$x, y \in \mathbb{R}^n, \rho_m(x, y) := \|x - y\| := \sqrt{(x_1 - y_1)^2 + (x_2 - y_2)^2 + \cdots + (x_n - y_n)^2} = \sqrt{\sum_{i=1}^{n} (x_i - y_i)^2}$$

▶ **BONUS EXERCISES:**

- Show that the Euclidian metric satisfies the properties of a metric function.
- Show that the function $\rho_m:V \times V \longrightarrow [0, \infty[$ defined as $\forall x, y \in V:\rho_m(x, y) := \|x - y\|$ satisfies the properties of a metric function.

1.1.6.c. Dot Product

We have thus far defined the sizes and distances between elements of a linear vector space. What about directions? In Euclidian vector spaces, $\mathbb{R}^2$ and $\mathbb{R}^3$ in particular, the "dot product" operation is used to relate lengths and directions between vectors. The dot product operation is usually preceded by the definition of an "inner product," which is presented here for completeness.

In general, an inner product is a way to multiply elements of a linear vector space:

Let V be a linear vector space over $\mathbb{R}$. An inner product is a mapping $\langle\cdot, \cdot\rangle:V \times V \rightarrow \mathbb{R}$ that satisfies the following properties $\forall x, y, z \in V$:

1. $\langle x + y, z\rangle = \langle x, z\rangle + \langle y, x\rangle$
2. $\langle \alpha x, z\rangle = \alpha\langle x, z\rangle$
3. $\langle x, z\rangle = \langle z, x\rangle$

4. $\langle x, x \rangle \geq 0$
5. $\langle x, x \rangle = 0 \Leftrightarrow x = \hat{0}$

The **Euclidian dot product** in $\mathbb{R}^n$ is defined as:

$$\cdot : \mathbb{R}^n \times \mathbb{R}^n \to \mathbb{R} \text{ such that } \forall x, y \in \mathbb{R}^n : x \cdot y = \sum_{i=1}^{n} x_i y_i$$

It is easily evident that the Euclidian norm of a vector is related to the dot product since: $\|x\|^2 = x \cdot x$

DEFINITION **Orthogonal (perpendicular) vectors.** $\forall x, y \in \mathbb{R}^n$, x and y are called orthogonal if $x \cdot y = 0$

▶ BONUS EXERCISE: Show that the Euclidian dot product satisfies the properties of the inner product.

1.1.7. Euclidian Vector Space $\mathbb{R}^n$

A Euclidian vector space is the $\mathbb{R}^n$ vector space equipped with the Euclidian norm, the Euclidian metric, and the Euclidian dot product functions. The norm function is related to the dot product function since $\forall x \in \mathbb{R}^n : \|x\|^2 = x \cdot x$.

1.1.7.a. Orthonormal Basis

Definition: Orthonormal basis. Consider a basis set in a Euclidian vector space. The basis set is called an orthonormal basis set if the following two conditions are satisfied. The first condition is that the elements in the basis set have to be orthogonal to each other. The second condition is that the norm (size) of each vector has to be equal to unity.

Notes:

- It can be shown by construction that, in any Euclidian vector space, there exists an orthonormal basis set. The construction procedure is termed the "Gram–Schmidt Process."
- Let $B \subset \mathbb{R}^n$ be an orthonormal basis set, $B = \{e_1, e_2 \ldots e_n\} = \{e_i\}_{i=1}^{n}$. Then, the following hold:
  - The basis vectors of B satisfy the following equality:

$$e_i \cdot e_j = \begin{cases} 1, & i = j \\ 0, & i \neq j \end{cases}$$

- $\forall a \in \mathbb{R}^n: a = \sum_{i=1}^{n} a_i e_i$. Then, the components a_i can be retrieved by the relation:

$$a_i = a \cdot e_i$$

These components can be called the Cartesian components of the vector a in the Cartesian coordinate system defined by the orthonormal basis set B.

Problem 1: Which of the following sets are orthonormal basis sets in the Euclidian vector space $\mathbb{R}^2$?

- $B_1 = \{e_1, e_2\},\ e_1 = \{1, 0\} e_2 = \{0, 1\}$
- $B_2 = \{e_1, e_2\},\ e_1 = \{1, 1\} e_2 = \{-1, 1\}$

Solution:

It was shown in section 1.1.4.a that the two $e_1 = \begin{pmatrix} 1 \\ 0 \end{pmatrix}$ and $e_2 = \begin{pmatrix} 0 \\ 1 \end{pmatrix}$ form a basis in $\mathbb{R}^2$. Considering $\mathbb{R}^2$ to be a Euclidian vector space and then, using the Euclidian norm and the Euclidian dot product: $\|e_1\| = \|e_2\| = \sqrt{1^2 + 0} = 1$ and $e_1 \cdot e_2 = 1 \times 0 + 0 \times 1 = 0$. Therefore, the set $B = \{e_1, e_2\}$ form an orthonormal basis in the Euclidian vector space $\mathbb{R}^2$.

It was also shown in section 1.1.4.a that the two vectors $e_1 = \begin{pmatrix} 1 \\ 1 \end{pmatrix}$ and $e_2 = \begin{pmatrix} -1 \\ 1 \end{pmatrix}$ form a basis in $\mathbb{R}^2$. Considering $\mathbb{R}^2$ to be a Euclidian vector space and then, using the Euclidian norm and the Euclidian dot product: $\|e_1\| = \|e_2\| = \sqrt{1^2 + 1^2} = \sqrt{2}$ and $e_1 \cdot e_2 = 1 \times -1 + 1 \times 1 = 0$. Therefore, the set $B = \{e_1, e_2\}$ form an orthogonal basis in the Euclidian vector space $\mathbb{R}^2$. However, the ele–ments are not normal (a normal vector is a vector whose norm equals to unity).

Problem 2: Let $u, v \neq \hat{0} \in \mathbb{R}^3$. Define $z := \frac{(v \cdot u)u}{\|u\|^2}$ to be the **orthogonal projection** of v on u. Define $\bar{z} := v - z$. Show that $\bar{z}$ is orthogonal to u.

Solution: $\bar{z} \cdot u = \left(v - \frac{(v \cdot u)u}{\|u\|^2}\right) \cdot u = (v \cdot u) - \frac{(v \cdot u)(u \cdot u)}{\|u\|^2} = (v \cdot u) - \frac{(u \cdot v)\|u\|^2}{\|u\|^2} = 0 \Longrightarrow \bar{z}$ is orthogonal to u. Figure 1–1 shows a schematic of the orthogonal projection.

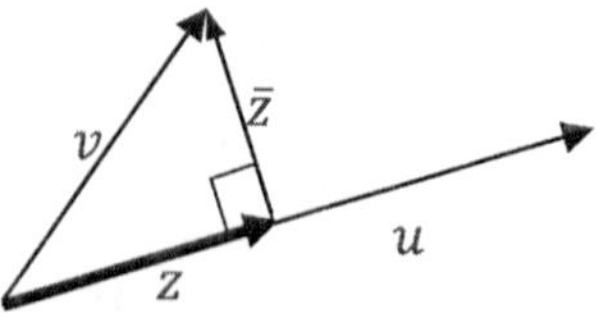

FIGURE 1-1 Orthogonal projection of v on u.

1.1.7.b. Cross Product

The traditional cross product that engineers are familiar with is a very special operation in the Euclidian vector space $\mathbb{R}^3$. The result of the cross product operation is a vector that is orthogonal to the original two vectors and whose magnitude is equal to the area of the parallelogram formed by the original two vectors. Thus, we can define the cross product as the operation $\times:\mathbb{R}^3 \times \mathbb{R}^3 \to \mathbb{R}^3$, satisfying the following properties $\forall u, v, w \in \mathbb{R}^3, \forall \alpha \in \mathbb{R}$: denote $z = (u \times v)$

1. $v \cdot z = u \cdot z = 0$
2. Skew symmetry: $u \times v = -v \times u$
3. $(\alpha u) \times v = \alpha(u \times v) = u \times (\alpha v)$
4. Distributivity: $u \times (v + w) = u \times v + u \times w$
5. $\|z\|^2 = z \cdot z = \|u\|^2\|v\|^2 - (u \cdot v)^2 = (u \cdot u)(v \cdot v) - (u \cdot v)^2$

The last property results in limiting the norm of the vector $z = (u \times v)$ to be equal to the area of the parallelogram formed by the original two vectors.

Note that the cross product as defined is unique for $\mathbb{R}^3$ and can be extended to higher dimen–sions, but it will require additional structure. The results in the following problem are direct con–sequences of the definition of the cross product presented in $\mathbb{R}^3$.

Problem: Consider the orthonormal basis set $B = \{e_1, e_2, e_3\}$ in $\mathbb{R}^3$ where:

$$e_1 = \begin{pmatrix} 1 \\ 0 \\ 0 \end{pmatrix}, \; e_2 = \begin{pmatrix} 0 \\ 1 \\ 0 \end{pmatrix}, \; e_3 = \begin{pmatrix} 0 \\ 0 \\ 1 \end{pmatrix}$$

Use the definition of the cross product to show the following results:

1. $e_1 \times e_2 = \pm e_3, \; e_2 \times e_3 = \pm e_1, \; e_3 \times e_1 = \pm e_2$
   This indicates that the definition of the cross product has to be augmented with an orientation (a choice of the positive or the negative sign in the above relationship), and in a right–handed system, the positive sign is chosen for the above relations.
2. $u \times v = \hat{0}$ if and only if either $u = \hat{0}$, $v = \hat{0}$ or u and v are linearly dependent.
3. The explicit representation of the cross product in an orthonormal basis in $\mathbb{R}^3$ is:

$$\begin{aligned} u \times v &= (u_1e_1 + u_2e_2 + u_3e_3) \times (v_1e_1 + v_2e_2 + v_3e_3) \\ &= (u_2v_3 - u_3v_2)e_1 + (u_3v_1 - u_1v_3)e_2 + (u_1v_2 - u_2v_1)e_3 \end{aligned}$$

4. $\forall u, v, w \in \mathbb{R}^3$: the operation $u \cdot (v \times w)$ is called the **triple product**. The triple product satisfies the following property:

$$u \cdot (v \times w) = w \cdot (u \times v) = v \cdot (w \times u) = -u \cdot (w \times v) = -v \cdot (u \times w) = -w \cdot (v \times u)$$

5. $\forall u, v, w \in \mathbb{R}^3$: $u \cdot (v \times w) = 0$ if and only if u, v and w are linearly dependent
6. Let $u, v, w \in \mathbb{R}^3$ such that u, v and w form an orthonormal basis set. Then, $u \cdot (v \times w) = \pm 1$

Solution:

1. Let $z = e_1 \times e_2 = \alpha e_1 + \beta e_2 + \gamma e_3$. Since $e_1 \cdot e_2 = e_1 \cdot e_3 = e_2 \cdot e_3 = 0$ and using property 1 of the cross product: $e_1 \cdot z = e_2 \cdot z = 0 \Longrightarrow \alpha = \beta = 0$. From property 5: $\|z\|^2 = \gamma^2 = 1 \Longrightarrow \gamma = \pm 1$

2. For u or $v = \hat{0}$, the proof is trivial using property 3.

   In the following, assume nonzero vectors u and v and let $z = u \times v$

   $\Rightarrow$ direction: Assume $z = \hat{0}$. From property 5: $\|z\|^2 = 0 \Longrightarrow (u \cdot v)^2 = \|u\|^2\|v\|^2$. Therefore $\left\| v - \frac{(u \cdot v)u}{\|u\|^2} \right\|^2 = \left(v - \frac{(u \cdot v)u}{\|u\|^2}\right) \cdot \left(v - \frac{(u \cdot v)u}{\|u\|^2}\right) = \|v\|^2 - \frac{(u \cdot v)^2}{\|u\|^2} = 0 \Longrightarrow v = \frac{(u \cdot v)u}{\|u\|^2} \Longrightarrow v = \alpha u \Longrightarrow v$ and u are linearly dependent.

   $\Leftarrow$ direction: assume $u = \alpha v$. Using 5: $\|z\|^2 = \|u\|^2\|v\|^2 - (u \cdot v)^2 = (\alpha^2 - \alpha^2)\|v\|^4 = 0 \Rightarrow z = \hat{0}$

3. Trivial

4. Using property 1 $\Rightarrow 0 = (u + v) \cdot ((u + v) \times w) = u \cdot (v \times w) + v \cdot (u \times w) = \ldots$ etc.

5. $\Leftarrow$ direction: Assume u, v and w are linearly dependent, then $u = \alpha v + \beta w \Rightarrow u \cdot (v \times w) = (\alpha v + \beta w) \cdot (v \times w) = 0$

   $\Rightarrow$ direction: assume $u \cdot (v \times w) = 0$. We will argue by contradiction. Assume u, v and w are linearly independent $\Rightarrow v \times w \neq \hat{0} \Rightarrow u$ along with v and w are orthogonal to $v \times w$. However, since $u, v, w \in \mathbb{R}^3$ and form a linearly independent set of three vectors, they form a basis in $\mathbb{R}^3$. Thus, every vector in $\mathbb{R}^3$ can be represented as a linear combination of u, v and w, which means that every vector is orthogonal to $v \times w$ (even the vector $v \times w$!). This would lead to $v \times w$ being orthogonal to itself, indicating that $v \times w = \hat{0}$, which is in contradiction with the linear independence of v and w.

6. Same proof as in 1.

1.1.8. Graphical Representation of $\mathbb{R}^2$ and $\mathbb{R}^3$

1.1.8.a. Addition and Multiplication by a Scalar

Every element in $\mathbb{R}^2$ or $\mathbb{R}^3$ can be represented in a natural way by a vector pointing from 0 to one point in the plane with two coordinate axes x and y or in a space with three coordinate axes x, y and z. Figure 1–2 illustrates the representation of the two vectors $v = \begin{pmatrix} 1 \\ 1 \end{pmatrix} \in \mathbb{R}^2$ and $w = \begin{pmatrix} 1 \\ 1 \\ 1 \end{pmatrix} \in \mathbb{R}^3$.

Addition of vectors and multiplication by a scalar in $\mathbb{R}^2$ or $\mathbb{R}^3$ can also be represented. Figure 1–3 the representation of the vectors: $v = \begin{pmatrix} 1 \\ 1 \end{pmatrix}$, $w = \begin{pmatrix} 1 \\ 0 \end{pmatrix}$, $2v$, $v + w$.

1.1.8.b. Norm, Distance and Dot product

The norm and the distances in $\mathbb{R}^n$ using Euclidian geometry give the geometric length of the vector and the geometric distance between two vectors. The dot product gives an indication of directions in $\mathbb{R}^n$ spaces and, in particular, for $\mathbb{R}^2$ and $\mathbb{R}^3$, the dot product is related to the geometric angle between two vectors:

Let $x, y \in \mathbb{R}^3$

$$x \cdot y = \sum_{i=1}^{3} x_i y_i = \|x\| \|y\| \cos \theta_{xy}$$

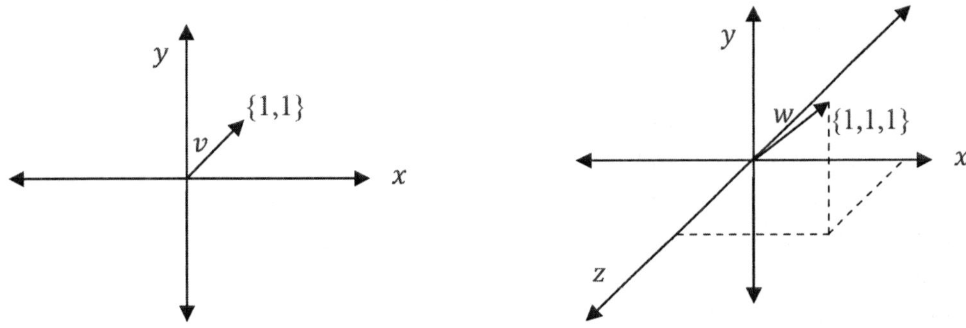

FIGURE 1-2 Graphical representation of vectors in $\mathbb{R}^2$ and $\mathbb{R}^3$.

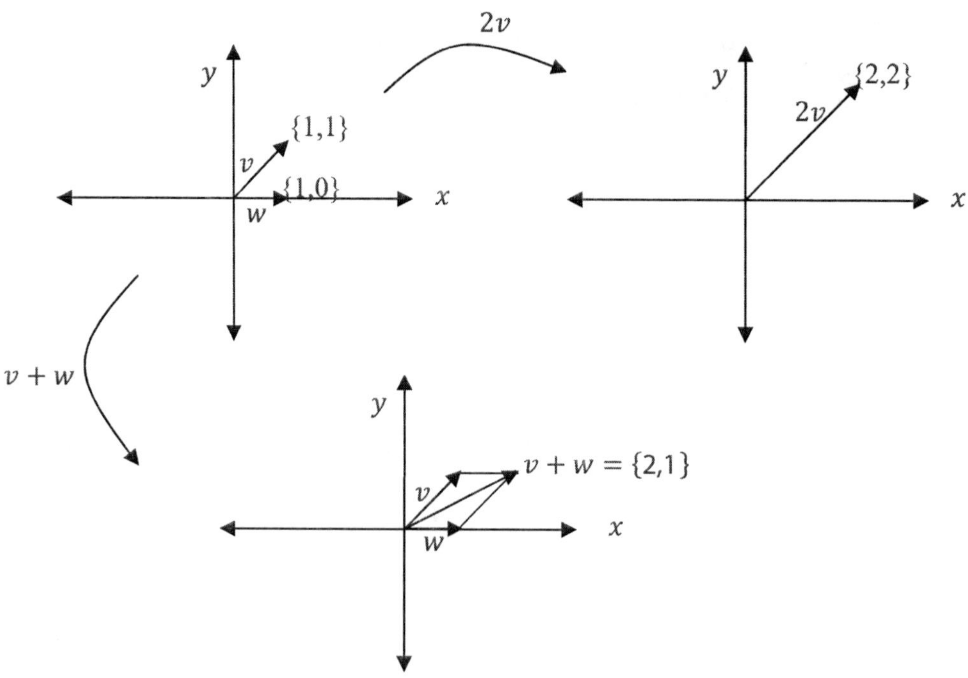

FIGURE 1-3 Graphical representation of vector addition and multiplication by a scalar in $\mathbb{R}^2$.

Where θ_{xy} is the geometric angle between x and y (Figure 1–4). Note that $\theta_{xy} = 90^{\circ}$ for orthogonal vectors

▶ **BONUS EXERCISE:** Verify the relationship between the geometric angle θ_{xy} and the dot product operation.

1.1.8.c. Cross Product

The cross product of two vectors in $\mathbb{R}^3$ is related to the geometric angle between those two vectors by the following relation:

Let $u, v \in \mathbb{R}^3$

$$u \times v = \|u\|\|v\| \sin \theta_{uv} n$$

where, n is a unit (normal) vector that is orthogonal to u and v and θ_{uv} is the geometric angle between the two vectors u and v (Figure 1–5).

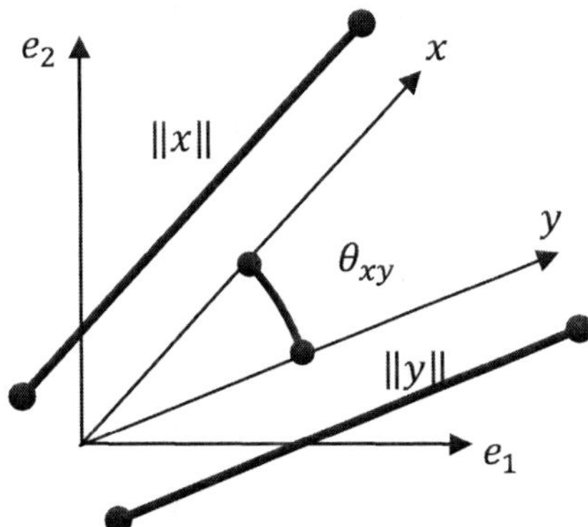

FIGURE 1-4 Geometric angle between two vectors and their length in $\mathbb{R}^2$.

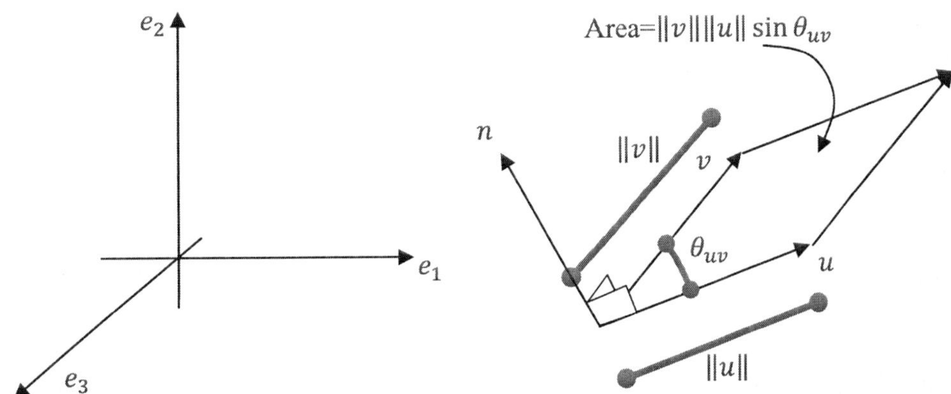

FIGURE 1-5 Graphical representation of the cross product in $\mathbb{R}^3$. (a) Right–handed orientation for the basis set, (b) right–handed result for $\boldsymbol{u} \times \boldsymbol{v}$.

▶ **BONUS EXERCISE:** Verify the relationship between the geometric angle θ_{uv} and the cross product operation.

Notice that the two vectors n and $-n$ are both orthogonal to u and v and, thus, the result is dependent on the definition of the operation according to the basis. If a right–handed geometric orientation is chosen for the basis, then the cross product also follows the right–handed geometric orientation (Figure 1–5). Another important aspect of the cross product is that the norm of the vector produced is equal to the size of the area of the parallelogram formed by the two vectors (Figure 1–5b).

1.1.9. Representing Elements of $\mathbb{R}^n$ and $\mathbb{M}^n$ in Mathematica

1.1.9.a. Direct Definition Using Braces {}

To define $x = \begin{pmatrix} 1 \\ 5 \end{pmatrix}$:

```
x={1,5}
```

To define $x = \begin{pmatrix} 1 \\ 5 \end{pmatrix}$ while suppressing the result, you need to add a semicolon:

```
x={1,5};
```

To define $X = \begin{pmatrix} 1 & 5 \\ 2 & 3 \end{pmatrix}$:

```
X={{1,5},{2,3}};
```

You can use //MatrixForm to display the results of the definition in matrix form. The following code defines $x = \begin{pmatrix} 1 \\ 2 \end{pmatrix}$ and $X = \begin{pmatrix} 1 & 5 \\ 2 & 3 \end{pmatrix}$ and then displays x and X in braces format and then displays x and X in matrix form.

```
x={1,2};
X={{1,5},{2,3}};
x
X
x//MatrixForm
X//MatrixForm
```

1.1.9.b. Definition Using Tables

To define $x = \begin{pmatrix} 1 \\ 2 \\ 3 \end{pmatrix}$ and $X = \begin{pmatrix} 1+1 & 1+2 & 1+3 \\ 2+1 & 2+2 & 2+3 \\ 3+1 & 3+2 & 3+3 \end{pmatrix} = \begin{pmatrix} 2 & 3 & 4 \\ 3 & 4 & 5 \\ 4 & 5 & 6 \end{pmatrix}$

```
x=Table[i,{i,1,3}];
X=Table[i+j,{i,1,3},{j,1,3}];
x//MatrixForm
X//MatrixForm
```

1.1.9.c. Operations

The Euclidian norm is the default when using the Norm[] function in Mathematica.

$$x = \begin{pmatrix} 1 \\ 2 \\ 3 \end{pmatrix}, \ \|x\| = \sqrt{1^2 + 2^2 + 3^2} = \sqrt{14}$$

```
x={1,2,3};
Norm[x]
```

The dot product between two vectors can be expressed using two methods. Notice the following example:

$$x = \begin{pmatrix} 1 \\ 2 \\ 3 \end{pmatrix}, \quad y = \begin{pmatrix} -1 \\ -5 \\ 2 \end{pmatrix}, \quad x \cdot y = -1 - 10 + 6 = -5$$

```
x={1,2,3};
y={-1,-5,2};
x.y
Dot[x,y]
```

The angle between two vectors can be obtained using two methods, as shown in the following example:

$$x = \begin{pmatrix} 1 \\ 2 \\ 3 \end{pmatrix}, \quad y = \begin{pmatrix} -1 \\ -5 \\ 2 \end{pmatrix}, x \cdot y = -5, \quad \|x\| = \sqrt{14}, \quad \|y\| = \sqrt{30}, \quad \theta_{xy} = acos\frac{(x \cdot y)}{\|x\|\,\|y\|} = 1.82^{rad}$$

```
x={1,2,3};
y={-1,-5,2};
theta1=ArcCos[x.y/Norm[x]/Norm[y]]
theta2=VectorAngle[x,y]
N[theta1]
N[theta2]
```

Notice that the function `N[]` returns the numerical value of the argument.

The cross product of two vectors is calculated using the function `Cross[]`. In the following code, $z = x \times y = \{19, -5, 3\}$. Notice that $z \cdot x = z \cdot y = 0$

```
x={1,2,3};
y={-1,-5,2};
z=Cross[x,y]
z.x
z.y
```

Note that the cross product for vectors with dimension n requires $n-1$ linearly independent vectors and produces a vector that is orthogonal to the set of vectors used in the argument:

```
x={1,2,3,4};
y={-1,-5,2,1};
z={1,1,1,1};
w=Cross[x,y,z]
x.w
y.w
z.w
```

```
New Mathematica Functions Explored in Section 1.1

ArcCos[x]
Cross[x,y]
Dot[x,y]
N[x]
Norm[x]
Table[]
VectorAngle[x,y]
//MatrixForm
```

▶ EXERCISES

The following apply to the Euclidian vector space $\mathbb{R}^2$

1. Show that the following sets are subspaces of $\mathbb{R}^2$. Represent each subspace graphically. Find two basis sets for each subspace.
   - $Y = \{\alpha x | x = \{1, 0\}, \alpha \in \mathbb{R}\} = \text{span}\{x\}$
   - $Z = \{\alpha z | z = \{1, 1\}, \alpha \in \mathbb{R}\} = \text{span}\{z\}$
2. Find three vectors that are orthogonal to $x = \{1, 1\}$
3. Choose a value for y_2 such that $x = \{1, 1\}$, $y = \{1, y_2\}$ are linearly independent
4. Verify that $x = \{1, 0\}$, $y = \{1, 2\}$ are linearly independent and then find the unique expansion of $v = \{5, 3\}$ in the basis set $B = \{x, y\}$
5. Find two different orthonormal basis sets and two different non–orthonormal basis sets for $\mathbb{R}^2$

6. Use the cross product to find the area of the parallelogram formed by the two vectors $x = \{1, 1\}, y = \{1, 2\}$

The following apply to the Euclidian vector space $\mathbb{R}^3$

7. Show that the following sets are subspaces of $\mathbb{R}^3$. Represent each subspace graphically. Find two basis sets for each subspace.
   - $Y = \{\alpha x | x = \{1, 0, 0\}, \alpha \in \mathbb{R}\} = \text{span}\{x\}$
   - $Z = \{\alpha x + \beta y | x = \{1, 0, 0\}, y = \{0, 1, 0\}, \alpha, \beta \in \mathbb{R}\} = \text{span}\{x, y\}$
8. Find three vectors that are orthogonal to $x = \{1, 1, 1\}$
9. Show that the following vectors are linearly dependent $x = \{1, 1, 1\}, y = \{1, 2, 1\}$, $z = \{0, -1, 0\}$. (Hint: use the triple product.)
10. Choose a value for z_3 such that $x = \{1, 1, 1\}, y = \{1, 2, 1\}, z = \{0, -1, z_3\}$ are linearly independent
11. Verify that $x = \{1, 0, 1\}$, $y = \{1, 2, 1\}$, $z = \{0, -1, 1\}$ are linearly independent and then find the unique expansion of $v = \{5, 3, 2\}$ in the basis set $B = \{x, y, z\}$
12. Find two different orthonormal basis sets and two different non–orthonormal basis sets for $\mathbb{R}^3$.
13. Consider $x = \{1, 1, 1\}$ and $y = \{1, 2, 0\}$. Use the cross product to find a vector orthogonal to both x and y. Also find the area of the parallelogram formed by the two vectors x and y.
14. For the shown cuboid, $A = \{0, 1, 0\}, B = \{0, 1, 1\}$ $C = \{2, 0, 1\}$ and $D = \{2, 0, 0\}$
   - Use the cross product to find *two* unit vectors orthogonal (perpendicular) to the plane $ABCD$.
   - Use the cross product to find the area of the parallelogram $ABCD$.
   - Find the angle between the vectors representing AC and BC.

Notice that a line starting at point A and ending at point B can be represented by a vector $AB = B - A$

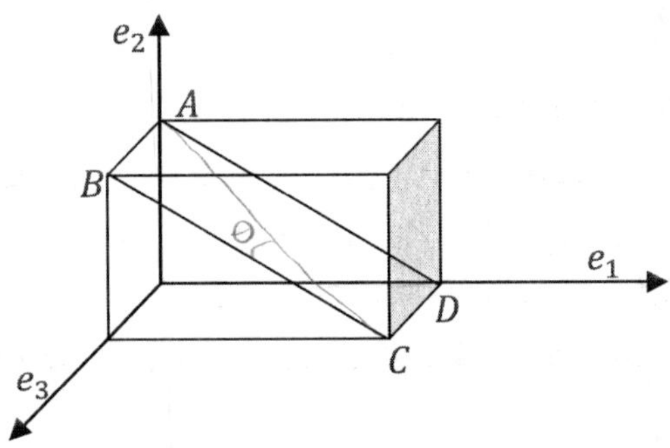

1.2. LINEAR MAPS BETWEEN VECTOR SPACES (MATRICES, TENSORS)

1.2.1. Introduction

The topic of matrices and/or tensors is the next ingredient for the study of linear algebra. Matrices and/or tensors are functions whose domain is one vector space, while the range is another (same or different) vector space. When dealing with Euclidian vector spaces, a matrix can be viewed as a list of several vectors, with each row representing a vector. When the matrix is multiplied by a vector, a new vector is produced. The components of the new vector are calculated by taking the dot product between each row vector in the matrix by the original vector. For example, consider the multiplication of the matrix $M = \begin{pmatrix} 11 & 12 \\ 21 & 22 \end{pmatrix}$ by the two dimensional vector $x = \begin{pmatrix} 1 \\ 2 \end{pmatrix}$ and denoting the resulting vector $y = Mx$. The first component in the new vector y is calculated by taking the dot product between the vector $\{11, 12\}$ and the vector $\{1, 2\}$, which is equal to $11 \times 1 + 12 \times 2 = 35$. The second component is calculated by taking the dot product between the vector $\{21, 22\}$ and the vector $\{1, 2\}$, which is equal to $21 \times 1 + 22 \times 2 = 65$. The linear operation $Mx = y$ is then:

$$\begin{pmatrix} 11 & 12 \\ 21 & 22 \end{pmatrix} \begin{pmatrix} 1 \\ 2 \end{pmatrix} = \begin{pmatrix} 35 \\ 65 \end{pmatrix}$$

The dot product as studied in section 1.1.6.c is a linear function and, thus, the matrix M in this example can be viewed as a linear map or a linear function from $\mathbb{R}^2$ to $\mathbb{R}^2$. In general, a matrix can be viewed as a linear map or a linear function from $\mathbb{R}^n$ to $\mathbb{R}^m$ with the number of columns equal to n and the number of rows equal to m. The matrix in this general case is considered a linear function with the argument being a vector in $\mathbb{R}^n$ and the resulting vector in $\mathbb{R}^m$. In a general sense, such matrix can be viewed as a combination of m vectors each having n components. In this section, we will study such linear maps, which, in solid mechanics, are traditionally termed tensors. We will particularly be concerned with "square matrices," which are linear functions from $\mathbb{R}^n$ to $\mathbb{R}^n$.

1.2.2. Basic Definitions

A linear map T between two vector spaces V and W is a function $T{:}V \to W$ such that $\forall x, y \in V$, $\forall \alpha, \beta \in \mathbb{R}$:

$$T(\alpha x + \beta y) = \alpha T(x) + \beta T(y)$$

As described in the previous section, matrices are linear maps between finite dimensional Euclidian vector spaces. It can be shown that any linear map between two finite dimensional

Euclidian vector spaces can be represented by a rectangular matrix; however, the proof of such statement is beyond the scope of this text. To illustrate this, however, an example will be used. Consider the linear map $T:\mathbb{R}^2 \to \mathbb{R}^2$ such that $\forall x \in \mathbb{R}^2: T(x) = \begin{pmatrix} x_1 + x_2 \\ x_1 - x_2 \end{pmatrix}$. We can associate the matrix $M = \begin{pmatrix} 1 & 1 \\ 1 & -1 \end{pmatrix}$ with the linear map T. Consider the vectors $x = \{1, 1\}$ and $y = \{0, 1\}$ and consider the scalars $\alpha = 1$ and $\beta = 3$. It is clear that the linearity of the matrix operation is evi– dent in the equality between $M(\alpha x + \beta y)$ and $\alpha Mx + \beta My$ which is independent of the order of the operations as follows:

$$M(\alpha x + \beta y) = \begin{pmatrix} 1 & 1 \\ 1 & -1 \end{pmatrix} \begin{pmatrix} 1 \\ 4 \end{pmatrix} = \begin{pmatrix} 5 \\ -3 \end{pmatrix}$$

$$\alpha Mx + \beta My = 1 \begin{pmatrix} 1 & 1 \\ 1 & -1 \end{pmatrix} \begin{pmatrix} 1 \\ 1 \end{pmatrix} + 3 \begin{pmatrix} 1 & 1 \\ 1 & -1 \end{pmatrix} \begin{pmatrix} 0 \\ 1 \end{pmatrix} = \begin{pmatrix} 5 \\ -3 \end{pmatrix}$$

The following Mathematica code illustrates the equivalence.

```
M={{1,1},{1,-1}};
x={1,1};
y={0,1};
alpha=1;
beta=3;
M.(alpha*x+beta*y)
alpha*M.x+beta*M.y
```

DEFINITION **Kernel:** Let T be a linear map between two vector spaces V and W. Then, $\ker T = \{x \in V | T(x) = \hat{0}\}$. The kernel of a linear map is the set of all the vectors that are "nullified" or are mapped to the zero vector when the linear map is applied to them. The terms "Null" or "Null Space" can also be used to mean the kernel of the linear map.

Problem: Find the kernel of the following matrix.

$$M = \begin{pmatrix} 1 & 1 \\ 2 & 2 \end{pmatrix}$$

Solution: Careful investigation of the given matrix indicates that any vector orthogonal to the vector $x = \{1, 1\}$ will be mapped to the zero vector. For example, consider the vector $y = \{-1, 1\}$:

$$My = \begin{pmatrix} 1 & 1 \\ 2 & 2 \end{pmatrix} \begin{pmatrix} -1 \\ 1 \end{pmatrix} = \begin{pmatrix} 0 \\ 0 \end{pmatrix}$$

The vector y is thus an element of the kernel of M. In fact, if the vector y is multiplied by any scalar, the resultant vector is still an element of the kernel of M. $\ker M = \{\alpha y \mid y = \{-1, 1\}, \alpha \in \mathbb{R}\} = \text{span}\{y\}$

▶ BONUS EXERCISE: Let T be a linear map between two vector spaces V and W. Use the prop–erties of a linear map to prove that the kernel of a linear map is a subspace of V.

1.2.3. Invertible Maps

The invertibility of linear maps is an important aspect in linear vector spaces, particularly when solving sets of linear equations. The following are important definitions in the study of linear maps:

The linear map $T{:}X \to Y$ is **injective** (one–to–one) if $\forall x \in X, \exists! y \in Y \, s.t. T(x) = y$.

The linear map $T{:}X \to Y$ is **surjective** (onto) if $\forall y \in Y, \exists x \in X \, s.t. T(x) = y$.

The linear map $T{:}X \to Y$ is **bijective** if it is one–to–one and onto. A bijective map is **inverible**.

An injective (one–to–one) map is a map in which each element in the domain has a unique image in the range. A surjective (onto) map is a map whose range covers all the codomain. A bijective map is one that can be inverted and, thus, it has to be both injective and surjective. Figure 1–6 illustrates the difference between these maps.

▶ BONUS EXERCISES:

- Prove that if T is an injective linear map, then $\ker T = \{\hat{0}\}$
- Prove that the linear $T{:}\mathbb{R}^n \to \mathbb{R}^n$ is invertible if and only if $\ker T = \{\hat{0}\}$ (Hint, it is required to show that T is both injective and surjective.)

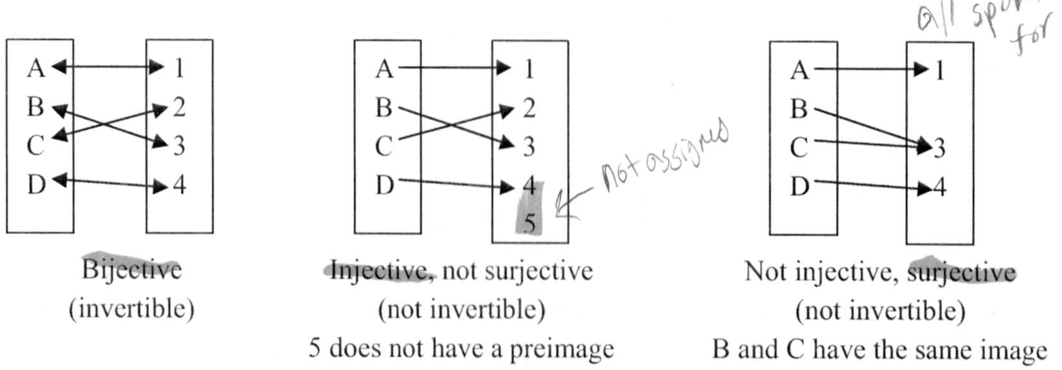

FIGURE 1-6 Difference between bijective, injective, and surjective maps.

1.2.4. Matrices from the Bottom Up

1.2.4.a. Rank-one Operators

A linear map that maps any multidimensional vector into an element of $\mathbb{R}$ is termed a rank–one operator:

Let $T:\mathbb{R}^n \to \mathbb{R}$ such that $\forall x \in \mathbb{R}^n: T(x) = v_1x_1 + v_2x_2 + \cdots + v_nx_n$ where $\forall i: v_i \in \mathbb{R}$.

We can associate the matrix $M = (v_1\ v_2 \ldots v_n)$ with the linear map T such that:

$$\forall x \in \mathbb{R}^n:$$

$$Mx = v_1x_1 + v_2x_2 + \cdots + v_nx_n$$

As a result of the choice of our Euclidian vector space, which is equipped with the Euclidian norm and the Euclidian dot product, the linear map T (or equivalently the matrix M) can be associated with a fixed vector $v = \{v_1\ v_2\ \ldots\ v_n\}$ and with an operation similar to the "dot product" operation. This "dot product" gives a measure of the angle between x and v. As a result, any vector in $\mathbb{R}^n$ that is orthogonal to v will be mapped to zero.

Figure 1–7 illustrates the linear map $T:\mathbb{R}^2 \to \mathbb{R}$ where $T(x) = v_1x_1 + v_2x_2$. Let $y \in \mathbb{R}^2$ be orthogonal to v, then $T(y) = My = v \cdot y = 0$. The kernel of T can be explicitly stated as: $\ker T = \{a \in \mathbb{R}^2 | v \cdot a = 0\} = \text{span}\{y\} = \{\alpha y | \alpha \in \mathbb{R}\}$. If $v \neq \hat{0}$, then the map T is surjective (onto) since the range covers all elements in the codomain; every element in $\mathbb{R}$ is the image of at least one vector $x \in \mathbb{R}^2$. However, the map T is not injective (one–to–one). The kernel of the map contains many elements, i.e., there is more than one element that will be mapped to 0. Thus, the map T is also not bijective, i.e., cannot be inverted.

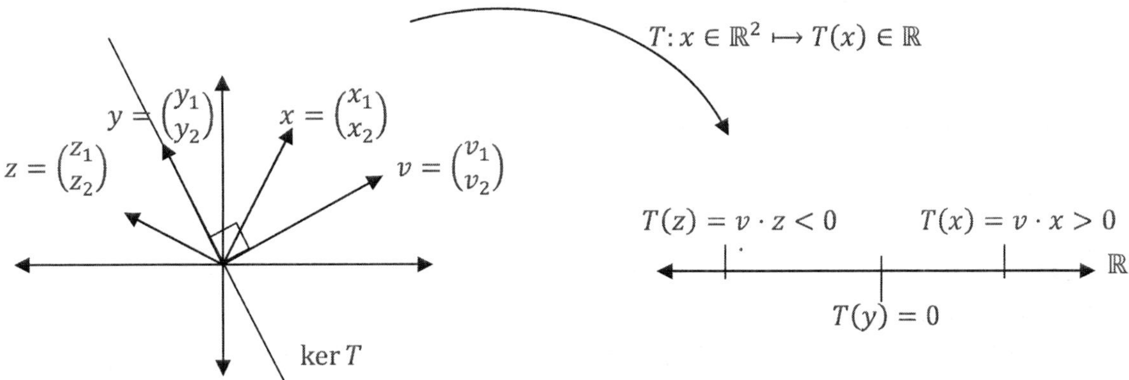

FIGURE 1-7 The action of the linear map T associated with the fixed vector v:

$$T(x) = v \cdot x > 0 \quad T(y) = v \cdot y = 0 \quad T(z) = v \cdot z < 0$$

1.2.4.b. General Operators

In general, a linear map is defined as $T{:}\mathbb{R}^n \rightarrow \mathbb{R}^m$, but in this section, we will restrict to the case when $n = m$. Let $T{:}\mathbb{R}^n \rightarrow \mathbb{R}^n$ such that $\forall x \in \mathbb{R}^n$:

$$T(x) = \begin{pmatrix} v_{11}x_1 + v_{12}x_2 + \cdots + v_{1n}x_n \\ v_{21}x_1 + v_{22}x_2 + \cdots + v_{2n}x_n \\ \cdots\cdots\cdots\cdots \\ v_{n1}x_1 + v_{n2}x_2 + \cdots + v_{nn}x_n \end{pmatrix}$$

where $\forall i, j{:}v_{ij} \in \mathbb{R}$. In this case, we can associate with the linear map the following matrix $M \in \mathbb{M}^n$

$$M = \begin{pmatrix} v_{11} & v_{12} & \ldots & v_{1n} \\ v_{21} & v_{22} & \ldots & v_{2n} \\ \ldots & \ldots & \ldots & \ldots \\ v_{n1} & v_{n2} & \ldots & v_{nn} \end{pmatrix}$$

The matrix is composed of n vectors $v_i = \{v_{i1} \quad v_{i2} \quad \ldots \quad v_{in}\}$ where i takes values from 1 to n. Then, $\forall x \in \mathbb{R}^n$:

$$Mx = \begin{pmatrix} v_1 \cdot x \\ v_2 \cdot x \\ \ldots \\ v_n \cdot x \end{pmatrix}$$

Notice that each row in the matrix represents one linear function that gives a scalar value when taking its dot product by the vector x. Thus, the linear map, or the associated matrix, is composed of n rank–one operators.

As an illustrative example, consider the linear map $T{:}\mathbb{R}^2 \rightarrow \mathbb{R}^2$ such that $\forall x \in \mathbb{R}^2$:

$$T(x) = \begin{pmatrix} v_{11}x_1 + v_{12}x_2 \\ v_{21}x_1 + v_{22}x_2 \end{pmatrix}$$

In this case, we can associate with the linear map a matrix $M = \begin{pmatrix} v_{11} & v_{12} \\ v_{21} & v_{22} \end{pmatrix}$ and two vectors $v_a = \begin{pmatrix} v_{11} \\ v_{12} \end{pmatrix}$ and $v_b = \begin{pmatrix} v_{21} \\ v_{22} \end{pmatrix}$ such that $\forall x \in \mathbb{R}^2$:

$$Mx = \begin{pmatrix} v_a \cdot x \\ v_b \cdot x \end{pmatrix} = \begin{pmatrix} v_{11}x_1 + v_{12}x_2 \\ v_{21}x_1 + v_{22}x_2 \end{pmatrix}$$

Let $e_1 = \begin{pmatrix} 1 \\ 0 \end{pmatrix}$, $e_2 = \begin{pmatrix} 0 \\ 1 \end{pmatrix}$, then:

$$Mx = \begin{pmatrix} v_a \cdot x \\ v_b \cdot x \end{pmatrix} = (v_a \cdot x)e_1 + (v_b \cdot x)e_2$$

This map is composed of two rank–one operators M_a and M_b associated with the vectors v_a and v_b. First, it will be investigated whether this linear map is injective or not. For this map to be injective, it is enough to show that ker$T$ has only one element, the zero element. This can be done by finding x that would satisfy: $Mx = \begin{pmatrix} v_a \cdot x \\ v_b \cdot x \end{pmatrix} = (v_a \cdot x)e_1 + (v_b \cdot x)e_2 = \begin{pmatrix} 0 \\ 0 \end{pmatrix}$. In other words, we seek a vector x that is orthogonal to v_a ($v_a \cdot x = 0$) and to v_b ($v_b \cdot x = 0$). If v_a and v_b are linearly dependent, then any vector that is orthogonal to v_a will be orthogonal to v_b as well, and therefore, **many** vectors will be mapped to $\hat{0}$ implying that the map M is not injective. In an injective map, v_a and v_b should be linearly independent, i.e., $\ker M_a \cap \ker M_b = \{\hat{0}\}$ (Figure 1–8). If v_a and v_b are linearly independent and are not equal to $\hat{0}$, then the map T is also surjective on $\mathbb{R}^2$ (using the results of the bonus exercises in section 1.2.3) and, therefore, it is invertible.

In general, associated with any linear map T:$\mathbb{R}^n \to \mathbb{R}^n$ is a matrix $M \in \mathbb{M}^n$ and a set of n vectors. The map T is invertible if and only if the n vectors are linearly independent.

▶ **BONUS EXERCISE:** Show that a linear map T:$\mathbb{R}^n \to \mathbb{R}^n$ is invertible (injective and surjective) if and only if the n vectors forming the square matrix associated with T are linearly independent.

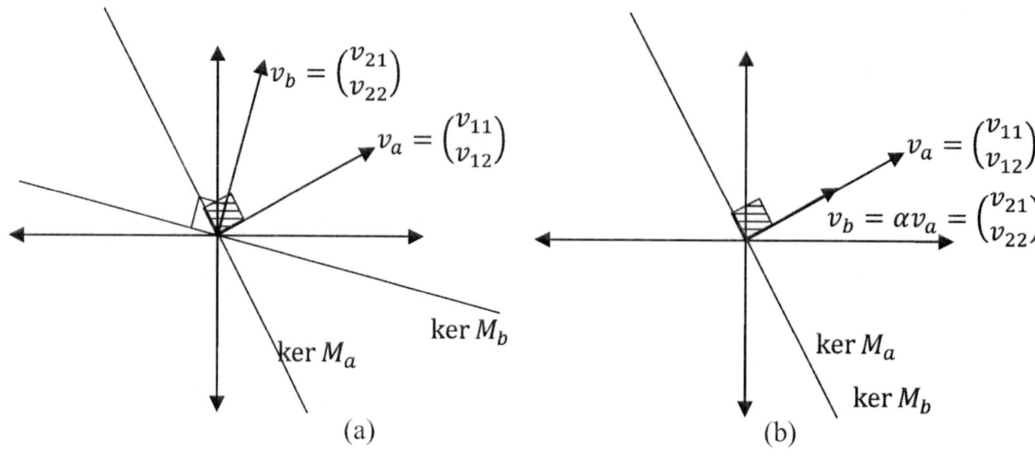

FIGURE 1-8 Graphical representation of the two vectors v_a and v_b associated with the a) injective linear map, b)non–injective linear map $T:\mathbb{R}^2 \to \mathbb{R}^2$.

1.2.4.c. Examples

Example 1: Let $T:\mathbb{R}^2 \to \mathbb{R}$ such that $\forall x \in \mathbb{R}^2: T(x) = 2x_1 + \frac{1}{2}x_2$. Consider the action of this map on the following vectors $x_a = \begin{pmatrix} 1 \\ 0 \end{pmatrix}$, $x_b = \begin{pmatrix} 0 \\ 1 \end{pmatrix}$, $x_c = \begin{pmatrix} 2 \\ \frac{1}{2} \end{pmatrix}$, $x_d = \begin{pmatrix} \frac{1}{2} \\ -2 \end{pmatrix}$, $x_e = \begin{pmatrix} -\frac{1}{2} \\ 2 \end{pmatrix}$ and $x_f = \begin{pmatrix} -1 \\ -1 \end{pmatrix}$. Sketch a graphical representation of the map. Why is this map not invertible?

Solution: Associated with this map is the matrix $M = \begin{pmatrix} 2 & \frac{1}{2} \end{pmatrix}$ and the vector $v = \begin{pmatrix} 2 \\ \frac{1}{2} \end{pmatrix}$.

$$T(x_a) = v \cdot x_a = 2 \quad T(x_b) = v \cdot x_b = 0.5 \quad T(x_c) = v \cdot x_c = 4.25$$

$$T(x_d) = T(x_e) = T(x_d - x_e) = 0 \qquad T(x_f) = v \cdot x_f = -2.5$$

As shown in Figure 1–9, ker T = span$\{x_e\}$. The kernel of this map contains more than one element, and therefore, the map cannot be inverted.

Example 2: Let $T:\mathbb{R}^2 \to \mathbb{R}^2$ such that $\forall x \in \mathbb{R}^2: T(x) = \begin{pmatrix} 2x_1 \\ 3x_2 \end{pmatrix}$. Show that this map is invertible.

Solution: Associated with this map is the matrix $M = \begin{pmatrix} 2 & 0 \\ 0 & 3 \end{pmatrix}$ and the two rank–one operators M_a and M_b associated with the vectors $v_a = \begin{pmatrix} 2 \\ 0 \end{pmatrix}$ and $v_b = \begin{pmatrix} 0 \\ 3 \end{pmatrix}$. Since the two vectors v_a and v_b

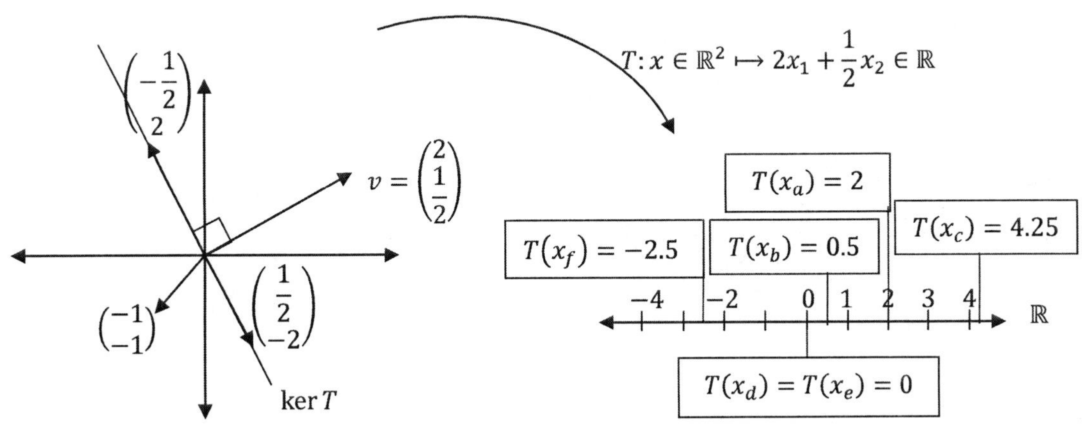

FIGURE 1-9 Graphical representation of Example 1.

are linearly independent, then the matrix M is invertible. $\ker M_a = \text{span}\left\{\begin{pmatrix} 0 \\ 1 \end{pmatrix}\right\}$, while $\ker M_b = \text{span}\left\{\begin{pmatrix} 1 \\ 0 \end{pmatrix}\right\}$. Therefore, $\ker M = \{\hat{0}\}$.

Note: Any vector in $\ker M_a$ will be mapped to a vector whose first component is zero. Let $x \in \ker M_a$, $x = \begin{pmatrix} 0 \\ x_2 \end{pmatrix}$, $Mx = \begin{pmatrix} 0 \\ 3x_2 \end{pmatrix}$. Similarly, any vector in $\ker M_b$ will be mapped to a vector whose second component is zero. Let $x \in \ker M_b x = \begin{pmatrix} x_1 \\ 0 \end{pmatrix}$, $Mx = \begin{pmatrix} 2x_1 \\ 0 \end{pmatrix}$. Figure 1–10 shows the action of this map on the vector $x = \begin{pmatrix} 1 \\ 1 \end{pmatrix}$.

Example 3: Let $T{:}\mathbb{R}^2 \to \mathbb{R}^2$ such that $\forall x \in \mathbb{R}^2{:}T(x) = \begin{pmatrix} 2x_1 \\ 3x_1 \end{pmatrix}$. Show that this map is not invertible and find its kernel.

Solution: Associated with this map is the matrix $M = \begin{pmatrix} 2 & 0 \\ 3 & 0 \end{pmatrix}$ and the two rank–one operators M_a and M_b associated with the vectors $v_a = \begin{pmatrix} 2 \\ 0 \end{pmatrix}$ and $v_b = \begin{pmatrix} 3 \\ 0 \end{pmatrix}$. Since the two vectors v_a and v_b

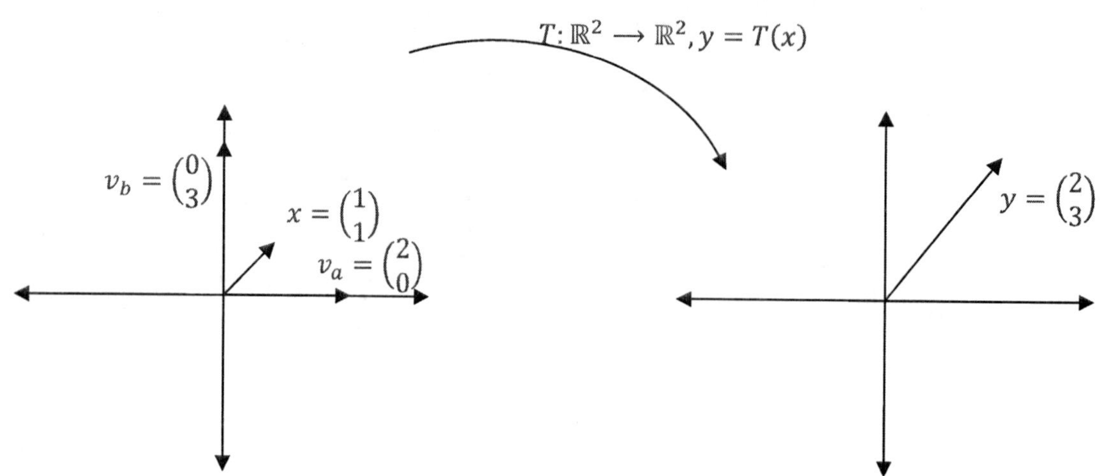

FIGURE 1-10 Graphical representation of Example 2.

are linearly dependent ($v_b = 1.5v_a$), then the matrix M is NOT invertible. $\ker M_a = \text{span}\left\{\begin{pmatrix} 0 \\ 1 \end{pmatrix}\right\}$, while $\ker M_b = \text{span}\left\{\begin{pmatrix} 0 \\ 1 \end{pmatrix}\right\}$. Therefore, $\ker M = \text{span}\left\{\begin{pmatrix} 0 \\ 1 \end{pmatrix}\right\}$.

Note: A matrix M is not invertible if $\exists x \neq \hat{0}$ such that $Mx = \hat{0}$; i.e., the map is not one–to–one. Clearly, any vector x whose second component is not zero will still be mapped to zero: $Mx = \begin{pmatrix} 2 & 0 \\ 3 & 0 \end{pmatrix}\begin{pmatrix} 0 \\ 5 \end{pmatrix} = \begin{pmatrix} 0 \\ 0 \end{pmatrix}$

Example 4: Let $T:\mathbb{R}^2 \rightarrow \mathbb{R}^2$ such that $\forall x \in \mathbb{R}^2: T(x) = \begin{pmatrix} 0 \\ 3x_1 + x_2 \end{pmatrix}$. Show that this map is not invert–ible, and find its kernel.

Solution: Associated with this map is the matrix $M = \begin{pmatrix} 0 & 0 \\ 3 & 1 \end{pmatrix}$ and the two rank–one opera–tors M_a and M_b associated with the vectors $v_a = \begin{pmatrix} 0 \\ 0 \end{pmatrix}$ and $v_b = \begin{pmatrix} 3 \\ 1 \end{pmatrix}$. $\ker M_a = \mathbb{R}^2$ while $\ker M_b = \text{span}\left\{\begin{pmatrix} -1 \\ 3 \end{pmatrix}\right\}$. Therefore, $\ker M = \ker M_a \cap \ker M_b = \text{span}\left\{\begin{pmatrix} -1 \\ 3 \end{pmatrix}\right\}$ (Figure 1–11).

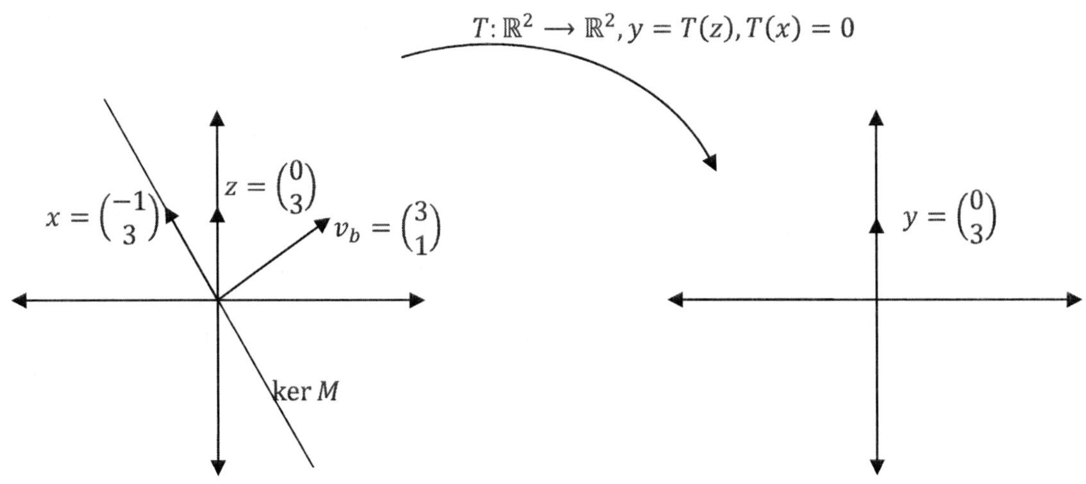

FIGURE 1-11 Graphical representation of Example 4.

Notice the following: Clearly, any vector x that is a multiple of $x = \begin{pmatrix} -1 \\ 3 \end{pmatrix}$ will still be mapped to zero:

$$Mx = \begin{pmatrix} 0 & 0 \\ 3 & 1 \end{pmatrix} \begin{pmatrix} -1 \\ 3 \end{pmatrix} = \begin{pmatrix} 0 \\ 0 \end{pmatrix}$$

Example 5: Let $T:\mathbb{R}^2 \to \mathbb{R}^2$such that $\forall x \in \mathbb{R}^2 : T(x) = \begin{pmatrix} 2x_1 + x_2 \\ x_2 \end{pmatrix}$. Show that this map is invertible.

Solution: Associated with this map is the matrix $M = \begin{pmatrix} 2 & 1 \\ 0 & 1 \end{pmatrix}$ and the two rank–one operators M_a and M_b associated with the vectors $v_a = \begin{pmatrix} 2 \\ 1 \end{pmatrix}$ and $v_b = \begin{pmatrix} 0 \\ 1 \end{pmatrix}$. $\ker M_a = \text{span}\left\{\begin{pmatrix} -1 \\ 2 \end{pmatrix}\right\}$ while $\ker M_b = \text{span}\left\{\begin{pmatrix} 1 \\ 0 \end{pmatrix}\right\}$. Therefore, $\ker M = \text{span}\left\{\begin{pmatrix} 0 \\ 0 \end{pmatrix}\right\} = \{\tilde{0}\}$. Therefore, the map T is invertible.

Example 6: Identity maps in Mathematica:

$T:\mathbb{R}^n \to \mathbb{R}^n$ is an **identity map** if $\forall x \in \mathbb{R}^n : T(x) = x$. Associated with this map is the identity matrix $I \in M^n$, which has unit diagonal entries and zero off–diagonal entries. The Mathematica function

`IdentityMatrix[n]` returns a square identity matrix with dimension n. In the following code, the identity matrix I in $\mathbb{M}^2$ and $\mathbb{M}^3$ are produced.

```
I1=IdentityMatrix[2];
I1//MatrixForm
I2=IdentityMatrix[3];
I2//MatrixForm
```

Example 7: Finding the Kernel using Mathematica:

The Mathematica function to find the kernel of a linear map is: `NullSpace[]`. The function gives a list of vectors that form a basis for the kernel. In the following code, the kernel of the five matrices defined in the previous five examples is obtained. Notice that Mathematica returns {}when the kernel contains only the zero element.

```
M1={{2,1/2}};
M2={{2,0},{0,3}};
M3={{2,0},{3,0}};
M4={{0,0},{3,1}};
M5={{2,1},{0,1}};
NullSpace[M1]
NullSpace[M2]
NullSpace[M3]
NullSpace[M4]
NullSpace[M5]
```

1.2.4.d. Composition of Linear Operators / Multiplying Matrices

Let $U:\mathbb{R}^n \to \mathbb{R}^k$ and $V:\mathbb{R}^k \to \mathbb{R}^m$ be linear maps. The composition map $T = V \circ U:\mathbb{R}^n \to \mathbb{R}^m$ is also a linear map since:

$$\forall x, y \in \mathbb{R}^n, \forall \alpha, \beta \in \mathbb{R}: T(\alpha x + \beta y) = (V \circ U)(\alpha x + \beta y) = V(\alpha U(x) + \beta U(Y)) = \alpha T(x) + \beta T(y)$$

As mentioned before, each linear map has an associated matrix when dealing with Euclidian vector spaces. When $n = m = k$, then associated with U and V are square matrices N and $L \in \mathbb{M}^n$. The composition map T is also associated with a square matrix $M \in \mathbb{M}^n$. The components of the matrix M can be obtained from the components of N and L by the equality $Mx = LNx$:

$$Mx = \begin{pmatrix} \sum_{j=1}^{n} M_{1j}x_j \\ \sum_{j=1}^{n} M_{2j}x_j \\ \dots \\ \sum_{j=1}^{n} M_{nj}x_j \end{pmatrix} = LNx = L \begin{pmatrix} \sum_{j=1}^{n} N_{1j}x_j \\ \sum_{j=1}^{n} N_{2j}x_j \\ \dots \\ \sum_{j=1}^{n} N_{nj}x_j \end{pmatrix} = \begin{pmatrix} \sum_{k=1}^{n}\sum_{j=1}^{n} L_{1k}N_{kj}x_j \\ \sum_{k=1}^{n}\sum_{j=1}^{n} L_{2k}N_{kj}x_j \\ \dots \\ \sum_{k=1}^{n}\sum_{j=1}^{n} L_{2k}N_{kj}x_j \end{pmatrix}$$

It follows that the components of the matrix M can be obtained as follows:

$$M_{ij} = \sum_{k=1}^{n} L_{ik} N_{kj}$$

Thus, the components of the matrix M are composed of the dot product between row vectors in the matrix L and column vectors in the matrix N. The order of the application of the linear map is important; the composition NL is not equal to LN.

Problem: Consider the matrices $LN \in \mathbb{M}^3$:

$$L = \begin{pmatrix} 1 & 2 & 3 \\ 2 & 2 & 2 \\ 3 & 2 & 1 \end{pmatrix}, \quad N = \begin{pmatrix} 5 & 5 & 5 \\ 1 & 2 & 3 \\ 3 & 2 & 1 \end{pmatrix}$$

Use Mathematica to show that the matrices $M_1 = LN$ and $M_2 = NL$ are not equal. Then consid– ering the vector $x = \{1, 1, 1\}$, show that the vector y obtained by the linear operation $y = M_1 x$ is equivalent to the vector l obtained as $l = Lz$ where $z = Nx$.

Solution: The matrix multiplication procedure described above produces the following two matrices, which are not equal:

$$M_1 = LN = \begin{pmatrix} 16 & 15 & 14 \\ 18 & 18 & 18 \\ 20 & 21 & 22 \end{pmatrix}, \quad M_2 = NL = \begin{pmatrix} 30 & 30 & 30 \\ 14 & 12 & 10 \\ 10 & 12 & 14 \end{pmatrix}$$

The vectors y and l are equal:

$$y = M_1 x \begin{pmatrix} 45 \\ 54 \\ 63 \end{pmatrix}, \quad z = Nx = \begin{pmatrix} 15 \\ 6 \\ 6 \end{pmatrix}, \quad l = Lz = \begin{pmatrix} 45 \\ 54 \\ 63 \end{pmatrix}$$

The following Mathematica code performs the required operations. The matrices and vectors operations are performed using the period character.

```
matL={{1,2,3},{2,2,2},{3,2,1}};
matN={{5,5,5},{1,2,3},{3,2,1}};
M1=matL.matN;
M2=matN.matL;
M1//MatrixForm
```

```
M2//MatrixForm
x={1,1,1};
y=M1.x;
y//MatrixForm
z=matN.x;
z//MatrixForm
l=matL.z;
l//MatrixForm
```

1.2.5. $\mathbb{M}^2$ and $\mathbb{M}^3$ Linear Maps

In solid and continuum mechanics, we primarily deal with linear maps $T:\mathbb{R}^2 \to \mathbb{R}^2$ and $T:\mathbb{R}^3 \to \mathbb{R}^3$. In this section, we will focus on those linear maps, bearing in mind that $\mathbb{M}^2$ represents all the possible linear maps $T:\mathbb{R}^2 \to \mathbb{R}^2$ while $\mathbb{M}^3$ represents all the possible linear maps $T:\mathbb{R}^3 \to \mathbb{R}^3$. In this section, we will use the term *matrix* freely as a representative of the corresponding linear map since we are dealing primarily with Euclidian vector spaces. Remember that an invertible matrix indicates that the associated linear map is invertible.

1.2.5.a. Notation and Summation Convention (Einstein Summation Convention)

Before we continue with introducing new material, it is necessary to introduce some notation tools used when dealing with $\mathbb{R}^3$ that would minimize the expansion terms used. In the following, $u, v, w \in \mathbb{R}^3$, $e_i (i = 1, 2, 3)$ form the standard orthonormal basis, I is the identity matrix in $\mathbb{M}^3$ and $M, N, K \in \mathbb{M}^3$.

- Any expression in which a suffix appears twice is understood to be summed over the range 1, 2, 3 of that suffix.

$$u = u_1 e_1 + u_2 e_2 + u_3 e_3 = \sum_{i=1}^{3} u_i e_i = u_i e_i$$

- The Kronecker delta is defined as:

$$\delta_{ij} = \begin{cases} 1, & i = j \\ 0, & i \neq j \end{cases}$$

So now we can write,

$$I_{ij} = e_i \cdot e_j = \delta_{ij}$$

- The Alternator is defined as:

$$\varepsilon_{ijk} = \begin{cases} 1, & i, j, k \text{ is a cyclic permutation of } 1, 2, 3 \\ -1, & i, j, k \text{ is a noncyclic permutation of } 1, 2, 3 \\ 0, & \textit{otherwise} \end{cases}$$

So now we can write

$$u \cdot (v \times w) = \varepsilon_{ijk} u_i v_j w_k, \quad e_i \times e_j = \varepsilon_{ijk} e_k$$

- Notice the following identities:

$$u \cdot e_i = u_i, \quad e_i \cdot M e_j = M_{ij}$$

- Examples:

$$\begin{aligned} Mu \in \mathbb{R}^3 \Rightarrow (Mu)_i &= M_{ij} u_j \\ MN \in \mathbb{M}^3 \Rightarrow (MN)_{ij} &= M_{ik} N_{kj} \\ MNK \in \mathbb{M}^3 \Rightarrow (MNK)_{ij} &= M_{ik} N_{kl} K_{lj} \\ v \cdot Mu \in \mathbb{R} \Rightarrow v \cdot Mu &= v_i M_{ij} u_j \end{aligned}$$

1.2.5.b. Determinant of a Matrix

The determinant of a two–dimensional or a three–dimensional matrix has two major roles. First, it is an indication of whether the matrix is invertible or not. A square matrix is invertible if and only if the determinant is not equal to zero. Second, the determinant of the matrix gives a geometric interpretation of the area (when $M \in \mathbb{M}^2$) and volume (when $M \in \mathbb{M}^3$) transformations associated with the map.

Definition of the Determinant of a Matrix in $\mathbb{M}^2$:

Let $M \in \mathbb{M}^2$. $M = \begin{pmatrix} a_1 & a_2 \\ b_1 & b_2 \end{pmatrix}$. The determinant of M is defined as:

$$\det M = a_1 b_2 - a_2 b_1$$

Results of the definition:

<u>$\det M = 0$ if and only if M is not invertible:</u>

Assume M is not invertible$\Longrightarrow$ section 1.2.4.b then asserts that the two vectors forming the matrix are linearly dependent $\Longrightarrow \exists \alpha$ s.t. $b_1 = \alpha a_1$ and $b_2 = \alpha a_2 \Longrightarrow \det M = 0$.

Assume M is invertible$\Longrightarrow$ section 1.2.4.b then asserts that the two vectors forming the matrix are linearly independent $\Longrightarrow \det M \neq 0$.

Therefore, $\det M = 0$ iff M is not invertible.

It was also shown in section 1.2.3 that a linear map is invertible if and only if its kernel set contains only the zero vector. Therefore, we reach the following important argument:

$$\det M = 0 \Longleftrightarrow M \text{ is not invertible} \Longleftrightarrow \text{the vectors associated with } M \text{ are linearly dependent}$$
$$\Longleftrightarrow \ker M \neq \left\{\hat{0}\right\} \Longleftrightarrow \exists x \in \mathbb{R}^2 x \neq \hat{0}\, s.t.\, Mx = \hat{0}$$

Similarly, the opposite argument is

$$\det M \neq 0 \Longleftrightarrow M \text{ is invertible} \Longleftrightarrow \text{the vectors associated with } M \text{ are linearly independent}$$
$$\Longleftrightarrow \ker M = \left\{\hat{0}\right\}$$

<u>Area transformation:</u>

Consider the two unit vectors $e_1 = \{1, 0\}$ and $e_2 = \{0, 1\}$. The area between the initial two vectors is a square with a unit area. Let x_1 and x_2 be their respective images under the transformation. $x_1 := Me_1 = \{a_1, b_1\}$. $x_2 := Me_2 = \left\{a_2, b_2\right\}$. The two image vectors form a parallelogram whose area is the determinant of the matrix M (Figure 1–12).

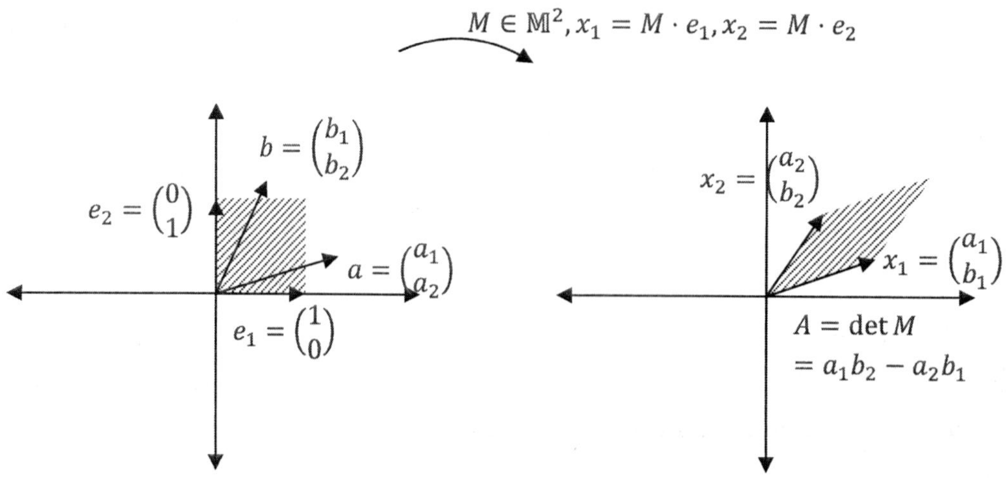

FIGURE 1-12 Area transformation under $\mathbb{M}^2$.

▶ **BONUS EXERCISE:** Show that the area of the parallelogram with vertices $\begin{pmatrix} 0 \\ 0 \end{pmatrix}$, $\begin{pmatrix} a_1 \\ b_1 \end{pmatrix}$, $\begin{pmatrix} a_1 + a_2 \\ b_1 + b_2 \end{pmatrix}$, $\begin{pmatrix} a_2 \\ b_2 \end{pmatrix}$, is equal to $a_1 b_2 - a_2 b_1$.

Definition of the Determinant of a Matrix in $\mathbb{M}^3$:

Let $M \in \mathbb{M}^3$. $M = \begin{pmatrix} a_1 & a_2 & a_3 \\ b_1 & b_2 & b_3 \\ c_1 & c_2 & c_3 \end{pmatrix}$. Let $a = \begin{pmatrix} a_1 \\ a_2 \\ a_3 \end{pmatrix}$, $b = \begin{pmatrix} b_1 \\ b_2 \\ b_3 \end{pmatrix}$, $c = \begin{pmatrix} c_1 \\ c_2 \\ c_3 \end{pmatrix}$, then the determinant of M is defined as:

$$\det M = a \cdot (b \times c) = a_1 (b_2 c_3 - b_3 c_2) + a_2 (b_3 c_1 - b_1 c_3) + a_3 (b_1 c_2 - b_2 c_1).$$

Results of the definition:

$\det M = 0$ iff M is not invertible:

Assume M is not invertible $\Longleftrightarrow$ section 1.2.4.b then asserts that the three vectors a, b and c forming the matrix are linearly dependent $\Longleftrightarrow$ The triple product described in section 1.1.7.b is equal to zero: $a \cdot (b \times c) = 0 \Longleftrightarrow \det M = 0$.

It was also shown in section 1.2.3 that a linear map is invertible if and only if its kernel set contains only the zero vector. Therefore, we reach the following important argument:

$$\det M = 0 \Longleftrightarrow M \text{ is not invertible} \Longleftrightarrow \text{the vectors associated with } M \text{ are linearly dependent}$$
$$\Longleftrightarrow \ker M \neq \{\hat{0}\} \Longleftrightarrow \exists x \in \mathbb{R}^3, x \neq \hat{0} \; s.t. Mx = \hat{0}$$

Similarly, the opposite argument is:

$$\det M \neq 0 \Longleftrightarrow M \text{ is invertible} \Longleftrightarrow \text{the vectors associated with } M \text{ are linearly independent}$$
$$\Longleftrightarrow \ker M = \{\hat{0}\}$$

Volume transformation:

Let's now consider the three unit vectors $e_1 = \{1, 0, 0\}$, $e_2 = \{0, 1, 0\}$ and $e_3 = \{0, 0, 1\}$. The volume between the initial three vectors is a cube with unit area. Let x_1, x_2 and x_3 be their respective images under the transformation.

$x_1 = M \cdot e_1 = \{a_1, b_1, c_1\}$, $\quad x_2 = M \cdot e_2 = \{a_2, b_2, c_2\}$, and $x_3 = M \cdot e_3 = \{a_3, b_3, c_3\}$.

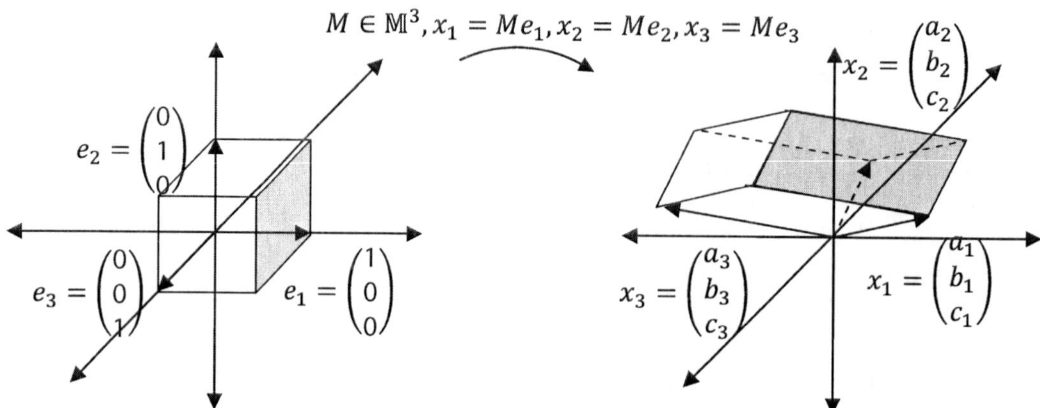

FIGURE 1-13 Volume transformation under $\mathbb{M}^3$.

The three image vectors form a parallelepiped whose volume is the determinant of the matrix M (Figure 1–13).

$$\begin{aligned}\det M &= a \cdot (b \times c) = a_1\left(b_2c_3 - b_3c_2\right) + a_2\left(b_3c_1 - b_1c_3\right) + a_3\left(b_1c_2 - b_2c_1\right)\\ &= a_1\left(b_2c_3 - b_3c_2\right) + b_1\left(a_3c_2 - a_2c_3\right) + c_1\left(a_2b_3 - b_2a_3\right)\\ &= x_1 \cdot (x_2 \times x_3)\\ &= Me_1 \cdot (Me_2 \times Me_3)\end{aligned}$$

The Alterator ε_{ijk} defined in section 1.2.5.a can be used to reach the following useful equality:

$$\det M = Me_1 \cdot (Me_2 \times Me_3) \Longrightarrow Me_i \cdot \left(Me_j \times Me_k\right) = \varepsilon_{ijk} \det M$$

General Definition of a Determinant:

The determinant definition can be extended to matrices of larger sizes. Let $M \in \mathbb{M}^n$. The determinant of M is by definition the recursive relationship: $\det M = \sum_{i=1}^{n} (-1)^{1+i} M_{1i} \det N$ where $N \in \mathbb{M}^{n-1}$ and is calculated by eliminating the 1st row and the ith column of the matrix M. It can be shown that $\det M = 0 \Longleftrightarrow M$ is not invertible.

1.2.5.c. Volume Transformation in $\mathbb{R}^3$ and Further Properties of the Determinant

Consider the three arbitrary, linearly independent vectors u, v and $w \in \mathbb{R}^3$. Consider the action of the linear map $M \in \mathbb{M}^3$. Then, the ratio between the volume of the transformed parallelepiped formed by Mu, Mv and Mw to the volume of parallelepiped formed by u, v and w is equal to $\det M$:

Let VOL denote the volume before the linear transformation:

$$VOL = u \cdot (v \times w) = \varepsilon_{ijk} u_i v_j w_k$$

Let vol denote the volume after the linear transformation and using the conventions and the results in sections 1.2.5.a and 1.2.5.b:

$$vol = Mu_i e_i \cdot \left(Mv_j e_j \times Mw_k e_k\right) = u_i v_j w_k \left(Me_i \cdot \left(Me_j \times Me_k\right)\right) = \left(\det M\right) \varepsilon_{ijk} u_i v_j w_k$$

Therefore,

$$\frac{vol}{VOL} = \det M = \frac{Mu \cdot (Mv \times Mw)}{u \cdot (v \times w)}$$

▶ **BONUS EXERCISE:** $\forall N, M \in \mathbb{M}^3$, $\forall \alpha \in \mathbb{R}$, I is the identity map in $\mathbb{M}^3$, use the above results to show the following:

$$\det NM = \det N \det M$$

$$\det \alpha M = \alpha^3 \det M$$

$$\det I = 1$$

1.2.5.d. Matrix Transpose

Let $M \in \mathbb{M}^n$, the transpose M^T of the matrix M is the unique matrix that satisfies $\forall u,\ v \in \mathbb{R}^n$:

$$Mu \cdot v = u \cdot M^T v$$

▶ **BONUS EXERCISE:** Use the above definition directly to show that $\forall M,\ N \in \mathbb{M}^n$:

$$(MN)^T = N^T M^T$$

$$(N + M)^T = N^T + M^T$$

Notes:

- Observe the component forms for the matrix transpose in the following two expressions:

$$\left(M^T\right)_{ij} = M_{ji}$$

$$\left(NM^T\right)_{ij} = N_{ik} M_{jk}$$

- The determinant of a matrix is equal to the determinant of its transpose:

$$\det M = \varepsilon_{ijk} M_{1i} M_{2j} M_{3k} = \varepsilon_{ijk} M_{i1} M_{j2} M_{k3} = \det M^T$$

1.2.5.e. Inverse of a Matrix

It was shown in section 1.2.4.b that a linear map $T:\mathbb{R}^n \to \mathbb{R}^n$ is invertible when the only element in the kernel is the zero vector. This is achieved when the square matrix associated with the linear map has a linearly independent set of vectors associated with it. In section 1.2.5.b, it was shown that the determinant of the matrix is equal to 0 if and only if it is associated with a linearly dependent set of vectors. It follows then that if a linear map is invertible, then the determinant is not equal to 0, and there exists an inverse that satisfies the following:

Let $M \dot{\in} \mathbb{M}^3,\ x \in \mathbb{R}^3, y := Mx, \det M \neq 0 \Longrightarrow \exists! M^{-1}$ s.t.

$$MM^{-1} = M^{-1}M = I$$

$$Mx = y \Rightarrow x = M^{-1}y$$

Using the results of section 1.2.5.c and 1.2.5.d:

$$\det\left(MM^{-1}\right) = \det I = 1 \Rightarrow \det M^{-1} = \frac{1}{\det M}$$

$$\left(M^T\right)^{-1} = \left(M^{-1}\right)^T = M^{-T}$$

The inverse of an invertible matrix $M \in \mathbb{M}^3$ can be constructed as follows:

Let $M = \begin{pmatrix} a_1 & a_2 & a_3 \\ b_1 & b_2 & b_3 \\ c_1 & c_2 & c_3 \end{pmatrix}$. Then, $M^{-1} = \dfrac{1}{a \cdot (b \times c)}\left((b \times c) \quad (c \times a) \quad (a \times b)\right)$

Where $(b \times c)$, $(c \times a)$ and $(a \times b)$ are input as column vectors in the matrix representation of M^{-1}.

1.2.5.f. General Area Transformation in $\mathbb{R}^3$

Consider the two arbitrary, linearly independent vectors v and $w \in \mathbb{R}^3$. Consider the action of the linear map $M \in \mathbb{M}^3$. Then, the cross product $v \times w = (A)N$ where $A \in \mathbb{R}$ is the area of the parallelogram formed by v and w, while $N \in \mathbb{R}^3$ is a unit vector orthogonal to both v and w. Consider the cross product $Mv \times MW = (a)n$ where $a \in \mathbb{R}$ is the area of the parallelogram formed by Mv and Mw while $n \in \mathbb{R}^3$ is a unit vector orthogonal to both Mv and Mw (Figure 1–14). The relationship between $(A)N$ and $(a)n$ can be obtained using the results of section 1.2.5.c. Let $u \in \mathbb{R}^3$ be an arbitrary vector, then:

$$Mu \cdot (Mv \times Mw) = \det M \,(u \cdot (v \times w))$$

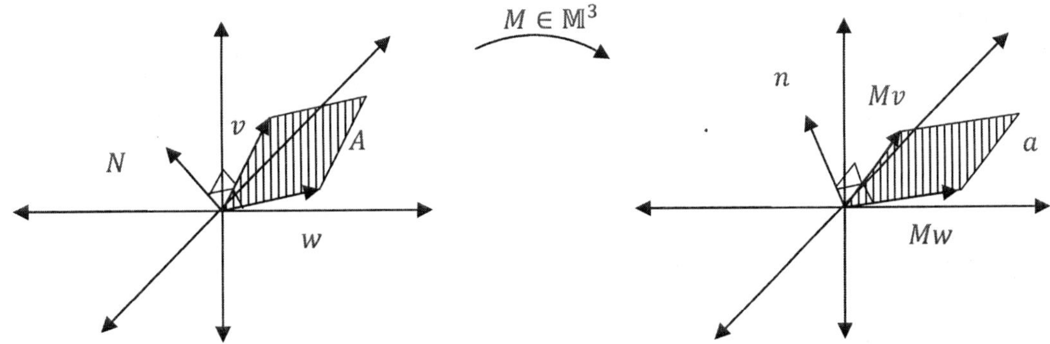

FIGURE 1-14 Area transformation under $\mathbb{M}^3$.

Using the definition of the transpose of the matrix (See section 1.2.5.d):

$$u \cdot M^T (Mv \times Mw) = u \cdot \det M\, (v \times w)$$

And since u is arbitrary it follows that:

$$M^T (Mv \times Mw) = \det M\, (v \times w)$$

$$M^T (a) n = \left(\det M\right) (A) N$$

If M is invertible, then:

$$(a) n = M^{-T} \left(\det M\right) (A) N$$

1.2.5.g. Eigenvalues and Eigenvectors

Let $M \in \mathbb{M}^n$. $\lambda \in \mathbb{R}$ is an eigenvalue of M if $\exists x \in \mathbb{R}^n$, $x \neq \hat{0}$ s.t. $Mx = \lambda x$. In this case x is an eigenvec–tor associated with the eigenvalue λ. Any nonzero multiplier of x is also an eigenvector associated with the eigenvalue λ.

The eigenvalue problem is the solution to the problem $(M - \lambda I)\, x = \hat{0}$ where I is the identity matrix and x is a nonzero vector in $\mathbb{R}^n$. In other words, the matrix $M - \lambda I$ is not invertible or $\det (M - \lambda I) = 0$. Notice that a very important result of the definition of the eigenvalues is that $\lambda = 0$ is an eigenvalue of any non–invertible matrix M (See exercises page 72.).

Remarks:

- Graphically, the eigenvector of a matrix is a vector that is parallel to its image due to the linear map produced by the matrix.

- The maximum number of eigenvalues of a matrix $M \in \mathbb{M}^3$ is three (**Why?**).
- It is possible that the eigenvalues are complex.

It will be shown in the next examples that the number of the eigenvalues and eigenvectors depend on the nature of the matrix.

Problem 1: Find the eigenvectors and eigenvalues of the identity matrix $I \in \mathbb{M}^3$.

Solution:

Eigenvalues:

$$(I - \lambda I)\,x = \hat{0} \Rightarrow \det\left((1-\lambda)\,I\right) = 0 \Rightarrow (1-\lambda)^3 = 0 \Rightarrow \lambda = 1$$

Eigenvectors: every nonzero vector is an eigenvector of I, $\forall x \in \mathbb{R}^3 : Ix = x$.

Problem 2: Find the eigenvectors and eigenvalues of the matrix $M = \begin{pmatrix} 1 & 0 & 0 \\ 0 & 2 & 0 \\ 0 & 0 & 3 \end{pmatrix}$.

Solution:

Eigenvalues:

$$(M - \lambda I)\,x = \hat{0} \Rightarrow \det(M - \lambda I) = 0 \Rightarrow (1-\lambda)(2-\lambda)(3-\lambda) = 0 \Rightarrow \lambda = 1, 2, 3$$

Eigenvectors:

Assume $\lambda = 1 \Rightarrow (M - I)\,x = \hat{0} \Rightarrow \begin{pmatrix} 0 & 0 & 0 \\ 0 & 1 & 0 \\ 0 & 0 & 2 \end{pmatrix}\begin{pmatrix} x_1 \\ x_2 \\ x_3 \end{pmatrix} = \begin{pmatrix} 0 \\ 0 \\ 0 \end{pmatrix} \Rightarrow x_1$ could have any value, x_2 and $x_3 = 0$.

Therefore, a nonzero UNIT eigenvector associated with $\lambda = 1$ is $x = \{1, 0, 0\}$

Similarly, $y = \{0, 1, 0\}$ and $z = \{0, 0, 1\}$ are the unit eigenvectors associated with $\lambda = 2$ and $\lambda = 3$, respectively.

Problem 3: Find the eigenvectors and eigenvalues of the matrix $M = \begin{pmatrix} 1 & 5 & 0 \\ 0 & 0 & 2 \\ 0 & 0 & 3 \end{pmatrix}$.

Solution:

Eigenvalues:

$$\det\begin{pmatrix}1-\lambda & 5 & 0\\ 0 & -\lambda & 2\\ 0 & 0 & 3-\lambda\end{pmatrix} = 0 \Rightarrow \lambda\,(\lambda-1)\,(3-\lambda) = 0 \Rightarrow \lambda = 0, 1, 3$$

Eigenvectors:

Assume $\lambda = 0 \Rightarrow Mx = \hat{0} \Rightarrow \begin{pmatrix}1 & 5 & 0\\ 0 & 0 & 2\\ 0 & 0 & 3\end{pmatrix}\begin{pmatrix}x_1\\ x_2\\ x_3\end{pmatrix} = \begin{pmatrix}0\\ 0\\ 0\end{pmatrix} \Rightarrow x_1 = -5x_2$ and $x_3 = 0$. Therefore, a nonzero eigenvector associated with $\lambda = 0$ is $x = \{-5, 1, 0\}$

Similarly, $y = \{1, 0, 0\}$ and $z = \{5, 2, 3\}$ are the eignevalues associated with $\lambda = 1$ and $\lambda = 3$ respec– tively.

Problem 4: Find the eigenvalues of the matrix $M = \begin{pmatrix}0 & 1 & 1\\ 0 & 0 & 1\\ 0 & -1 & 0\end{pmatrix}$

Solution:

Eigenvalues: (Notice that the eigenvalues are complex.)

$$\det\begin{pmatrix}-\lambda & 1 & 1\\ 0 & -\lambda & 1\\ 0 & -1 & -\lambda\end{pmatrix} = 0 \Rightarrow -\lambda\left(\lambda^2+1\right) = 0 \Rightarrow \lambda = 0, i, -i$$

1.2.5.h. Mathematica Functions

The following Mathematica code clears the assignment for the variables a, b, c and d, and creates the matrix $M = \begin{pmatrix}a & b\\ c & d\end{pmatrix}$. Then, it calculates its inverse, its determinant, its eigenvalues and eigenvectors, and its transpose.

```
Clear[a,b,c,d]
M={{a,b},{c,d}}
Inverse[M]
Det[M]
Eigensystem[M]
Transpose[M]
```

There are various methods to create the identity matrix. In the following code, the function `IdentityMatrix[3]` is used to create $I \in \mathbb{M}^3$. Also, the function `KroneckerDelta[i,j]` is used to create the same matrix but using the table command:

```
I1=IdentityMatrix[3]
I2=Table[KroneckerDelta[i,j],{i,1,3},{j,1,3}]
I1//MatrixForm
I2//MatrixForm
```

1.2.6. Special Matrices in $\mathbb{M}^2$ and $\mathbb{M}^3$

In this section, families of matrices that have special roles in solid mechanics will be considered. The first family is that of orthogonal matrices. These matrices play an important role in rotations, reflections, and changing of the basis set of a linear vector space. Secondly, since the majority of the matrices in solid mechanics are symmetric, some important properties of symmetric matrices will be studied. Some of the definitions are given for the general family of matrices $\mathbb{M}^n$ but the examples are given for $n = 2$ and/or 3.

1.2.6.a. Orthogonal Matrices

DEFINITION A matrix $Q \in \mathbb{M}^2$ $(\mathbb{M}^3)$ is orthogonal if the two (three) associated vectors are orthonormal. The following properties of orthogonal matrices follow from the definition. Let a, b and c form a set of three orthonormal vectors associated with Q such that:

$$Q = \begin{pmatrix} a_1 & a_2 & a_3 \\ b_1 & b_2 & b_3 \\ c_1 & c_2 & c_3 \end{pmatrix}, \quad Q^T = \begin{pmatrix} a_1 & b_1 & c_1 \\ a_2 & b_2 & c_2 \\ a_3 & b_3 & c_3 \end{pmatrix}$$

Then,

$$QQ^T = \begin{pmatrix} a \cdot a & a \cdot b & a \cdot c \\ b \cdot a & b \cdot b & b \cdot c \\ c \cdot a & c \cdot b & c \cdot c \end{pmatrix} = \begin{pmatrix} 1 & 0 & 0 \\ 0 & 1 & 0 \\ 0 & 0 & 1 \end{pmatrix} = I$$

It follows that Q is invertible and that $Q^T = Q^{-1}$. Hence,

$$QQ^T = Q^T Q = I$$

Let $Q \in \mathbb{M}^3$ be an orthogonal matrix, then:

- $\det Q = \det Q^T = \det Q^{-1} = a \cdot (b \times c) = \pm 1$
- Q preserves the norm and the dot product, $\forall u, v \in \mathbb{R}^3$:

$$Qu \cdot Qv = u \cdot Q^T Qv = u \cdot v$$

$$\|Qu\|^2 = Qu \cdot Qu = u \cdot Q^T Qu = u \cdot u = \|u\|^2$$

- The identities $QQ^T = Q^T Q = I$ can be written in component form as:

$$Q_{ik} Q_{jk} = Q_{ki} Q_{kj} = \delta_{ij}$$

- If Q_1, Q_2 are orthogonal matrices, then their product $Q_1 Q_2$ is an orthogonal matrix:

$$Q_1 Q_2 (Q_1 Q_2)^T = Q_1 Q_2 Q_2^T Q_1^T = I$$

Rotation Matrices: When $\det Q = 1$ then Q is considered a rotation. In this case, the associated row vectors of Q follow the right–handed orientation. When $Q \in \mathbb{M}^2$, it represents a planar rotation of an angle θ, while if $Q \in \mathbb{M}^3$, it is associated with three angles of rotation θ_x, θ_y and θ_z.

Examples of Rotation Matrices:

- The identity matrix is a rotation matrix associated with zero rotation angles.
- $Q = \begin{pmatrix} \cos\theta & -\sin\theta \\ \sin\theta & \cos\theta \end{pmatrix}$ is the general rotation matrix in $\mathbb{M}^2$ that rotates every vector $u \in \mathbb{R}^2$ counter–clockwise by an angle θ (Figure 1–15). Notice that $Qe_1 = \begin{pmatrix} \cos\theta \\ \sin\theta \end{pmatrix}$, $Qe_2 = \begin{pmatrix} -\sin\theta \\ \cos\theta \end{pmatrix}$.
- The following are three counter–clockwise rotation matrices Q_x, Q_y and Q_z around the three mutually perpendicular axes x, y and z aligned with e_1, e_2 and e_3 in $\mathbb{R}^3$ respectively. Notice that the rotation matrix $Q_1 = Q_z Q_y Q_x$ is different from the rotation matrix $Q_2 = Q_x Q_y Q_z$

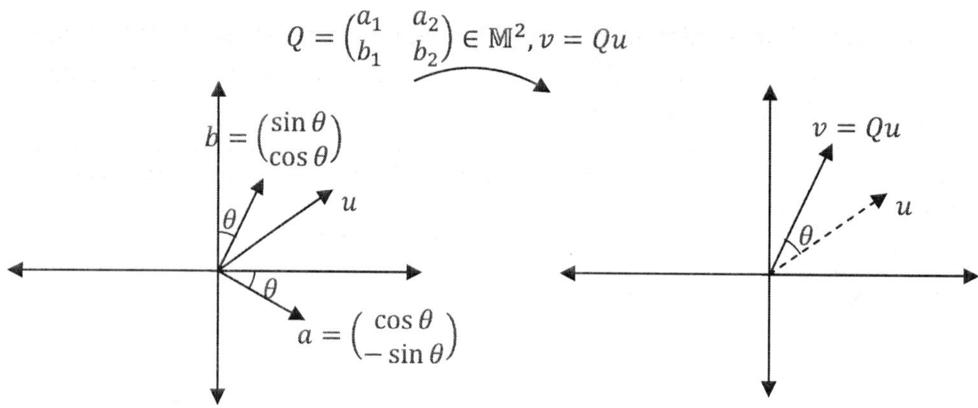

FIGURE 1-15 Counter–clockwise rotation in $\mathbb{R}^2$ by an angle θ.

$$Q_x = \begin{pmatrix} 1 & 0 & 0 \\ 0 & \cos\theta_x & -\sin\theta_x \\ 0 & \sin\theta_x & \cos\theta_x \end{pmatrix}, \; Q_y = \begin{pmatrix} \cos\theta_y & 0 & \sin\theta_y \\ 0 & 1 & 0 \\ -\sin\theta_y & 0 & \cos\theta_y \end{pmatrix}, \; Q_z = \begin{pmatrix} \cos\theta_z & -\sin\theta_z & 0 \\ \sin\theta_z & \cos\theta_z & 0 \\ 0 & 0 & 1 \end{pmatrix}$$

Reflection Matrices: When $\det Q = -1$ then Q is considered a reflection. In this case, the associated row vectors of Q follow the left–handed orientation. When $Q \in \mathbb{M}^2$, it represents a planar reflection across a vector, while if $Q \in \mathbb{M}^3$, it is associated with a reflection across a plane. In continuum mechanics, reflection matrices are usually avoided since reflection of bodies is not physically allowed.

Examples of Reflection Matrices:

- The following matrices are reflections:

$$I_1 = \begin{pmatrix} -1 & 0 & 0 \\ 0 & 1 & 0 \\ 0 & 0 & 1 \end{pmatrix}, \; I_2 = \begin{pmatrix} 1 & 0 & 0 \\ 0 & -1 & 0 \\ 0 & 0 & 1 \end{pmatrix}, \; I_3 = \begin{pmatrix} 1 & 0 & 0 \\ 0 & 1 & 0 \\ 0 & 0 & -1 \end{pmatrix}$$

1.2.6.b. Orthogonal Matrices in Mathematica

Example 1. Q preserves the norm and the dot product: In the following code, the rotation matrix $Q \in \mathbb{M}^2$ is defined using the function `RotationMatrix[angle]`. Notice that the angle input is automatically set at radians unless the option "Degree" is used after the angle. Because of the symbolic notation of Mathematica, the function `FullSimplify[]` is used to show that the rotation matrix preserves the norm of the vector $u = \{2, 2\}$ after rotation and that the rotation angle between u and its image v is exactly 20 degrees.

```
(*The following is a rotation matrix using the angle Pi/6 in radians*)
Q=RotationMatrix[Pi/6];
(*The following is a rotation matrix using 20 degrees*)
Q=RotationMatrix[20Degree];
u={2,2};
v=Q.u;
Norm[u]
Norm[v]
FullSimplify[Norm[v]]
VectorAngle[u,v]
FullSimplify[VectorAngle[u,v]]
```

Example 2. Creating a triangle and rotating it by 60 degrees: (Figure 1–16) In the following code, an array of 4 points is defined. Then, a new array is calculated by applying a rotation of 60 degrees to every point. Finally, two triangles are created using the `Polygon` function and viewed using the `Graphics` function.

```
th=60Degree;
Q=RotationMatrix[th];
Points={{0,1},{5,1},{2,3},{0,1}};
Points2=Table[Q.Points[[i]],{i,1,4}];
c=Polygon[Points];
c2=Polygon[Points2];
(*Without any graphics formatting*)
Graphics[{c,c2},Axes->True,AxesOrigin->{0,0}]
(*Applying graphics formatting*)
Graphics[{Opacity[0.6],c,c2},Axes->True,AxesOrigin->{0,0},AxesStyle->
{Directive[Bold,17],Directive[Bold,17]}
```

The function `Options[]` in Mathematica can be used to identify the different options that could be changed in an expression, in particular, in Graphics functions. For example, to identify the different options that a user can change in `Graphics[]` or `Plot[]`, use the following code:

```
Options[Graphics]
Options[Plot]
```

Example 3. Creating a cube and rotating it by 60 degrees: (Figure 1–17) You can also use the function `RotationMatrix[angle,vector]` to create $Q \in \mathbb{M}^3$ that is a rotation around a certain vector. You can use two– or three–dimensional graphics objects to visualize the effect of the rotation matrices. For example, the following code creates a rotation matrix Q_x as defined above with an angle of 60 degrees. Then it creates a unit cube and creates another cube by applying a geometric transformation to the original cube. Then, the two cubes before and after rotation are

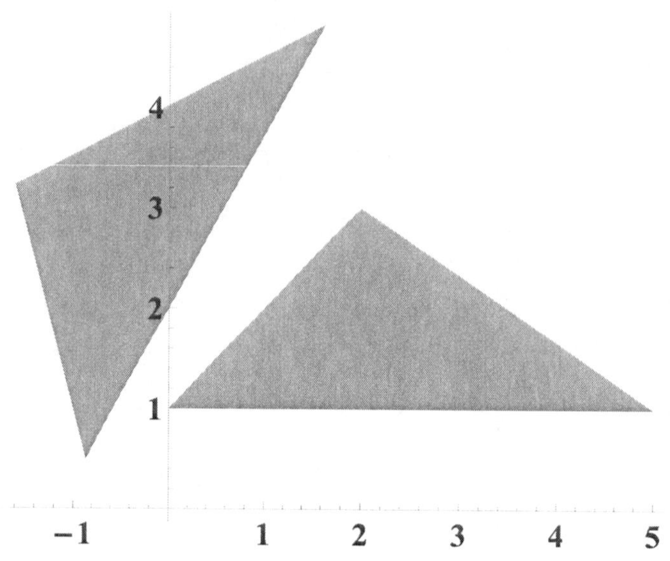

FIGURE 1-16 Rotating a polygon by 60 degrees using Mathematica.

viewed in the same plot. Notice the chosen options for the function `Graphics3D` to differentiate between the two objects and to view the result in gray.

```
thx=60Degree
Qx=RotationMatrix[thx,{1,0,0}];
c=Cuboid[{0,0,0},{1,1,1}];
c1=GeometricTransformation[c,Qx]
(*Without any graphics formatting*)
Graphics3D[{c,,c1},Axes->True,AxesOrigin->{0,0,0}]
(*Applying graphics formatting*)
Graphics3D[{GrayLevel[.3,0.4],Specularity[White,2],EdgeForm[Thickness[0.005]],c,
Specularity[White,6],EdgeForm[Thickness[0.01]],c1},Axes->True,AxesOrigin
->{0,0,0},Lighting->"Neutral",AxesStyle->{Directive[Bold,13],Directive[Bold,13],
Directive[Bold,13]}]
```

Example 4. Visualizing three dimensional rotation of a cuboid: (Figure 1–18) The following code creates the rotation matrices Q_x, Q_y, Q_z defined in the examples above. A cuboid is then defined with opposite corners at {0,0,0} and {3,2,1}. The function `Manipulate[]` is used to view the cuboid before and after applying the rotations $Q_1 = Q_xQ_yQ_z$ and $Q_2 = Q_zQ_yQ_x$. Copy and paste the code into your Mathematica software, and move the sliders to change the values of the rotation angles. The order of consecutive rotations is important. The application of Q_1 on a vector is equivalent to rotating the vector first by Q_z, then by Q_y and finally by Q_x. The resulting vector is different if Q_2 is applied.

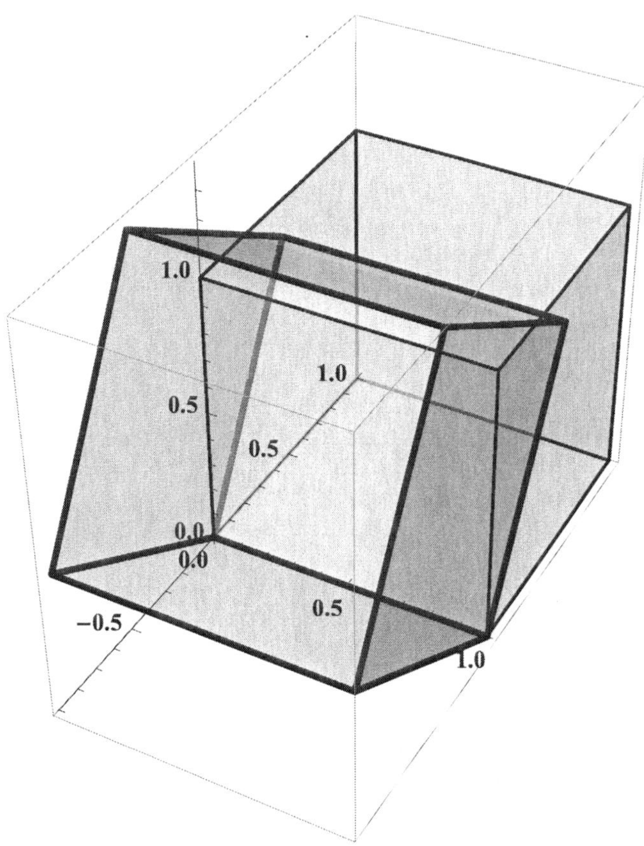

FIGURE 1-17 Rotating a cube by 60 degrees using Mathematica.

```
Clear[thx,thy,thz]
Manipulate[
Qx=RotationMatrix[thx,{1,0,0}];Qy=RotationMatrix[thy,{0,1,0}];
Qz=RotationMatrix[thz,{0,0,1}];c=Cuboid[{0,0,0},{3,2,1}];
c1=GeometricTransformation[c,Qx.Qy.Qz]; c2=GeometricTransformation[c,Qz.Qy.Qx];
Graphics3D[{c,{Gray,c1},{GrayLevel[0.2],c2}},AxesStyle->{Directive[Bold,13],
Directive[Bold,13],Directive[Bold,13]},AxesLabel->{"x","y","z"},Axes->True,
AxesOrigin->{0,0,0},Lighting->"Neutral"],{thx,0,2Pi},{thy,0,2Pi},{thz,0,2Pi}]
```

Example 5. Reflection of a Triangle: (Figure 1–19) In the following code, two reflection matri– ces are created using the `ReflectionMatrix[vector]` function that returns a reflection matrix. Reflection is applied across the plane perpendicular to the vector specified. The dark and lighter triangles are reflections in the dark and lighter arrow directions, respectively.

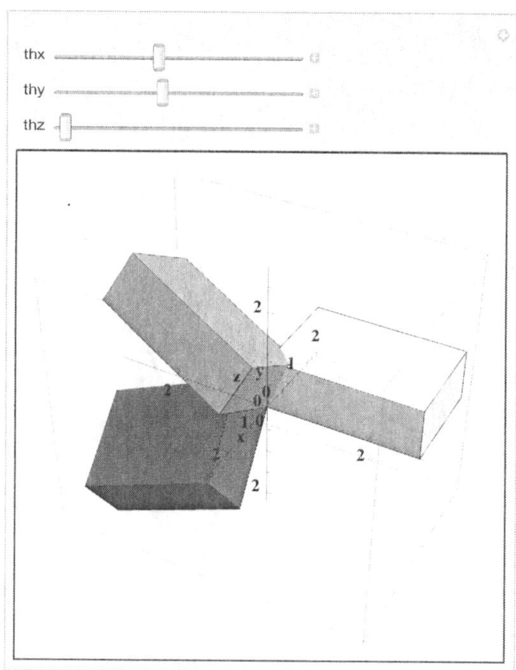

FIGURE 1-18 Using `Manipulate[]` function to visualize 3–dimensional rotation using Mathematica. The light and dark cuboids are those rotated using Q_1 and Q_2, respectively.

```
Q1=ReflectionMatrix[{1,0}]
Q2=ReflectionMatrix[{1,1}]
Points={{0,1},{5,1},{2,3},{0,1}};
Points1=Table[Q1.Points[[i]],{i,1,4}];
Points2=Table[Q2.Points[[i]],{i,1,4}];
L1=Arrow[{{-5,0},{5,0}}]
L2=Arrow[{{-5,-5},{5,5}}]
c=Polygon[Points];
c1=Polygon[Points1];
c2=Polygon[Points2];
(*In gray tones*)
Graphics[{c,{GrayLevel[0.2],c1},{Arrowheads[{-.1,.1}],{GrayLevel[0.2],
L1}},{Arrowheads[{-.1,.1}],{GrayLevel[0.5],L2}},{GrayLevel[
0.5],c2}},Axes->True,AxesStyle->{Directive[Bold,13],Directive[Bold,13]},
AxesOrigin->{0,0}]
(*In colour*)
Graphics[{c,{Blue,c1},{Arrowheads[{-.1,.1}],{Blue,L1}},{Arrowheads[{-.1,.1}],
{Red,L2}},{Red,c2}},Axes->True,AxesOrigin->{0,0}]
```

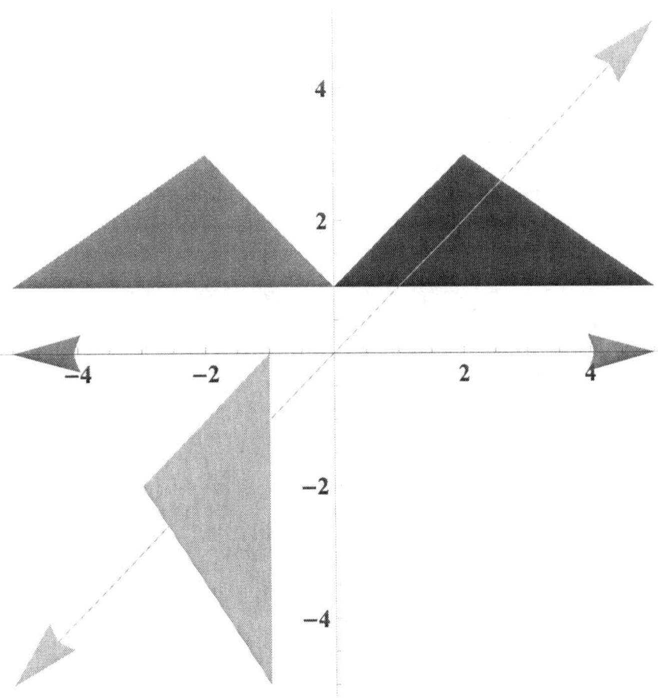

FIGURE 1-19 Using Mathematica to apply reflections of the original (black) triangle along the dark and light lines.

1.2.6.c. Orthogonal Change of Basis (Coordinate Transformation)

In the previous section, rotation matrices were shown to be important in describing how an object can rotate in its space by simply applying a rotation matrix to the vectors representing its points. Rotation matrices are also important in describing how the components of a vector and the components of a matrix change when the orthonormal set of basis rotates.

Statements of Orthogonal Change of Basis: Let $B = \{e_i\}_{i=1}^n$ be an orthonormal set of basis vectors in $\mathbb{R}^n$ and let $B' = \{e'_i\}_{i=1}^n$ be another orthonormal set of basis vectors in $\mathbb{R}^n$. Let $u \in \mathbb{R}^n$. Then, $u = \sum_{i=1}^n u_i e_i = u_i e_i = \sum_{i=1}^n u'_i e'_i = u'_i e'_i$. The relationship between the new components u'_i and u_i is according to:

$$u' = Qu, \quad u = Q^T u'$$

where, $u' = \{u'_1, u'_2, \ldots, u'_n\}$, $u = \{u_1, u_2, \ldots, u_n\}$ and Q is an orthogonal matrix whose components are $Q_{ij} = e'_i \cdot e_j$.

Also, let $M \in \mathbb{M}^n$ be the matrix associated with a linear map defined using the basis set B then:

$$M' = QMQ^T, \quad M = Q^T M' Q$$

where, M' is the matrix associated with the same linear map defined using the basis set B'.

Proof of the Orthogonal Change of Basis Statements:

First, we show that Q defined above is indeed an orthogonal matrix.

$$Q = \left(\begin{bmatrix} e'_1 \cdot e_1 & \cdots & e'_1 \cdot e_n \\ \vdots & \ddots & \vdots \\ e'_n \cdot e_1 & \cdots & e'_n \cdot e_n \end{bmatrix}\right)$$

$\forall i \leq n$:e'_i can be expressed in the basis set B as (summation convention is employed):

$$e'_i = \sum_{j=1}^{n} (e'_i \cdot e_j)\, e_j = Q_{ij} e_j$$

Also notice that $e_i = \sum_{j=1}^{n} \left(e_i \cdot e'_j\right) e'_j = Q_{ji} e'_j$.

Since the sets B and B' are orthonormal:

$$\begin{aligned} \delta_{ij} &= e'_i \cdot e'_j \\ &= Q_{ik} e_k \cdot Q_{jl} e_l \\ &= Q_{ik} Q_{jl} \delta_{kl} \\ &= Q_{ik} Q_{jk} \end{aligned}$$

In matrix form:

$$QQ^T = I$$

Clearly, the vectors associated with the matrix Q are linearly independent, and thus, Q is an orthogonal matrix and if $\det Q = 1$ then Q is a rotation matrix.

The relationship between the components of an arbitrary vector in the basis set B and its compo– nents in the basis set B' can be shown as follows. $\forall u \in \mathbb{R}^n$:

$$u = u_i e_i = u'_i e'_i$$

e'_i can be replaced by $Q_{ij}e_j$ as shown above:

$$u = u_i e_i = u'_i Q_{ij} e_j$$

Finally, the indices i and j on the right–hand side can be replaced to reach the following equality:

$$u_i e_i = u'_j Q_{ji} e_i$$

Since the multipliers of e_i are equal on both sides, then:

$$u_i = Q_{ji} u'_j$$

Therefore,

$$u = Q^T u', \quad u' = Qu$$

Notice that, unlike section 1.2.6.a, the vector u itself has not rotated, but rather, the basis set is changed. u' is not considered a new vector, but rather, the vector expressed with new components in the new basis set B'.

Let $M \in \mathbb{M}^n$ be the matrix associated with a linear map expressed in the basis set B, and consider v the image vector of u under the linear map, then:

$$v = Mu$$

The vectors v and u can be replaced by their representation in the new coordinate system:

$$Q^T v' = MQ^T u'$$

Multiplying both sides by Q yields:

$$v' = QMQ^T u'$$

Comparing this equality to the equality $v' = M'u'$ yields:

$$M' = QMQ^T$$

Since u is arbitrary, therefore, the relationship between the components of the linear map expressed in the basis set B and those expressed in the basis set B' is:

$$M = Q^T M' Q, \; M' = QMQ^T$$

Figure 1–20 shows a schematic of the orthogonal change of basis in $\mathbb{R}^2$. u is a vector expressed in the coordinate system defined by e_1 and e_2. The vector v is the image of the vector u when the linear map represented by the matrix M is applied. If a new orthonormal set of basis e'_1 and e'_2 is used, the components of the vectors u and v and the matrix M change according to a rotation matrix Q to become u', v' and M' respectively.

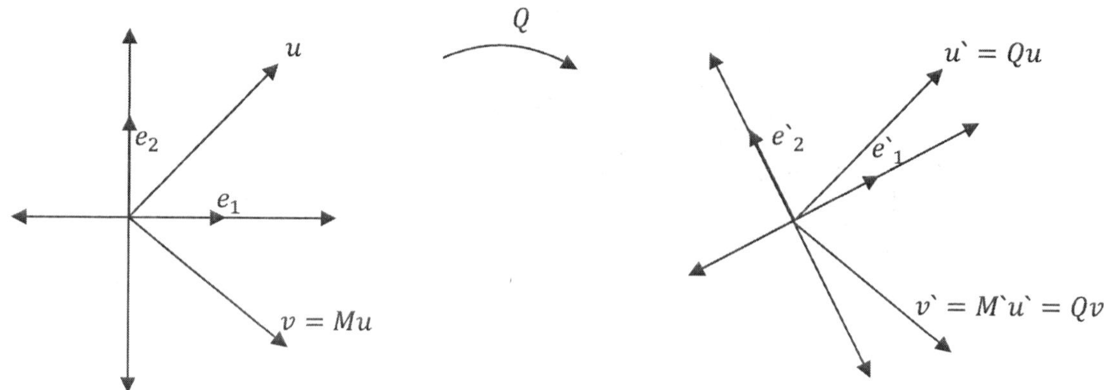

FIGURE 1-20 A schematic showing the orthogonal change of basis.

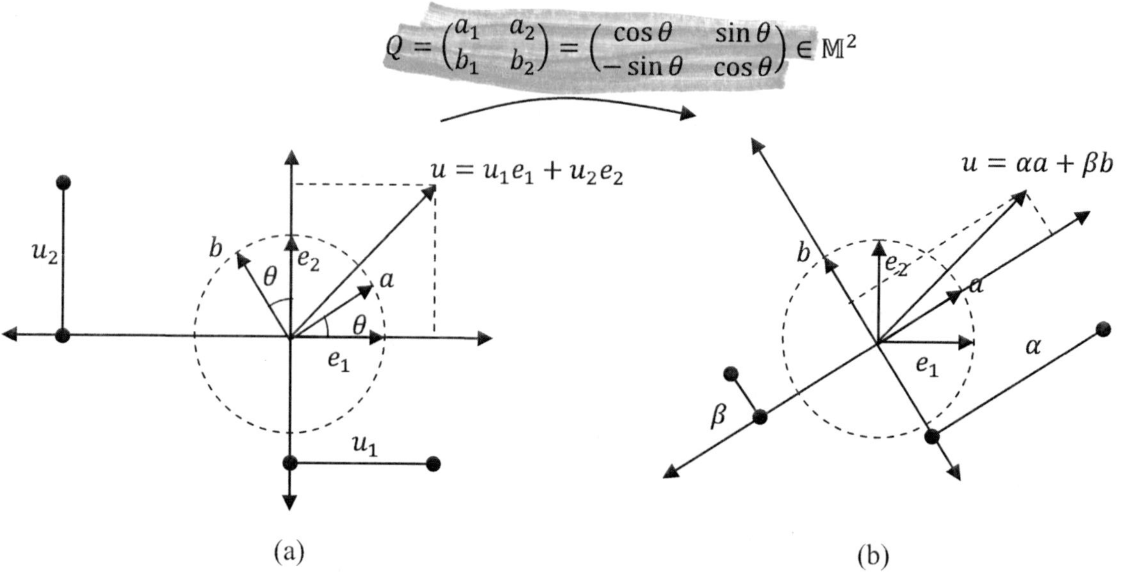

FIGURE 1-21 Counter–clockwise rotation of a basis set by an angle θ in $\mathbb{R}^2$. (a) components in original basis, (b) components in new basis.

Problem 1: (Figure 1–21) Let $\{e_1, e_2\}$ be the traditional set of orthonormal basis in a Euclidian vector space $\mathbb{R}^2$, therefore, $\forall u \in \mathbb{R}^2$, $u = u_i e_i$. Let $\{a, b\}$ be a new set of orthonormal basis such that $\forall u \in \mathbb{R}^2$, $u = \alpha a + \beta b$. Write α and β as functions of $\{u_1, u_2\}$ and the components of the vectors a and b.

Solution:

The rotation matrix associated with the change of basis is:

$$Q = \begin{pmatrix} a \cdot e_1 & a \cdot e_2 \\ b \cdot e_1 & b \cdot e_2 \end{pmatrix} = \begin{pmatrix} a_1 & a_2 \\ b_1 & b_2 \end{pmatrix}, Q^T = \begin{pmatrix} a_1 & b_1 \\ a_2 & b_2 \end{pmatrix}, QQ^T = Q^T Q = I$$

Notice that the relationship between the basis vectors in the original coordinate system can be obtained using the rotation matrices defined above, as follows:

$$a = Q^T e_1, \quad b = Q^T e_2$$
$$e_1 = Qa, \quad e_2 = Qb$$

Since the vector u itself does not rotate, it follows that:

$$\{\alpha, \beta\} \begin{pmatrix} a \\ b \end{pmatrix} = \{u_1, u_2\} \begin{pmatrix} e_1 \\ e_2 \end{pmatrix}$$

$$\{\alpha, \beta\} \begin{pmatrix} a_1 & a_2 \\ b_1 & b_2 \end{pmatrix} \begin{pmatrix} e_1 \\ e_2 \end{pmatrix} = \{u_1, u_2\} \begin{pmatrix} e_1 \\ e_2 \end{pmatrix}$$

Taking the transpose of the two sides of the above equation:

$$\{e_1, e_2\} \begin{pmatrix} a_1 & b_1 \\ a_2 & b_2 \end{pmatrix} \begin{pmatrix} \alpha \\ \beta \end{pmatrix} = \{e_1, e_2\} \begin{pmatrix} u_1 \\ u_2 \end{pmatrix}$$

This leads to the relationship between the different components:

$$\begin{pmatrix} u_1 \\ u_2 \end{pmatrix} = \begin{pmatrix} a_1 & b_1 \\ a_2 & b_2 \end{pmatrix} \begin{pmatrix} \alpha \\ \beta \end{pmatrix}, \begin{pmatrix} \alpha \\ \beta \end{pmatrix} = \begin{pmatrix} a_1 & a_2 \\ b_1 & b_2 \end{pmatrix} \begin{pmatrix} u_1 \\ u_2 \end{pmatrix}$$

Setting $u_e = \{u_1, u_2\}$, $u_a = \{\alpha, \beta\}$ we get:

$$a = Q^T e_1, \ b = Q^T e_2, \ u_a = Qu_e$$
$$e_1 = Qa, \ e_2 = Qb, \ u_e = Q^T u_a$$

Notice that the components are transformed using the transpose of the rotation matrix used for the basis vectors. Also, compare with Figure 1–15 to realize the difference between rotating a vector and rotating the basis.

Problem 2: Consider the vectors $x = \{1, 1, 1\}$, $u = \{0.6, 0.8, 0\}$, $v = \{-0.8, 0.6, 0\}$ and $w = \{0, 0, 1\}$. Consider the linear map represented by the matrix M:

$$M = \begin{pmatrix} 0.25 & 0.5 & 0.1 \\ 0.5 & -0.5 & 0.5 \\ -0.25 & -0.5 & 0.5 \end{pmatrix}$$

- Find the components of the vector y, the image of the vector x under the linear map M.
- Show that the vectors u, v and w form an orthonormal basis set in $\mathbb{R}^3$.
- Find the components of x, y and M if the set $B = \{u, v, w\}$ is chosen as the orthonormal basis set.

Solution:

The components of y are:

$$y = Mx = \begin{pmatrix} 0.85 \\ 0.5 \\ -0.25 \end{pmatrix}$$

The norm of each of the vectors u, v and w is equal to 1. In addition, $u \cdot v = u \cdot w = w \cdot v = 0$. Therefore, they form an orthonormal basis set in $\mathbb{R}^3$.

The rotation matrix associated with the orthogonal change of basis is:

$$Q = \begin{pmatrix} u \cdot e_1 & u \cdot e_2 & u \cdot e_3 \\ v \cdot e_1 & v \cdot e_2 & v \cdot e_3 \\ w \cdot e_1 & w \cdot e_2 & w \cdot e_3 \end{pmatrix} = \begin{pmatrix} u_1 & u_2 & u_3 \\ v_1 & v_2 & v_3 \\ w_1 & w_2 & w_3 \end{pmatrix} = \begin{pmatrix} 0.6 & 0.8 & 0 \\ -0.8 & 0.6 & 0 \\ 0 & 0 & 1 \end{pmatrix}$$

The components of the vectors y and x in the new coordinate system are:

$$x' = Qx = \begin{pmatrix} 0.6 & 0.8 & 0 \\ -0.8 & 0.6 & 0 \\ 0 & 0 & 1 \end{pmatrix} \begin{pmatrix} 1 \\ 1 \\ 1 \end{pmatrix} = \begin{pmatrix} 1.4 \\ -0.2 \\ 1 \end{pmatrix}, \quad y' = Qy = \begin{pmatrix} 0.91 \\ -0.38 \\ -0.25 \end{pmatrix}$$

The components of the matrix M in the new coordinate system are:

$$M' = QMQ^T = \begin{pmatrix} 0.25 & -0.5 & 0.46 \\ -0.5 & -0.5 & 0.22 \\ -0.55 & -0.1 & 0.5 \end{pmatrix}$$

The components of y' can also be obtained as the image of x' using the linear map M' as follows:

$$y' = M'x' = \begin{pmatrix} 0.25 & -0.5 & 0.46 \\ -0.5 & -0.5 & 0.22 \\ -0.55 & -0.1 & 0.5 \end{pmatrix} \begin{pmatrix} 1.4 \\ -0.2 \\ 1 \end{pmatrix} = \begin{pmatrix} 0.91 \\ -0.38 \\ -0.25 \end{pmatrix}$$

The following Mathematica code was used for the above calculations:

```
x={1,1,1}
M={{0.25,0.5,0.1},{0.5,-0.5,0.5},{-0.25,-0.5,0.5}};
y=M.x
u={3/5,4/5,0}
v={-4/5,3/5,0}
w={0,0,1.}
Norm[u]
Norm[v]
Norm[w]
u.v
u.w
v.w
Q={u,v,w}
xd=Q.x;
xd//MatrixForm
yd=Q.y;
yd//MatrixForm
Md=Q.M.Transpose[Q];
Md//MatrixForm
Md.xd//MatrixForm
```

1.2.6.d. Vector and Matrix Invariants

In continuum and solid mechanics, we are concerned with quantities associated with vectors and matrices that do not change when a coordinate transformation is applied. Such quantities are termed **invariants**. For instance, if a person wishes to qualify or quantify the state of stress at a point in a certain coordinate system, the quantity used should be independent of the chosen coordinate system.

In general, the **dot product** and the **norm** operations are **invariants** under orthonormal coordinate transformations. Let $Q \in \mathbb{M}^3$ be a coordinate transformation orthogonal matrix. Let $u, v \in \mathbb{R}^3$ with components in the orthonormal basis set $B = \{e_1, e_2, e_3\}$ and let u' and v' represent the components of u and v in the orthonormal basis set $B' = \{e_1', e_2', e_3'\}$. Then, we have:

$$\|v'\|^2 = v' \cdot v' = Qv \cdot Qv = v \cdot Q^T Qv = v \cdot v = \|v\|^2$$
$$u' \cdot v' = Qu \cdot Qv = u \cdot Q^T Qv = u \cdot v$$

There are the three major invariants of a general matrix $M \in \mathbb{M}^3$. These invariants, along with the eigenvalues and the eigenvectors of a matrix, do not change when an orthogonal change of basis is applied. In the following, $Q \in \mathbb{M}^3$ is an orthogonal matrix representing a coordinate transformation and $M' = QMQ^T$ is the representation of the matrix in the new coordinate system.

First Invariant—Matrix Trace:

The sum of the diagonal terms of a matrix is invariant under any orthonormal coordinate transformation.

$$\begin{aligned} I_1(M) &= \operatorname{trace} M := \sum_{i=1}^{3} M_{ii} = M_{ii} = \operatorname{trace} M' = \operatorname{trace} QMQ^T \\ &= \sum_{i,j,k=1}^{3} Q_{ij} M_{jk} Q_{ik} = \sum_{j,k=1}^{3} M_{jk} \delta_{jk} = \sum_{j}^{3} M_{jj} \end{aligned}$$

Second Invariant:

The second invariant is defined as:

$$\begin{aligned} I_2(M) &= \frac{1}{2}\left((I_1(M))^2 - I_1(MM)\right) \\ &= (M_{22}M_{33} + M_{11}M_{22} + M_{11}M_{33} - M_{12}M_{21} - M_{13}M_{31} - M_{23}M_{32}) \end{aligned}$$

The second invariant is independent of the coordinate transformation:

$$\begin{aligned} I_2(M') &= \frac{1}{2}\left((I_1(M'))^2 - I_1(M'M')\right) = \frac{1}{2}\left((I_1(M))^2 - I_1\left(QMMQ^T\right)\right) \\ &= \frac{1}{2}\left((I_1(M))^2 - I_1(MM)\right) = I_2(M) \end{aligned}$$

Third Invariant—Matrix Determinant:

The determinant of a matrix is invariant under any orthonormal coordinate transformation (See section 1.2.5.c for the properties of the determinant.):

$$I_3(M) = \det M' = \det QMQ^T = \det Q \det M \det Q^T = \det M$$

These three invariants are closely related to the eigenvalue problem of a matrix $M \in \mathbb{M}^3$. The eigenvalues of the matrix M can be obtained by requiring that the matrix $M - \lambda I$ is not invertible:

$$\det(M - \lambda I) = 0$$

Therefore, the eigenvalues of the matrix M are the roots of the following cubic polynomial with the coefficients being the invariants of the matrix M:

$$\begin{gathered}(M_{11} - \lambda)\left((M_{22} - \lambda)(M_{33} - \lambda) - M_{23}M_{32}\right) - M_{12}\left(M_{21}(M_{33} - \lambda) - M_{23}M_{31}\right) \\ + M_{13}\left(M_{12}M_{32} - M_{13}(M_{22} - \lambda)\right) = 0 \\ -\lambda^3 + \lambda^2 \operatorname{trace} M - \lambda\left(M_{22}M_{33} + M_{11}M_{22} + M_{11}M_{33} - M_{12}M_{21} - M_{13}M_{31} - M_{23}M_{32}\right) + \det M = 0 \\ \lambda^3 - I_1(M)\lambda^2 + I_2(M)\lambda - I_3(M) = 0\end{gathered}$$

▶ **BONUS EXERCISES:**

- Show that the eigenvalues and the eigenvectors are invariant under a coordinate transformation.
- Show the following relations for the first invariant given that: $M, N \in \mathbb{M}^3$, $\alpha \in \mathbb{R}$ and $I \in \mathbb{M}^3$ is the identity map:

$$I_1(\alpha M) = \alpha I_1(M), \quad I_1(I) = 3, \quad I_1(M + N) = I_1(M) + I_1(N)$$

1.2.6.e. Alternative Forms of Matrix Invariants

The matrix invariants in the previous section are presented as functions of the components of the matrix, which are tied to a certain choice of an orthonormal coordinate system. However, it is possible to write the matrix invariants in forms that are not tied to any coordinate system. Let M represent a linear map between three dimensional Euclidian vector spaces and let a, b and $c \in \mathbb{R}^3$ be three arbitrary linearly independent vectors. Consider the components represented in the orthonormal basis set $B = \{e_1, e_2, e_3\}$. Then, the matrix invariants can be written in the following forms as functions of the orthonormal basis vectors or as functions of the three arbitrary linearly independent vectors a, b and c.

$$\begin{aligned} I_1(M) = \operatorname{trace} M = M_{ii} &= Me_1 \cdot e_1 + Me_2 \cdot e_2 + Me_2 \cdot e_2 + Me_3 \cdot e_3 \\ &= Me_1 \cdot (e_2 \times e_3) + e_1 \cdot (Me_2 \times e_3) + e_1 \cdot (e_2 \times Me_3) \\ &= \frac{Ma \cdot (b \times c) + a \cdot (Mb \times c) + a \cdot (b \times Mc)}{a \cdot (b \times c)} \end{aligned}$$

$$\begin{aligned} I_2(M) &= Me_1 \cdot (Me_2 \times e_3) + e_1 \cdot (Me_2 \times Me_3) + Me_1 \cdot (e_2 \times Me_3) \\ &= \frac{Ma \cdot (Mb \times c) + a \cdot (Mb \times Mc) + Ma \cdot (b \times Mc)}{a \cdot (b \times c)} \end{aligned}$$

The determinant of a matrix M as given by the results of section 1.2.5.c has the following forms:

$$I_3(M) = \det M = \frac{(Me_1 \cdot (Me_2 \times Me_3))}{e_1 \cdot (e_2 \times e_3)} = \frac{(Ma \cdot (Mb \times Mc))}{a \cdot (b \times c)}$$

▶ **BONUS EXERCISE:** Use the components of a, b and c to prove the given equalities for both the first and the second invariants.

1.2.6.f. Symmetric Matrices

Definition: Let $S \in \mathbb{M}^n$. S is a symmetric matrix if $S^T = S$.

The following are direct consequences of the definition:

- In component form $S_{ij} = S_{ji}$
- $\forall u, v \in \mathbb{R}^n : u \cdot Sv = Su \cdot v$
- $\forall M \in \mathbb{M}^n : M^T SM$ is symmetric. In particular, if M is an orthogonal matrix associated with a coordinate transformation, then the representation of S in any coordinate system stays symmetric.

Symmetric matrices form a very important class of matrices for many fields especially solid and continuum mechanics. For example, the stresses and strains are themselves, in most cases, assumed to be symmetric matrices. A very important result for symmetric matrices is that any symmetric matrix possesses three real eigenvalues and three major perpendicular directions (eigenvectors). For example, as will be shown later, the stress matrix possesses three principal stresses and three principal directions. The same statement applies for the strain matrix since it is symmetric as well. Thus, the proof of the existence of real eigenvalues and perpendicular eigen–vectors for a general symmetric matrix will first be introduced in this section.

Statement for the Eigenvalues and Eigenvectors of Symmetric Matrices:

A matrix $S \in \mathbb{M}^3$ is symmetric if and only if it possesses three REAL eigenvalues and it is possible to choose three corresponding orthonormal eigenvectors.

Proof: Note that the same proof can be applied for $S \in \mathbb{M}^n$.

First, we show that if S is symmetric, then it possesses three REAL eigenvalues. Let $S \in \mathbb{M}^3$ be a symmetric matrix. We will argue by contradiction and that is by assuming that S is a symmetric matrix and that it possesses complex valued eigenvalues. Assume λ is a complex eigenvalue of S, then $\exists x \neq \hat{0}, x \in \mathbb{C}^3$ s.t. (where $\mathbb{C}$ is the space of complex numbers).

$$Sx = \lambda x$$

Taking the conjugate of the above relation:

$$S\bar{x} = \overline{\lambda x} = \bar{\lambda}\bar{x}$$

If we then take the dot product of the first equation with $\bar{x}$ and the second equation with x we get:

$$\bar{x} \cdot Sx = \lambda\bar{x} \cdot x, \quad x \cdot S\bar{x} = \bar{\lambda}x \cdot \bar{x}$$

But, since S is symmetric we have:

$$\bar{x} \cdot Sx = x \cdot S\bar{x} \Rightarrow \lambda\bar{x} \cdot x - \bar{\lambda}x \cdot \bar{x} = \left(\lambda - \bar{\lambda}\right)\bar{x} \cdot x = 0$$

But $\bar{x} \cdot x > 0$ therefore, $\lambda = \bar{\lambda}$, thus, $\lambda \in \mathbb{R}$.

Remark: if $\lambda = a + bi,$ s.t. $a, b \in \mathbb{R},\ i = \sqrt{-1}$, then $\bar{\lambda} = a - bi$ and $\lambda\bar{\lambda} = a^2 + b^2 \in [0, \infty[$

Therefore, S possesses three real eigenvalues.

Now it will be shown that it is possible to choose three corresponding orthonormal eigenvectors.

Case 1: The three eigenvalues of S are distinct:

Let $\lambda_1 \neq \lambda_2$ be eigenvalues of S with x_1 and x_2 the corresponding eigenvectors:

$$x_1 \cdot Sx_2 = x_2 \cdot Sx_1 \Rightarrow \lambda_1 x_1 \cdot x_2 - \lambda_2 x_2 \cdot x_1 = (\lambda_1 - \lambda_2)\, x_2 \cdot x_1 = 0$$

So, $x_2 \cdot x_1 = 0$ and by normalizing $x_1 = \frac{x_1}{\|x_1\|}$ and $x_2 = \frac{x_2}{\|x_2\|}$ we get that x_1 and x_2 are orthonormal. The same applies for λ_3 and x_3.

Case 2: Two of the eigenvalues of S are equal:

Let S possess one distinct eigenvalue λ_3 and two equal eigenvalues λ_1 and λ_2 with corresponding eigenvectors x_3, x_2 and x_1. Then, $x_3 \cdot x_2 = 0$ and $x_3 \cdot x_1 = 0$. In addition, it can be shown that we can find two linearly independent eigenvectors x_1 and x_2 that correspond to λ_1 but this proof is beyond the scope of this text. Therefore, any linear combination of x_2 and x_1 is also an eigenvector of S. Hence, it is possible to choose an orthonormal set of vectors: x_3 and two vectors in the plane perpendicular to x_3 to form an orthonormal eigenvectors set for S.

Case 3: The three eigenvalues of S are equal:

Finally, if S possesses one distinct eigenvalue λ_1, then it has at least three corresponding eigenvectors x_1, x_2 and x_3. In addition, it can be shown that we can find three linearly independent eigenvectors $x_1,\ x_2$ and x_3 that correspond to λ_1 but this proof is beyond the scope of this text. Therefore, any linear combination of x_1, x_2 and x_3 is also an eigenvector of S. Hence, it is possible to choose an orthonormal set of vectors to form an orthonormal eigenvectors set for S.

It is now required to show the opposite statement that if a matrix has three real eigenvalues and three orthonormal eigenvectors, then it is symmetric. Let p, q and $r \in \mathbb{R}^3$ be the three orthonormal eigenvectors of S and λ_1, λ_2 and λ_3 be the corresponding eigenvalues. Then, any two vectors $x, y \in \mathbb{R}^3$ can be expressed as a linear combination of p, q and r. Let $x = \alpha_1 p + \beta_1 q + \delta_1 r$ and $y = \alpha_2 p + \beta_2 q + \delta_2 r$ with $\alpha_1, \beta_1, \delta_1, \alpha_2, \beta_2$ and $\delta_2 \in \mathbb{R}$. Then, using the orthogonality of p, q and r:

$$x \cdot Sy = y \cdot Sx = \lambda_1\alpha_1\alpha_2 + \lambda_2\beta_1\beta_2 + \lambda_3\delta_1\delta_2$$

Therefore, S is symmetric.

Remark: It is important to clarify the wording: "it is possible to choose...". That is because the eigenvectors are not unique. The following two examples are used to clarify this concept.

Example 1: Consider the matrix $M = pI$ where $p \in \mathbb{R}$ and I is the identity matrix in $\mathbb{M}^3$. Then, the three eigenvalues of the matrix M are all equal to the value of p. Moreover, any nonzero vector is an eigenvector of M; therefore, in this case, any three orthonormal vectors are eigenvectors, e.g., the standard set $B = \{e_1, e_2, e_3\}$.

Example 2: Consider the matrix $M = \begin{pmatrix} p & 0 & 0 \\ 0 & p & 0 \\ 0 & 0 & q \end{pmatrix}$, where $p \neq q \in \mathbb{R}$. Then, the eigenvalues are $\lambda_1 = p$, $\lambda_2 = p$, and $\lambda_3 = q$. Clearly, any arbitrary vector in the form $u = \{u_1, u_2, 0\}$ is an eigen–vector associated with λ_1 and λ_2. Also, $v = \{0, 0, v_3\}$ is an arbitrary eigenvector associated with λ_3. In this case, it is also possible to choose the standard set $B = \{e_1, e_2, e_3\}$ as the set of eigen–vectors associated with the eigenvalues. It is also possible to choose $B' = \{e'_1, e'_2, e'_3\}$ such that $e'_1 = \left\{\frac{1}{\sqrt{2}}, \frac{1}{\sqrt{2}}, 0\right\}$, $e'_2 = \left\{\frac{-1}{\sqrt{2}}, \frac{1}{\sqrt{2}}, 0\right\}$ and $e'_3 = \{0, 0, 1\}$ as the set of orthonormal eigenvectors of M.

Spectral decomposition of a symmetric matrix:

The spectral decomposition of a symmetric matrix is a remarkable result. **It states that you can always find a coordinate system in which a symmetric matrix is a diagonal matrix.** Let $S \in \mathbb{M}^3$ s.t. $S = S^T$. Then, as seen in the previous proof, there exists three orthonormal eigenvectors for S. Let p, q and $r \in \mathbb{R}^3$ be the orthonormal eigenvectors of S associated with the three eigenvalues λ_1, λ_2 and λ_3 respectively. Construct $Q \in \mathbb{M}^3$ such that:

$$Q = \begin{pmatrix} p_1 & p_2 & p_3 \\ q_1 & q_2 & q_3 \\ r_1 & r_2 & r_3 \end{pmatrix}$$

Since $QQ^T = I$, then Q is an orthogonal matrix and thus can represent a coordinate transformation to the new orthonormal coordinate system whose bases set $B' = \{p, q, r\}$ as shown in section 1.2.6.c.

Let $S' = QSQ^T$

$$S' = QS\begin{pmatrix} p_1 & q_1 & r_1 \\ p_2 & q_2 & r_2 \\ p_3 & q_3 & r_3 \end{pmatrix} = Q\begin{pmatrix} \lambda_1 p_1 & \lambda_2 q_1 & \lambda_3 r_1 \\ \lambda_1 p_2 & \lambda_2 q_2 & \lambda_3 r_2 \\ \lambda_1 p_3 & \lambda_2 q_3 & \lambda_3 r_3 \end{pmatrix} = \begin{pmatrix} \lambda_1 (p \cdot p) & 0 & 0 \\ 0 & \lambda_2 (q \cdot q) & 0 \\ 0 & 0 & \lambda_3 (r \cdot r) \end{pmatrix}$$

$$S' = \begin{pmatrix} \lambda_1 & 0 & 0 \\ 0 & \lambda_2 & 0 \\ 0 & 0 & \lambda_3 \end{pmatrix}$$

Therefore, any symmetric matrix is represented by a diagonal matrix in the coordinate system defined by the directions of its eigenvectors.

Principal Invariants of a Symmetric Matrix:

The spectral decomposition of a symmetric matrix $S \in \mathbb{M}^3$ enables a simple expression for its principle invariants in terms of its eigenvalues λ_1, λ_2 and λ_3:

$$I_1(S) = I_1(S') = \lambda_1 + \lambda_2 + \lambda_3$$

$$I_2(S) = I_2(S') = \frac{1}{2}\left(I_1(S')^2 - I_1(S'S')\right) = \frac{1}{2}\left((\lambda_1 + \lambda_2 + \lambda_3)^2 - \left(\lambda_1^2 + \lambda_2^2 + \lambda_3^2\right)\right)$$

$$= \lambda_1\lambda_2 + \lambda_2\lambda_3 + \lambda_1\lambda_3$$

$$I_3(S) = I_3(S') = \lambda_1\lambda_2\lambda_3$$

Graphical Representation:

The results of this section show that a symmetric matrix is associated with extension or contraction of its principal directions without any rotation. Figure 1–22 shows that when $S \in \mathbb{M}^2$ is applied to a circular shape, the resulting shape is an ellipse with its axes being the eigenvectors of S. These eigenvectors do not change direction, but rather, increase or decrease in length. As long as S is symmetric, then the existence of perpendicular eigenvectors is guaranteed by the theorem in this section.

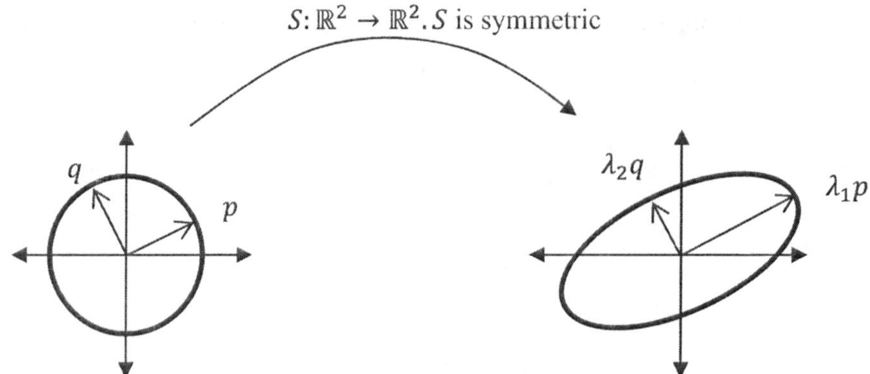

FIGURE 1-22 Graphical representation of symmetric matrices in two–dimensional spaces. $p, q \in \mathbb{R}^2$ are the eigenvectors of S while λ_1 and λ_2 are the corresponding real eigenvalues.

Problem: Find a coordinate system in which the following symmetric matrix is diagonal.

$$M = \begin{pmatrix} 1 & 2 & 3 \\ 2 & 1 & -1 \\ 3 & -1 & 2 \end{pmatrix}$$

Solution:

The eigenvalues, the eigenvectors, and the normalized eigenvectors of the matrix M are given by:

$$\lambda_1 = 4.63897, \quad x_1 = \begin{pmatrix} 0.964812 \\ 0.255463 \\ 1 \end{pmatrix}, \quad \frac{x_1}{\|x_1\|} = \begin{pmatrix} 0.682887 \\ 0.180815 \\ 0.707793 \end{pmatrix}$$

$$\lambda_2 = -2.79696, \quad x_2 = \begin{pmatrix} -1.28549 \\ 0.940485 \\ 1 \end{pmatrix}, \quad \frac{x_2}{\|x_2\|} = \begin{pmatrix} -0.683521 \\ 0.500074 \\ 0.531719 \end{pmatrix}$$

$$\lambda_3 = 2.15799, \quad x_3 = \begin{pmatrix} -0.554319 \\ -1.82095 \\ 1 \end{pmatrix}, \quad \frac{x_3}{\|x_3\|} = \begin{pmatrix} -0.257806 \\ -0.846896 \\ 0.465085 \end{pmatrix}$$

Consider a new coordinate system with the three orthonormal vectors e'_1, e'_2 and e'_3 (Verify that they are orthogonal!):

$$e'_1 = \frac{x_1}{\|x_1\|}, \quad e'_2 = \frac{x_2}{\|x_2\|}, \quad e_3 = \frac{x_3}{\|x_3\|} \tag{1.1}$$

Associated with the coordinate transformation is the rotation matrix Q:

$$Q = \begin{pmatrix} 0.682887 & 0.180815 & 0.707793 \\ -0.683521 & 0.500074 & 0.531719 \\ -0.257806 & -0.846896 & 0.465085 \end{pmatrix}$$

The matrix M in the new coordinate system is indeed diagonal and is given by:

$$M' = QMQ^T = \begin{pmatrix} 4.63897 & 0 & 0 \\ 0 & -2.79696 & 0 \\ 0 & 0 & 2.15799 \end{pmatrix}$$

The following is the Mathematica code used for the above calculations. Notice that the function `Chop[]` replaces very small values such as 10^{-16} with zeros:

```
M={{1,2,3},{2,1,-1},{3,-1,2}}
s=N[Eigensystem[M]]
ev1=s[[2,1]]/Norm[s[[2,1]]]
ev2=s[[2,2]]/Norm[s[[2,2]]]
ev3=s[[2,3]]/Norm[s[[2,3]]]
Q={ev1,ev2,ev3}
Q//MatrixForm
Chop[Q.M.Transpose[Q]]
```

1.2.6.g. Positive Definite and Semi-positive Definite Symmetric Matrices

A symmetric matrix $S \in \mathbb{M}^n$ is positive definite if $\forall a \neq \hat{0},\ a \in \mathbb{R}^n,\ a \cdot Sa > 0$

A symmetric matrix $S \in \mathbb{M}^n$ is semi–positive definite if $\forall a \in \mathbb{R}^n,\ a \cdot Sa \geq 0$

Problem: Let $S \in \mathbb{M}^n$ be a symmetric matrix and denote λ as any eigenvalue of S. Show that S is positive definite (semi–positive) if and only if $\lambda > 0$ ($\lambda \geq 0$).

Solution:

Let $S \in \mathbb{M}^n$ be a positive definite symmetric matrix and λ be an eigenvalue associated with the eigenvector $x \in \mathbb{R}^n$. $x \neq \hat{0} \Longrightarrow (x \cdot x) > 0$. Then:

$$x \cdot Sx = \lambda\,(x \cdot x) > 0 \Rightarrow \lambda > 0$$

Note, if $S \in \mathbb{M}^n$ is a semipositive definite symmetric matrix, then:

$$x \cdot Sx = \lambda (x \cdot x) \geq 0 \Rightarrow \lambda \geq 0$$

The opposite statement is left as a bonus exercise with the proof similar to that given in section 1.2.6.f for the statement of eigenvalues and eigenvectors of a symmetric matrix.

1.2.6.h. Skew-symmetric (Anti-symmetric) Matrices

Definition: Let $W \in \mathbb{M}^n$. W is an antisymmetric matrix if $W^T = -W$.

The following are direct consequences of the definition:

- In component form: $W_{ij} = -W_{ji}$. Diagonal entries are therefore equal to zero!
- $\forall u, v \in \mathbb{R}^n : u \cdot Wv = -Wu \cdot v$
- $\forall M \in \mathbb{M}^n : M^T WM$ is antisymmetric. In particular, if M is an orthogonal matrix associated with a coordinate transformation, then the representation of W in any coordinate system stays antisymmetric.
- $\forall M \in \mathbb{M}^n$: M can be written as the sum of two unique matrices, a symmetric matrix $S = \frac{1}{2}\left(M + M^T\right)$ and an antisymmetric matrix $W = \frac{1}{2}\left(M - M^T\right)$.
- $\forall u \in \mathbb{R}^n : Wu$ is orthogonal to u. (**Bonus exercise:** show this relation.)

Problem 1: If W is a skew–symmetric matrix in a three–dimensional Euclidian vector space, then show that 0 is an eigenvalue of W and thus W is not invertible.

Solution: Since W is a matrix in a three–dimensional Euclidian vector space, it has one or three real eigenvalues. Let λ be an eigenvalue of W associated with the eigenvector $r \in \mathbb{R}^3, r \neq \hat{0}$. Then:

$$r \cdot Wr = \lambda (r \cdot r)$$

Also, since W is a skew symmetric matrix, then:

$$r \cdot Wr = -Wr \cdot r = -\lambda (r \cdot r)$$

By equating the above two expressions: $\lambda = 0$. Therefore, $r \neq \hat{0} \in \ker W$; therefore, W is not invertible.

Problem 2: If W is a skew–symmetric matrix in a three–dimensional Euclidian vector space and e_1 the unit eigenvector associated with the eigenvalue $\lambda = 0$, then show that $\forall a \in \mathbb{R}^3 : Wa = \omega e_1 \times a$, where $\omega = e_3 \cdot We_2$ and $B = \{e_1, e_2, e_3\}$ is an orthonormal basis set formed for $\mathbb{R}^3$.

Solution: Since W is a skew–symmetric matrix, then:

$$\omega = e_3 \cdot We_2 = -We_3 \cdot e_2$$

Assume that $We_2 = \alpha e_1 + \beta e_2 + \gamma e_3$. Then, by noticing that e_1 is an eigenvector of W:

$$We_2 \cdot e_1 = -e_2 \cdot We_1 = 0 \Longrightarrow \alpha = 0$$

Also, by using the identity $We_2 \cdot e_2 = 0 \Longrightarrow \beta = 0$. Thus, $We_2 = \gamma e_3 \Longrightarrow \gamma = \omega \Longrightarrow We_2 = \omega e_3$. Similarly, $We_3 = -\omega e_2$. Note that if $\omega = 0$ then $W = \hat{0}$.

Let $a \in \mathbb{R}^3 : a = a_1 e_1 + a_2 e_2 + a_3 e_3$, then:

$$Wa = a_2 We_2 + a_3 We_3 = \omega a_2 e_3 - \omega a_3 e_2$$

We also have:

$$\omega e_1 \times a = \omega a_2 e_3 - \omega a_3 e_2 \Longrightarrow Wa = \omega e_1 \times a$$

The vector ωe_1 is called the axial vector of W and is denoted by $w \in \mathbb{R}^3$.

The results of the previous problem are independent of the chosen coordinate system since the eigenvector e_1 of W is invariant (See section 1.2.6.d.). In a general coordinate system $B = \{e_1, e_2, e_3\}$ where the basis vectors are arbitrary, W has the form:

$$W = \begin{pmatrix} 0 & W_{12} & W_{13} \\ -W_{12} & 0 & W_{23} \\ -W_{13} & -W_{23} & 0 \end{pmatrix}$$

and the axial vector w of W in component form is:

$$w = \begin{pmatrix} -W_{23} \\ W_{13} \\ -W_{12} \end{pmatrix}$$

Skew–symmetric Matrices and the Instantaneous Rate of Rotation:

In solid mechanics, a skew symmetric matrix $W \subset \mathbb{M}^3$ usually represents the instantaneous rota– tion of a three–dimensional body around its axial vector $w \in \mathbb{R}^3$. $\forall a \in \mathbb{R}^3 : Wa = w \times a$ gives a vec– tor in the direction of rotation of the tip a and whose norm (magnitude) is the angle of rotation of the vector a around the axis of rotation w (Figure 1–23).

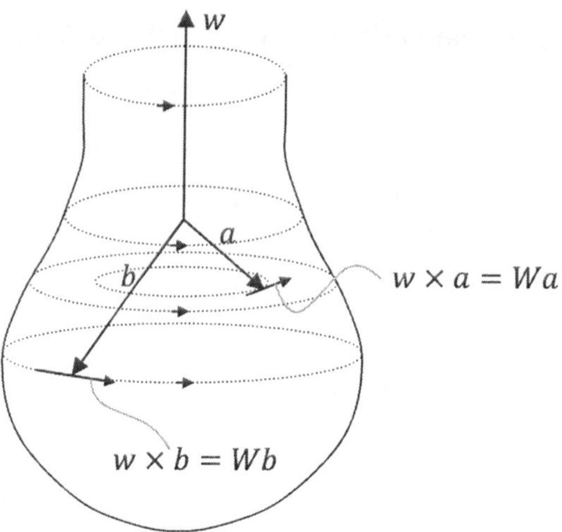

FIGURE 1-23 Axial vector w of a skew–symmetric matrix W.

Problem 3: Show that at **small rotations:**

- The instantaneous rate of rotation $\dot{Q}$ is represented by a skew–symmetric matrix, and
- Any skew–symmetric matrix represents a rate of rotation.

Solution: A rotation matrix satisfies:

$$QQ^T = I$$

Assuming that a body is rotating in space such that the rotation matrix Q is a function of time, then the rate of rotation is represented by the matrix $\dot{Q} = dQ/dt$:

$$\frac{d\left(QQ^T\right)}{dt} = \frac{dI}{dt} \Longrightarrow \dot{Q}Q^T + Q\dot{Q}^T = \hat{0} \Longrightarrow \dot{Q}Q^T = -Q\dot{Q}^T$$

At small rotations, $Q = I$ and thus $\dot{Q}$ is a skew–symmetric matrix:

$$\dot{Q} = -\dot{Q}^T$$

Let W be a skew–symmetric matrix. At small rotations, showing that W represents a rate of rotation is equivalent to showing that the matrix $Q = (I + Wdt)$ is a rotation matrix. Indeed, Q is a rotation matrix:

$$(I + Wdt)\left(I + W^T dt\right) = I + \left(W + W^T\right)dt - \left(WW^T\right)dt^2 \approx I$$

Problem 4: If W, S are skew–symmetric and symmetric matrices, respectively, in a three–dimensional Euclidian vector space and $B = \{e_1, e_2, e_3\}$ is an orthonormal basis set for $\mathbb{R}^3$, then show that tr$SW = S_{ij}W_{ji} = 0$.

Solution:

Using the symmetry of S and the antisymmetry of W:

$$S_{ij}W_{ji} = \frac{S_{ij}}{2}\left(W_{ji} + W_{ji}\right) = \frac{S_{ij}}{2}\left(W_{ji} - W_{ij}\right) = \frac{S_{ij}W_{ji}}{2} - \frac{S_{ij}W_{ij}}{2} = \frac{S_{ij}W_{ji}}{2} - \frac{S_{ji}W_{ij}}{2} = 0$$

Problem 5: A cone has a circular base of radius $r = 2mm$ that is concentric with the origin of the coordinate system. The tip of the cone is situated at $x = \{0, 0, 5\}\,mm$. The instantaneous rate of rotation at $t = 0$ of the cone in radians per minute is represented by the skew–symmetric matrix:

$$W = \begin{pmatrix} 0 & -5 & 5 \\ 5 & 0 & -5 \\ -5 & 5 & 0 \end{pmatrix}$$

Find the axial vector around which the cone is rotating and the angular velocity. Find the instantaneous velocity vector of a point on the base of the cylinder situated at $x = \{2, 0, 0\}$ at $t = 0$. Use Mathematica to draw the cone before and during rotation, assuming that the speed is constant.

Solution: The axial vector around which the cone is rotating is:

$$w = \begin{pmatrix} 5 \\ 5 \\ 5 \end{pmatrix}$$

The angular velocity is:

$$\omega = \|w\| = 5\sqrt{3}rad/min = \frac{1}{4\sqrt{3}}rad/sec$$

The velocity vector of the point situated at $x = \{2, 0, 0\}$ is:

$$Wx = \{0, 10, -10\}\,mm/min$$

The function `RotationMatrix` in Mathematica is used to rotate the cone at intervals of 5 seconds around the axis of rotation. The function `Manipulate` is used to visualize the rotated

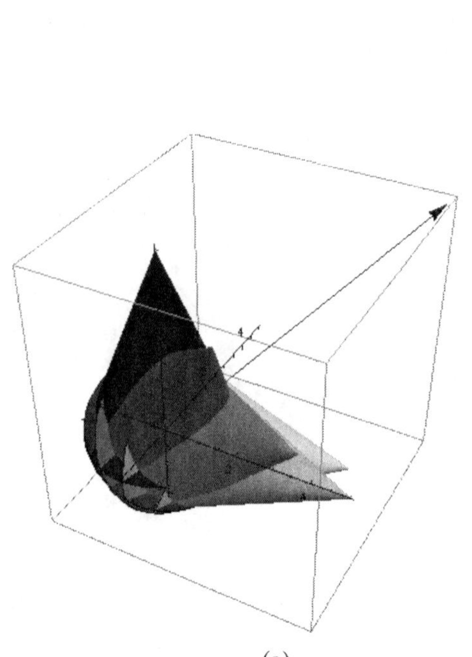

(a)

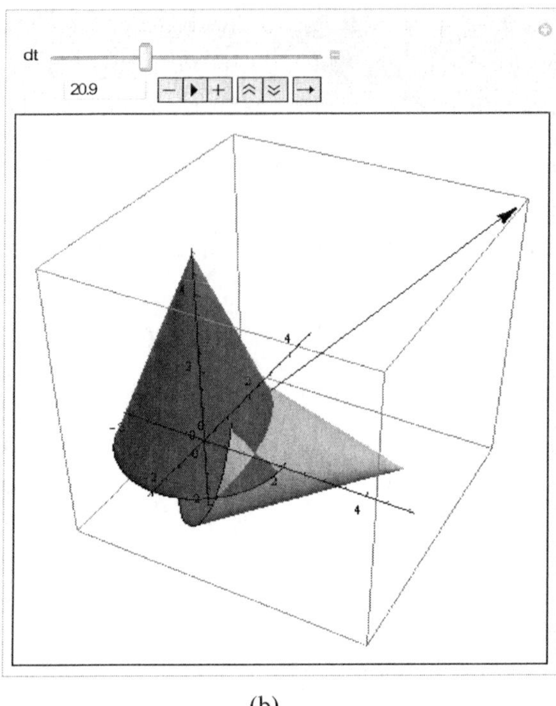

(b)

FIGURE 1-24 Cone rotation at a) $t = 0, 5, 10, 15, 20$ **and 25sec. b) user–specified** t**. Arrow pointing in the direction of the axial vector.**

cone at different intervals (Figure 1–24). Copy and paste the code into your Mathematica to visu–alize the rotation of the cone.

```
a=Cone[{{0,0,0},{0,0,5}},2];
W={{0,-5,5},{5,0,-5},{-5,5,0}};
w={-W[[2,3]],W[[1,3]],-W[[1,2]]};
axialvector=Arrow[{{0,0,0},w}];
Norm[w]
deltatheta=Norm[w]/60
x={2,0,0};
velocityvector=W.x
Q=Table[RotationMatrix[deltatheta*(5*i),w],{i,1,4}];
b=Table[GeometricTransformation[a,Q[[i]]],{i,1,4}];
Graphics3D[{axialvector,b,Green,a},Axes->True,AxesOrigin->{0,0,0}]

Manipulate[Q=RotationMatrix[deltatheta*(dt),w];
b=GeometricTransformation[a,Q];
Graphics3D[{axialvector,b,Green,a},Axes->True,AxesOrigin->{0,0,0}],{dt,1,60}]
```

1.2.7. Higher-Order Tensors

We will restrict the discussion here to three–dimensional vector spaces, although the same applies for higher dimensions. Let $\mathbb{R}^3$ be a Euclidian vector space and let $B = \{e_1, e_2, e_3\}$ and $B' = \{e'_1, e'_2, e'_3\}$ be two orthonormal basis sets. Then, as seen in section 1.2.6.c, the rotation matrix $Q \in \mathbb{M}^3$ with $Q_{ij} = e'_i \cdot e_j$ can be used for the orthogonal change of basis operation relating the components of the vectors and the matrices using both orthonormal basis sets. Considering the orthonormal basis set B, a vector $u \in \mathbb{R}^3$ has three independent components denoted $u_i = u \cdot e_i$, while a square matrix $M: \mathbb{R}^3 \to \mathbb{R}^3$ contains 3×3 independent components denoted $M_{ij} = e_i \cdot Me_j$. The components u'_i and u_i of u' and u considering B' and B respectively, are related by the relation:

$$u' = Qu, \quad u'_i = Q_{ij}u_j$$

The components M'_{ij} and M_{ij} of M' and M considering B' and B, respectively, are related by the relation:

$$M' = QMQ^T, \quad M'_{ij} = Q_{ik}M_{kl}Q_{jl}$$

The above results motivated the terminology that a vector is termed a **first–order tensor**, while a square matrix is termed a **second–order tensor**. Higher–order tensors can be defined similarly:

A linear map $D: \mathbb{M}^3 \to \mathbb{R}^3$ is considered a **third–order tensor** since it has $3 \times 3 \times 3$ independent components denoted D_{ijk} such that $\forall M \in \mathbb{M}^3, \exists u \in \mathbb{R}^3$ such that:

$$D_{ijk}M_{jk} = u_i$$

The relationship between the components D'_{ijk} and D_{ijk} considering B' and B, respectively, can be obtained as follows:

$$DM = u \Longrightarrow QDM = Qu \Longrightarrow Q_{il}D_{lmn}M_{mn} = Q_{ij}u_j \Longrightarrow Q_{il}D_{lmn}Q_{jm}Q_{kn}M'_{jk} = u'_i \Longrightarrow$$
$$D'_{ijk} = Q_{il}Q_{jm}Q_{kn}D_{lmn}$$

A linear map $C: \mathbb{M}^3 \to \mathbb{M}^3$ is considered a **fourth–order tensor** since it has $3 \times 3 \times 3 \times 3$ independent components denoted C_{ijkl} such that $\forall M \in \mathbb{M}^3, \exists N \in \mathbb{M}^3$:

$$C_{ijkl}M_{kl} = N_{ij}$$

The components C'_{ijkl} and C_{ijkl} considering B' and B, respectively, can be obtained as follows:

$$CM = N \Longrightarrow QCMQ^T = QNQ^T \Longrightarrow Q_{im}C_{mnop}M_{op}Q_{jn}$$
$$= Q_{ik}N_{kl}Q_{jl} \Longrightarrow Q_{im}Q_{jn}C_{mnop}Q_{ko}Q_{lp}M'_{kl} = N'_{ij} \Longrightarrow C'_{ijkl} = Q_{im}Q_{jn}Q_{ko}Q_{lp}C_{mnop}$$

```
New Mathematica Functions Explored in Section 1.2
(*Notes are typed between the two bracketed stars*)
Arrow[]
Chop[]
Clear[]
Cuboid[]
Det[]
Eigensystem[]
FullSimplify[]
GeometricTransformation[Objects,Q]
Graphics[{objects},AxesLabel->{},Axes->True,AxesOrigin->{}]
Graphics3D[{objects},AxesLabel->{},Axes->True,AxesOrigin->{}]
IdentityMatrix[]
Inverse[]
KroneckerDelta[i,j]
Manipulate[]
Options[]
Polygon[]
ReflectionMatrix[]
RotationMatrix[Angle,vector]
Table[]
Transpose[]
```

EXERCISES

1. Let $M \in \mathbb{M}^n$. Show that if $\det M = 0$, then $\lambda = 0$ is an eigenvalue for M and that any nonzero vector $x \in \ker M$ is an eigenvector associated with the eigenvalue $\lambda = 0$.

2. Find the determinant of the following matrices. If a matrix is not invertible, then find its kernel. If a matrix is invertible, then find its inverse.

$$\begin{pmatrix} 1 & 1 \\ -1 & -1 \end{pmatrix}, \begin{pmatrix} 1 & 1 \\ 0 & -1 \end{pmatrix}, \begin{pmatrix} 5 & 0 & 0 \\ 0 & 2 & -1 \\ 1 & 2 & 5 \end{pmatrix}, \begin{pmatrix} 5 & 0 & 0 \\ 0 & 2 & -1 \\ 10 & 2 & -1 \end{pmatrix}$$

3. Consider the vectors $x = \{1, 0, 0\}$, $y = \{0, 1, 0\}$ and $z = \{1, 1, 1\}$. Find the volume of the parallelepiped formed by the three vectors. Also, find the volume of the parallelepiped formed when the three vectors are linearly mapped using the matrices M_1 and M_2 defined below. Comment on the results.

$$M_1 = \begin{pmatrix} 1.1 & 0.1 & 0 \\ 0 & 0.9 & 0 \\ 0 & 0 & 0.9 \end{pmatrix}, \quad M_2 = \begin{pmatrix} 0 & 1.1 & 0 \\ 0 & 1 & 0 \\ 0 & 0 & 1 \end{pmatrix}$$

4. Find the eigenvalues and eigenvectors of the matrices in question 2.

5. Choose a value for M_{11} in the following matrices so that the matrix is not invertible, and then find their kernels.

$$\begin{pmatrix} M_{11} & 5 & 5 \\ -3 & 5 & -1 \\ 0 & 2 & -1 \end{pmatrix} \quad \begin{pmatrix} M_{11} & 2 & 3 & 1 \\ -1 & -2 & 1 & -2 \\ 1 & 3 & 1 & -5 \\ 5 & 2 & 1 & 7 \end{pmatrix}$$

6. Use Mathematica Software to draw a two–dimensional polygon with five sides, and then use the table command to create 10 copies of the polygon rotated with an angle $\frac{2\pi}{10}$ between each copy. Finally, use the graphics command to view the 10 polygons you drew.

7. A rectangle with length 2 units and height 4 units is situated as shown in the figure. Find the coordinates of its vertices in the orthonormal basis set $B = \{e_1, e_2\}$. Find the coordinates of its vertices if the coordinate system defined by the orthonormal basis set $B' = \{e'_1, e'_2\}$ is chosen instead.

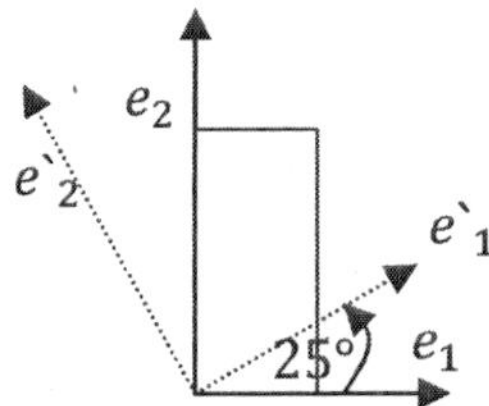

8. Verify that the following matrix is an orthogonal matrix, and specify whether it is a rotation or a reflection.

$$Q = \begin{pmatrix} 3/4 & -\sqrt{3}/4 & 1/2 \\ 3\sqrt{3}/8 & 5/8 & -\sqrt{3}/4 \\ -1/8 & 3\sqrt{3}/8 & 3/4 \end{pmatrix}$$

9. Find the value of m so that the following matrix is a rotation matrix:

$$Q = \begin{pmatrix} \frac{\sqrt{3}}{4} & m & \frac{\sqrt{3}}{2} \\ \frac{3}{8} + \frac{\sqrt{3}}{4} & \frac{3}{4} - \frac{\sqrt{3}}{8} & -\frac{1}{4} \\ \frac{1}{4} - \frac{3\sqrt{3}}{8} & \frac{3}{8} + \frac{\sqrt{3}}{4} & \frac{\sqrt{3}}{4} \end{pmatrix}$$

10. Consider the orthonormal basis set $B = \{e_1, e_2, e_3\}$. Consider the matrix M:

$$M = \begin{pmatrix} 1 & 5 & -2 \\ 2 & 2 & 3 \\ -5 & 2 & 1 \end{pmatrix}$$

- Let $a = \{1, 1, 2\}$ $b = \{-1, 1, 0\}$. Verify that $a \cdot b = 0$ and find a vector c that is orthogonal to both a and b. Then, find a new orthonormal basis set B' using the three normalized a, b and c vectors.
- Find the components of M' in the new orthonormal basis set B'.
- Find the three invariants of M and M' and verify that they are equal.
- Find the eigenvalues of M and M' and verify that they are equal.

11. Consider the symmetric matrix M:

$$M = \begin{pmatrix} 1 & 0 & -2 \\ 0 & 2 & 0 \\ -2 & 0 & 1 \end{pmatrix}$$

- Find the eigenvalues and eigenvectors of M.
- Find a new coordinate system (new orthonormal basis set) B' where M' (the components of M in the new coordinate system) is a diagonal matrix.

12. Identify the positive definite and the semi–positive definite symmetric matrices in the following:

$$\begin{pmatrix} 1 & 0 & -2 \\ 0 & 2 & 0 \\ -2 & 0 & 1 \end{pmatrix}, \begin{pmatrix} 1 & 0 & 2 \\ 0 & 2 & 0 \\ 2 & 0 & 1 \end{pmatrix}, \begin{pmatrix} 1 & 0 & -2 \\ 0 & 2 & 0 \\ -2 & 0 & 0 \end{pmatrix}, \begin{pmatrix} -1 & 0 & -2 \\ 0 & 2 & 0 \\ -2 & 0 & 1 \end{pmatrix}$$

13. Let a, b and $c \in \mathbb{R}^3$ such that a, b and c are linearly independent nonzero vectors and that $\mathbb{R}^3$ is a Euclidean vector space equipped with the dot product operation. Using the properties of the dot product, construct an orthonormal basis as a function of a, b and c. (Such construction is referred to as: Gram–Schmidt Process.)

14. Using Einstein summation convention described in section 1.2.5.a, and assuming that the underlying space is the Euclidian space $\mathbb{R}^3$, show that the following relations hold true for the Kronecker Delta δ_{ij} and the Alternator ε_{ijk}:

    a. $\delta_{ii} = 3$

b. $\delta_{ik}\varepsilon_{ikm} = 0$

c. $\varepsilon_{iks}\varepsilon_{mks} = 2\delta_{im}$

d. $\varepsilon_{ikm}\varepsilon_{ikm} = 6$

e. $\varepsilon_{iks}\varepsilon_{mps} = \delta_{im}\delta_{kp} - \delta_{ip}\delta_{km}$

15. Write the following expressions out in full, assuming that the underlying space is the Euclidian space $\mathbb{R}^3$:

a. $c = d_{ijkl}\delta_{jk}\delta_{il}$

b. $b_i = (a_{ij} + a_{ji})\, c_j$

c. $a_i = \varepsilon_{ijk}c_j d_k$

16. Write the following expressions in a more compact form using index notation and the summation convention.

a.

$$
\begin{aligned}
c_{11}x_1 + c_{12}x_2 + c_{13}x_3 &= d_1 \\
c_{21}x_1 + c_{22}x_2 + c_{23}x_3 &= d_2 \\
c_{31}x_1 + c_{32}x_2 + c_{33}x_3 &= d_3
\end{aligned}
$$

b.

$$a = c_{11}x_1^2 + c_{22}x_2^2 + c_{33}x_3^2 + (c_{12} + c_{21})\, x_1x_2 + (c_{13} + c_{31})\, x_1x_3 + (c_{23} + c_{32})\, x_2x_3$$

c.

$$
\begin{aligned}
\sigma_{11} &= 2\mu\varepsilon_{11} + \lambda\varepsilon \\
\sigma_{22} &= 2\mu\varepsilon_{22} + \lambda\varepsilon \\
\sigma_{33} &= 2\mu\varepsilon_{33} + \lambda\varepsilon \\
\sigma_{12} &= 2\mu\varepsilon_{12} \\
\sigma_{13} &= 2\mu\varepsilon_{13} \\
\sigma_{23} &= 2\mu\varepsilon_{23}
\end{aligned}
$$

where $\varepsilon = \varepsilon_{11} + \varepsilon_{22} + \varepsilon_{33}$

17. Show that symmetric and antisymmetric matrices stay symmetric and antisymmetric, respectively, under any orthogonal coordinate transformation (orthogonal change of basis):

a. Directly, using the definitions of symmetric and antisymmetric matrices and using the orthogonal transformation rules without reference to components.

b. Using components; i.e., show that if A_{ij} and B_{ij} are the components of two second–order tensors that are symmetric and antisymmetric, respectively, when referred to a particular orthonormal basis set, then, the same will be true when they are referred to any other orthonormal basis set.

18. Let $S \in \mathbb{M}^3$ be a symmetric matrix. Let $P \in \mathbb{M}^3$ be an invertible matrix. The matrices S and $T = PSP^{-1}$ are called similar. Show that similar matrices have the same eigenvalues. What is the relationship between the eigenvectors of S and T?

19. In a three–dimensional Euclidian vector space, how many independent components will the following tensors have?

    a. A completely symmetric* third–order tensor.

    b. A completely symmetric fourth–order tensor

    *The term completely symmetric means that components having indices with the same numerical values are equal, regardless of the order of the indices. For example $C_{1213} = C_{1132} = C_{1231} = C_{1231} = C_{3211}$ etc.

20. Show that if A and B are symmetric matrices, then AB is not necessarily a symmetric matrix. (Hint: Find a counter example)

21. Show that if $A \in \mathbb{M}^3$ is a symmetric matrix, then, $\forall B \in \mathbb{M}^3$ the matrix $C = BAB^T$ is symmetric.

22. Let A, B, $C \in \mathbb{M}^n$ and I be the identity matrix in $\mathbb{M}^n$. Show that if $BA = I$ and $AC = I$, then, $B = C$. (The matrix B is referred to as the **inverse** of A and is denoted A^{-1})

23. Let $S \in \mathbb{M}^n$ be a symmetric matrix. Let $\lambda_{\min}$ and $\lambda_{\max}$ be the minimum and maximum eigenvalues of S. Show that $\forall a \in \mathbb{R}^n$: $\lambda_{\min} \leq a \cdot Aa \leq \lambda_{\max}$. (The quantity $a \cdot Aa$ is referred to as the **Rayleigh Quotient**).

1.3. VECTOR CALCULUS

1.3.1. Fields and their Derivatives

Consider the set $D \subset \mathbb{R}^n$. The following are scalar, vector, and tensor fields, or functions defined on the set D, respectively:

$$\phi: D \to \mathbb{R}, \quad u: D \to \mathbb{R}^n, \quad T: D \to \mathbb{M}^n$$

For simplicity, in this section, tensors are restricted to second–order tensors. We remind the reader that a second–order tensor can be represented by a square matrix tied to a specific coordinate system. The arguments of the functions ϕ, u and T are vectors in the set D. The functions ϕ, u and T are scalar–valued, vector–valued, and tensor–valued functions, respectively (the outputs are scalars, vectors and tensors, respectively).

Considering a three–dimensional Euclidian vector space $\mathbb{R}^3$ with an orthonormal basis set $B = \{e_1, e_2, e_3\}$ in which $\forall x \in D$: $\exists! x_i \in \mathbb{R},\ i = 1, 2, 3$ such that $x = x_i e_i$, then, the arguments of the functions ϕ, u and T could be chosen to be the Cartesian coordinates x_i:

$$\phi(x) = \phi(x_1, x_2, x_3), \quad u(x) = u(x_1, x_2, x_3), \quad T(x) = T(x_1, x_2, x_3)$$

In the following, unless otherwise noted, we will consider that any scalar–, vector–, or tensor–valued functions are continuous and smooth (differentiable), but the definition of continuity and smoothness is omitted. Also, we will consider an orthonornal basis set, even though in the abstract definitions, components do not appear.

1.3.1.a. Gradient of a Scalar Field

Consider a three–dimensional Euclidian vector space $\mathbb{R}^3$ with an orthonormal basis set $B = \{e_1, e_2, e_3\}$. Consider the set $D \subset \mathbb{R}^3$. Let $\phi: D \to \mathbb{R}$ be a scalar–valued smooth function. The gradient of ϕ, denoted by $\nabla\phi$ or gradϕ at any element $x \in D$, is the vector with the components:

$$\nabla\phi = \begin{pmatrix} \dfrac{\partial\phi}{\partial x_1} \\ \dfrac{\partial\phi}{\partial x_2} \\ \dfrac{\partial\phi}{\partial x_3} \end{pmatrix}$$

Notice that this vector transforms by the usual orthogonal transformation matrix if a new basis set $B' = \{e'_1, e'_2, e'_3\}$ is considered:

$$(\nabla\phi)' = \begin{pmatrix} \dfrac{\partial\phi}{\partial x_1}\dfrac{\partial x_1}{\partial x'_1} + \dfrac{\partial\phi}{\partial x_2}\dfrac{\partial x_2}{\partial x'_1} + \dfrac{\partial\phi}{\partial x_3}\dfrac{\partial x_3}{\partial x'_1} \\ \dfrac{\partial\phi}{\partial x_1}\dfrac{\partial x_1}{\partial x'_2} + \dfrac{\partial\phi}{\partial x_2}\dfrac{\partial x_2}{\partial x'_2} + \dfrac{\partial\phi}{\partial x_3}\dfrac{\partial x_3}{\partial x'_2} \\ \dfrac{\partial\phi}{\partial x_1}\dfrac{\partial x_1}{\partial x'_3} + \dfrac{\partial\phi}{\partial x_2}\dfrac{\partial x_2}{\partial x'_3} + \dfrac{\partial\phi}{\partial x_3}\dfrac{\partial x_3}{\partial x'_3} \end{pmatrix} = Q\nabla\phi$$

Directional derivative:

The scalar field $\phi: D \to \mathbb{R}$ varies with respect to the different coordinates x_1, x_2 and x_3. The derivative of a scalar field at a point in a specified direction is termed the directional derivative. The directional derivative of ϕ in the direction of the unit vector $n \in \mathbb{R}^3$ can be obtained as follows:

$$\text{directional derivative} = \lim_{h \to 0} \frac{\phi(x + hn) - \phi(x)}{h}$$

The term $\phi(x + hn)$ can be expanded using a Taylor series, as follows:

$$\phi(x + hn) = \phi(x) + \frac{\partial \phi}{\partial x} \cdot hn + \cdots$$

Then, the directional derivative can be evaluated as:

$$\text{directional derivative} = \frac{\partial \phi}{\partial x} \cdot n = \nabla \phi \cdot n$$

▶ **BONUS EXERCISE:** Show that the maximum value of the directional derivative is attained in the direction of the vector $\nabla\phi$. Also, indicate what it means that a vector $n \in \mathbb{R}^3$ is orthogonal to $\nabla\phi$.

Problem: Consider $\mathbb{R}^2$. The distribution of the scalar field $\phi : D \to \mathbb{R}$ on the set $D = \{(x_1, x_2) \,|\, 0 \le x_1 \le 3 \text{ and } 0 \le x_2 \le 3\}$ is according to the equation: $\phi = 5x_1^2 + 3x_1x_2 - 5x_2^2$. Draw the contour plot of ϕ and the vector plot of gradϕ. Find the directional derivative of ϕ evaluated at $x \in D, x = \{2, 2\}$ in the direction of the unit vector $z = \{1, 0\}$. Also, find a vector pointing to the direction of constant ϕ evaluated at $x \in D, x = \{2, 2\}$.

Solution:

The contour plot of ϕ and the vector plot of the gradient of ϕ are shown in Figure 1–25. The arrows in the vector plot are orthogonal to the contour lines in the contour plot (Why?). The gradient of the scalar field is the vector field:

$$\text{grad}\,\phi = \begin{pmatrix} 10x_1 + 3x_2 \\ 3x_1 - 10x_2 \end{pmatrix}$$

The value of the vector field at $x = \{2, 2\}$ is:

$$\text{grad}\,\phi = \begin{pmatrix} 26 \\ -14 \end{pmatrix}$$

The directional derivative of ϕ in the direction of z is:

$$\text{grad}\,\phi \cdot z = 26$$

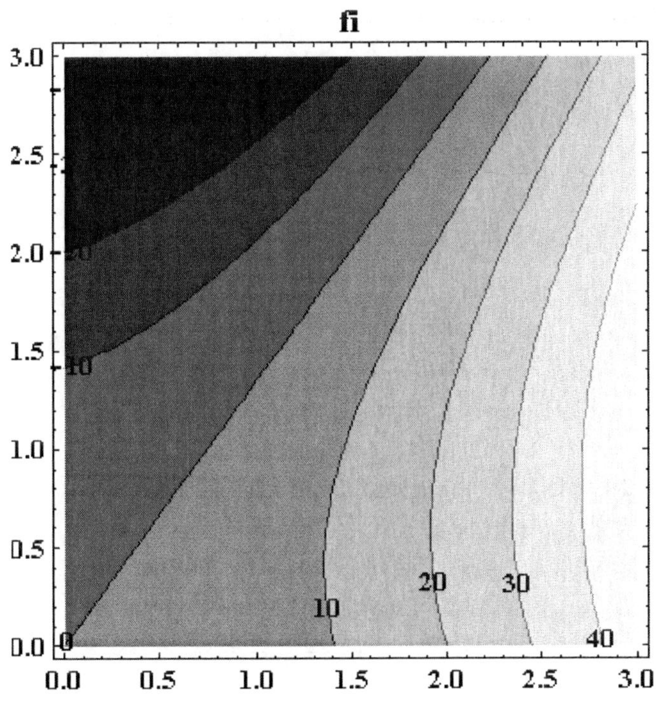

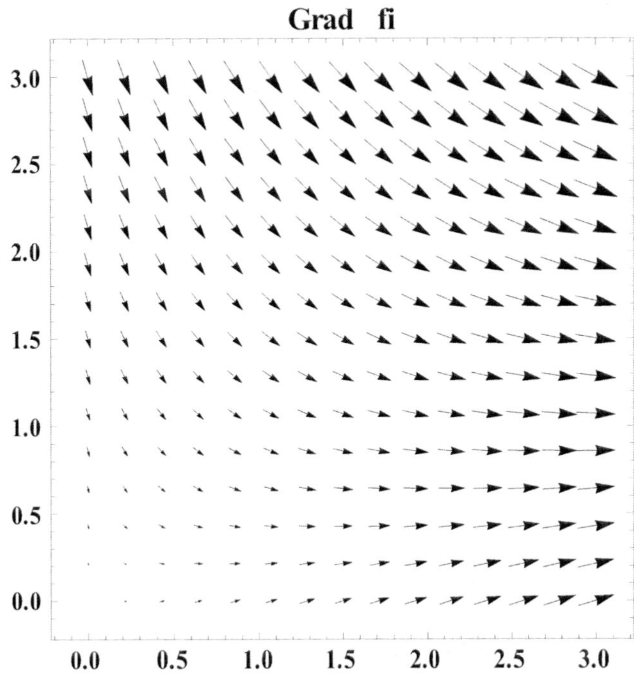

FIGURE 1-25 Contour plot of a scalar field and vector plot of the gradient of the scalar field over a set in $\mathbb{R}^2$.

A vector pointing to the direction of constant ϕ evaluated at x is:

$$n = \begin{pmatrix} 14 \\ 26 \end{pmatrix}$$

A unit vector pointing to the direction of constant ϕ evaluated at x is:

$$n = \begin{pmatrix} 0.4741 \\ 0.8805 \end{pmatrix}$$

Notice that in Mathematica, a rule can be used to evaluate an expression at certain values for the variables within the expression by using `/.rule`. For example, the following code evaluates the value of $y = 5x^2 + x^3$ by replacing x with the value of 3. This rule assignment does not assign any value to the variable x, but only evaluates y by replacing x with 3.

```
Clear[y,x];
y=5*x^2+x^3;
y/.x->3
x
```

The following Mathematica code illustrates the above calculations (Notice that the x and y vari– ables are used for x_1 and x_2):

```
Clear[x,y];
fi=5*x^2+3*x*y-5*y^2;
ContourPlot[fi,{x,0,3},{y,0,3},BaseStyle->Directive[Bold,15],ColorFunction->
"GrayTones",AspectRatio->Automatic,ContourLabels->All,PlotLabel->"fi"]
gradfi={D[fi,x],D[fi,y]};
VectorPlot[gradfi,{x,0,3},{y,0,3}, BaseStyle->Directive[Bold,15],
VectorStyle->Black,AspectRatio->Automatic,PlotLabel->"Grad-fi"]
gradfiatx=gradfi/.{x->2,y->2}
dirderivative=gradfiatx.{1,0}
v={-gradfiatx[[2]],gradfiatx[[1]]};
v=N[v/Norm[v]]
v.gradfiatx
```

1.3.1.b. Gradient of a Vector Field

Consider a three–dimensional Euclidian vector space $\mathbb{R}^3$ with an orthonormal basis set $B = \{e_1, e_2, e_3\}$, $\forall x \in \mathbb{R}^3 : \exists! x_i \in \mathbb{R},\ i = 1, 2, 3$ such that $x = x_i e_i$. Consider the set $D \subset \mathbb{R}^3$. Let $u{:}D \to \mathbb{R}^3$ be a vector–valued smooth function. The gradient of u, denoted by ∇u or grad u at any element

$x \in D$, is the tensor with the components:

$$\nabla u = \begin{pmatrix} \dfrac{\partial u_1}{\partial x_1} & \dfrac{\partial u_1}{\partial x_2} & \dfrac{\partial u_1}{\partial x_3} \\ \dfrac{\partial u_2}{\partial x_1} & \dfrac{\partial u_2}{\partial x_2} & \dfrac{\partial u_2}{\partial x_3} \\ \dfrac{\partial u_3}{\partial x_1} & \dfrac{\partial u_3}{\partial x_2} & \dfrac{\partial u_3}{\partial x_3} \end{pmatrix}.$$

This tensor transforms by the usual orthogonal transformation matrix if a new basis set $B' = \{e'_1, e'_2, e'_3\}$ is considered (adopting Einstein summation convention):

$$(\nabla u)'_{ij} = \frac{\partial u'_i}{\partial x'_j} = \frac{\partial Q_{ik} u_k}{\partial x_l} \frac{\partial x_l}{\partial x'_j} = Q_{ik} \frac{\partial u_k}{\partial x_l} Q_{jl}$$

Therefore:

$$(\nabla u)' = Q \nabla u Q^T$$

Problem: Consider $\mathbb{R}^2$. The distribution of the vector field $u: D \to \mathbb{R}^2$ on the set $D = \{(x_1, x_2) \mid 0 \le x_1 \le 3 \text{ and } 0 \le x_2 \le 3\}$ is according to the equation: $u = \{5x_1^2 + 2x_2, 3x_2 + x_1\}$. Find grad u.

Solution: The gradient of the vector field u is the tensor field:

$$\text{grad}\, u = \begin{pmatrix} 10x_1 & 2 \\ 1 & 3 \end{pmatrix}$$

1.3.1.c. Divergence of Vector and Tensor fields

Consider a three–dimensional Euclidian vector space $\mathbb{R}^3$ with an orthonormal basis set $B = \{e_1, e_2, e_3\}$, $\forall x \in \mathbb{R}^3 : \exists! x_i \in \mathbb{R},\ i = 1, 2, 3$ such that $x = x_i e_i$. Given a set $D \subset \mathbb{R}^3$. Let $u: D \to \mathbb{R}^3$ be a smooth vector–valued function and $T: D \to \mathbb{M}^3$ be a smooth tensor–valued function. Then, div u and div T are defined by:

$$\text{div}\, u = \frac{\partial u_1}{\partial x_1} + \frac{\partial u_2}{\partial x_2} + \frac{\partial u_3}{\partial x_3}, \qquad \text{div}\, T = \begin{Bmatrix} \dfrac{\partial T_{11}}{\partial x_1} + \dfrac{\partial T_{21}}{\partial x_2} + \dfrac{\partial T_{31}}{\partial x_3} \\ \dfrac{\partial T_{12}}{\partial x_1} + \dfrac{\partial T_{22}}{\partial x_2} + \dfrac{\partial T_{32}}{\partial x_3} \\ \dfrac{\partial T_{13}}{\partial x_1} + \dfrac{\partial T_{23}}{\partial x_2} + \dfrac{\partial T_{33}}{\partial x_3} \end{Bmatrix}$$

The divergence of a vector field is a scalar field, while the divergence of a tensor field is a vector field. Also, div u and div T are sometimes denoted by div $u = \nabla \cdot u$ and div $T = \nabla \cdot T$, respectively.

1.3.1.d. Curl

Consider a three–dimensional Euclidian vector space $\mathbb{R}^3$ with an orthonormal basis set $B = \{e_1, e_2, e_3\}, \forall x \in \mathbb{R}^3 : \exists! x_i \in \mathbb{R},\ i = 1, 2, 3$ such that $x = x_i e_i$. Consider the set $D \subset \mathbb{R}^3$. Let $u: D \to \mathbb{R}^3$ be a vector–valued smooth function. The curl of u, denoted by curl u, at any element $x \in D$, is the vector:

$$\operatorname{curl} u = \begin{pmatrix} \frac{\partial u_2}{\partial x_3} - \frac{\partial u_3}{\partial x_2} \\ \frac{\partial u_3}{\partial x_1} - \frac{\partial u_1}{\partial x_3} \\ \frac{\partial u_1}{\partial x_2} - \frac{\partial u_2}{\partial x_1} \end{pmatrix}$$

or in component form:

$$(\operatorname{curl} u)_i = \varepsilon_{ijk} \left(\frac{\partial u_j}{\partial x_k} \right)$$

1.3.1.e. Laplacian

Consider a three–dimensional Euclidian vector space $\mathbb{R}^3$ with an orthonormal basis set $B = \{e_1, e_2, e_3\}, \forall x \in \mathbb{R}^3 : \exists! x_i \in \mathbb{R},\ i = 1, 2, 3$ such that $x = x_i e_i$. Consider the set $D \subset \mathbb{R}^3$. Let $\phi: D \to \mathbb{R}$ be a scalar–valued smooth function. The Laplacian of ϕ, denoted by $\nabla \cdot \nabla\phi$ or $\nabla^2\phi$ at any element $x \in D$, is the divergence of the gradient of ϕ:

$$\nabla^2\phi = \frac{\partial^2\phi}{\partial x_1^2} + \frac{\partial^2\phi}{\partial x_2^2} + \frac{\partial^2\phi}{\partial x_3^2}$$

1.3.1.f. Comma Notation

For ease of notation, partial differentiation is sometimes replaced by a comma notation. Consider a three–dimensional Euclidian vector space $\mathbb{R}^3$ with an orthonormal basis set $B = \{e_1, e_2, e_3\}, \forall x \in \mathbb{R}^3 : \exists! x_i \in \mathbb{R},\ i = 1, 2, 3$ such that $x = x_i e_i$. Consider the set $D \subset \mathbb{R}^3$. Let $\phi: D \to \mathbb{R}$, $u: D \to \mathbb{R}^3$, $T: D \to \mathbb{M}^3$ be scalar–valued, vector–valued, and tensor–valued smooth functions, respectively. The following notations are equivalent:

$$(\nabla\phi)_i = (\operatorname{grad}\phi)_i = \phi_{,i} = \frac{\partial\phi}{\partial x_i}$$

$$(\nabla u)_{ij} = (\operatorname{grad} u)_{ij} = u_{i,j} = \frac{\partial u_i}{\partial x_j}$$

$$\operatorname{div} u = u_{i,i} = \frac{\partial u_i}{\partial x_i}$$

$$(\operatorname{div} T)_i = T_{ji,j} = \frac{\partial T_{ji}}{\partial x_j}$$

$$(\operatorname{curl} u)_i = \varepsilon_{ijk} u_{j,k} = \varepsilon_{ijk} \left(\frac{\partial u_j}{\partial x_k} \right)$$

$$\nabla^2 \phi = \frac{\partial \frac{\partial \phi}{\partial x_i}}{\partial x_i} = \phi_{,ii}$$

Problem: If $\phi = 5x_1^3 x_2 + 7x_2^3$ is a scalar field defined over $\mathbb{R}^2$, then evaluate the gradient of ϕ, the Laplacian of ϕ, the gradient of grad ϕ, and the divergence of the gradient of grad ϕ.

Solution:

$$\operatorname{grad} \phi = \begin{pmatrix} 15x_1^2 x_2 \\ 21x_2^2 + 5x_1^3 \end{pmatrix}$$

$$\nabla^2 \phi = \operatorname{div} (\operatorname{grad} \phi) = 30x_1 x_2 + 42x_2$$

$$\operatorname{grad} (\operatorname{grad} \phi) = \begin{pmatrix} 30x_1 x_2 & 15x_1^2 \\ 15x_1^2 & 42x_2 \end{pmatrix}$$

$$\operatorname{div} (\operatorname{grad} (\operatorname{grad} \phi)) = \begin{pmatrix} 30x_2 \\ 42 + 30x_1 \end{pmatrix}$$

The following Mathematica code is used to construct the required tables:

```
fi=5*x1^3*x2+7*x2^3;
xx={x1,x2};
gradfi=Table[D[fi,xx[[i]]],{i,1,2}]
divgradfi=Sum[D[gradfi[[i]],xx[[i]]],{i,1,2}]
gradgradfi=Table[D[gradfi[[i]],xx[[j]]],{i,1,2},{j,1,2}]
divgradgradfi=Table[Sum[D[gradgradfi[[j,i]],xx[[j]]],{j,1,2}],{i,1,2}]
```

1.3.2. Divergence Theorem

The divergence theorem, also known as the Gauss theorem, is a mathematical tool that relates volume integrals of the gradients of functions to the surface integrals of those functions. It has many applications in engineering. In physical terms, the divergence theorem states that the change of a quantity inside a closed volume is due to this quantity entering or exiting through the boundary, as long as there are no sinks or sources inside the volume. The mathematical theorem is as follows:

Given a volume represented by a closed set $V \subset \mathbb{R}^3$ with a boundary $S = \partial V$.

Let $n \in \mathbb{R}^3$ define the outward normal to the boundary S,

$\phi: V \to \mathbb{R}$ be a smooth scalar–valued function,

X_1, X_2 and X_3 represent the coordinates in the orthonormal basis set $B = \{e_1, e_2, e_3\}$.

Then the divergence theorem states:

$$\int \frac{\partial \phi}{\partial X_1} dV = \int \phi n_1 dS, \quad \int \frac{\partial \phi}{\partial X_2} dV = \int \phi n_2 dS, \quad \int \frac{\partial \phi}{\partial X_3} dV = \int \phi n_3 dS$$

The proof of the divergence theorem is a straightforward technical application of the fundamental theorem of calculus and will be omitted. An interested reader should consult a book on vector calculus.

▶ **BONUS EXERCISE:** Given a volume represented by a closed set $V \subset \mathbb{R}^3$ with a boundary $S = \partial V$.

Let $n \in \mathbb{R}^3$ define the outward normal to the boundary S,

$\phi: V \to \mathbb{R}$ be a smooth scalar–valued function,

$u: V \to \mathbb{R}^3$ be a smooth vector–valued function,

$T: V \to \mathbb{M}^3$ be a smooth tensor–valued function,

X_1, X_2 and X_3 represent the coordinates in the orthonormal basis set $B = \{e_1, e_2, e_3\}$.

$\operatorname{div} u$ and $\operatorname{div} T$ are as per their definitions:

$$\operatorname{div} u = \frac{\partial u_1}{\partial X_1} + \frac{\partial u_2}{\partial X_2} + \frac{\partial u_3}{\partial X_3}, \quad \operatorname{div} T = \begin{Bmatrix} \frac{\partial T_{11}}{\partial X_1} + \frac{\partial T_{21}}{\partial X_2} + \frac{\partial T_{31}}{\partial X_3} \\ \frac{\partial T_{12}}{\partial X_1} + \frac{\partial T_{22}}{\partial X_2} + \frac{\partial T_{32}}{\partial X_3} \\ \frac{\partial T_{13}}{\partial X_1} + \frac{\partial T_{23}}{\partial X_2} + \frac{\partial T_{33}}{\partial X_3} \end{Bmatrix}$$

Then, use the divergence theorem to show the following equalities:

$$\int \operatorname{grad}\phi dV = \int (\phi)\, n dS, \quad \int \operatorname{div} u\, dV = \int u \cdot n\, dS, \quad \int \operatorname{div} T\, dV = \int T^T n\, dS$$

Problem 1: Consider $\mathbb{R}^2$. Let V be the area inside the circle $X_1^2 + X_2^2 = 1$. Let $f = \{2X_2, 5X_1\}$. Verify that $\int \text{div} f dV = \int f \cdot n dS$

Solution: For the left–hand side of the equality:

$$\text{div} f = \frac{\partial f_1}{\partial X_1} + \frac{\partial f_2}{\partial X_2} = 0 \Rightarrow \int \text{div} f dV = 0$$

The right–hand side of the equality applies on the boundary of V (on a circle of radius 1) at which:

$$n = \{\cos\theta, \sin\theta\}, \quad f = \{2\sin\theta, 5\cos\theta\} \Rightarrow f \cdot n = 7\sin\theta\cos\theta \Rightarrow \int_0^{2\pi} 7\sin\theta\cos\theta d\theta = 0$$

The last integral can be verified using Mathematica:

```
Integrate[7*Sin[th]*Cos[th], {th, 0, 2*Pi}]
```

The function f can be visualized using Mathematica in the region of the circle by using the function `VectorPlot[]` and utilizing the `RegionFunction` option. The following code draws the distri–bution of f inside the region specified by the circle equation (Figure 1–26):

```
Clear[x2,x1];
f={2x2,5x1};
VectorPlot[f,{x1,-1.1,1.1},{x2,-1.1,1.1},BaseStyle
->Directive[Bold,15],VectorStyle->Black,RegionFunction
->Function[{x1,x2},x1^2+x2^2<=1]]
```

Problem 2: Consider $\mathbb{R}^3$. Let V represent the closed region bounded by a sphere with radius of 1 unit. The equation of the surface of the sphere is $X_1^2 + X_2^2 + X_3^2 = 1$. Let $v = \{2X_1^2, 5X_2, 3X_3\}$ be the velocity of air particles inside the sphere. Find the amount of air leaving or entering the ball surface per unit time first using a volume integral and second using a surface integral.

Solution:

The velocity vectors of the air particles can be viewed on the region of interest in Mathematica using the following code (Figure 1–27):

```
Clear[f,x1,x2,x3,a,b,n];
f={2*x1^2,5*x2,3*x3};
VectorPlot3D[f,{x1,-1.1,1.1},{x2,-1.1,1.1},{x3,-1.1,1.1},BaseStyle
->Directive[Bold,15],VectorStyle->Black,RegionFunction
->Function[{x1,x2,x3},x1^2+x2^2+x3^2<=1],AxesLabel->{"x1","x2","x3"}]
```

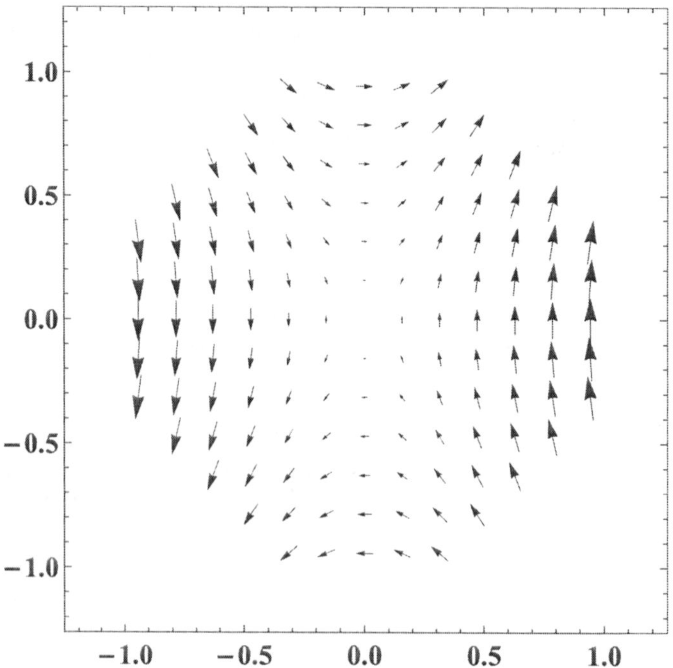

FIGURE 1-26 Vector plot of the function f on the circular region of radius 1 unit.

Using a surface integral:

Let S be the set representing the surface of the sphere and n represent the outward normal. The amount of air leaving or entering the surface per unit time is:

$$\int v \cdot ndS$$

If a spherical coordinate system is adopted with ϕ being the angle between any vector and the X_3 axis and θ is the angle between the X_1 axis and the projection of a vector onto the X_1 and X_2 plane, then on the surface of the unit sphere:

$$X_1 = \sin\phi\cos\theta, \quad X_2 = \sin\phi\sin\theta, \quad X_3 = \cos\phi, \quad dS = \sin\phi\, d\phi d\theta, \quad n = \{X_1, X_2, X_3\}$$

Then:

$$\int v \cdot ndS = \int_{\theta=0}^{2\pi}\int_{\phi=0}^{\pi} v \cdot n\sin\phi\, d\phi d\theta = \frac{32\pi}{3}$$

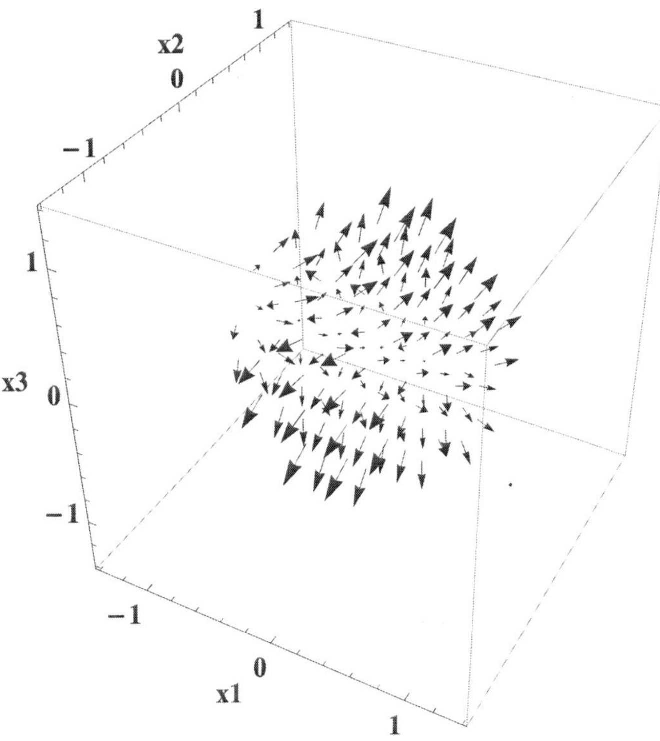

FIGURE 1-27 Vector plot of the function v on the spherical region of radius 1 unit.

Using a volume integral:

Using the divergence theorem the volume integral is:

$$\int v \cdot n dS = \int \left(\frac{\partial v_1}{\partial X_1} + \frac{\partial v_2}{\partial X_2} + \frac{\partial v_3}{\partial X_3} \right) dV = \int 8 + 4X_1 dV = \frac{32\pi}{3}$$

The following Mathematica code is used for the calculations:

```
Clear[f,x1,x2,x3,a,b,n];
f={2*x1^2,5*x2,3*x3};
df=D[f[[1]],x1]+D[f[[2]],x2]+D[f[[3]],x3]
a=Integrate[df*Boole[x1^2+x2^2+x3^2<=1],{x1,-1,1},{x2,-1,1},{x3,-1,1}]
x1=Cos[th]*Sin[fi];
x2=Sin[th]*Sin[fi];
x3=Cos[fi];
n={x1,x2,x3};
b=Integrate[f.n*Sin[fi],{fi,0,Pi},{th,0,2Pi}]
```

New Mathematica Functions Explored in Section 1.3

```
/.var->val
ContourPlot[,{},AspectRatio-> ,ContourLabels-> ,PlotLabel->""]
D[]
Integrate[]
Sum[]
VectorPlot[,{},AspectRatio->Automatic,PlotLabel->""]
VectorPlot3D[,{},RegionFunction->Function[],AxesLabel->{}]
```

▶ EXERCISES:

1. If $\phi = 5x_1^3 + 3x_1x_2 - 15x_2^3$ is a scalar field defined over the set $D \subset \mathbb{R}^2$: $D = \{(x_1, x_2)\,|0 \le x_1 \le 3, 0 \le x_2 \le 1\}$. Then, evaluate and indicate the order (scalar, vector, tensor) of the gradient of ϕ, the Laplacian of ϕ, the gradient of grad ϕ, the divergence of the gradient of grad ϕ. Use Mathematica software to visualize ϕ and gradϕ. Why is gradient of grad ϕ always symmetric?
2. Given a volume represented by a closed set $V \subset \mathbb{R}^3$ with a boundary $S = \partial V$.
   Let $n \in \mathbb{R}^3$ define the outward normal to the boundary S,
   $\phi: V \to \mathbb{R}$ be a smooth scalar–valued function,
   $u: V \to \mathbb{R}^3$ be a smooth vector–valued function,
   $T: V \to \mathbb{M}^3$ be a smooth tensor–valued function,
   X_1, X_2 and X_3 represent the coordinates in the orthonormal basis set $B = \{e_1, e_2, e_3\}$.

Then use the definitions in this section and the divergence theorem to show the following equal–ities:

$$\text{div}\, Tu = u \cdot \text{div}\, T + I_1\left(T\,\text{grad}\, u\right)$$

$$\text{div}\, \phi T = T^T \text{grad}\, \phi + \phi\, \text{div}\, T$$

$$\int \text{curl}\, u\, dV = \int u \times n\, dV$$

$$\int \text{div}\, Tu\, dV = \int T^T n \cdot u\, dS$$

$$\int \text{div}\, \phi u\, dV = \int n \cdot \phi u\, dS$$

CHAPTER 2
Stress

The mechanical design of loadbearing components involves ensuring that such components can sustain various loads. This design process involves calculating stresses inside the component being designed and comparing those stresses with the maximum stresses that the material used in manufacturing the component can withstand. This process is similar, whether the component is a bridge, a rocket, or even an artificial knee joint. The aim of this chapter is to familiarize the student with the concepts behind the word *stress*.

Figure 2–1 shows a bar with a cross–sectional area A subjected to a constant total force vector $F = \{f, 0, 0\}$ equally distributed on both sides. For the bar to be in equilibrium, the external forces have to sum up to zero (indeed, this is the case). It is clear that the stress inside the bar is constant, i.e., the value of the stress is independent of the point under consideration. The magnitude of the force per unit area over the surface on each side is denoted $s = f/A$. The simplified notion of the stress inside the bar is traditionally thought of as the force per unit area, i.e., the stress inside the bar in this example is equal to s. However, it will be shown that the general state of stress at a point inside this bar cannot be fully described by the scalar quantity s.

Considering different planes inside the material, the magnitude and direction of the force per unit area on each plane depends on its orientation. For example, the plane with normal $n_3 = \{1, 0, 0\}$ has a force vector per unit area that is equal to $t_3 = \{s, 0, 0\}$, while the plane with normal $n_4 = \{\cos 45, -\cos 45, 0\}$ has a force vector per unit area that is equal to $t_4 = \{s \cos 45^\circ, 0, 0\}$. It is also important to notice that the force vector per unit area is perpendicular to the plane with normal n_3 while this is not the case for the plane with normal n_4. In addition, Newton's third law predicts that the material on the left of a plane exerts a force vector on the material to the right of the plane that is exactly equal and opposite to the force exerted by the material to the right of the plane on the material to the left of the plane. If we denote t_n as the force vector per unit area on the plane with normal vector n, then the previous argument dictates that $t_n = -t_{-n}$. The force vectors per unit area are termed **Traction Vectors**. The following table lists the different planes shown in Figure 2–1, along with the traction vector applied to each plane.

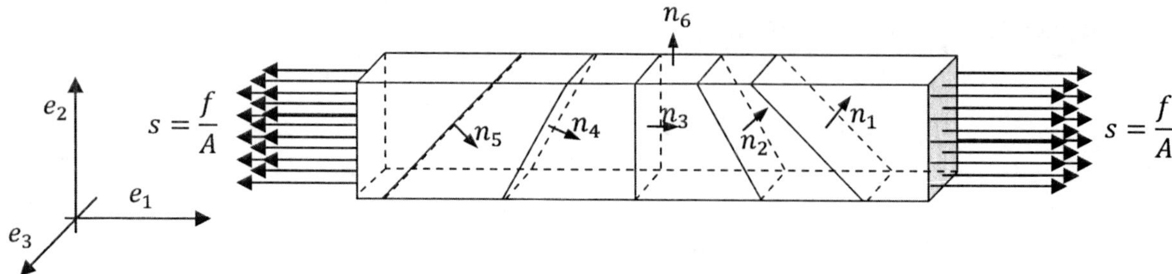

FIGURE 2-1 A bar of cross–sectional area A subjected to a distributed force f on each end. n_i represents the normal vector on a plane after removing the material on the right of the plane.

TABLE 2-1 Traction vector (force/unit area) on the different planes inside the bar shown in Figure 2–1:

Plane angle to the horizontal	Normal unit vector	Traction vector (force/unit area)
45°	$n_1 = \{\cos 45, \cos 45, 0\}$	$t_1 = \left\{\frac{f}{A}\cos 45^\circ, 0, 0\right\}$
60°	$n_2 = \{\cos 30, \cos 60, 0\}$	$t_2 = \left\{\frac{f}{A}\cos 30^\circ, 0, 0\right\}$
90°	$n_3 = \{1, 0, 0\}$	$t_3 = \left\{\frac{f}{A}, 0, 0\right\}$
-45°	$n_4 = \{\cos 45, -\cos 45, 0\}$	$t_4 = \left\{\frac{f}{A}\cos 45^\circ, 0, 0\right\}$
-60°	$n_5 = \{\cos 30, -\cos 60, 0\}$	$t_5 = \left\{\frac{f}{A}\cos 30^\circ, 0, 0\right\}$
0	$n_6 = \{0, 1, 0\}$	$t_6 = \{0, 0, 0\}$

Table 2–1 indicates that the traction vector (force per unit area) depends on the orientation of the plane under investigation. The stress measure at a point inside the bar should predict the value of the force vector per unit area on any given plane with normal n. This implies that the measure of the stress should be a function whose argument is n and the result is t_n.

Now let's consider the matrix $\sigma = \begin{pmatrix} s & 0 & 0 \\ 0 & 0 & 0 \\ 0 & 0 & 0 \end{pmatrix} \in \mathbb{M}^3 : \mathbb{R}^3 \to \mathbb{R}^3$. If this matrix is multiplied by any of the normal unit vectors in Table 2–1, the resulting vector is in fact the corresponding traction vector shown in the table. In particular, this matrix is a linear transformation from the Euclidian vector space of area vectors (normal vectors to the areas under consideration) to the Euclidian vector space of force vectors per unit area:

$$\sigma : n \in \mathbb{R}^3 \mapsto t_n \in \mathbb{R}^3 \quad \text{such that} \quad t_n = \sigma^T n$$

In this chapter, we will study stress as a linear transformation between Euclidian vector spaces. In particular, the stress matrix will be shown to be a symmetric matrix and thus will inherit all the properties of a symmetric matrix described previously.

2.1. THE STRESS TENSOR (MATRIX)

2.1.1. Traction Vector

The traction vector at a point inside a body can be defined after removing the adjacent material and assuming that a force vector $\Delta p \in \mathbb{R}^3$ is exerted by the removed material on an area Δa surrounding that point. The traction vector $t_n \in \mathbb{R}^3$ on this arbitrary surface with the normal vector $n \in \mathbb{R}^3$ is then (Figure 2–2):

$$t_n = \lim_{\Delta a \to 0} \frac{\Delta p}{\Delta a}$$

Newton's third law asserts that every action has a reaction that is equal in magnitude and opposite in direction (Figure 2–2). Thus,

$$t_n = -t_{-n}$$

2.1.2. Cauchy Stress Tetrahedron

In the introductory section, we presented, without proof, that the stress at a point can be rep–resented by a matrix $\sigma \in \mathbb{M}^3 : n \in \mathbb{R}^3 \mapsto t \in \mathbb{R}^3$ s.t. $t = \sigma^T n$. The proof that will be presented here is attributed to the French mathematician Augustin–Louis Cauchy (1789–1857), who had major contributions in complex analysis, mathematics, and mathematical physics. The ultimate goal of the proof is to show that the force vector per unit area (Traction vector defined above) $t_n \in \mathbb{R}^3$ on an arbitrary surface is given by the formula:

$$t_n = \sigma^T n$$

First, we need to clearly define the stresses on the planes perpendicular to our chosen set of basis. Let σ_{ij} indicates the stress on the face with normal i in the direction j (Figure 2–3). Notice that in many textbooks, this definition could be reversed.

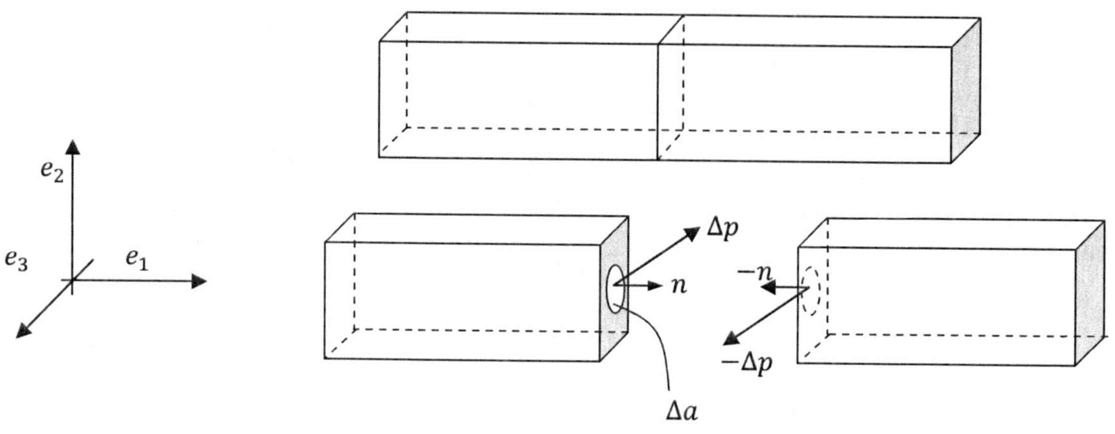

FIGURE 2-2 Traction vector $t_n = \lim_{\Delta_a \to 0} \frac{\Delta p}{\Delta a}$.

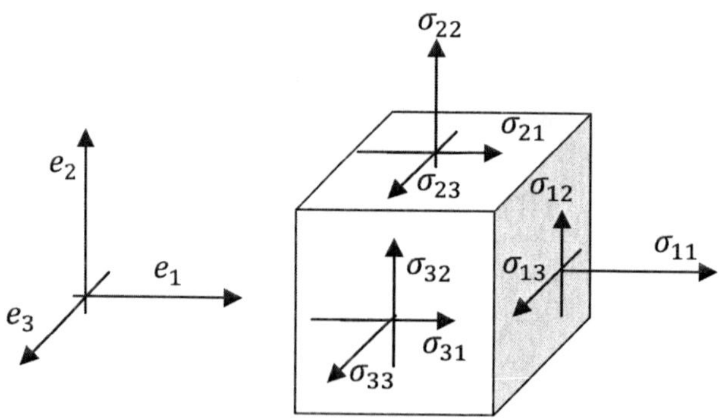

FIGURE 2-3 Definition of the components of the stress matrix (tensor).

The basic assumption behind the existence of this stress matrix is that each block of material can be separated from its surroundings by simply cutting it and considering the effect on the cut sur–face as a force vector. Thus, we can consider the tetrahedron shown in Figure 2–4 as cut from its surrounding. Let A_n be the area of the surface with normal $n \in \mathbb{R}^3$, then:

$$nA_n = A_1e_1 + A_2e_2 + A_3e_3$$

where A_1, A_2 and A_3 are the areas of the triangular faces perpendicular to e_1, e_2 and e_3 respectively.

Multiplying each side of the previous formula by $1/A_n$:

$$n = n_1e_1 + n_2e_2 + n_3e_3, \text{ where } \forall i : n_i = \frac{A_i}{A_n}$$

▶ **BONUS EXERCISE:** Derive the formula $nA_n = A_1e_1 + A_2e_2 + A_3e_3$ using the properties and the geometric representation of the cross product presented in sections 1.1.7.b and 1.1.8.c.

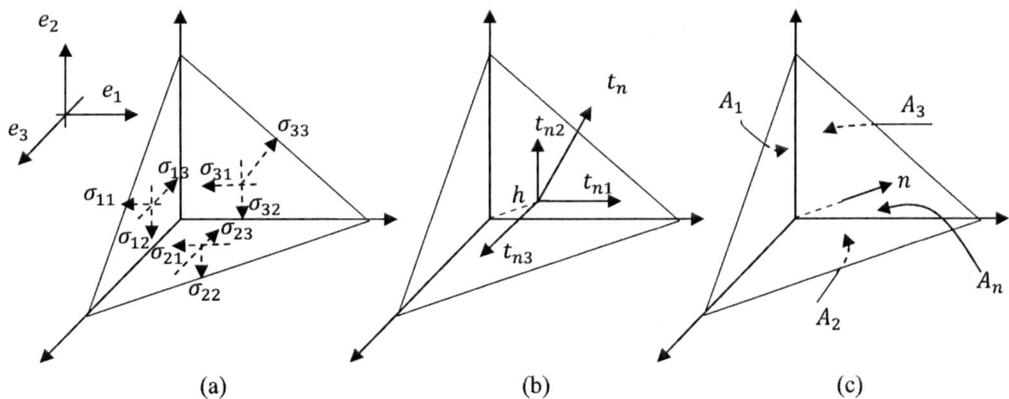

FIGURE 2-4 **Cauchy stress tetrahedron. (a) Traction components on the faces perpendicular to the basis vectors. (b) Components of the traction vector t_n on the inclined face. (c) Areas of the different faces.**

Newton's equilibrium principles state that the sum of the external forces (including body forces $g = g_1e_1 + g_2e_2 + g_3e_3$) acting on the tetrahedron is equal to its mass m multiplied by its acceler–ation vector $a = a_1e_1 + a_2e_2 + a_3e_3$. Newton's equilibrium principle applied in the direction of e_1 produces:

$$-\sigma_{11}A_1 - \sigma_{21}A_2 - \sigma_{31}A_3 + t_{n1}A_n + mg_1 = ma_1$$

The mass m can be expressed as $m = \left(\frac{\rho h A_n}{3}\right)$ where ρ is the density, h is the height of the tetrahedron perpendicular to A_n:

$$-\sigma_{11}A_1 - \sigma_{21}A_2 - \sigma_{31}A_3 + t_{n1}A_n + \left(\frac{\rho h A_n}{3}\right) g_1 = \left(\frac{\rho h A_n}{3}\right) a_1$$

Multiplying by $1/A_n$ throughout produces:

$$-\sigma_{11}n_1 - \sigma_{21}n_2 - \sigma_{31}n_3 + t_{n1} = \left(\frac{\rho h}{3}\right)(a_1 - g_1)$$

Similarly, in the directions of e_2 and e_3:

$$-\sigma_{12}n_1 - \sigma_{22}n_2 - \sigma_{32}n_3 + t_{n2} = \left(\frac{\rho h}{3}\right)(a_2 - g_2)$$

$$-\sigma_{13}n_1 - \sigma_{23}n_2 - \sigma_{33}n_3 + t_{n3} = \left(\frac{\rho h}{3}\right)(a_3 - g_3)$$

Taking the limit on both sides as the volume of the tetrahedron goes to zero (i.e., as $h \to 0$) leads to the linear relationship between the general traction vector t_n and the unit normal vector n:

$$\begin{pmatrix} t_{n1} \\ t_{n2} \\ t_{n3} \end{pmatrix} = \begin{pmatrix} \sigma_{11} & \sigma_{21} & \sigma_{31} \\ \sigma_{12} & \sigma_{22} & \sigma_{32} \\ \sigma_{13} & \sigma_{23} & \sigma_{33} \end{pmatrix} \begin{pmatrix} n_1 \\ n_2 \\ n_3 \end{pmatrix}$$

The stress matrix can then be introduced as:

$$\sigma = \begin{pmatrix} \sigma_{11} & \sigma_{12} & \sigma_{13} \\ \sigma_{21} & \sigma_{22} & \sigma_{23} \\ \sigma_{31} & \sigma_{32} & \sigma_{33} \end{pmatrix}$$

Thus, the relationship between t_n and n is:

$$t_n = \sigma^T n$$

This asserts that the stress is a linear mapping (a matrix) from the three–dimensional (area) vector space into the three–dimensional (traction) vector space. Thus, in this form, the stress tensor has nine independent components. The term "Tensor" is often used in continuum mechanics to describe linear maps between vector spaces of the same dimensions. Tensors, or linear maps, are abstract operators and relate vectors independent of the chosen coordinate system. Matrices, however, are dependent on the choice of the coordinate system. In this text, however, the terms *stress matrix* and *stress tensor* are freely used to denote σ.

2.1.3. Symmetry of the Cauchy Stress Tensor

The stress tensor belongs to the family of symmetric matrices. As will be shown in this section, this is a direct consequence of the angular momentum equilibrium equation, irrespective of whether the body is in static or dynamic equilibrium. According to Euler's laws of motion, the rate of change of the angular momentum L in the material is equal to the moment of the external forces. Euler's laws can be applied to an infinitesimal cuboid with external forces acting on its faces as shown in Figure 2–5. The moment around the axis perpendicular to the plane shown and passing through the bottom left corner (black dot), neglecting smaller quantities, can be evaluated as follows:

$$M_{external} = (\sigma_{12} - \sigma_{21})\Delta x_1 \Delta x_2 \Delta x_3$$

Assuming the cuboid rotates around the bottom left corner with angular velocity ω, then $\dot{L}$ has the following form:

$$\text{Rate of change of angular momentum} = \frac{dL}{dt} = \dot{L} = \rho\dot{\omega}I = \rho\dot{\omega}\left(\frac{1}{3}\Delta x_2 \Delta x_3 \Delta x_1 (\Delta x_2^2 + \Delta x_1^2)\right)$$

where, ρ is the density, $\dot{\omega}$ is the angular acceleration and I is the polar moment of inertia. Equating $M_{external}$ with $\frac{dL}{dt}$ results in the following expression:

$$(\sigma_{12} - \sigma_{21}) = \rho\dot{\omega}\left(\frac{1}{3}(\Delta x_2^2 + \Delta x_1^2)\right)$$

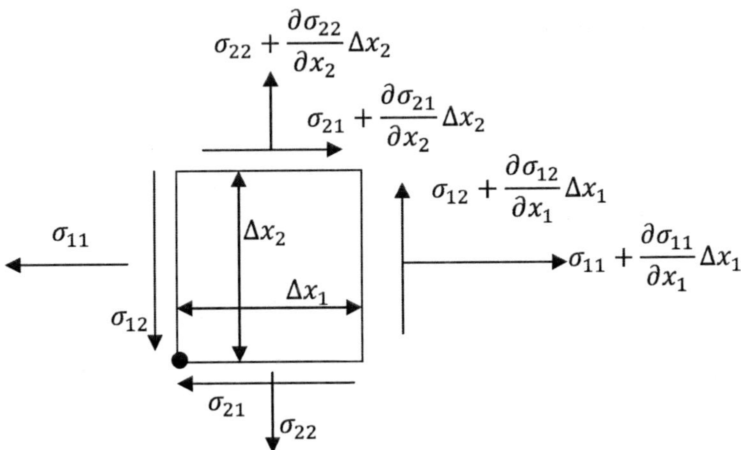

FIGURE 2-5 View perpendicular to e_3 of an infinitesimal cuboid with dimensions Δx_1, Δx_2 and Δx_3.

In the limit, as Δx_1 and Δx_2 go to zero, the stress components σ_{12} and σ_{21} are equal. Similarly, $\sigma_{13} = \sigma_{31}$ and $\sigma_{23} = \sigma_{32}$. This implies that the Cauchy stress matrix is symmetric and thus has six independent components.

2.1.4. Principal Stresses of the Cauchy Stress Tensor:

As shown in the above sections, the Cauchy stress $\sigma \in \mathbb{M}^3$ is a symmetric matrix. Using the results from section 1.2.6.f regarding symmetric matrices, associated with σ are a set of three real eigen–values and a set of three orthonormal eigenvectors. The eigenvalues $\sigma_1, \sigma_2, \sigma_3$ of σ are called the **principal stresses**, and the eigenvectors are called the **principal directions.**

Problem: The plane state of stress ($\sigma_{33} = \sigma_{13} = \sigma_{23} = 0$) shown in the figure exists at a point inside a body. Find the principal stresses and the principal directions. All units are in MPa.

Solution: The stress matrix associated with the above state is:

$$\sigma = \begin{pmatrix} 4 & 2 & 0 \\ 2 & 1 & 0 \\ 0 & 0 & 0 \end{pmatrix} MPa$$

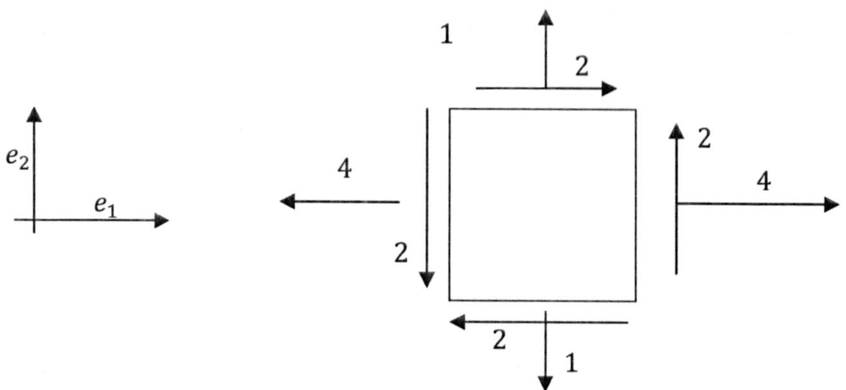

We can consider only the first two directions since the third direction is already a principal direction with stress = 0 units:

$$\sigma = \begin{pmatrix} 4 & 2 \\ 2 & 1 \end{pmatrix} MPa$$

Using Mathematica, the function `Eigensystem` returns a list of the eigenvalues and the eigen– vectors. Notice, however, that the eigenvectors are not necessarily normalized. In the following code, the variable sol contains the output, which is a list of eigenvalues and a list of eigenvec– tors. The angle between the eigenvectors and the basis vectors can be obtained using the function `VectorAngle`

```
s={{4,2},{2,1}};
sol=Eigensystem[s]
ev1=sol[[2,1]]
ev2=sol[[2,2]]
VectorAngle[ev1,{1,0}]
N[VectorAngle[ev1,{1,0}]/Degree]
VectorAngle[ev2,{-1,0}]
N[VectorAngle[ev2,{-1,0}]/Degree]
```

The output of the variable `sol` contains the list {5,0}, the eigenvalues, and the list of the associated eigenvectors {2,1} and {−1,2}:

```
{{5,0},{{2,1},{-1,2}}}
```

The output indicates that the first eigenvalue (first principal stress) is equal to $5MPa$ while the second principal stress is 0. The first principal stress is associated with the direction of the vector $ev1 = \{2,1\}$ while the second principal stress is associated with the direction of the vector $ev2 = \{-1,2\}$. The angle between the first eigenvector and $e1 = \{1,0\} = 26.6^{\circ}$

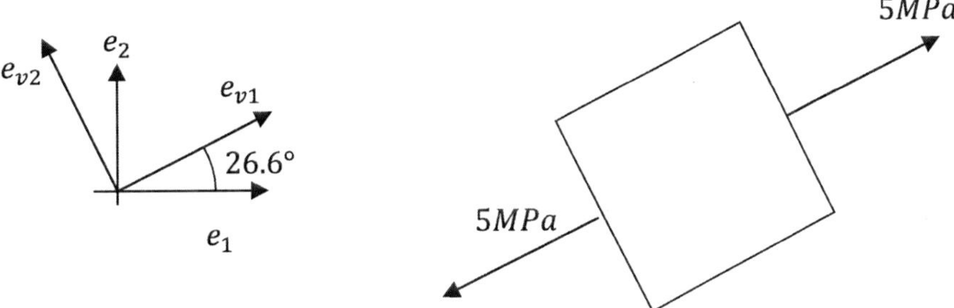

Using hand calculations, we can solve the following quadratic function to find the eigenvalues:

$$\det(\sigma - \lambda I) = 0 \Longrightarrow \det\begin{pmatrix} 4-\lambda & 2 \\ 2 & 1-\lambda \end{pmatrix} = (4-\lambda)(1-\lambda) - 4 = 0$$

$$\lambda^2 - 5\lambda = 0 \Rightarrow \lambda = \{5, 0\}$$

Let $\lambda = 5$; therefore:

$$\begin{pmatrix} 4-\lambda & 2 \\ 2 & 1-\lambda \end{pmatrix}\begin{pmatrix} x_1 \\ x_2 \end{pmatrix} = \begin{pmatrix} -1 & 2 \\ 2 & -4 \end{pmatrix}\begin{pmatrix} x_1 \\ x_2 \end{pmatrix} = \begin{pmatrix} 0 \\ 0 \end{pmatrix}$$

$$ev1 = \begin{pmatrix} x_1 \\ x_2 \end{pmatrix} = \begin{pmatrix} 2 \\ 1 \end{pmatrix}$$

Let $\lambda = 0$; therefore:

$$\begin{pmatrix} 4-\lambda & 2 \\ 2 & 1-\lambda \end{pmatrix}\begin{pmatrix} x_1 \\ x_2 \end{pmatrix} = \begin{pmatrix} 4 & 2 \\ 2 & 1 \end{pmatrix}\begin{pmatrix} x_1 \\ x_2 \end{pmatrix} = \begin{pmatrix} 0 \\ 0 \end{pmatrix}$$

$$ev2 = \begin{pmatrix} x_1 \\ x_2 \end{pmatrix} = \begin{pmatrix} -2 \\ 1 \end{pmatrix}$$

2.1.5. Normal and Shear Stresses

Let $\sigma \in \mathbb{M}^3$ be the stress tensor at a point, and let $n \in \mathbb{R}^3$ be a unit vector perpendicular to an area A inside a body. The traction vector on the area is $t_n = \sigma^T n$. The **normal stress component** $\sigma_n \in \mathbb{R}$ is the component of the traction vector t_n in the direction of the normal vector n (positive σ_n indicates tension and negative indicates compression). The **shear stress** $\tau_n \in \mathbb{R}$ is the component

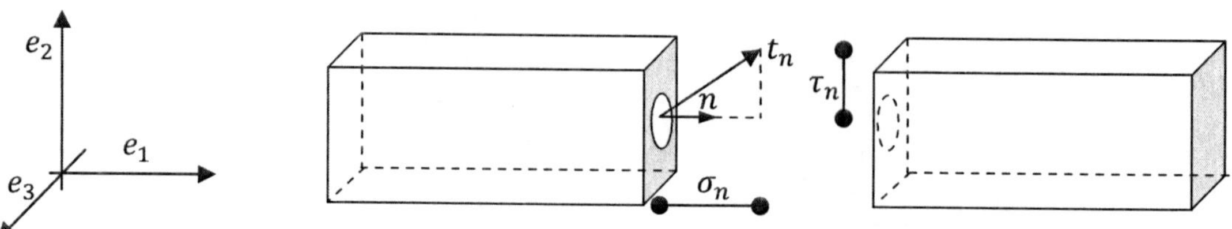

FIGURE 2-6 Normal and shear stress components.

of the traction vector t_n parallel to the area A (Figure 2–6). σ_n and τ_n can be obtained using the following vector operations (See orthogonal projection in section 1.1.7.a):

$$\sigma_n = t_n \cdot n$$

$$\tau_n = \|t_n - \sigma_n n\|$$

Problem 1: The stress at a point inside a continuum body is given by the stress matrix $\sigma = \begin{pmatrix} 1 & -1 & 0 \\ -1 & -5 & 0 \\ 0 & 0 & 4 \end{pmatrix}$. Find the normal and shear stress components on an area whose normal vector is in the direction of the vector $u = \{1, 1, 1\}$.

Solution: The vector normal to the area is not normalized, so, the first step is to normal–ize u and then directly apply the equations from the definition of the normal and shear stress components.

$$n = \frac{1}{\sqrt{3}} \begin{pmatrix} 1 \\ 1 \\ 1 \end{pmatrix}$$

$$t_n = \frac{1}{\sqrt{3}} \begin{pmatrix} 1 & -1 & 0 \\ -1 & -5 & 0 \\ 0 & 0 & 4 \end{pmatrix} \begin{pmatrix} 1 \\ 1 \\ 1 \end{pmatrix} = \frac{1}{\sqrt{3}} \begin{pmatrix} 0 \\ -6 \\ 4 \end{pmatrix}$$

$$\sigma_n = t_n \cdot n = \frac{1}{3}(-6+4) = -\frac{2}{3}$$

$$t_n - \sigma_n n = \frac{1}{\sqrt{3}}\begin{pmatrix}0\\-6\\4\end{pmatrix} - \frac{-2}{3\sqrt{3}}\begin{pmatrix}1\\1\\1\end{pmatrix} = \frac{1}{3\sqrt{3}}\begin{pmatrix}2\\-16\\14\end{pmatrix}$$

$$\tau_n = \|t_n - \sigma_n n\| = \frac{1}{3\sqrt{3}}\sqrt{2^2+16^2+14^2} = \frac{2\sqrt{38}}{3}$$

The following Mathematica code gives the above answer:

```
s={{1, -1, 0},{-1, -5, 0},{0, 0, 4}}
u={1,1,1};
n=u/Norm[u];
tn=Transpose[s].n
sn=tn.n
FullSimplify[tn-sn*n]
tawn=FullSimplify[Norm[tn-sn*n]]
```

Problem 2: The stress at a point inside a continuum body is given by the stress matrix $\sigma = pI$ where $p \in \mathbb{R}$ and I is the identity matrix. Show that the normal stress component on any surface is equal to p while the shear stress component is equal to zero.

Solution: Let $n \in \mathbb{R}^3$ be an arbitrary normal (unit) vector:

$$t_n = \sigma^T n = pn$$

$$\sigma_n = t_n \cdot n = p(n \cdot n) = p$$

$$\tau_n = \|t_n - pn\| = 0 \qquad .$$

Problem 2 shows that if the stress tensor at a point is a multiple of the identity matrix, then all the planes at this point would have zero shear components. This state of stress is usually termed **pressure** or a **hydrostatic stress state**.

Problem 3: Find the new basis set (coordinate system) at which the stress matrix $\sigma = \begin{pmatrix}1&2&0\\2&4&0\\0&0&1\end{pmatrix}$ is diagonal.

Solution: The set A of eigenvalues of the stress matrix is $A = \{5, 1, 0\}$ corresponding to the set B of eigenvectors $B = \{\{1, 2, 0\}, \{0, 0, 1\}, \{-2, 1, 0\}\}$. The normalized eigenvectors are:

$$ev_1 = \begin{pmatrix} 1/\sqrt{5} \\ 2/\sqrt{5} \\ 0 \end{pmatrix}, \quad ev_2 = \begin{pmatrix} -2/\sqrt{5} \\ 1/\sqrt{5} \\ 0 \end{pmatrix}, \quad ev_3 = \begin{pmatrix} 0 \\ 0 \\ 1 \end{pmatrix}$$

The coordinate system defined by the directions of the eigenvectors of σ can be obtained using the following rotation matrix:

$$Q = \begin{pmatrix} 1/\sqrt{5} & 2/\sqrt{5} & 0 \\ -2/\sqrt{5} & 1/\sqrt{5} & 0 \\ 0 & 0 & 1 \end{pmatrix}$$

When choosing an order for the eigenvectors in the orthogonal matrix, care has to be taken so that the matrix produced is a rotation rather than reflection. Notice that such matrix rotates the current basis system into the new basis system defined by the three orthonormal vectors:

$$e'_1 = \begin{pmatrix} 1/\sqrt{5} \\ 2/\sqrt{5} \\ 0 \end{pmatrix}, \quad e'_2 = \begin{pmatrix} -2/\sqrt{5} \\ 1/\sqrt{5} \\ 0 \end{pmatrix}, \quad e'_3 = \begin{pmatrix} 0 \\ 0 \\ 1 \end{pmatrix}$$

The stress matrix in the new coordinate system is given by:

$$\sigma' = Q\sigma Q^T = \begin{pmatrix} 5 & 0 & 0 \\ 0 & 0 & 0 \\ 0 & 0 & 1 \end{pmatrix}$$

The following Mathematica code shows the above calculations:

```
s={{1,2,0},{2,4,0},{0,0,1}};
sol1=Eigensystem[s]
ev1=sol1[[2,1]]/Norm[sol1[[2,1]]]
ev2=sol1[[2,3]]/Norm[sol1[[2,3]]]
ev3=sol1[[2,2]]/Norm[sol1[[2,2]]]
Q=FullSimplify[{ev1,ev2,ev3}]
FullSimplify[Q.s.Transpose[Q]]//MatrixForm
```

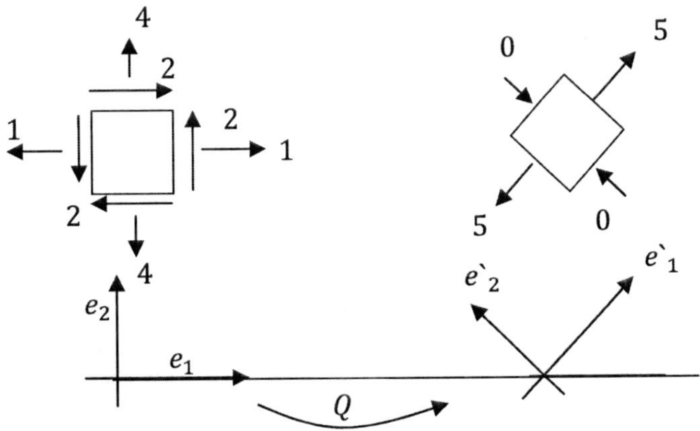

View perpendicular to e_3 (and $e`_3$)

Problem 4: Show that the pure shear stress state $\sigma = \begin{pmatrix} 0 & \sigma_{12} & 0 \\ \sigma_{12} & 0 & 0 \\ 0 & 0 & 0 \end{pmatrix}$ is dependent on the choice of the basis set, i.e., it is possible to find a different basis set in which there are no shear stress components (off diagonal components) in the stress matrix.

Solution: The set A of eigenvalues is $A = \{0, -\sigma_{12}, \sigma_{12}\}$ and the corresponding set B of eigenvectors is $B = \{\{0, 0, 1\}, \{-1, 1, 0\}, \{1, 1, 0\}\}$. The normalized eigenvectors can be used to form the following rotation matrix:

$$Q = \begin{pmatrix} \frac{1}{\sqrt{2}} & \frac{1}{\sqrt{2}} & 0 \\ \frac{-1}{\sqrt{2}} & \frac{1}{\sqrt{2}} & 0 \\ 0 & 0 & 1 \end{pmatrix}$$

By performing a coordinate transformation, the stress matrix in the new basis set is diagonal and thus has no shear components:

$$\sigma' = Q\sigma Q^T = \begin{pmatrix} \sigma_{12} & 0 & 0 \\ 0 & -\sigma_{12} & 0 \\ 0 & 0 & 0 \end{pmatrix}$$

```
s={{0,s12,0},{s12,0,0},{0,0,0}};
sol1=Eigensystem[s]
ev1=sol1[[2,3]]/Norm[sol1[[2,3]]]
ev2=sol1[[2,2]]/Norm[sol1[[2,2]]]
ev3=sol1[[2,1]]/Norm[sol1[[2,1]]]
Q={ev1,ev2,ev3};
Q.s.Transpose[Q]//MatrixForm
```

2.2. STRESS MEASURES AND STRESS INVARIANTS

The concept of stress invariants is very important for any mechanical design. We have shown in the previous section that the stress is a symmetric matrix with six independent components. Stress invariants are important to answer the following two questions:

- At which combination of those six independent components will a material fail?
- What is the measure of stress that would not be affected by a change of coordinate systems?

The stress matrix is simply an example of the symmetric matrices studied in section 1.2.6.f. and thus has the following properties:

- The stress tensor has three eigenvalues associated with three orthonormal eigenvectors. These eigenvalues and eigenvectors are termed the principal stresses and the principal directions, respectively.
- The invariants of the stress matrix are functions of the principal stresses.
- There is a coordinate system at which the stress matrix is diagonal.

In this section, three stress invariants will be presented; the hydrostatic stress, the von Mises stress, and the maximum shear stress. These invariants are used in the design of different mechanical components. In addition, the deviatoric stress tensor, from which the von Mises stress is derived, will be presented.

2.2.1. Hydrostatic Stress

DEFINITION Given a stress matrix $\sigma \in \mathbb{M}^3$, σ is symmetric. Let σ_1, σ_2 and σ_3 denote the eigenvalues of σ and let σ_{ij} denote the components of σ. The **Hydrostatic Stress** $p \in \mathbb{R}$ is defined as:

$$p = \frac{I_1(\sigma)}{3} = \frac{(\sigma_1 + \sigma_2 + \sigma_3)}{3} = \frac{(\sigma_{11} + \sigma_{22} + \sigma_{33})}{3}$$

where $I_1(\sigma)$ is the first invariant of the stress matrix (Section 1.2.6.d).

Remarks:

- As the definition implies, the hydrostatic stress is a measure of the average stress on a point in the continuum. Also, as will be shown later, it is the measure of the stress at a point that is responsible for volumetric changes in linear elastic materials.
- **Alternative names**: The names "Volumetric Stress" and "Pressure" exist for the same quantity. However, the pressure is defined as the negative of the hydrostatic stress.
- If the average stress at a point is positive, then the material point is under an average tensile stress; in this case, the hydrostatic stress is positive, while the pressure is negative.
- If the average stress at a point is negative, then the material point is under an average compressive stress; in this case, the hydrostatic stress is negative, while the pressure is positive.

Problem: The stress at a point is represented by the following stress matrices. Find the value of the hydrostatic stress in each case.

$$\sigma = \begin{pmatrix} 1 & 0 & 0 \\ 0 & 2 & 0 \\ 0 & 0 & 3 \end{pmatrix}, \begin{pmatrix} 5 & 2 & 1 \\ 2 & -2 & 1 \\ 1 & 1 & -3 \end{pmatrix}, \begin{pmatrix} -4 & 2 & 1 \\ 2 & -2 & 1 \\ 1 & 1 & -3 \end{pmatrix}$$

Solution: In the first case, the material point is under an average tensile stress or a hydrostatic stress of 2 units:

$$\sigma = \begin{pmatrix} 1 & 0 & 0 \\ 0 & 2 & 0 \\ 0 & 0 & 3 \end{pmatrix} \Rightarrow p = \frac{(1+2+3)}{3} = 2$$

In the second case, the material point is under zero average stress:

$$\sigma = \begin{pmatrix} 5 & 2 & 1 \\ 2 & -2 & 1 \\ 1 & 1 & -3 \end{pmatrix} \Rightarrow p = \frac{(5-2-3)}{3} = 0$$

In the third case, the material point is under an average compressive stress of 3 units:

$$\sigma = \begin{pmatrix} -4 & 2 & 1 \\ 2 & -2 & 1 \\ 1 & 1 & -3 \end{pmatrix} \Rightarrow p = \frac{(-4-2-3)}{3} = -3$$

2.2.2. Deviatoric Stress Tensor

DEFINITION Given a stress matrix $\sigma \in \mathbb{M}^3$, σ is symmetric. Let p denote the hydrostatic stress of σ. The **Deviatoric Stress Tensor** $S \in \mathbb{M}^3$ is defined as:

$$S = \sigma - pI$$

The components of the deviatoric stress tensor can be represented as:

$$S_{ij} = \sigma_{ij} - p\delta_{ij}$$

Remarks:

- The deviatoric stress has zero trace:

$$\text{trace}\,(S) = I_1(S) = I_1(\sigma - pI) = I_1(\sigma) - pI_1(I) = I_1(\sigma) - \frac{I_1(\sigma)}{3}3 = 0$$

- As the name implies, the deviatoric stress is a measure of the deviation of the stress state at a point in the continuum from the average stress at that point. Also, as will be shown later, it is the measure of the stress at a point that is responsible for changes of the shape of linear elastic materials.
- The stress matrix can be decomposed into two additive components, one representing a hydrostatic stress state and the other representing a deviatoric stress state.

$$\sigma = S + pI$$

The logic behind such decomposition lies in the physics of the failure of different materials. For example, it has been found that the deviatoric stress S is more relevant when discussing the yielding of metals, while the value of p does not positively or negatively affect the onset of failure of metals. In soils, however, the material is naturally more stable when it is under a high compressive (negative value) of the hydrostatic stress. Thus, the value of p is important when discussing the failure of soils.

Problem: Decompose the following stress matrices into their deviatoric and hydrostatic stress components

$$\sigma = \begin{pmatrix} 1 & 0 & 0 \\ 0 & 2 & 0 \\ 0 & 0 & 3 \end{pmatrix}, \begin{pmatrix} 5 & 2 & 1 \\ 2 & -2 & 1 \\ 1 & 1 & -3 \end{pmatrix}, \begin{pmatrix} -4 & 2 & 1 \\ 0 & -2 & 1 \\ 2 & 0 & -3 \end{pmatrix} MPa$$

Solution:

$$\sigma = \begin{pmatrix} 1 & 0 & 0 \\ 0 & 2 & 0 \\ 0 & 0 & 3 \end{pmatrix} \Rightarrow p = \frac{(1+2+3)}{3} = 2MPa \Rightarrow \sigma = S + pI = \begin{pmatrix} -1 & 0 & 0 \\ 0 & 0 & 0 \\ 0 & 0 & 1 \end{pmatrix} + \begin{pmatrix} 2 & 0 & 0 \\ 0 & 2 & 0 \\ 0 & 0 & 2 \end{pmatrix} MPa$$

$$\sigma = \begin{pmatrix} 5 & 2 & 1 \\ 2 & -2 & 1 \\ 1 & 1 & -3 \end{pmatrix} \Rightarrow p = \frac{(5-2-3)}{3} = 0 \Rightarrow \sigma = S = \begin{pmatrix} 5 & 2 & 1 \\ 2 & -2 & 1 \\ 1 & 1 & -3 \end{pmatrix} MPa$$

$$\sigma = \begin{pmatrix} -4 & 2 & 1 \\ 2 & -2 & 1 \\ 1 & 1 & -3 \end{pmatrix} \Rightarrow p = \frac{-4-2-3}{3} = -3MPa \Rightarrow \sigma = S + pI$$

$$= \begin{pmatrix} -1 & 2 & 1 \\ 2 & 1 & 1 \\ 1 & 1 & 0 \end{pmatrix} + \begin{pmatrix} -3 & 0 & 0 \\ 0 & -3 & 0 \\ 0 & 0 & -3 \end{pmatrix} MPa$$

2.2.3. Second Invariant of the Deviatoric Stress Tensor (Von Mises Stress)

One of the very important measures of stress that plays a major role in the prediction of yielding of metals is the von Mises stress: a measure of the second invariant of the deviatoric stress. As shown in section 1.2.6.d, the second invariant of $S \in \mathbb{M}^3$ is:

$$I_2(S) = \frac{1}{2}(I_1(S)^2 - I_1(SS))$$

As seen in section 2.2.2, the first invariant of S, namely the trace of S is equal to zero. The **von Mises stress** is an invariant measure of the deviatoric stress and is equal to:

$$Von\ Mises\ stress = \sqrt{-3I_2(S)} = \sqrt{\frac{3I_1(SS)}{2}}$$

In component form:

$$\text{Von Mises stress} = \sqrt{\frac{3}{2}\sum_{i,j=1}^{3} S_{ij}S_{ji}} = \sqrt{\frac{3}{2}\sum_{i,j=1}^{3} S_{ij}S_{ij}}$$

2.2.3.a. Von Mises Stress in the Principal Directions Coordinate System

If the coordinate system for $\mathbb{R}^3$ coincides with the principal directions of $\sigma \in \mathbb{M}^3$, then σ, $S \in \mathbb{M}^3$ and $p \in \mathbb{R}$ have the following forms (where σ_1, σ_2 and σ_3 are the eigenvalues of σ):

$$\sigma = \begin{pmatrix} \sigma_1 & 0 & 0 \\ 0 & \sigma_2 & 0 \\ 0 & 0 & \sigma_3 \end{pmatrix}, \quad p = \frac{\sigma_1 + \sigma_2 + \sigma_3}{3}, \quad S = \begin{pmatrix} \sigma_1 - p & 0 & 0 \\ 0 & \sigma_2 - p & 0 \\ 0 & 0 & \sigma_3 - p \end{pmatrix}$$

In this case, the von Mises Stress can be expressed in terms of the principal stresses (eigenvalues of σ) as follows:

$$\text{Von Mises stress} = \sqrt{\frac{3}{2}((\sigma_1 - p)^2 + (\sigma_2 - p)^2 + (\sigma_3 - p)^2)} = \sqrt{\frac{(\sigma_1 - \sigma_2)^2 + (\sigma_2 - \sigma_3)^2 + (\sigma_3 - \sigma_1)^2}{2}}$$

2.2.3.b. Von Mises Stress in a General Coordinate System

If a general orthonormal basis system is chosen for $\mathbb{R}^3$, then σ, $S \in \mathbb{M}^3$ and $p \in \mathbb{R}$ have the following forms (where σ_{ij} are the components of σ):

$$\sigma = \begin{pmatrix} \sigma_{11} & \sigma_{12} & \sigma_{13} \\ \sigma_{21} & \sigma_{22} & \sigma_{23} \\ \sigma_{31} & \sigma_{32} & \sigma_{33} \end{pmatrix}, \quad p = \frac{\sigma_{11} + \sigma_{22} + \sigma_{33}}{3}, \quad S = \begin{pmatrix} \sigma_{11} - p & \sigma_{12} & \sigma_{13} \\ \sigma_{21} & \sigma_{22} - p & \sigma_{23} \\ \sigma_{31} & \sigma_{32} & \sigma_{33} - p \end{pmatrix}$$

In this case, the von Mises stress can be expressed as follows:

$$\text{Von Mises stress} = \sqrt{\frac{3}{2}((\sigma_{11} - p)^2 + (\sigma_{22} - p)^2 + (\sigma_{33} - p)^2 + \sigma_{12}^2 + \sigma_{13}^2 + \sigma_{21}^2 + \sigma_{23}^2 + \sigma_{31}^2 + \sigma_{32}^2)}$$

Using the symmetry of σ:

$$\text{Von Mises stress} = \sqrt{\frac{(\sigma_{11} - \sigma_{22})^2 + (\sigma_{22} - \sigma_{33})^2 + (\sigma_{33} - \sigma_{11})^2 + 6\sigma_{12}^2 + 6\sigma_{13}^2 + 6\sigma_{23}^2}{2}}$$

Remarks:

- **Alternative names:** The von Mises stress is known as the equivalent stress.
- The mathematical theory describing the physical behaviour of metals asserts that a metal will yield once the von Mises stress reaches a critical value. The von Mises stress measure is invariant, i.e., it is independent of the coordinate system at which it is calculated.
- **Uniaxial state of stress**: When there is only one nonzero diagonal component (σ_{11}) in σ, then,

$$\textit{Von Mises stress} = \sqrt{\frac{(\sigma_{11} - 0)^2 + (0 - \sigma_{11})^2 + 0}{2}} = \sigma_{11}$$

- **Pure shear state of stress:** When there are only two nonzero off–diagonal components ($\sigma_{12} = \sigma_{21}$) in σ, then,

$$\textit{Von Mises stress} = \sqrt{\frac{0 + 6\sigma_{12}^2}{2}} = \sqrt{3}\sigma_{12}$$

- The von Mises stress is the same whether a material is under compression or tension!

Problem 1: Show that the following two stress states are equivalent.

$$\sigma = \begin{pmatrix} 2 & 2 & 0 \\ 2 & 5 & 0 \\ 0 & 0 & -5 \end{pmatrix}, \quad \sigma = \begin{pmatrix} 1 & 0 & 0 \\ 0 & -5 & 0 \\ 0 & 0 & 6 \end{pmatrix}$$

Solution: A quick glance at the stress matrix on the left shows a shear stress of 2 units on one plane and three normal stresses of 2, 5 and –5 units. The stress matrix on the right shows no shear stresses with three normal stresses of 1, –5 and 6 units. A careless observer could assert that the right stress matrix is "more severe" since it shows a plane with a normal stress component of 6 units. However, using the stress invariants, we can easily show that both stress matrices are equivalent in terms of severity:

Considering the left stress matrix:

$$\sigma = \begin{pmatrix} 2 & 2 & 0 \\ 2 & 5 & 0 \\ 0 & 0 & -5 \end{pmatrix} \Rightarrow p = \frac{2}{3}$$

$$\textit{Von Mises stress} = \sqrt{\frac{(2-5)^2 + (5+5)^2 + (-5-2)^2 + 6(2)^2 + 0}{2}} = \sqrt{91}$$

Considering the right stress matrix:

$$\sigma = \begin{pmatrix} 1 & 0 & 0 \\ 0 & -5 & 0 \\ 0 & 0 & 6 \end{pmatrix} \Rightarrow p = \frac{2}{3}$$

$$\textit{Von Mises stress} = \sqrt{\frac{(1+5)^2 + (-5-6)^2 + (6-1)^2 + 0}{2}} = \sqrt{91}$$

In fact, we can use Mathematica to show that both matrices represent the same state of stress, but at two different coordinate systems. The following code utilizes the diagonalization procedure shown in section 1.2.6.f for symmetric matrices. The eigenvalues and eigenvectors of the left matrix are obtained. Then, a transformation matrix is formed using the three normalized eigenvectors. The final matrix is shown to be exactly the right stress matrix above.

```
S1={{2,2,0},{2,5,0},{0,0,-5}};
ss=Eigensystem[S1];
eigenvector1=ss[[2,1]]/Norm[ss[[2,1]]];
eigenvector2=ss[[2,2]]/Norm[ss[[2,2]]];
eigenvector3=ss[[2,3]]/Norm[ss[[2,3]]];
Q={eigenvector3,eigenvector2,eigenvector1};
S2=FullSimplify[Q.S1.Transpose[Q]];
S2//MatrixForm
```

It is worth noting that stress invariants are only valid for materials whose strength is not orientation dependent, as will be shown in section 2.3.

Problem 2: Create a function in Mathematica that returns the value of the von Mises stress of a general stress matrix $\sigma \in \mathbb{M}^3$ and test it.

Solution:

```
VonMises[sigma_]:=Sqrt[1/2*((sigma[[1,1]]-sigma[[2,2]])^2+(sigma[[2,2]]-
sigma[[3,3]])^2+(sigma[[3,3]]-
sigma[[1,1]])^2+6*(sigma[[1,2]]^2+sigma[[1,3]]^2+sigma[[2,3]]^2))]
stress1={{2, 2, 0},{2, 5, 0},{0, 0,-5}};
stress2={{2, 2, 1},{2, 4, -3},{1, -3, 7}};
Q=RotationMatrix[30Degree,{2,1,1}];
Qt=Transpose[Q];
VonMises[stress1]
FullSimplify[VonMises[Q.stress1.Qt]]
VonMises[stress2]
FullSimplify[VonMises[Q.stress2.Qt]]
```

Problem 3: Find the von Mises Stress in a general plane state of stress ($\sigma_{33} = \sigma_{13} = \sigma_{23} = 0$)

Solution: The von Mises stress when $\sigma_{33} = \sigma_{13} = \sigma_{23} = 0$ is:

$$\textit{Von Mises Stress} = \sqrt{\frac{(\sigma_{11} - \sigma_{22})^2 + (\sigma_{22})^2 + (\sigma_{11})^2 + 6\sigma_{12}^2}{2}} = \sqrt{\sigma_{11}^2 - \sigma_{11}\sigma_{22} + \sigma_{22}^2 + 3\sigma_{12}^2}$$

In terms of the principal stresses, the von Mises stress is:

$$\textit{Von Mises Stress} = \sqrt{\sigma_1^2 - \sigma_1\sigma_2 + \sigma_2^2}$$

2.2.4. Maximum Shear Stress

Another invariant measure of the stress matrix is the maximum shear stress. Let $\sigma \in \mathbb{M}^3$, where, σ_{ij} are the components of σ in an arbitrary orthonormal basis set (arbitrary Cartesian coordinate system). The **maximum shear stress** is half the maximum difference between the principal stresses (where σ_1, σ_2 and σ_3 are the eigenvalues (principal stresses) of σ):

$$\max \sigma_{ij}(i \neq j) = \frac{\max(|\sigma_1 - \sigma_2|, |\sigma_1 - \sigma_3|, |\sigma_2 - \sigma_3|)}{2}$$

The maximum shear stress is obtained on a plane that is inclined by 45° to the two principal planes having the maximum difference in the principal stresses.

This can be shown by considering a plane problem aligned with the principal directions. The stress matrix in this case is given by:

$$\sigma = \begin{pmatrix} \sigma_1 & 0 \\ 0 & \sigma_2 \end{pmatrix}$$

Consider a general anti–clockwise coordinate transformation matrix (as shown in section 1.2.6.c) of the form:

$$Q = \begin{pmatrix} \cos\theta & \sin\theta \\ -\sin\theta & \cos\theta \end{pmatrix}$$

The stress matrix σ' in this new coordinate system defined by the coordinate transformation matrix Q has the form:

$$\sigma' = Q\sigma Q^T = \begin{pmatrix} \sigma_1\cos^2\theta + \sigma_2\sin^2\theta & (\sigma_2 - \sigma_1)\cos\theta\sin\theta \\ (\sigma_2 - \sigma_1)\cos\theta\sin\theta & \sigma_2\cos^2\theta + \sigma_1\sin^2\theta \end{pmatrix}$$

The absolute value of the maximum shear is obtained when $(\sigma_2 - \sigma_1)\cos\theta\sin\theta$ is maximum (or minimum), i.e.:

$$\frac{d(\sigma_2 - \sigma_1)\cos\theta\sin\theta}{d\theta} = 0 \Rightarrow \cos^2\theta = \sin^2\theta \Rightarrow \theta = \textit{multiples of}\ 45^\circ$$

Then,

$$|\sigma_{12}| = |(\sigma_2 - \sigma_1)\cos 45^\circ \sin 45^\circ| = \frac{|(\sigma_2 - \sigma_1)|}{2}$$

Notice that, in this case, the diagonal terms (the normal stresses on the respective planes) are equal to $\frac{\sigma_1+\sigma_2}{2}$

The following code utilizes Mathematica to show the above result. The variables `ss` and `news` are the stress matrices in the principal coordinate system and in the rotated coordinate system, respectively. The variable `eq1` is equal to $\frac{d(\sigma_2-\sigma_1)\cos\theta\sin\theta}{d\theta}$. The function `Solve` finds the solution to the equation (`eq1==0`):

```
Clear[s1,ss,s2, theta]
ss={{s1,0},{0,s2}};
Q=RotationMatrix[-theta];
news=FullSimplify[Q.ss.Transpose[Q]];
Q//MatrixForm
news//MatrixForm
eq1=D[news[[1,2]],theta];
Solve[eq1==0,theta]
```

Problem 1: Find the maximum shear stress for the following state of stress represented by the matrix below (units are in MPa). Also, find the coordinate transformation matrix that would pro– duce such value:

$$\sigma = \begin{pmatrix} 29 & 72 \\ 72 & 71 \end{pmatrix}$$

Solution: The eigenvalues and eigenvectors of the above stress matrix are:

$$\sigma_1 = \lambda_1 = 125, \quad ev_1 = \begin{pmatrix} 0.6 \\ 0.8 \end{pmatrix}, \quad \sigma_2 = \lambda_2 = -25, \quad ev_2 = \begin{pmatrix} -0.8 \\ 0.6 \end{pmatrix}$$

Thus, to transform into the new principal coordinate system, the following rotation matrix is used:

$$Q = \begin{pmatrix} 0.6 & 0.8 \\ -0.8 & 0.6 \end{pmatrix}$$

In the new coordinate system:

$$\sigma' = Q\begin{pmatrix} 29 & 72 \\ 72 & 71 \end{pmatrix}Q^T = \begin{pmatrix} 125 & 0 \\ 0 & -25 \end{pmatrix}$$

The maximum shear stress is $\frac{|125+25|}{2} = 75$ units.

The maximum shear stress is obtained after the application of a 45° rotation given by Q_2. The new stress matrix σ'' after rotation by Q_2 is:

$$\sigma'' = Q_2\sigma' Q_2^T = \begin{pmatrix} \cos 45^\circ & \sin 45^\circ \\ -\sin 45^\circ & \cos 45^\circ \end{pmatrix}\begin{pmatrix} 125 & 0 \\ 0 & -25 \end{pmatrix}\begin{pmatrix} \cos 45^\circ & -\sin 45^\circ \\ \sin 45^\circ & \cos 45^\circ \end{pmatrix} = \begin{pmatrix} 50 & -75 \\ -75 & 50 \end{pmatrix}$$

Therefore, the total rotation is Q_2Q (as shown in the figure below) and the relationship between the final stress matrix in the final coordinate system and the original stress matrix in the original coordinate system is:

$$\sigma'' = Q_2 Q \sigma Q^T Q_2^T$$

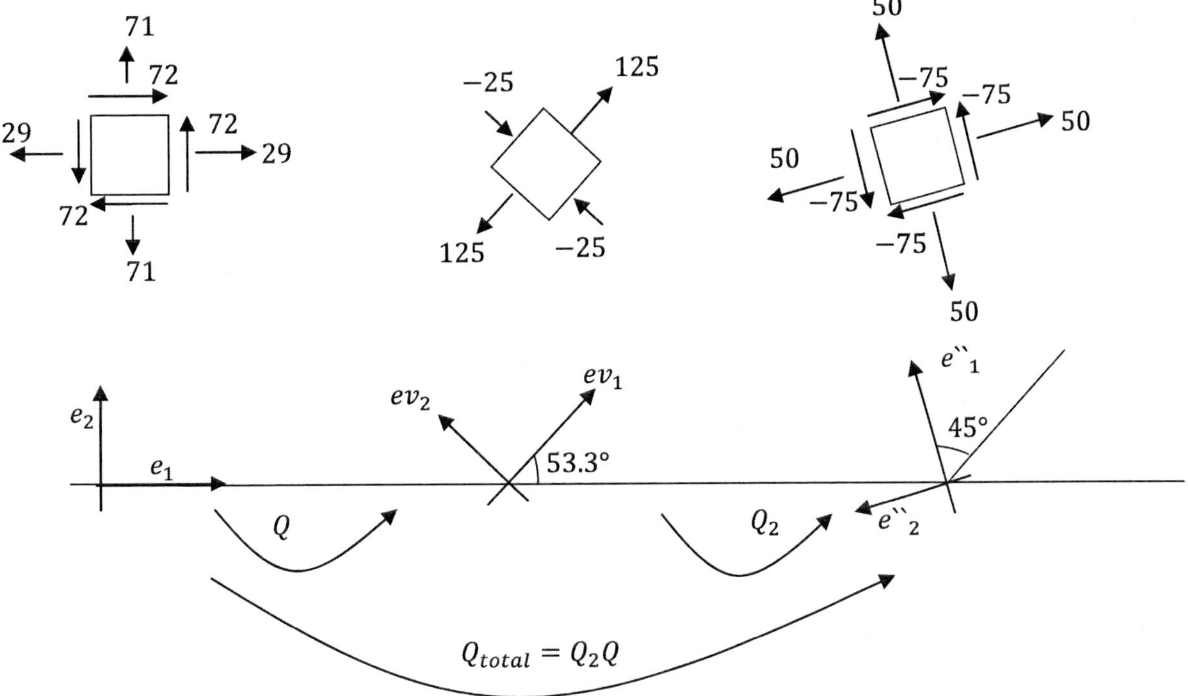

The following Mathematica code illustrates the above example:

```
s={{29,72},{72,71}};
Sol=Eigensystem[s]
ev1=Sol[[2,1]]/Norm[Sol[[2,1]]];
ev2=Sol[[2,2]]/Norm[Sol[[2,2]]];
Q={ev1,ev2};
Qt=Transpose[Q];
Q2=RotationMatrix[-45Degree];
Q2t=Transpose[Q2];
Sfinal=FullSimplify[Q2.Q.s.Qt.Q2t];
Sfinal//MatrixForm
s={{29,72},{72,71}};
Sol=Eigensystem[s]
ev1=Sol[[2,1]]/Norm[Sol[[2,1]]];
ev2=Sol[[2,2]]/Norm[Sol[[2,2]]];
Q={ ev1,ev2};
Qt=Transpose[Q];
Q2=RotationMatrix[45Degree];
Q2t=Transpose[Q2];
Sfinal=FullSimplify[Q2.Q.s.Qt.Q2t];
Sfinal//MatrixForm
```

Problem 2: Find the maximum shear stress for the following state of stress represented by the matrix. Also, find the coordinate transformation matrix that would produce such value:

$$\sigma = \begin{pmatrix} 17.5 & 36 & -48 \\ 36 & 33.7 & 38.4 \\ -48 & 38.4 & 11.3 \end{pmatrix}$$

Solution: The first step is to obtain the principal stresses and directions:

```
s={{17.5,36,-48},{36,33.7,38.4},{-48,38.4,11.3}};
sol=Eigensystem[s]
```

The output is:

```
{62.5,62.5,-62.5},{{0.624695,0.780869,0.},{-0.499756,0.399805,0.768375},
{0.6,-0.48,0.64}}
```

Thus, the maximum principal stresses are 62.5, 62.5 and –62.5 respectively aligned with the prin–cipal directions:

$$ev_1 = \begin{pmatrix} 0.6247 \\ 0.78087 \\ 0 \end{pmatrix}, \quad ev_2 = \begin{pmatrix} -0.49976 \\ 0.399805 \\ 0.768375 \end{pmatrix}, \quad ev_3 = \begin{pmatrix} 0.6 \\ -0.48 \\ 0.64 \end{pmatrix}$$

As a result, the transformation matrix that would rotate the coordinates into the principal coordinate system is:

$$Q = \begin{pmatrix} 0.6247 & 0.78087 & 0 \\ -0.49976 & 0.399805 & 0.768375 \\ 0.6 & -0.48 & 0.64 \end{pmatrix}$$

The stress matrix in the new principal coordinate system is:

$$\sigma' = Q \begin{pmatrix} 17.5 & 36 & -48 \\ 36 & 33.7 & 38.4 \\ -48 & 38.4 & 11.3 \end{pmatrix} Q^T = \begin{pmatrix} 62.5 & 0 & 0 \\ 0 & 62.5 & 0 \\ 0 & 0 & -62.5 \end{pmatrix}$$

The maximum shear stress is $(62.5 + 62.5)/2 = 62.5$ units, and this occurs by rotating the principal directions ev_2 and ev_3 by 45° through the rotation matrix Q_2. Note that the maximum shear could also be obtained by rotating the principal directions ev_1 and ev_3 by 45°

$$Q_2 = \begin{pmatrix} 1 & 0 & 0 \\ 0 & \cos 45^\circ & \sin 45^\circ \\ 0 & -\sin 45^\circ & \cos 45^\circ \end{pmatrix}, \quad Q_2^T = \begin{pmatrix} 1 & 0 & 0 \\ 0 & \cos 45^\circ & -\sin 45^\circ \\ 0 & \sin 45^\circ & \cos 45^\circ \end{pmatrix}$$

Thus, the transformation from the original coordinate system to the new coordinate system with the maximum shear is obtained using the following rotation matrix:

$$Q_{total} = Q_2 Q = \begin{pmatrix} 0.6247 & 0.78087 & 0. \\ -0.77764 & 0.6221 & 0.090775 \\ 0.07088 & -0.0567 & 0.99587 \end{pmatrix}$$

The following code illustrates the procedure above:

```
s={{17.5,36,-48},{36,33.7,38.4},{-48,38.4,11.3}};
sol=Eigensystem[s]
e1=sol[[2,1]]/Norm[sol[[2,1]]]
e2=sol[[2,2]]/Norm[sol[[2,2]]]
e3=sol[[2,3]]/Norm[sol[[2,3]]]
Q={e1,e2,e3};
Qt=Transpose[Q];
news=Q.s.Qt;
news//MatrixForm
Qnew=RotationMatrix[45Degree,{ 1,0,0}]
Qtnew=Transpose[Qnew];
```

```
maxshears=Qnew.news.Qtnew;
maxshears//MatrixForm
Qtotal=Qnew.Q;
Qtotal//MatrixForm
Qtotal.s.Transpose[Qtotal]//MatrixForm
```

2.3. STRESS-BASED FAILURE CRITERIA

In most load–bearing applications, a material is designed such that a measure of the stress applied is less than or equal to a maximum strength of the material. If the material is isotropic (e.g., steel and traditional concrete), i.e., the strength is independent of the choice of the coordinate system, then one of the stress invariants is often used as a measure of the stress that a material exhibits. In this section, we will discuss the most common isotropic failure criteria that are successfully used for metals. However, to illustrate the difference between such failure criteria and a general non–isotropic failure criteria (e.g., for wood), an example is presented where the strength of the material is dependent on the plane of consideration.

Problem: A wood beam can withstand:

- A maximum tensile stress of $25MPa$ in the directions of the fibres $u_{fibres} = \{1, 1, 0\}$
- A maximum tensile stress of $10MPa$ on the surface with normal $u = \{-1, 1, 0\}$
- A maximum shear stress of $10MPa$ on the surface with normal $u = \{-1, 1, 0\}$

Given the stress state at a point inside the beam is represented by stress matrix (units are in MPa):

$$\sigma = \begin{pmatrix} 25 & 10 & 10 \\ 10 & 10 & -3 \\ 10 & -3 & 5 \end{pmatrix}$$

Show whether this wood beam has reached its maximum capacity or not.

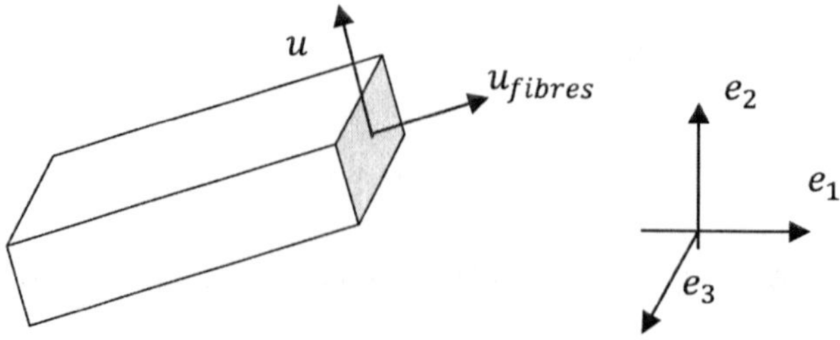

Solution: The normal and shear stress components presented in section 2.1.5 will be used to show whether the material has reached its capacity or not.

The unit normal vector for a surface perpendicular to the fibers:

$$n_{fibres} = \left\{\frac{1}{\sqrt{2}}, \frac{1}{\sqrt{2}}, 0\right\}$$

The traction vector on this surface is (units are in MPa):

$$t_{nfibres} = \sigma^T n_{fibres} = \{24.75, 14.14, 4.94\}$$

The normal stress component can be obtained as follows:

$$\sigma_{nfibres} = t_{nfibres} \cdot n_{fibres} = 27.5MPa > 25MPa$$

Thus, the beam has reached its capacity in terms of the tensile stresses along the fibers.

The unit normal vector for a surface perpendicular to u:

$$n = \left\{\frac{-1}{\sqrt{2}}, \frac{1}{\sqrt{2}}, 0\right\}$$

The traction vector on this surface is (units are in MPa):

$$t_n = \sigma^T n = \{-10.61, 0, -9.19\}$$

The normal stress component on the surface perpendicular to u:

$$\sigma_n = t_n \cdot n = 7.5MPa < 10MPa$$

Therefore, at this point, the beam has *not* reached its tensile stress capacity in the direction of u.

The shear stress component on the surface perpendicular to u:

$$\tau_n = \|t_n - \sigma_n n\| = 11.86MPa > 10MPa$$

Consequently, at this point, the beam has reached its capacity in terms of the *shear* stresses on a surface with normal u.

The following Mathematica code performs the above calculations:

```
s={{25,10,10},{10,10,-3},{10,-3,5}};
nfibers={1,1,0}/Norm[{1,1,0}];
n={-1,1,0}/Norm[{-1,1,0}];
tnfibres=s.nfibers;
tn=s.n;
snfibres=N[tnfibres.nfibers]
sn=N[tn.n]
tawn=Norm[tn-sn*n]
```

2.3.1. Maximum Shear Stress (Tresca)

The Tresca or maximum shear stress criterion states that a metal will yield when the maximum shear stress at any point reaches a maximum (critical) value termed τ_{max}. As shown in section 2.2.4, given a state of stress, the value of the maximum shear stress is half the maximum difference between the eigenvalues (principal stresses) of the stress matrix. Given $\sigma \in \mathbb{M}^3$ with eigenvalues σ_1, σ_2 and σ_3, then according to Tresca yield criterion, the material will yield when:

$$\frac{\max(|\sigma_1 - \sigma_2|, |\sigma_1 - \sigma_3|, |\sigma_2 - \sigma_3|)}{2} = \tau_{max}$$

where τ_{max} is a constant dependent on the type of material.

Remarks:

- The Tresca yield or failure criterion is independent of the choice of the coordinate system, since, as shown in section 1.2.6.d, the eigenvalues of a matrix are invariant.
- The material constant τ_{max} can be determined experimentally by performing a uniaxial state of stress test. In this case, σ_{11} is the only nonzero component in σ. When the material reaches yield or failure, the value of $\sigma_{11} = \sigma_Y$, then:

$$\sigma = \begin{pmatrix} \sigma_Y & 0 & 0 \\ 0 & 0 & 0 \\ 0 & 0 & 0 \end{pmatrix}$$

In this case, the eigenvalues of σ are $\lambda \in \{\sigma_Y, 0, 0\}$ and the material constant can be determined as:

$$\tau_{max} = \frac{\max(|\sigma_1 - \sigma_2|, |\sigma_1 - \sigma_3|, |\sigma_2 - \sigma_3|)}{2} = \frac{\sigma_Y}{2}$$

Problem: A metal that follows the Tresca yield criterion yields in a uniaxial state of stress when the uniaxial stress reaches $200MPa$. Will the metal yield when the stress is described by the matrix

$$\sigma = \begin{pmatrix} 25 & 40 & 0 \\ 40 & 20 & 30 \\ 0 & 30 & 20 \end{pmatrix}$$

Solution: The first step is to use the uniaxial state of stress test to determine the material constant τ_{max}. The material constant is related to the uniaxial failure load using the formula:

$$\tau_{max} = \frac{\sigma_Y}{2} = \frac{200}{2} = 100MPa$$

The stress state represented by the given matrix can now be investigated. The eigenvalues of the stress matrix are $\lambda \in \{71.7, -28.5, 21.8\}$. The maximum shear in this case is less than τ_{max}:

$$\frac{\max(|\sigma_1 - \sigma_2|, |\sigma_1 - \sigma_3|, |\sigma_2 - \sigma_3|)}{2} = \frac{(71.7 - (-28.5))}{2} = 50.1MPa < \tau_{max}$$

Thus, the material will not yield under the effect of the stress state represented by the given matrix.

Graphical Representation:

Plane stress state: Let $\sigma \in \mathbb{M}^3$ be such that $\sigma_1, \sigma_2 \neq 0$ while $\sigma_3 = 0$. Therefore, the material will yield when:

$$\max(|\sigma_1 - \sigma_2|, |\sigma_1|, |\sigma_2|) = 2\tau_{max}$$

The above relationship can be viewed graphically, as shown in Figure 2–7. If a stress state is represented by a point inside the polygon, then the material has not reached yield. The material will reach yield once the stress state is represented by a point on the polygon. A stress state outside the polygon cannot physically exist for the material since failure occurs when a point is on the polygon surface.

Three dimensional stress state: Let $\sigma \in \mathbb{M}^3$ be such that $\sigma_1, \sigma_2, \sigma_3 \neq 0$. Therefore, the material will yield when:

$$\max(|\sigma_1 - \sigma_2|, |\sigma_1 - \sigma_3|, |\sigma_2 - \sigma_3|) = 2\tau_{max}$$

The above relationship can be viewed graphically, as shown in Figure 2–8. If a stress state is represented by a point inside the surface shown, then the material has not reached yield. The shown surface is thus termed "the yield surface." The material will reach yield once the stress state is represented by a point on the yield surface. Notice that when viewed along the vector $u = \{1, 1, 1\}$ the yield surface has the shape of a hexagon.

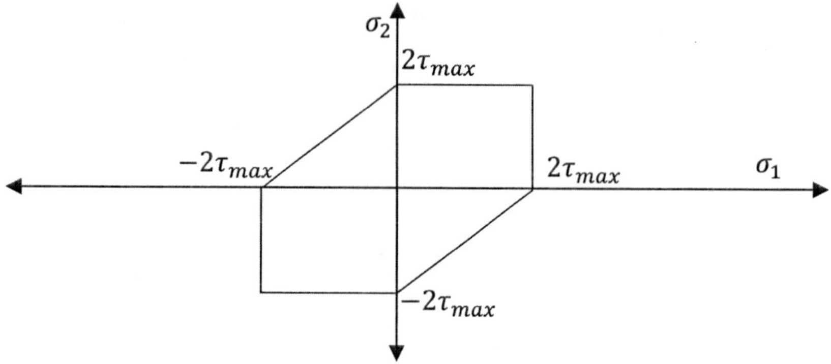

FIGURE 2-7 Graphical representation of the Tresca yield criterion in a state of plane stress with only two nonzero eigenvalues of the stress matrix.

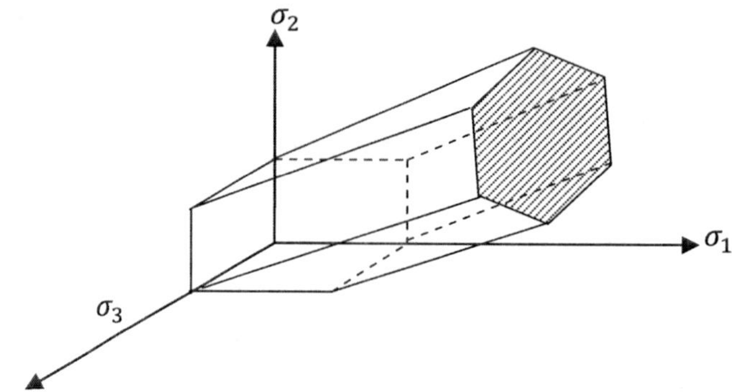

FIGURE 2-8 Graphical representation of the Tresca yield criterion in a three–dimensional state of stress with three axes representing the three eigenvalues of the stress matrix.

2.3.2. Von Mises Yield Criteria (J2 Flow Theory)

The von Mises yield criterion states that a metal will yield when the von Mises stress (a function of the second invariant of the deviatoric stress tensor) at any point reaches a maximum (critical) value termed σ_{max}. Given $\sigma \in \mathbb{M}^3$, then according to von Mises yield criterion, the material will yield when:

$$\textit{Von Mises stress} = \sqrt{\frac{(\sigma_{11}-\sigma_{22})^2+(\sigma_{22}-\sigma_{33})^2+(\sigma_{33}-\sigma_{11})^2+6\sigma_{12}^2+6\sigma_{13}^2+6\sigma_{23}^2}{2}} = \sigma_{max}$$

where σ_{max} is a constant dependent on the type of the material and the expression for the von Mises stress is according to section 2.2.3.b.

Remarks:

- The von Mises yield or failure criterion is independent of the choice of the coordinate system (**Why?**).
- The von Mises yield criterion does not differentiate between tension and compression.
- The von Mises yield criterion is sometimes called the maximum distortion energy criterion.
- Similar to the Tresca yield criterion, the von Mises yield criterion predicts that the material will not fail under a hydrostatic stress state where $\sigma = pI$ (**Why?**).
- The material constant σ_{max} can be experimentally determined by performing a uniaxial state of stress test. In this case, σ_{11} is the only nonzero component in σ. When the material reaches yield or failure, the value of $\sigma_{11} = \sigma_Y$, then:

$$\sigma = \begin{pmatrix} \sigma_Y & 0 & 0 \\ 0 & 0 & 0 \\ 0 & 0 & 0 \end{pmatrix}$$

In this case,

$$Von\ Mises\ stress = \sigma_{max} = \sigma_Y$$

Problem: A metal that follows the von Mises yield criterion yields in a uniaxial state of stress when the uniaxial stress reaches $200MPa$. Will the metal yield when the stress reaches a state described by the matrix?

$$\sigma = \begin{pmatrix} 25 & 40 & 0 \\ 40 & 20 & 30 \\ 0 & 30 & 20 \end{pmatrix}$$

Solution: The material constant $\sigma_{max} = \sigma_Y = 200MPa$

The eigenvalues of the stress matrix are $\lambda = \{71.7, -28.5, 21.8\}$. Using the expression in section 2.2.3.a for the von Mises stress:

$$Von\ Mises\ stress = \sqrt{\frac{(71.7 + 28.5)^2 + (71.7 - 21.8)^2 + (21.8 + 28.5)^2}{2}} = 86.75MPa < \sigma_{max}$$

Therefore, under the given stress state, the material has *not* reached yield.

Graphical Representation:

Plane stress state: Let $\sigma \in \mathbb{M}^3$ be such that $\sigma_1, \sigma_2 \neq 0$ while $\sigma_3 = 0$. Therefore, the material will yield when (according to section 2.2.3.a):

$$\sqrt{\frac{(\sigma_1 - \sigma_2)^2 + (\sigma_2)^2 + (\sigma_1)^2}{2}} = \sqrt{\sigma_1^2 - \sigma_1\sigma_2 + \sigma_2^2} = \sigma_{max}$$

The above relationship is represented by an ellipse, as shown graphically in Figure 2–9. If a stress state is represented by a point inside the surface, then the material has not reached yield. The material will reach yield once the stress state is *represented* by a point on the surface.

Three–dimensional stress state: Let $\sigma \in \mathbb{M}^3$ be such that $\sigma_1, \sigma_2, \sigma_3 \neq 0$. Therefore, the material will yield when:

$$\sqrt{\frac{(\sigma_1 - \sigma_2)^2 + (\sigma_2 - \sigma_3)^2 + (\sigma_3 - \sigma_1)^2}{2}} = \sigma_{max}$$

The above relationship is represented graphically by a cylinder whose radius is equal to $\sqrt{\frac{2}{3}}\sigma_Y$, as shown in Figure 2–10. The axis of the cylinder is oriented along the direction of the vector $u = \{1,\ 1,\ 1\}$. If a stress state is represented by a point inside the surface shown, then the material has

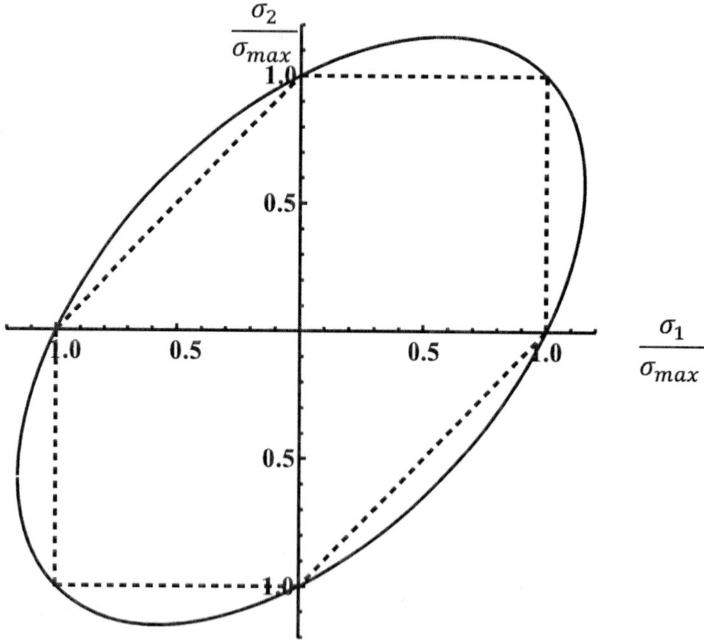

FIGURE 2-9 Graphical representation of the von Mises yield criterion in a state of plane stress (with only two nonzero eigenvalues for the stress matrix). The dotted line represents the Tresca yield criterion.

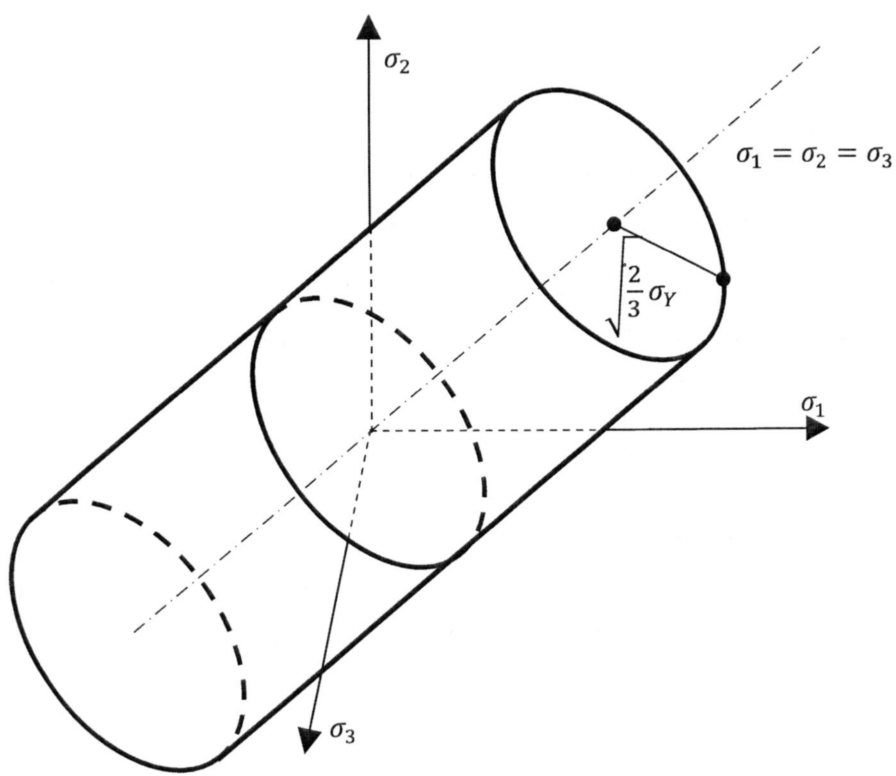

FIGURE 2-10 Graphical representation of the von Mises yield criterion in a three–dimensional state of stress with three axes representing the three eigenvalues of the stress matrix.

not reached yield. The shown surface is thus termed the yield surface. The material will reach yield once the stress state is represented by a point on the yield surface.

▶ **BONUS EXERCISE:** Show that the radius of the von Mises cylinder is equal to $\sqrt{\frac{2}{3}}\sigma_Y$ where σ_y is the uniaxial yield stress

The Mathematica code used to draw Figure 2–9 is given below:

```
Clear[s1,s2,t,a1,a2]
a1=Piecewise[{{t,0<=t<=1},{1,1<t<=2},{3-t,2< t<=3},{-t+3,3<t<=4},
{-1,4< t<=5},{-6+t,5<t<=6}}]
a2=Piecewise[{{1,0<=t<=1},{2-t,1<t<=2},{-t+2,2<t<=3},{-1,3< t<=4},
{-5+t,4<t<=5},{t-5,5<t<=6}}]
ParametricPlot[{a1,a2},{t,0,6},PlotStyle->{Black,Thick},AxesStyle-
>{{Black,Thick},{Black,Thick}}]

s1=2*Sin[t]
s=Solve[s1^2-s2*s1+s2^2==1,{s2}]
s21=s2/.s[[2]]
```

```
s22=s2/.s[[1]]

ParametricPlot[{{s1,s21},{s1,s22},{a1,a2}},{t,0,2*Pi},BaseStyle-
>Directive[Bold,15],PlotStyle-
>{{Black,Thick},{Black,Thick},{Black,Dashed,Thick}},AxesStyle-
>{{Black,Thick},{Black,Thick}}]
```

2.3.3. Other Isotropic Yield Criteria

Tresca and von Mises yield criteria have been very successful in describing the behavior of met– als. However, the failure of other materials, such as soils and rubber, is also dependent on the hydrostatic stress. For example, in soil materials, the greater confining compression applied to the material, the higher its shear strength. The following are two examples of isotropic yield criteria that incorporate the hydrostatic stress in the expression for the failure criterion and thus are more appropriate for such materials:

- Mohr–Coulomb yield criterion: The material will fail when the maximum shear stress reaches a critical value that is dependent on the average stress:

$$\left|\frac{\sigma_i - \sigma_j}{2}\right| = -\frac{(\sigma_i + \sigma_j)}{2}\sin\phi + c\cos\phi$$

  where ϕc, are material constants and σ_i, σ_j are the principal stresses and $i \neq j$

- Drucker–Prager yield criterion: The material will fail when the von Mises stress reaches a critical value that is dependent on the average stress:

$$Von\ Mises\ Stress = A + BI_1(\sigma)$$

  where A, B are material constants and σ is the stress matrix

2.4. MOHR'S CIRCLE FOR PLANE STRESS

Mohr's circle, attributed to the German engineer Christian Otto Mohr, is a realization that the graph representing the relationship between the normal stress component and the shear stress component in a plane stress state on any plane passing through a point can be represented as a circle (Figure 2–11). Let $\sigma \in \mathbb{M}^2$ be a general stress tensor with a general orthonormal basis system $B = \{e_1, e_2\}$ for $\mathbb{R}^2$. Let $B' = \{e'_1, e'_2\}$ be another orthonormal basis system that is rotated counter– clockwise by an angle θ. As per section 1.2.6.c, associated with this coordinate transformation is the matrix $Q \in \mathbb{M}^2$ defined by:

$$Q = \begin{pmatrix} e'_1 \cdot e_1 & e'_1 \cdot e_2 \\ e'_2 \cdot e_1 & e'_2 \cdot e_2 \end{pmatrix} = \begin{pmatrix} \cos\theta & \sin\theta \\ -\sin\theta & \cos\theta \end{pmatrix}$$

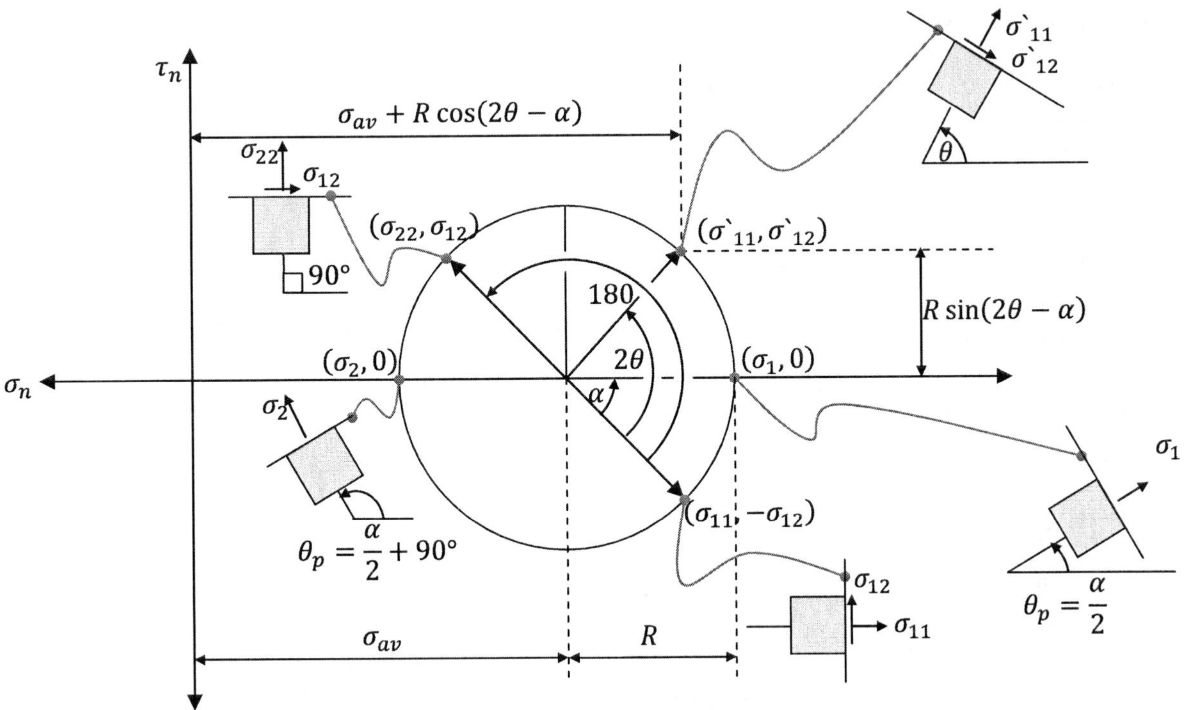

FIGURE 2-11 Construction of Mohr's circle.

Then, using the results of section 1.2.6.c, the stress matrix σ' when B' is the orthonormal basis system is:

$$\sigma' = Q\sigma Q^T$$

Using the symmetry of the stress tensor the relationship above has the following form:

$$\begin{pmatrix} \sigma'_{11} & \sigma'_{12} \\ \sigma'_{12} & \sigma'_{22} \end{pmatrix} = \begin{pmatrix} \cos\theta & \sin\theta \\ -\sin\theta & \cos\theta \end{pmatrix} \begin{pmatrix} \sigma_{11} & \sigma_{12} \\ \sigma_{12} & \sigma_{22} \end{pmatrix} \begin{pmatrix} \cos\theta & -\sin\theta \\ \sin\theta & \cos\theta \end{pmatrix}$$

Using the trigonometric relationships $\sin 2\theta = 2\sin\theta\cos\theta$ and $\cos 2\theta = 2\cos^2\theta - 1 = 1 - 2\sin^2\theta$, the following relationship is obtained:

$$\begin{pmatrix} \sigma'_{11} & \sigma'_{12} \\ \sigma'_{12} & \sigma'_{22} \end{pmatrix} = \begin{pmatrix} \frac{(\sigma_{22} + \sigma_{11})}{2} - \frac{(\sigma_{22} - \sigma_{11})}{2}\cos 2\theta + \sigma_{12}\sin 2\theta & \frac{(\sigma_{22} - \sigma_{11})}{2}\sin 2\theta + \sigma_{12}\cos 2\theta \\ \frac{(\sigma_{22} - \sigma_{11})}{2}\sin 2\theta + \sigma_{12}\cos 2\theta & \frac{(\sigma_{22} + \sigma_{11})}{2} + \frac{(\sigma_{22} - \sigma_{11})}{2}\cos 2\theta - \sigma_{12}\sin 2\theta \end{pmatrix}$$

Mohr's circle can be constructed by first defining:

$$R = \sqrt{\left(\frac{\sigma_{22} - \sigma_{11}}{2}\right)^2 + \sigma_{12}^2}, \quad \sigma_{av} = \frac{(\sigma_{22} + \sigma_{11})}{2}, \quad \alpha = \text{atan}\frac{2\sigma_{12}}{(\sigma_{11} - \sigma_{22})}$$

Then, the relationship between the components can be written in the following form:

$$\begin{pmatrix} \sigma'_{11} & \sigma'_{12} \\ \sigma'_{12} & \sigma'_{22} \end{pmatrix} = \begin{pmatrix} \sigma_{av} + R\cos(2\theta - \alpha) & -R\sin(2\theta - \alpha) \\ -R\sin(2\theta - \alpha) & \sigma_{av} - R\cos(2\theta - \alpha) \end{pmatrix}$$

2.4.1. Orientation and Magnitude of Principal Stresses

The principal directions (the eigenvectors of the stress matrix) are oriented at an angle θ_p with respect to the basis vector e_1. θ_p can be obtained using the relationships defined above by noticing that when B' is aligned with the eigenvectors of σ then the component $\sigma'_{12} = 0$:

$$\sigma'_{12} = -R\sin(2\theta_p - \alpha) = 0 \Longrightarrow \theta_p = \frac{\alpha}{2} \text{ (or) } \theta_p = \frac{\alpha}{2} + 90$$

Thus, the values of the principal stresses (σ_1 and σ_2) are:

$$\sigma_1 = \sigma_{av} + R, \quad \sigma_2 = \sigma_{av} - R$$

2.4.2. Orientation and Magnitude of Maximum Shear Stresses

As shown in section 2.2.4, the maximum shear stress in a plane stress problem is located on a plane oriented 45° to the principal planes. Thus, the maximum shear stresses are located on the plane oriented by $45^\circ + \theta_p$ to the plane perpendicular to e_1.

2.4.3. Construction of Mohr's Circle

If the horizontal axis represents the normal stress component σ_n while the vertical axis represents the shear component τ_n on any plane, then the center of the circle lies at the point $(\sigma_{av}, 0)$, while the radius of the circle is equal to R. The sign convention is such that positive values for the normal stress indicate tension, while the positive values for the shear components indicate that the shear components cause clockwise rotation. Mohr's circle can then be used to find the normal and shear components of the stress acting on any plane oriented with a counter–clockwise angle θ from the plane perpendicular to e_1. To find those components, one traverses on the circle a distance sub–tending a counter–clockwise angle of 2θ (Figure 2–11). Can you identify the points representing the principal planes and the planes of the maximum shear?

2.4.4. Illustrative Example for Mohr's Circle

The following example constructs the Mohr's circle using Mathematica for a state of stress repre–sented by a stress matrix $\sigma \in \mathbb{M}^2$ when each component ranges from -5 to 5 (Figure 2–12). The

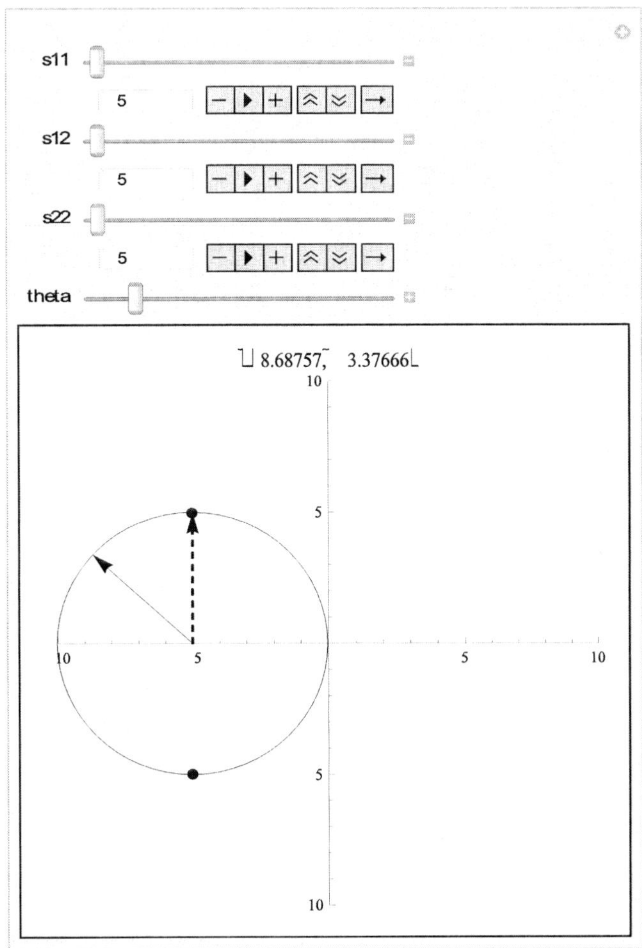

FIGURE 2-12 Construction of Mohr's circle using Mathematica.

dashed arrow represents the plane perpendicular to e_1. The black point on the circle represents the plane perpendicular to e_2, while the solid arrow represents the plane whose normal forms a counter–clockwise angle θ with e_1. The values on top of the graph are σ_n and τ_n on the plane whose normal forms a counter–clockwise angle θ with e_1. Use the code to identify the states of stress for which the Mohr's circle reduces to a point. What is the name of such states of stress?

```
Manipulate[
s={{s11,s12},{s12,s22}};
R=Sqrt[((s[[2,2]]-s[[1,1]])/2)^2+s[[1,2]]^2];
sav=1/2*(s[[2,2]]+s[[1,1]]);
alpha=ArcTan[(s[[1,1]]-s[[2,2]])/2,s[[1,2]]];
Centre={sav,0};
```

```
a=Circle[Centre,R];
P1=Point[{s[[1,1]],-s[[1,2]]}];
P2=Point[{s[[2,2]],s[[1,2]]}];
ar1=Arrow[{Centre,P1[[1]]}];
sn=sav+R*Cos[2theta-alpha];
tn=-R*Sin[2theta-alpha];
P3=Point[{sn,-tn}];
ar3=Arrow[{Centre,P3[[1]]}];
Graphics[{{PointSize[Large],P1},{PointSize[Large],P2},a,{Dashed,ar1},ar3},Ax
es->True,PlotRange->{{-10,10},{-10,10}},PlotLabel->{P3[[1,1]],-
P3[[1,2]]}],{s11,-5,5},{s12,-5,5},{s22,-5,5},{theta,0,Pi}]
```

2.5. OTHER MEASURES OF STRESS

2.5.1. Engineering Stress versus True Stress

During material uniaxial tests, the value of the applied stress is obtained by dividing the applied force F by the measured initial cross–sectional area of the specimen A_0. In this case, the stress is termed the "Engineering Stress." If excessive decrease (or increase) in the cross–sectional area occurs, then the force should be divided by the actual area A (area after deformation) to obtain the "True Stress":

$$\sigma_{eng} = \frac{F}{A_0}, \qquad \sigma_{true} = \frac{F}{A}$$

Similarly, the "Engineering Strain" is obtained by dividing the elongation Δ by the original length L_0, while the "True Strain" is obtained by taking the natural logarithm $\ln \frac{L_0+\Delta}{L_0}$. (More discussion on the strain will be presented in Chapter 3.):

$$\varepsilon_{eng} = \frac{\Delta}{L_0}, \qquad \varepsilon_{true} = \ln \frac{L_0 + \Delta}{L_0} = \ln(1 + \varepsilon_{eng})$$

Considering the behavior of metals, at higher strains, the deformation of a metal is volume pre–serving and thus:

$$A_0 L_0 = A(L_0 + \Delta) \Longrightarrow A = \frac{A_0}{1 + \varepsilon_{eng}}$$

Then, the true stress can be obtained as follows:

$$\sigma_{true} = \frac{F}{A} = \frac{F}{A_0}(1 + \varepsilon_{eng})$$

Remarks: The Cauchy stress matrix presented in section 2.1.2 is obviously the three–dimensional representation of the "True Stress."

Problem: The engineering stress vs. the engineering strain of a metal is given by the following table:

$\varepsilon_{eng}(\%)$	σ_{eng} (MPa)
0	0
0.001	200
0.002	400
0.004	600
0.005	630
0.01	690
0.02	750
0.03	790
0.04	810

Compute the true stress and the true strain, and plot the relationships

Solution:

ε_{eng} (mm/mm)	σ_{eng} (MPa)	$\varepsilon_{true} = \ln(1 + \varepsilon_{eng})$ (mm/mm)	$\sigma_{true} = \sigma_{eng}(1 + \varepsilon_{eng})$ (MPa)
0	0	0	0
0.001	200	0.0009995	200.2
0.002	400	0.001998	400.8
0.004	600	0.00399	602.4
0.005	630	0.004988	633.15
0.01	690	0.00995	696.9
0.02	750	0.0198	765
0.03	790	0.02956	813.7
0.04	810	0.0392	842.4

As shown in Figure 2–13, in metals, the value of the true stress is usually slightly higher than the value of the engineering stress.

2.5.2. First and Second Piola-Kirchhoff Stress Tensors

The Cauchy stress tensor, as defined in this chapter, related area vectors to traction vectors in the current state of deformation of a material object. The first and second Piola–Kirchhoff stress

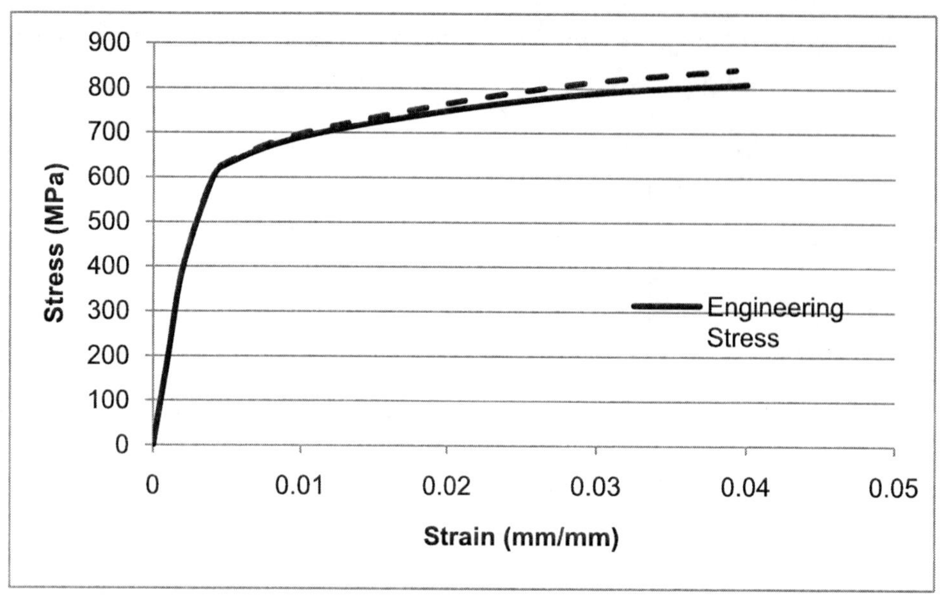

FIGURE 2-13 Difference between engineering and true stress for a metallic specimen.

tensors extend the concept of "true" and "engineering" stress to the three–dimensional case and operate on area vectors in the undeformed state of the material. Assume that the area vector before any forces are applied to a continuum is equal to $(A)N$ where $A \in \mathbb{R}$ is the magnitude of the area and $N \in \mathbb{R}^3(undeformed)$ is a vector perpendicular to the area. Assume that the forces applied lead to a linear mapping $F \in \mathbb{M}^3 : u \in \mathbb{R}^3(undeformed) \longmapsto Fu \in \mathbb{R}^3(deformed)$. Let the area vector after deformation be $(a)n$ where $a \in \mathbb{R}$ is the magnitude of the area and $n \in \mathbb{R}^3(deformed)$ is a vector perpendicular to the deformed area. Then, using the results of section 1.2.5.f:

$$(a)n = (\det F)(A)F^{-T}N$$

Let $f_n = (a)t_n = (a)\sigma^T n \in \mathbb{R}^3$ be the force vector on the deformed area with unit normal n. The First Piola–Kirchoff stress tensor P is defined as the matrix that would produce the same force vector f_n when applied to the undeformed area vector $(A)N$:

$$f_N = f_n = (A)PN$$

Equating the force vector obtained using $(A)PN$ and $(a)\sigma^T n$ and replacing $(a)n$ with the formula above leads to the relationship between P and σ^T:

$$(a)\sigma^T n = (A)PN \Longrightarrow (\det F)(A)\sigma^T F^{-T}N = (A)PN \Longrightarrow P = (\det F)\sigma^T F^{-T}$$

The second Piola–Kirchhoff stress tensor S is defined as the matrix that would produce a force vector $f2_N = F^{-1}f$ when applied to the undeformed area vector $(A)N$. Notice that the actual force vector f_n is obtained by applying the mapping F to the fictitious force vector $f2_N$. The relationship between S and σ^T can be obtained as follows:

$$F^{-1}f_n = f2_N \Longrightarrow (a)F^{-1}\sigma^T n = (A)SN \Longrightarrow (\det F)(A)F^{-1}\sigma^T F^{-T}N = (A)SN$$

Thus, S can be written in terms of σ^T and F or in terms of P and F as follows:

$$S = (\det F)F^{-1}\sigma^T F^{-T} = F^{-1}P$$

Remarks:

- The first Piola–Kirchhoff stress $P \in \mathbb{M}^3$ is a three–dimensional generalization of the concept of engineering stress.
- The first Piola–Kirchhoff stress is not necessarily a symmetric matrix. The second Piola–Kirchhoff stress, however, is a symmetric matrix.
- The term: "Nominal Stress Tensor" is sometimes used in the literature in reference to the first Piola–Kirchhoff stress or to its transpose.

▶ BONUS EXERCISE: Show that the first Piola–Kirchhoff stress is not necessarily symmetric, while the second Piola–Kirchhoff stresss is always symmetric.

Problem: Consider $\mathbb{R}^2$. Assume that a region inside a material deforms according to the linear transformation described by:

$$F = \begin{pmatrix} 1.2 & 0.4 \\ 0.4 & 1.2 \end{pmatrix}$$

The Cauchy stress of the deformed body is described by the matrix:

$$\sigma = \begin{pmatrix} 5 & 2 \\ 2 & 3 \end{pmatrix}$$

Find:

- The area vectors after deformation of the two original unit areas with area vectors $N1 = \{1, 0\}$ and $N2 = \{0, 1\}$.
- The first and second Piola–Kirchhoff stress matrices.
- The force vectors on the planes with the original unit area vectors $N1 = \{1, 0\}$ and $N2 = \{0, 1\}$ using the Cauchy stress matrix.
- The action of the first Piola–Kirchhoff stress matrix on the unit area vectors $N1 = \{1, 0\}$ and $N2 = \{0, 1\}$.

- The action of the second Piola–Kirchhoff stress matrix on the unit area vectors $N1 = \{1, 0\}$ and $N2 = \{0, 1\}$.

Solution: Consider a square of unit length in the original material before deformation. The unit areas ($A = 1$) with the area vectors $N1 = \{1, 0\}$ and $N2 = \{0, 1\}$ deform such that the new area vectors are:

$$(a_1)n1 = (\det F)(A)F^{-T}N1 = \begin{pmatrix} 1.2 \\ -0.4 \end{pmatrix}$$

$$(a_2)n2 = (\det F)(A)F^{-T}N2 = \begin{pmatrix} -0.4 \\ 1.2 \end{pmatrix}$$

Where a_1 and a_2 are the magnitudes of the "deformed" areas and $n1$ and $n2$ are the corresponding normal vectors to the areas whose original normal vectors before deformation are $N1$ and $N2$, respectively.

The first and second Piola–Kirchhoff stress matrices P and S are:

$$P = (\det F)\sigma^T F^{-T} = \begin{pmatrix} 5.2 & 0.4 \\ 1.2 & 2.8 \end{pmatrix}, \quad S = F^{-1}P = \begin{pmatrix} 4.5 & -0.5 \\ -0.5 & 2.5 \end{pmatrix}$$

The force vectors on the deformed area vectors $(a_1)n1$ and $(a_2)n2$ are:

$$f_{n1} = (a_1)t_{n1} = (a_1)\sigma^T n1 = \begin{pmatrix} 5.2 \\ 1.2 \end{pmatrix}, \quad f_{n2} = (a_2)t_{n2} = (a_2)\sigma^T n2 = \begin{pmatrix} 0.4 \\ 2.8 \end{pmatrix}$$

The force vectors on the original area vectors $N1$ and $N2$ obtained using the first Piola–Kirchoff stress are exactly the same:

$$f_{N1} = (A)PN1 = \begin{pmatrix} 5.2 \\ 1.2 \end{pmatrix}, \quad f_{N2} = (A)PN2 = \begin{pmatrix} 0.4 \\ 2.8 \end{pmatrix}$$

The force vectors on the original area vectors $N1$ and $N2$ when using the Second Piola–Kirchhoff stress ($A = 1$):

$$f2_{N1} = (A)SN1 = \begin{pmatrix} 4.5 \\ -0.5 \end{pmatrix}, \quad f2_{N2} = (A)SN2 = \begin{pmatrix} -0.5 \\ 2.5 \end{pmatrix}$$

Figure 2–14 shows the deformation of an originally square object of unit length and the corresponding deformation of areas. The force vectors are also shown on the different areas before and after deformation. The Cauchy stress and the first Piola–Kirchhoff stress produce the same force

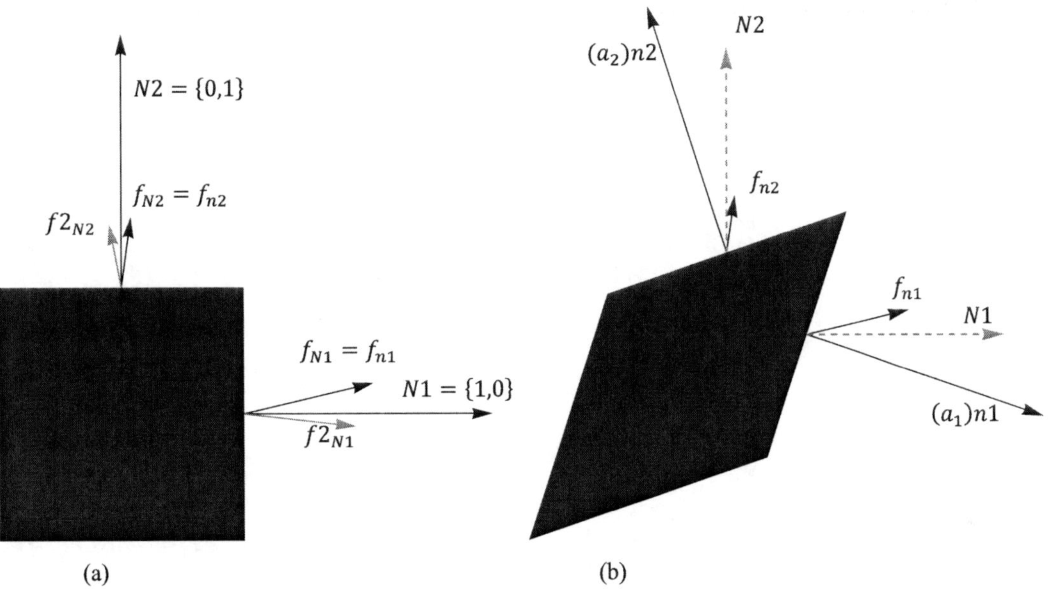

FIGURE 2-14 The (a) undeformed versus the (b) deformed configurations and the respective area vectors and force vectors.

vectors. The Cauchy stress matrix is intuitively viewed as the force in the deformed configuration per unit area of the deformed configuration. The first Piola–Kirchhoff stress matrix is intuitively viewed as the force in the deformed configuration per unit area of the undeformed configuration, while the second Piola–Kirchhoff stress is intuitively viewed as the force in the undeformed con–figuration per unit area of the undeformed configuration. Notice that all these stress measures are equivalent, but they are useful for different applications.

The following Mathematica code was used to produce the above results and figure:

```
F={{1.2,0.4},{0.4,1.2}};
Bx=Rectangle[{0,0}];
Bx2=GeometricTransformation[Bx,F];
Sigma={{5,2},{2,3}};
N1={1,0};
N2={0,1};
Farea=Det[F]*Transpose[Inverse[F]];
an1=Farea.N1;
an2=Farea.N2;
tn1=Transpose[Sigma].an1
tn2=Transpose[Sigma].an2
FirstPiola=Transpose[Sigma].Inverse[Transpose[F]]*Det[F];
SecondPiola=Inverse[F].FirstPiola;
TN1=FirstPiola.N1;
TN2=FirstPiola.N2;
```

```
STN1=SecondPiola.N1;
STN2=SecondPiola.N2;
Pt11=F.N1+1/2*F.N2;
Pt12=Pt11+an1;
Pt21=1/2*F.N1+F.N2;
Pt22=Pt21+an2;
ar1=Arrow[{Pt11,Pt12}];
ar2=Arrow[{Pt21,Pt22}];
A1dotted=Arrow[{Pt11,Pt11+{1,0}}];
A2dotted=Arrow[{Pt21,Pt21+{0,1}}];
A1=Arrow[{{1,0.5},{2,0.5}}];
A2=Arrow[{{0.5,1},{0.5,2}}];
artn1=Arrow[{Pt11,Pt11+1/10*tn1}];
artn2=Arrow[{Pt21,Pt21+1/10*tn2}];
arTN1=Arrow[{{1,0.5},{1,0.5}+1/10*TN1}];
arTN2=Arrow[{{0.5,1},{0.5,1}+1/10*TN2}];
arSTN1=Arrow[{{1,0.5},{1,0.5}+1/10*STN1}];
arSTN2=Arrow[{{0.5,1},{0.5,1}+1/10*STN2}];
Graphics[{Black,Bx,A1,A2,GrayLevel[0.1],arTN1,arTN2,GrayLevel[0.5],arSTN1,ar
STN2}]
Graphics{Black,Bx2,ar1,ar2,GrayLevel[0.1],artn1,artn2,Dashed,GrayLevel[0.5],
A1dotted,A2dotted}
```

New Mathematica Functions Explored in Chapter 2

```
ArcTan[]
Circle[]
Cos[]
Manipulate[]
ParametricPlot[]
Piecewise[]
Point[]
Rectangle[]
Sin[]
Solve[{Equations},{variables}]
Sqrt[]

- Defining functions in Mathematica:
f[x_]:= expression in x
```

▶ EXERCISES

(Answers are given in round brackets for some problems.):

1. A 1m average diameter pressure vessel is constructed by butt–welding a 10mm plate with a spiral seam weld, as shown in the figure. A constant pressure P of $1.3MPa$ is maintained inside the vessel. Determine:
    a. The traction vector on the spiral seam weld plane in the coordinate system shown in the figure.

b. The normal stress on the spiral seam weld. (**Answer:** 48.75 MPa).

c. The shearing stress on the spiral seam weld. (**Answer:** 16.25 MPa).

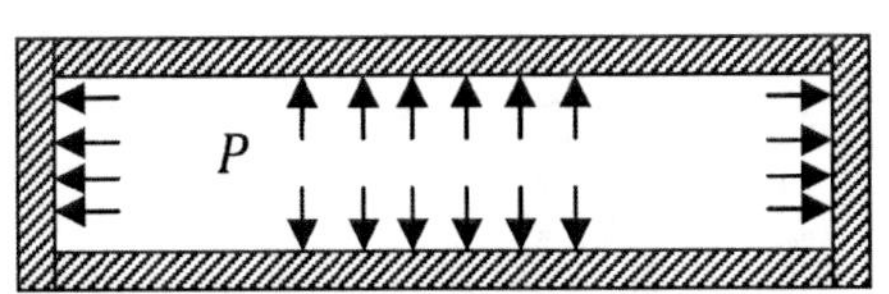

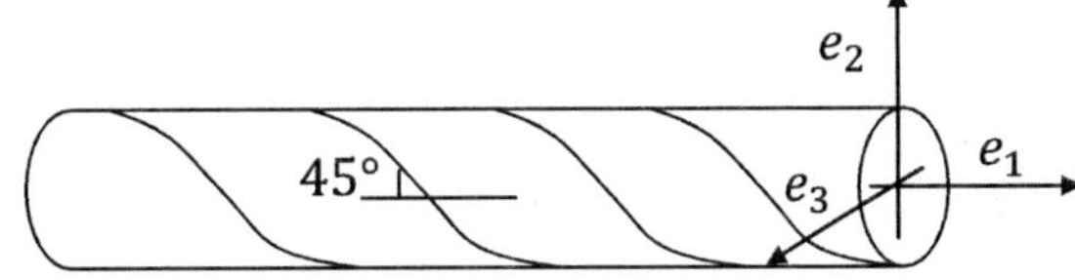

2. The shown cuboid is under the state of stress represented by the Cauchy stress matrix below. The coordinates of points C, B and D are {3,0,0}, {3,0,2} and {0,1,0}, respectively. The cuboid is composed of two halves glued together on the plane $ABCD$. The maximum shear stress that the glue can take is 100MPa, while the maximum normal tensile stress is 200MPa. Will the glue fail in shear and/or in tension?
(**Answer**: $\sigma_n = 40$, $\tau_n = 223.25MPa$)

$$\sigma = \begin{pmatrix} 100 & 200 & 30 \\ 200 & -100 & -50 \\ 30 & -50 & 100 \end{pmatrix} MPa$$

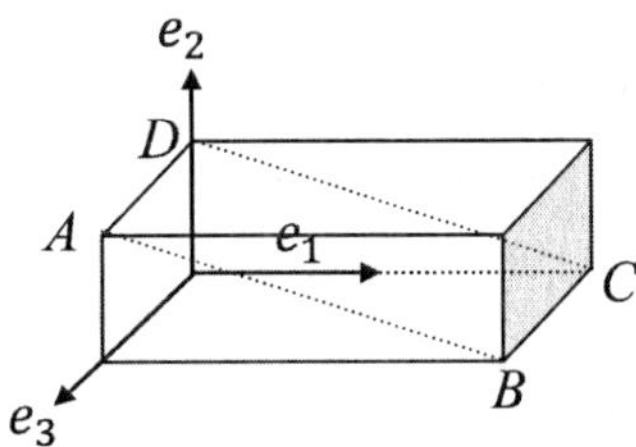

3. A material point is subjected to a state of plane stress, as shown in the figure. Determine:

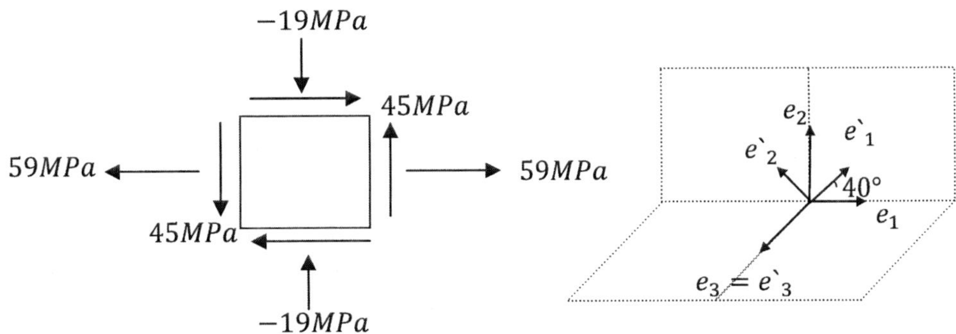

a. The Cauchy stress tensor σ in the orthonormal basis set $B = \{e_1, e_2, e_3\}$

b. The Cauchy stress tensor σ' in the orthonormal basis set $B' = \{e_1', e_2', e_3'\}$

c. The coordinate system in which the Cauchy stress matrix is diagonal. Also find the Cauchy stress matrix in that coordinate system. (**Answer:** $e_1'' = \{0.91, 0.42, 0\} e_2'' = \{-0.42, 0.91, 0\} e_3'' = \{0, 0, 1\}$)

d. The hydrostatic stress, the von Mises stress and the maximum shear stress and show that they are equal for both σ and σ'. (**Answer**: $13.33MPa$, $105.06MPa$, $59.55MPa$)

4. A machine component was analyzed by engineer A in the orthonormal basis set $B = \{e_1, e_2, e_3\}$ using a finite element analysis package. The critical stress state was found to be represented by the Cauchy stress matrix σ shown below. Another engineer (engineer B) conducted a separate analysis in which the machine component was oriented as shown in the figure with the orthonormal basis set $B' = \{e_1', e_2', e_3'\}$. Notice that point C is the new origin of the coordinate system chosen by engineer B. Assuming that the loads and the boundary conditions of both parts have the same orientation with respect to the machine component, find the critical Cauchy stress matrix σ' obtained by engineer B and show that the hydrostatic stress and the von Mises stress are equal in both cases. (Hint: Find the transformation matrix between the coordinate systems.) (**Answer**: *Hydrostatic stress* = $33.3MPa$, Von Mises stress = $412.55MPa$).

$$\sigma = \begin{pmatrix} 100 & 200 & 30 \\ 200 & -100 & -50 \\ 30 & -50 & 100 \end{pmatrix} MPa$$

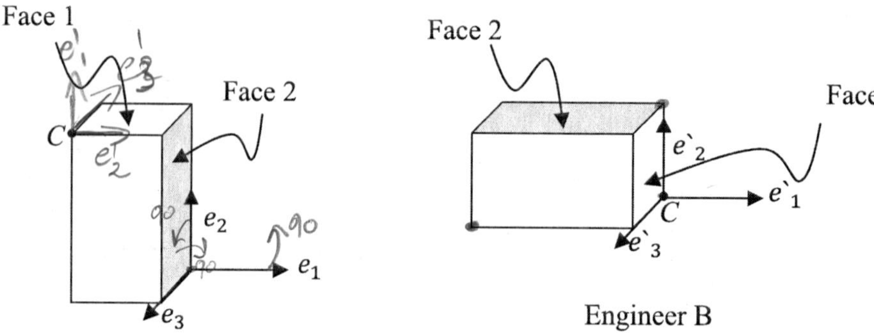

5. The shown cylinder is subjected to a confined compression test in which the horizontal stress is kept constant at –200MPa, while the magnitude of the compressive vertical stress is increased above the value of –200MPa. Find the magnitude of the compressive vertical stress at which the material will fail in the following two separate cases:

   a. If the maximum shear stress that the sample can withstand is 200MPa. (**Answer**: $\sigma_{22} = -600MPa$)

b. If the maximum von Mises stress that the sample can withstand is 400MPa. (**Answer**: $\sigma_{22} = -600MPa$)

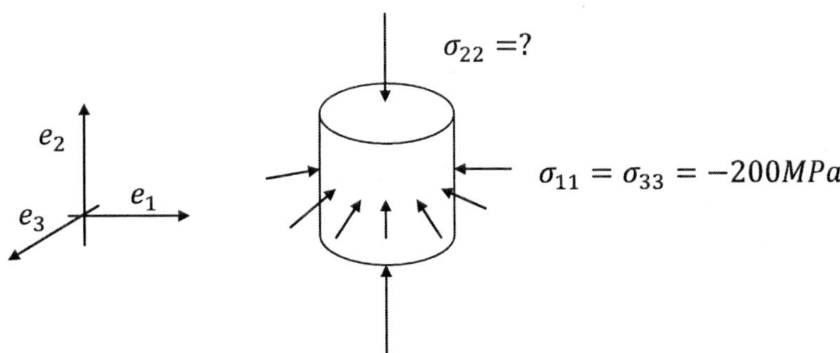

6. A metal plate is subjected to a shear state of stress in a certain coordinate system represented by the matrix below.

$$\sigma = \begin{pmatrix} 0 & t & 0 \\ t & 0 & 0 \\ 0 & 0 & 0 \end{pmatrix} MPa$$

If the yield stress of the material is σ_y, show that the maximum value for t at which the material yields is:

a. $t = \frac{\sigma_y}{\sqrt{3}}$ if the von Mises yield criterion is used

b. $t = \frac{\sigma_y}{2}$ if the Tresca yield criterion is used

7. The two plates are connected together using a bolt through the hole shown in the figure. If the diameter of the bolt is 20mm, find the force F that will cause yielding in the bolt if the yield stress of the bolt material is 400MPa in the following two cases:

a. Using the von Mises yield criterion. (**Answer**: $F = 72.55\ kN$)

b. Using the Tresca yield criterion. (**Answer**: $F = 62.83\ kN$)

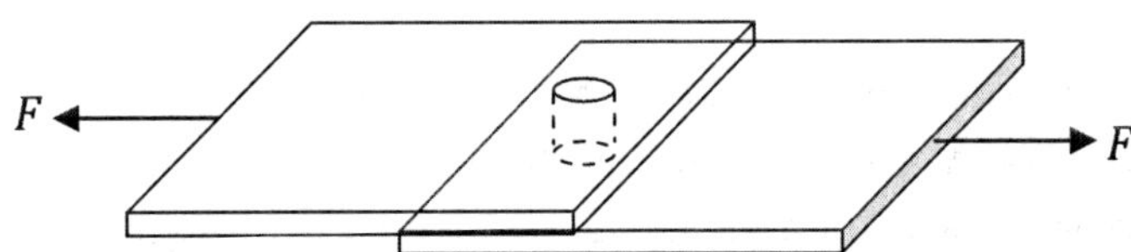

8. Find the principal stresses and the principal directions using Mohr's circle and verify your answer using Mathematica for the shown plane state of stress. Also, find the

orientation at which the maximum shear stress is attained. Find the value of the maximum shear stress. (**Answer:** $\sigma_1 = 70MPa, \sigma_2 = -30MPa$, Maximum shear stress = $50MPa$)

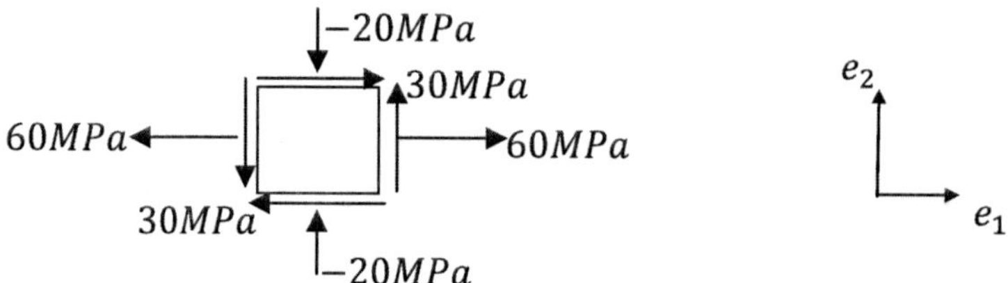

9. Let a and $b \in \mathbb{R}^3$ be two unit vectors defining distinct directions at a point P in a body under stress. The vectors t_a and t_b are the traction vectors at P associated with areas having unit normals a and b respectively.
   a. Show that the following relation is true: $t_a \cdot b = t_b \cdot a$
   b. Explain the physical meaning of the right– and left–hand sides of the expression in Part (a).
10. If the components of the Cauchy stress tensor at some point in a body are given by:

$$\sigma = \begin{pmatrix} 45 & -12 & 16 \\ -12 & 8 & 23 \\ 16 & 23 & -10 \end{pmatrix}$$

   Determine the following:
   a. the components of the traction vector t_n acting on an area with unit normal vector $n = \left\{-\frac{2}{3}, \frac{1}{3}, -\frac{2}{3}\right\}$.
   b. the norm (magnitude) of t_n.
   c. the component of t_n normal to the area.
   d. the component of t_n in the plane of the area.
   e. the angle between t_n and n.
11. If S is a scalar having the units of stress and $n \in \mathbb{R}^3$ is a unit vector, show that the stress tensor whose components are given by: $\sigma_{ij} = S\, n_i n_j$ represents a state of uniaxial tension S in the direction n. (Hint: Use a coordinate transformation into an orthonormal basis set $B = \{n, a, b\}$ with a and b being vectors orthogonal to n.)

12. The stress at a point is given by the following Cauchy stress tensor.

$$\sigma = \begin{pmatrix} 12 & 0 & 43 \\ 0 & 20 & -14 \\ 43 & -14 & 25 \end{pmatrix}$$

If the material point deforms under the action of the following linear mapping:

$$F = \begin{pmatrix} 1.1 & 0 & 0 \\ 0.2 & 1.2 & 0 \\ 0.1 & 0 & 1.3 \end{pmatrix}$$

Determine:

a. The three principal stresses and the unit vectors in the principal directions
b. The hydrostatic stress
c. The von Mises stress
d. The first and second Piola–Kirchhoff stress tensors
e. The area vectors after deformation of the three original unit areas with area vectors $N1 = \{1, 0, 0\}$, $N2 = \{0, 1, 0\}$ and $N3 = \{0, 0, 1\}$
f. The force vectors after deformation on the planes with the original unit area vectors $N1 = \{1, 0, 0\}$, $N2 = \{0, 1, 0\}$ and $N3 = \{0, 0, 1\}$ using the Cauchy stress matrix
g. The action of the first Piola–Kirchhoff stress matrix on the unit area vectors $N1 = \{1, 0, 0\}$, $N2 = \{0, 1, 0\}$ and $N3 = \{0, 0, 1\}$
h. The action of the second Piola–Kirchhoff stress matrix on the unit area vectors $N1 = \{1, 0, 0\}$, $N2 = \{0, 1, 0\}$ and $N3 = \{0, 0, 1\}$

CHAPTER 3

Deformation and Strain

The mathematical description of deformation is a very exciting topic that relies heavily on concepts from geometry. A deformed continuum body can be viewed as an image of an undeformed body or object. The measures of deformation quantify the difference between the deformed and the undeformed versions of the body. The different deformation measures rely heavily on concepts in Euclidian geometry, where distances and angles can be calculated as per Euclidian vector spaces. Simple concepts in differential geometry are also utilized in which tangents to material curves before deformation are related to tangents to material curves after deformation. In this chapter, the mathematical description of continuum bodies is presented, along with the description of deformation. The deformation gradient and the displacement gradient are introduced as measures that relate undeformed vectors to deformed vectors. Strain measures, which give an account of how much material points are displaced with respect to each other, are then introduced. At the end of this chapter, deformations that vary in time are discussed, along with the corresponding measures of the rate of deformation.

3.1. DESCRIPTION OF MOTION

3.1.1. The Mathematical Description of a Body and its Displacement

A material (continuum) body is assumed to be placed or embedded in the Euclidian Vector Space $\mathbb{R}^3$ such that every point inside the body is represented by a vector. The set of vectors corresponding to the different points inside a body is denoted Ω ($\Omega \subset \mathbb{R}^3$). Such embedding of a material body in $\mathbb{R}^3$ is called a *configuration*. Naturally, a body can have multiple configurations representing different states of deformation. However, we are usually concerned with comparing a *deformed* configuration Ω with a *reference* configuration Ω_0 (Figure 3–1). Let $X \in \Omega_0$ be a vector representing a material point in the reference configuration and $x \in \Omega$ be a vector representing the same material point but in the deformed configuration. The **position function** is the mapping

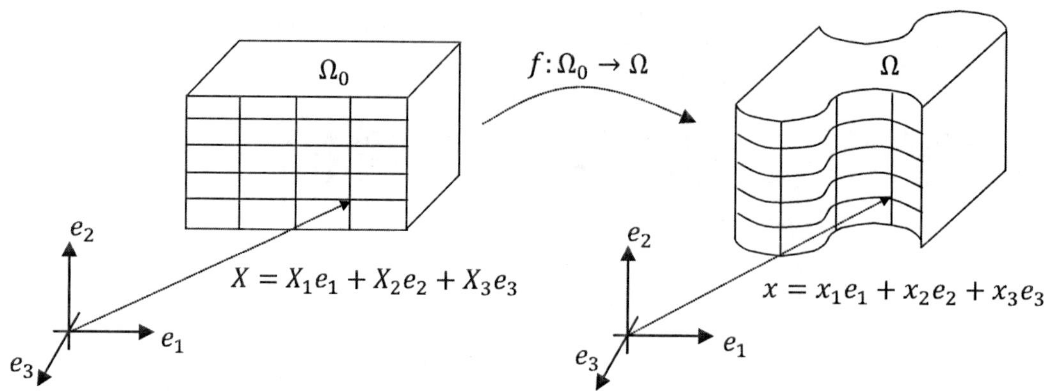

FIGURE 3-1 **A body is embedded in $\mathbb{R}^3$: Ω_0 is the set of vectors representing the reference configuration, while Ω is the set representing the deformed configuration.**

from material points (represented by vectors) in the reference configuration to material points (represented by vectors) in the deformed configuration $f{:}X \in \Omega_0 \longmapsto x \in \Omega$ such that $x = f(X)$. The **displacement function** or displacement field is a vector valued function $u{:}X \in \Omega_0 \longmapsto u(X) \in \mathbb{R}^3$ that assigns a displacement vector to each material point and is given by:

$$u(X) = x - X = f(X) - X$$

A solid mechanics problem usually involves finding the **position function** $f{:}X \in \Omega_0 \longmapsto x \in \Omega$. In most cases, the unknown position function f is assumed to be continuous, smooth (differen–tiable), and bijective. In solid mechanics, each of these assumptions is important. The continuity of f ensures that the body is left intact; in other words, no cracks or holes appear within the body. The smoothness of f, as will be shown later, is a requirement for the definition of any *strain* mea–sure and ensures that if a smooth curve is drawn within the reference configuration, the curve stays smooth in the deformed configuration. The bijectivity of f ensures that no material points are lost by the deformation.

The assumptions of smoothness and bijectivity pose a significant restriction on the motions that can be described by the formulation this text discusses. The following are examples of motions that cannot be described by a smooth bijective function (Figure 3–2). An example of a deformation function that is neither continuous nor smooth can be illustrated by imagining a box contain–ing balls. If the balls inside the box are rearranged, then it might not be possible to find a con–tinuous and smooth function that relates the position before arrangement to the position after arrangement (Figure 3–2a). Another example of a deformation function that could be continu–

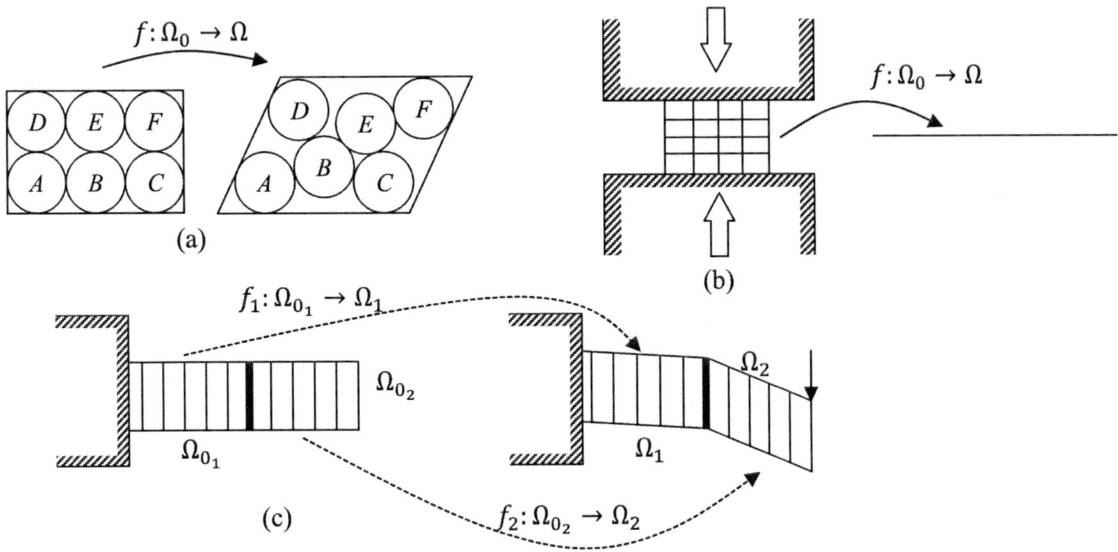

FIGURE 3-2 **Motions that contradict the a) smoothness, b) bijectivity assumptions, while c) a motion that can be described by two functions that are smooth and continuous except at the boundary between two objects where the smoothness is not satisfied.**

ous and smooth but not bijective is when a metal block is flattened. During the flattening process, cross sections of the block deform into single lines on the flat sheet (Figure 3–2b). Note that the smoothness of the position function can be relaxed at boundaries of different objects. For example, Figure 3–2c shows the motion of a body made up of two different materials. The function $x = \begin{cases} f_1(X), & X \in \Omega_{0_1} \\ f_2(X), & X \in \Omega_{0_2} \end{cases}$ is not smooth (differentiable) at the boundary between Ω_{0_1} and Ω_{0_2}.

In the following sections, some examples of simple displacement functions will be presented. Common assumptions for those displacement functions are:

- $\Omega_0 \subset \mathbb{R}^3$ is a set representing a reference configuration.
- $\Omega \subset \mathbb{R}^3$ is a set representing a deformed configuration.
- $f{:}X \in \Omega_0 \longmapsto x \in \Omega$ is a bijective, smooth, and continuous function.

The displacement function $u \in \mathbb{R}^3$ at each point is defined by:

$$u = x - X = f(X) - X$$

3.1.1.a. Rigid-Body Displacement (Translation)

A rigid–body displacement is represented by a constant displacement vector ($c \in \mathbb{R}^3$) at every point:

$$x = f(X) = X + c$$
$$u = c$$

If the reference frame is assumed to be fixed, i.e., the basis vectors are the same in both the reference and deformed configuration, then the following relationship is obtained:

$$\begin{pmatrix} x_1 \\ x_2 \\ x_3 \end{pmatrix} = \begin{pmatrix} X_1 \\ X_2 \\ X_3 \end{pmatrix} + \begin{pmatrix} c_1 \\ c_2 \\ c_3 \end{pmatrix}$$

In the following Mathematica example (Figure 3–3), a cuboid is drawn and then a copy is transformed by a displacement vector without a rotation. Then, the `Manipulate` function is used to visually see the effect of changing each coordinate.

```
box=Cuboid[{0,0,0},{3,2,1}];
c={2,2,2};
Graphics3D[{box,GeometricTransformation[box,{IdentityMatrix[3],c}]},Axes-
>True,AxesOrigin->{0,0,0}]
Manipulate[
Graphics3D[{GrayLevel[0.4],EdgeForm[Thickness[0.005]],box,
Specularity[White,6],EdgeForm[Thickness[0.01]],
GeometricTransformation[box,{IdentityMatrix[3],{cx,cy,cz}}]}],
BaseStyle->Directive[Bold,15],Axes->True,
Lighting->"Neutral",AxesOrigin->{0,0,0}],{cx,0,4},{cy,0,
4},{cz,0,4}]
```

3.1.1.b. Rigid-Body Rotation

A rigid–body rotation is represented by a rotation matrix ($Q \in \mathbb{M}^3$) at every point (See section 1.2.6.a.):

$$x = f(X) = QX$$
$$u = x - X = (Q - I)X$$

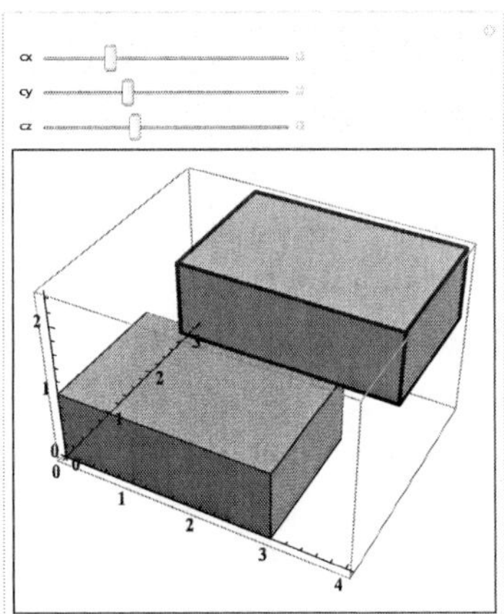

FIGURE 3-3 Rigid–body displacement of a cuboid.

If the reference frame is assumed to be fixed, i.e., the basis vectors are the same in both the reference and deformed configuration, then the following relationship in matrix form is obtained:

$$\begin{pmatrix} x_1 \\ x_2 \\ x_3 \end{pmatrix} = \begin{pmatrix} Q_{11} & Q_{12} & Q_{13} \\ Q_{21} & Q_{22} & Q_{23} \\ Q_{31} & Q_{32} & Q_{33} \end{pmatrix} \begin{pmatrix} X_1 \\ X_2 \\ X_3 \end{pmatrix}$$

Section 1.2.6.b shows examples of such deformations.

3.1.1.c. Rigid-Body Motion

A rigid–body motion is a combination of a rigid–body displacement and a rigid–body rotation, i.e. $\exists Q \in \mathbb{M}^3, \exists c \in \mathbb{R}^3$ s.t.:

$$x = f(X) = QX + c$$

$$u = x - X = (Q - I)X + c$$

The Mathematica Function `GeometricTransformation` applied on any geometric object can be used to specify both the rotation matrix and the translation matrix. In the following example, the different components of rigid–body motion can be visualized using the `Manipulate` function (Figure 3–4):

```
Clear[thx,thy,thz,ux,uy,uz]
Qx=RotationMatrix[thx,{1,0,0}];
Qy=RotationMatrix[thy,{0,1,0}];
Qz=RotationMatrix[thz,{0,0,1}];
```

```
box=Cuboid[{0,0,0},{3,2,1}];
Manipulate[
Graphics3D[{GrayLevel[0.4],EdgeForm[Thickness[0.005]],box,
Specularity[White,6],EdgeForm[Thickness[0.01]],
GeometricTransformation[
box,{RotationMatrix[thx,{1,0,0}].RotationMatrix[
thy,{0,1,0}].RotationMatrix[thz,{0,0,1}],{cx,cy,
cz}}]},BaseStyle->Directive[Bold,15],Lighting->"Neutral",
Axes->True,AxesOrigin->{0,0,0}],{thx,0,2Pi},{thy,0,
2Pi},{thz,0,2Pi},{cx,0,4},{cy,0,4},{cz,0,4}]
```

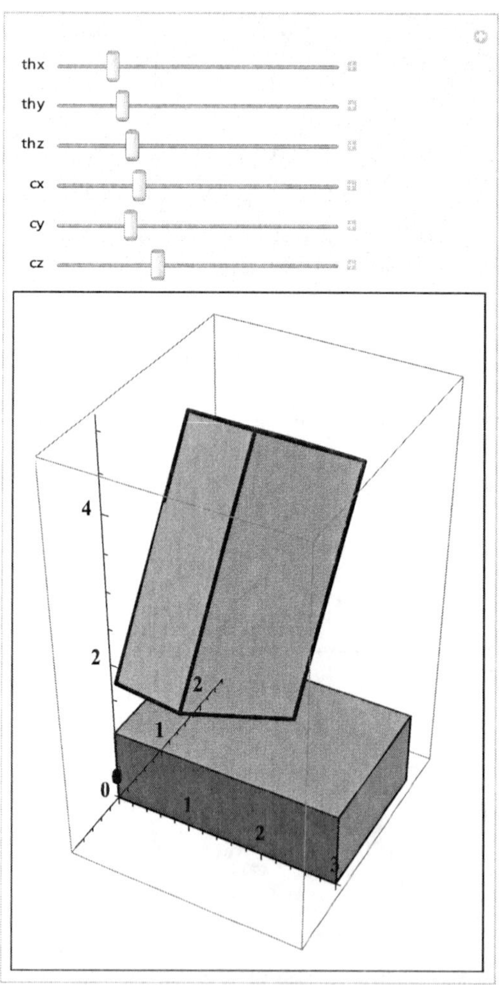

FIGURE 3-4 Rigid–body motion of a cuboid.

3.1.1.d. Uniform Extension/Contraction

A uniform extension/contraction is represented by three positive constants k_1, k_2 and $k_3 \in \mathbb{R}$ such that:

$$\begin{pmatrix} x_1 \\ x_2 \\ x_3 \end{pmatrix} = \begin{pmatrix} k_1 X_1 \\ k_2 X_2 \\ k_3 X_3 \end{pmatrix}$$

Such motion is described by the matrix $M \in \mathbb{M}^3$:

$$x = f(X) = MX = \begin{pmatrix} k_1 & 0 & 0 \\ 0 & k_2 & 0 \\ 0 & 0 & k_3 \end{pmatrix} \begin{pmatrix} X_1 \\ X_2 \\ X_3 \end{pmatrix}$$

Notice that an extension in a specified direction is represented by a constant $k_i > 1$, while a con– traction is represented by a constant $k_i < 1$. A value of $k_i = 1$ indicates no change in length. $k_i = 0$ is not allowed (**Why?**).

The following Mathematica code uses the `Manipulate` function to visualize the extension/ contraction of a cube of unit dimensions in the different directions (Figure 3–5):

```
Clear[kx,ky,kz];
box=Cuboid[{0,0,0},{1,1,1}];
Manipulate[Graphics3D[{GrayLevel[0.4],EdgeForm[Thickness[0.005]],box,Specula
rity[White,6],EdgeForm[Thickness[0.01]],GeometricTransformation[box,{{kx,0,0
},{0,ky,0},{0,0,kz}}]},BaseStyle->Directive[Bold,15],Lighting-
>"Neutral",Axes->True,AxesOrigin->{0,0,0}],{kx,0.1,4},{ky,0.1,4},{kz,0.1,4}]
```

3.1.1.e. Uniform Shear

A uniform shearing motion is described by a shearing angle along a certain direction and perpen– dicular to another direction. For example, a uniform shearing of an angle θ in the direction of e_1 and perpendicular to e_2 is described by the following position function:

$$\begin{pmatrix} x_1 \\ x_2 \\ x_3 \end{pmatrix} = \begin{pmatrix} X_1 + \tan\theta X_2 \\ X_2 \\ X_3 \end{pmatrix}.$$

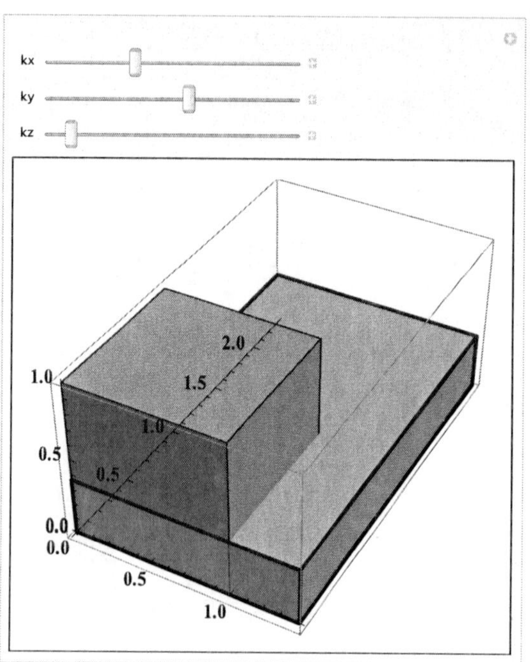

FIGURE 3-5 Uniform extension/contraction of a cube with unit length.

Such motion is described by the matrix $M \in \mathbb{M}^3$:

$$x = f(X) = MX = \begin{pmatrix} 1 & \tan\theta & 0 \\ 0 & 1 & 0 \\ 0 & 0 & 1 \end{pmatrix} \begin{pmatrix} X_1 \\ X_2 \\ X_3 \end{pmatrix}$$

The following Mathematica code uses the `Manipulate` function to visualize the shearing of a cube of unit dimensions, as described by the matrix M defined above (Figure 3–6). The function `ShearingMatrix[]` in Mathematica can also be used to generate a general uniform shear matrix.

```
Clear[theta];
box=Cuboid[{0,0,0},{1,1,1}];
Manipulate[Graphics3D[{GrayLevel[0.4],EdgeForm[Thickness[0.0055]],box,Specul
arity[White,6],EdgeForm[Thickness[0.01]],GeometricTransformation[box,{{1Tan
[theta],0}{0,1,0}{0,0,1}}]},Lighting->"Neutral",BaseStyle-
>Directive[Bold,15],Axes->True,AxesOrigin->{0,0,}],{theta,0Pi/2}]
```

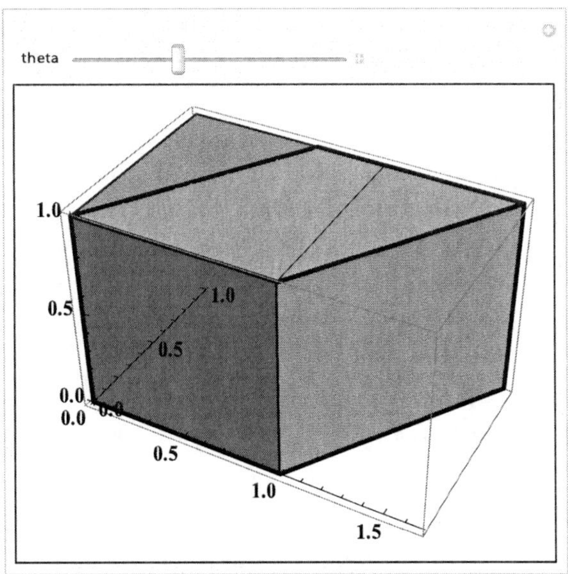

FIGURE 3-6 Uniform shear.

3.2. THE DEFORMATION GRADIENT

3.2.1. The Deformation Gradient Tensor

To measure the local deformation of a body embedded in the Euclidian vector space $\mathbb{R}^3$, it is imperative to be able to follow different material lines in different configurations. Let Ω_0 and Ω represent a reference and a deformed configuration of a body in $\mathbb{R}^3$ with the set of orthonormal basis $B = \{e_1, e_2, e_3\}$ and with the bijective smooth function $f: \Omega_0 \to \Omega$.[1] A material line inside the material can be represented by a one–parameter family of vectors $X(\xi)$ and $x(\xi, t)$, respectively, where t refers to time:

$$X = \begin{pmatrix} X_1(\xi) \\ X_2(\xi) \\ X_3(\xi) \end{pmatrix}, \quad x = f(X(\xi), t) = \begin{pmatrix} x_1(X(\xi), t) \\ x_2(X(\xi), t) \\ x_3(X(\xi), t) \end{pmatrix}$$

[1]Notice that f and Ω may vary in time for a continuously moving continuum body. It will be assumed that the orthonormal basis set B is the coordinate system for both the reference and the deformed configuration. However, a more general treatment of the material can be done by assuming a different basis set for the different configurations.

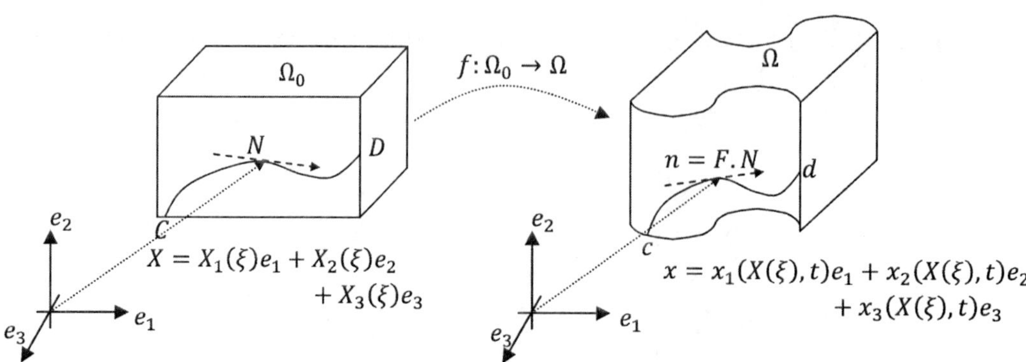

FIGURE 3-7 **Tangents to material curves in reference and deformed configurations. Points *C* and *D* in the reference configuration correspond to points *c* and *d* in the deformed configuration.**

At any point in time, the tangents to the material curves at a point in the reference and deformed configurations are equal to N and n, respectively (Figure 3–7):

$$N = \frac{\partial X}{\partial \xi} = \begin{pmatrix} \frac{\partial X_1}{\partial \xi} \\ \frac{\partial X_2}{\partial \xi} \\ \frac{\partial X_3}{\partial \xi} \end{pmatrix}, \quad n = \frac{\partial x}{\partial \xi} = \frac{\partial x}{\partial X}\frac{\partial X}{\partial \xi} = \begin{pmatrix} \frac{\partial x_1}{\partial X_1} & \frac{\partial x_1}{\partial X_2} & \frac{\partial x_1}{\partial X_3} \\ \frac{\partial x_2}{\partial X_1} & \frac{\partial x_2}{\partial X_2} & \frac{\partial x_2}{\partial X_3} \\ \frac{\partial x_3}{\partial X_1} & \frac{\partial x_3}{\partial X_2} & \frac{\partial x_3}{\partial X_3} \end{pmatrix} \begin{pmatrix} \frac{\partial X_1}{\partial \xi} \\ \frac{\partial X_2}{\partial \xi} \\ \frac{\partial X_3}{\partial \xi} \end{pmatrix} = FN$$

The matrix F is termed the **deformation gradient** and maps all the tangents to the material curves in the reference configuration to the corresponding tangents to the material curves in the deformed configuration. In fact, the matrix F contains all the required information regarding the local changes in lengths, volumes, and angles due to the deformation exhibited by every material point in the set Ω_0.

Remarks:

- **Local Volume Transformation:** As per section 1.2.5.c, the ratio of the local volumes at any point inside the material body is embedded in F. The local rate of change of the deformed configuration volume v with respect to the reference configuration volume V can be evaluated through:

$$\frac{dv}{dV} = \det F$$

- A deformation is called *isochoric* or volume preserving if $\det F = 1$ at every point.
- As per section 1.2.5.f, if an infinitesimal area vector is equal to $(da)n$ and $(dA)N$ in the deformed and reference configurations, respectively, with $da, dA \in R$ being the

magnitudes of the areas and $n, N \in \mathbb{R}^3$ being the unit vectors perpendicular to the corresponding areas, then the area vectors transform according to the equation:

$$(da)n = (\det F)(dA)F^{-T}N$$

- A deformation is called homogeneous if F is not a function of the position X of the material body. A deformation is called non–homogeneous if F varies according to the position X inside the material body.

Physical Restrictions on the Deformation Gradient:

For the matrix F to adequately describe physical deformations, it has to abide by the restriction that $\det F > 0$. If $\det F = 0$, then F is not invertible and $\exists N \neq \hat{0} \in \mathbb{R}^3$ s.t. $FN = \hat{0}$; i.e., there will be a direction where the material will be squished to zero thickness. Such deformation is not acceptable and thus $\det F$ cannot be equal to 0 (See sections 1.2.3 and 1.2.5.b.). In addition, if $\det F < 0$, then the material is said to have turned inside out. Section 1.2.5.c indicates that in such cases, the vol–ume of the deformed configuration would be negative! Thus, to preserve the material orientation and to ensure that the material does not diminish, the condition $\det F > 0$ is required.

Problem: A continuum body deforms such that the vectors $dX = \{1, 0, 0\}$, $dY = \{1, 1, 0\}$ and $dZ = \{1, 1, 1\}$ deform to the vectors $dx = \{1, 0, 0\}$, $dy = \{0, 1, 1\}$ and $dz = \{0, 0, 1\}$. Find the deformation gradient, and state whether this deformation is physically possible or not.

Solution: The nine components of the deformation gradient in the traditional orthonormal basis set can be obtained by solving the linear system of nine equations:

$$F\,dX = dx, \quad F\,dY = dy, \quad F\,dZ = dz$$

Using Mathematica to solve the above system of equations, F and $\det F$ are:

$$F = \begin{pmatrix} 1 & -1 & 0 \\ 0 & 1 & -1 \\ 0 & 1 & 0 \end{pmatrix}, \quad \det F = 1$$

Since $\det F = 1$, the deformation is physically possible. Notice that this example illustrates that F is fully defined by the map between three linearly independent vectors in the undeformed configu–ration and their images in the deformed configuration. Naturally, if the deformation is physically possible, the three images of the original three linearly independent vectors have to be linearly independent in the deformed configuration as well (**Why?**).

The following is the Mathematica code utilized in this problem. Notice the introduction of the new function `Flatten[]` that converts the nine components of the matrix F into a 9 component vector.

```
F = Table[Subscript[Ff, i, j], {i, 3}, {j, 3}];
dX={1,0,0};
dY={1,1,0};
dZ={1,1,1};
dx={1,0,0};
dy={0,1,1};
dz={0,0,1};
alist=Flatten[F]
a=Solve[{F.dX==dx,F.dY==dy,F.dZ==dz},alist]
F=F/.a[[1]]
Det[F]
```

3.2.2. The Displacement Gradient Tensor

Following the same assumptions in the previous section, the displacement vector–valued function $u:\Omega_0 \to \mathbb{R}^3$ is defined as the difference between the position vector after and before deformation:

$$u = x - X$$

Or in component form:

$$\begin{pmatrix} u_1 \\ u_2 \\ u_3 \end{pmatrix} = \begin{pmatrix} x_1 \\ x_2 \\ x_3 \end{pmatrix} - \begin{pmatrix} X_1 \\ X_2 \\ X_3 \end{pmatrix}$$

The gradient of the displacement function (the displacement gradient tensor) is:

$$\text{Grad } u = \nabla u = F - I$$

Or in component form:

$$\begin{pmatrix} \frac{\partial u_1}{\partial X_1} & \frac{\partial u_1}{\partial X_2} & \frac{\partial u_1}{\partial X_3} \\ \frac{\partial u_2}{\partial X_1} & \frac{\partial u_2}{\partial X_2} & \frac{\partial u_2}{\partial X_3} \\ \frac{\partial u_3}{\partial X_1} & \frac{\partial u_3}{\partial X_2} & \frac{\partial u_3}{\partial X_3} \end{pmatrix} = \begin{pmatrix} \frac{\partial x_1}{\partial X_1} & \frac{\partial x_1}{\partial X_2} & \frac{\partial x_1}{\partial X_3} \\ \frac{\partial x_2}{\partial X_1} & \frac{\partial x_2}{\partial X_2} & \frac{\partial x_2}{\partial X_3} \\ \frac{\partial x_3}{\partial X_1} & \frac{\partial x_3}{\partial X_2} & \frac{\partial x_3}{\partial X_3} \end{pmatrix} - \begin{pmatrix} 1 & 0 & 0 \\ 0 & 1 & 0 \\ 0 & 0 & 1 \end{pmatrix}$$

3.2.2.a. Decomposition of the Displacement Gradient Tensor

Assuming that tangents at a point to a material curve in the reference and deformed configurations are given by the vectors $dX, dx \in \mathbb{R}^3$, then the relationship between the vectors in the different configurations is:

$$dx = FdX = (I + \nabla u)dX = dX + \nabla u\, dX$$

Section 1.2.6.h shows that any tensor can be uniquely decomposed into a symmetric and a skew–symmetric matrix. The displacement gradient tensor can, thus, be uniquely decomposed into a symmetric matrix denoted ε and termed the **infinitesimal strain tensor** and a skew–symmetric matrix denoted W_{inf} and termed the **infinitesimal rotation tensor**:

$$\varepsilon = \frac{1}{2}(\nabla u + \nabla u^T), \quad W_{inf} = \frac{1}{2}(\nabla u - \nabla u^T)$$

The deformed material vector dx can now be written in terms of ε and W_{inf}, as follows:

$$dx = dX + \varepsilon dX + W_{inf} dX$$

The above relationship shows that the deformed vector dx is equal to the undeformed vector dX, in addition to two new vectors given by εdX and $W_{inf}dX$ described by the linear mappings ε and W_{inf} applied to dX. This relationship is shown schematically in Figure 3–8. At small displacements, the change in the direction of the infinitesimal material vector dX is thought to be due to the action of: a skew–symmetric matrix W_{inf} which, as shown in section 1.2.6.h, represents a rotation, and a symmetric matrix ε, which, as shown in section 1.2.6.f, has three orthogonal eigenvectors and thus represents extension without rotation. Thus, if dX happens to be the direction of an eigenvector of ε, then the vector εdX would be in the same direction as dX. In this case, the change in direction of the vector dX would be given by $W_{inf}dX$.

Both the infinitesimal strain tensor and infinitesimal rotation tensor are physically meaningful when the gradient of the displacement tensor ∇u has very small components such that $\frac{\partial u_i}{\partial X_j} \ll 1$. In the following section, small displacements will be assumed.

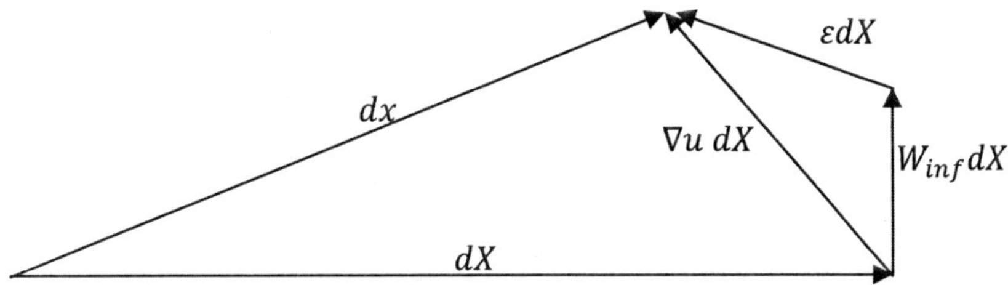

FIGURE 3-8 Decomposition of the gradient of the displacement tensor into a symmetric and a skew–symmetric matrix and their action on the material curve represented by the vector dX.

3.2.2.b. Physical Interpretation of the Infinitesimal Strain Tensor

The infinitesimal strain tensor, in component form, is:

$$\varepsilon = \frac{1}{2}\left(\nabla u + \nabla u^T\right) = \frac{1}{2}\begin{pmatrix} 2\frac{\partial u_1}{\partial X_1} & \frac{\partial u_1}{\partial X_2} + \frac{\partial u_2}{\partial X_1} & \frac{\partial u_1}{\partial X_3} + \frac{\partial u_3}{\partial X_1} \\ \frac{\partial u_1}{\partial X_2} + \frac{\partial u_2}{\partial X_1} & 2\frac{\partial u_2}{\partial X_2} & \frac{\partial u_2}{\partial X_3} + \frac{\partial u_3}{\partial X_2} \\ \frac{\partial u_1}{\partial X_3} + \frac{\partial u_3}{\partial X_1} & \frac{\partial u_2}{\partial X_3} + \frac{\partial u_3}{\partial X_2} & 2\frac{\partial u_3}{\partial X_3} \end{pmatrix}$$

Let two material curves $dX^{(1)}$ and $dX^{(2)}$ in the reference configuration be in the direction of the basis vectors e_1 and e_2, respectively. Assuming that the lengths of $dX^{(1)}$ and $dX^{(2)}$ are dL_1 and dL_2, respectively, then:

$$dX^{(1)} = \begin{pmatrix} dL_1 \\ 0 \\ 0 \end{pmatrix}, \quad dX^{(2)} = \begin{pmatrix} 0 \\ dL_2 \\ 0 \end{pmatrix}$$

If $dx^{(1)}$ and $dx^{(2)}$ are the corresponding images in the deformed configuration, then:

$$dx^{(1)} = (I + \nabla u)dX^{(1)} = \begin{pmatrix} 1 + \frac{\partial u_1}{\partial X_1} \\ \frac{\partial u_2}{\partial X_1} \\ \frac{\partial u_3}{\partial X_1} \end{pmatrix} dL_1, \quad dx^{(2)} = \begin{pmatrix} \frac{\partial u_1}{\partial X_2} \\ 1 + \frac{\partial u_2}{\partial X_2} \\ \frac{\partial u_3}{\partial X_2} \end{pmatrix} dL_2$$

If lengths of the vectors $dx^{(1)}$ and $dx^{(2)}$ are dl_1 and dl_2, respectively, then the lengths in the deformed configuration and the lengths in the reference configuration can be related using the diagonal components of the infinitesimal strain tensor:

$$dl_1^2 = dx^{(1)} \cdot dx^{(1)} \approx \left(1 + \frac{\partial u_1}{\partial X_1}\right)^2 dL_1^2, \quad dl_2^2 = dx^{(2)} \cdot dx^{(2)} \approx \left(1 + \frac{\partial u_2}{\partial X_2}\right)^2 dL_2^2$$

$$dl_1 \approx (1 + \varepsilon_{11})dL_1, \qquad dl_2 \approx (1 + \varepsilon_{22})dL_2$$

$$\varepsilon_{11} \approx \frac{dl_1 - dL_1}{dL_1}, \qquad \varepsilon_{22} \approx \frac{dl_2 - dL_2}{dL_2}$$

The physical interpretation of the off–diagonal components can be shown by taking the dot product of the two image vectors $dx^{(1)}$ and $dx^{(2)}$ with θ being the angle between these vectors

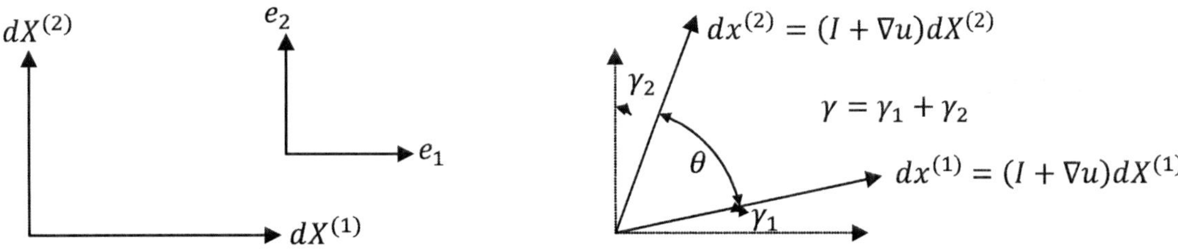

FIGURE 3-9 Geometric interpretation of the components of the infinitesimal strain tensor.

(Figure 3–9):

$$dl_1 dl_2 \cos\theta = dx^{(1)} \cdot dx^{(2)} \approx dL_1 dL_2 \left(\frac{\partial u_1}{\partial X_2} + \frac{\partial u_2}{\partial X_1} \right) = dL_1 dL_2 2\varepsilon_{12}$$

The assumptions of small displacements enable the introduction of a **shearing strain angle** $\gamma = \gamma_1 + \gamma_2$. Utilizing the relationships $\cos\theta = \sin\gamma \approx \gamma$, γ can be used to replace $\cos\theta$ in the above relationship:

$$dl_1 dl_2 \gamma \approx 2\varepsilon_{12} dL_1 dL_2$$

Again, if the displacements are very small, then the product of the lengths before deformation is almost equal to the product of lengths after deformation, and, therefore, the shearing strain angle γ is directly related to the off diagonal components ε_{12} by the relationship:

$$\varepsilon_{12} \approx \frac{\gamma}{2}$$

Therefore, the diagonal components of the infinitesimal strain tensor carry information on the changes in length of vectors originally aligned with the coordinate system vectors, while the diag–onal components carry information on the change in angles or the shearing angles between the same vectors.

3.2.2.c. Physical Interpretation of the Infinitesimal Rotation Tensor

The infinitesimal rotation tensor W_{inf}, in component form is:

$$W_{inf} = \frac{1}{2}(\nabla u - \nabla u^T) = \frac{1}{2} \begin{pmatrix} 0 & \frac{\partial u_1}{\partial X_2} - \frac{\partial u_2}{\partial X_1} & \frac{\partial u_1}{\partial X_3} - \frac{\partial u_3}{\partial X_1} \\ -\frac{\partial u_1}{\partial X_2} + \frac{\partial u_2}{\partial X_1} & 0 & \frac{\partial u_2}{\partial X_3} - \frac{\partial u_3}{\partial X_2} \\ -\frac{\partial u_1}{\partial X_3} + \frac{\partial u_3}{\partial X_1} & -\frac{\partial u_2}{\partial X_3} + \frac{\partial u_3}{\partial X_2} & 0 \end{pmatrix}$$

As shown in the schematic in Figure 3–10, the components $-W_{12}$, W_{13} and $-W_{23}$ represent the average rotation of a small material element around e_3, e_2 and e_1, respectively.

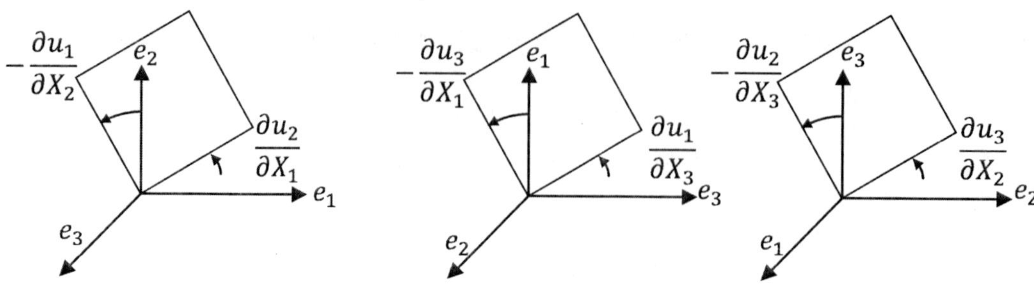

FIGURE 3-10 Geometric interpretation of the components of the infinitesimal rotation matrix.

Problem 1: A cube of unit side length has one vertex situated at the origin of the coordinate system and three sides situated along the three orthonormal basis vectors e_1, e_2 and e_3. A deformation is applied such that the cube rotates around the axis of rotation $u = \{1, 1, 1\}$ with an angle of rotation θ. Find the infinitesimal instantaneous rotation tensor and the infinitesimal strain tensor at $\theta = 1^\circ$ and 45°. From the infinitesimal rotation tensor, recalculate the angle of rotation and the axial vector (See section 1.2.6.h.).

Solution: The position function is:

$$x = QX$$

with $Q \in \mathbb{M}^3$ being the rotation matrix. Thus, the deformation gradient in this case is equal to the rotation matrix:

$$F = Q$$

The gradient of the displacement is:

$$\nabla u = F - I = Q - I$$

The infinitesimal rotation and infinitesimal strain tensors are:

$$W_{inf} = \frac{1}{2}\left(\nabla u - \nabla u^T\right), \quad \varepsilon = \frac{1}{2}\left(\nabla u + \nabla u^T\right)$$

The infinitesimal strain tensors at $\theta = 1^\circ$ and 45° are, respectively:

$$\varepsilon = \begin{pmatrix} -0.0001 & 0.00005 & 0.00005 \\ 0.00005 & -0.0001 & 0.00005 \\ 0.00005 & 0.00005 & -0.0001 \end{pmatrix}, \quad \varepsilon = \begin{pmatrix} -0.2 & 0.1 & 0.1 \\ 0.1 & -0.2 & 0.1 \\ 0.1 & 0.1 & -0.2 \end{pmatrix}$$

The infinitesimal rotation tensors at $\theta = 1^\circ$ and 45° are, respectively:

$$W_{inf} = \begin{pmatrix} 0 & -0.01 & 0.01 \\ 0.01 & 0 & -0.01 \\ -0.01 & 0.01 & 0 \end{pmatrix}, \quad W_{inf} = \begin{pmatrix} 0 & -0.41 & 0.41 \\ 0.41 & 0 & -0.41 \\ -0.41 & 0.41 & 0 \end{pmatrix}$$

The axial vectors retrieved from W at $\theta = 1^\circ$ and 45° respectively are:

$$w = \begin{pmatrix} 0.01 \\ 0.01 \\ 0.01 \end{pmatrix}, \quad w = \begin{pmatrix} 0.41 \\ 0.41 \\ 0.41 \end{pmatrix}$$

The angle of rotation at $\theta = 1^\circ$ and 45°:

$$\omega = \|w\| = 0.017^{rad} = 1^\circ, \quad \omega = \|w\| = 0.707^{rad} = 40.5^\circ$$

Figure 3–11 shows the cube in the reference and deformed configuration after 45° rotation. Physically, the pure rotation applied should not predict any changes in angles or lengths of material vectors; i.e., the components of ε should be identically zero. When $\theta = 1^\circ$ the strain components

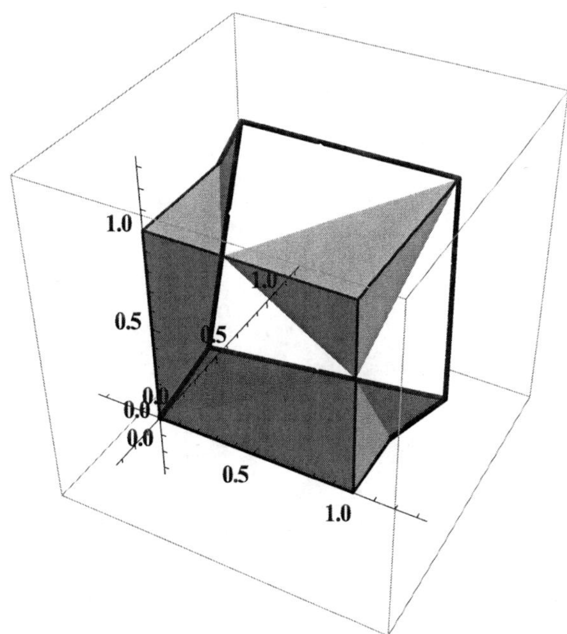

FIGURE 3-11 The reference and the deformed configuration of a cube after a 45° rotation around the vector $u = \{1, 1, 1\}$.

are very close to zero while at $\theta = 45^\circ$ the infinitesimal strain tensor produces erroneous results predicting high strains. Thus, ε is not an appropriate deformation measure at large angles of rota– tion. In addition, at a small angle of rotation, the infinitesimal rotation tensor is able to retrieve the rotation of 1° while the angle of rotation, as indicated by the infinitesimal rotation tensor, is underpredicted when the actual rotation is 45°. The following Mathematica code is used for the above calculations when $\theta = 45^\circ$:

```
a=Cuboid[{0,0,0},{1,1,1}];
rotationaxis={1,1,1};
Q=RotationMatrix[45Degree,rotationaxis];
b=GeometricTransformation[a,Q];
F=Q;
Gradu=F-IdentityMatrix[3];
esmall=N[1/2(Gradu+Transpose[Gradu])];
Wsmall=N[1/2(Gradu-Transpose[Gradu])];
esmall//MatrixForm
Wsmall//MatrixForm
sol=Eigensystem[Wsmall];
axialvector=N[{-Wsmall[[2,3]],Wsmall[[1,3]],-Wsmall[[1,2]]}]
Norm[axialvector]/Pi*180
Graphics3D[{GrayLevel[0.4],EdgeForm[Thickness[0.0055]]a,Specularity[White,6
],EdgeForm[Thickness[0.01]],b},Lighting->"Neutral",BaseStyle-
>Directive[Bold,15],AxesOrigin->{0,0,0},Axes->True]
```

Problem 2: A plate has dimensions of 2 units by 2 units and thickness t. The motion of the plate is restricted to the e_1 and e_2 plane, and the bottom left corner of the plate is situated at the origin. The displacement function is given by:

$$u_1 = 0.2X_1$$

$$u_2 = 0.09X_2X_1$$

If the plate does not move in the third direction, find the deformation gradient and the infinitesimal rotation tensor. Also, draw a vector plot showing the infinitesimal rotation vector (described by the linear map W_{inf}) and the infinitesimal extension vector (described by the linear map ε) of unit vectors in the direction of e_1 and in the direction of e_2 throughout the plate.

Solution: The third displacement component $u_3 = 0$ and, thus, the displacement gradient tensor is:

$$\nabla u = \begin{pmatrix} 0.2 & 0 & 0 \\ 0.09X_2 & 0.09X_1 & 0 \\ 0 & 0 & 0 \end{pmatrix}$$

The deformation gradient is:

$$F = \nabla u + I = \begin{pmatrix} 1.2 & 0 & 0 \\ 0.09X_2 & 1.09X_1 & 0 \\ 0 & 0 & 1 \end{pmatrix}$$

The infinitesimal rotation tensor and the infinitesimal strain tensor are:

$$W_{inf} = \frac{1}{2}\left(\nabla u - \nabla u^T\right) = \begin{pmatrix} 0 & -0.045X_2 & 0 \\ 0.045X_2 & 0 & 0 \\ 0 & 0 & 0 \end{pmatrix}, \quad \varepsilon = \frac{1}{2}\left(\nabla u + \nabla u^T\right) = \begin{pmatrix} 0.2 & 0.045X_2 & 0 \\ 0.045X_2 & 0.09X_1 & 0 \\ 0 & 0 & 0 \end{pmatrix}$$

The axial vector around which the instantaneous rotation occurs is:

$$w = \begin{pmatrix} 0 \\ 0 \\ 0.045X_2 \end{pmatrix}$$

The rotation vector r and the extension vector ext at every point of a unit vector $dX = \{1, 0, 0\}$ are equal to:

$$r = W_{inf}dX = \begin{pmatrix} 0 \\ 0.045X_1 \\ 0 \end{pmatrix}, \quad ext = \varepsilon dX = \begin{pmatrix} 0.2 \\ 0.045X_2 \\ 0 \end{pmatrix}$$

The rotation vector r and the extension vector ext at every point of a unit vector $dX = \{0, 1, 0\}$ are equal to:

$$r = W_{inf}dX = \begin{pmatrix} -0.045X_2 \\ 0 \\ 0 \end{pmatrix}, \quad ext = \varepsilon dX = \begin{pmatrix} 0.045X_2 \\ 0.09X_1 \\ 0 \end{pmatrix}$$

To draw the required plots in Mathematica, the two–dimensional part of the above two vectors are drawn on the specified domain (Figure 3–12). The difference between the skew–symmetric part and the symmetric part of the gradient of the displacement tensor can be visualized in Figure 3–12 by comparing corner 1 ($X_1 = 0, X_2 = 2$) with corner 2 ($X_1 = 2, X_2 = 0$). The vectors in corner 1

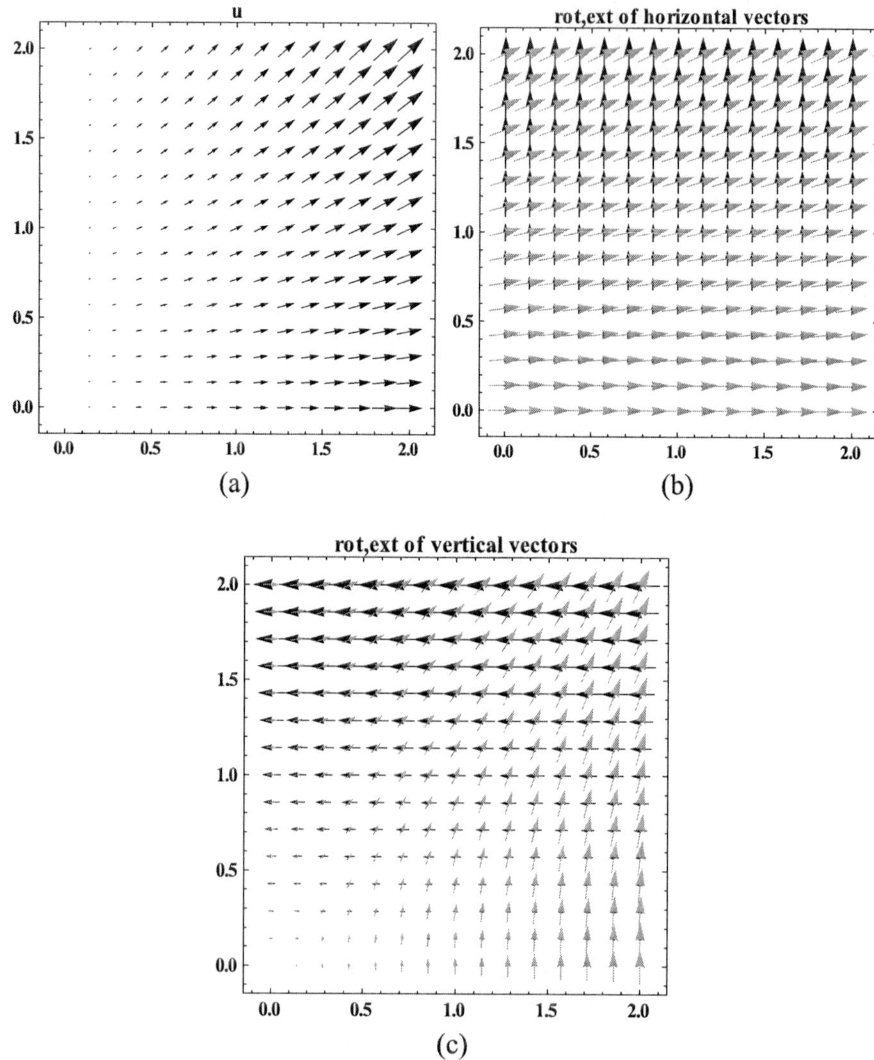

FIGURE 3-12 a) The displacement vector; b) the decomposition of the change of horizontal material vectors due to the rotation (dark) and the extension (gray); and c) the decomposition of the change of vertical material vectors due to the rotation (dark) and the extension (gray).

primarily rotate with small extension, while vectors in corner 2 primarily extend without rotation. The following is the Mathematica code used:

```
Clear[x1,x2,X1,X2,X3,x3]
X={X1,X2,X3};
u={2/10*X1,9/100*X2*X1,0};
x=u+X;
Gradu=Table[D[u[[i]],X[[j]]],{i,1,3},{j,1,3}];
```

```
F=Gradu+IdentityMatrix[3];
W=1/2*(Gradu-Transpose[Gradu]);
esmall=1/2*(Gradu+Transpose[Gradu]);
N[W]
N[esmall]
X={1,0,0}
rot=W.X;
ext=esmall.X;
rot2d={rot[[1]],rot[[2]]};
ext2d={ext[[1]],ext[[2]]};
N[rot2d]
N[ext2d]
VectorPlot[{u[[1]],u[[2]]},{X1,0,2},{X2,0,2},BaseStyle-
>Directive[Bold,15],VectorStyle->Black,AspectRatio->Automatic,PlotLabel-
>"u"]
VectorPlot[{rot2d,ext2d},{X1,0,2},{X2,0,2},BaseStyle-
>Directive[Bold,15],VectorStyle->{Black,GrayLevel[0.5]},AspectRatio-
>Automatic,PlotLabel->"rot,ext of horizontal vectors"]
X={0,1,0}
rot=W.X;
ext=esmall.X;
rot2d={rot[[1]],rot[[2]]};
ext2d={ext[[1]],ext[[2]]};
N[rot2d]
N[ext2d]
VectorPlot[{rot2d,ext2d},{X1,0,2},{X2,0,2},BaseStyle-
>Directive[Bold,15],VectorStyle->{Black,GrayLevel[0.5]},AspectRatio-
>Automatic,PlotLabel->"rot,ext of vertical vectors"]
```

3.2.3. The Polar Decomposition of the Deformation Gradient[2]

It was shown in sections 3.2.2 and 3.2.2.a that the symmetric and skew–symmetric parts of the displacement gradient are not sufficient to differentiate between rotations and stretching when a continuum body undergoes large rotations. The deformation gradient tensor, however, can be decomposed into two linear operators such that any large motion can be separated into stretching and rotation. Naturally, such decomposition is needed so that strains would only be associated with the stretching part and not the rotation part. In this section, it will be shown that the defor–mation gradient can be uniquely decomposed such that:

$$F = RU = VR$$

where, F is the deformation gradient tensor with the restriction $\det F > 0$.

[2]Ciarlet, P.G. (2004). Mathematical Elasticity. Volume 1: Three Dimensional Elasticity. *Elsevier Ltd.* and Chadwick, P. (1999). Continuum Mechanics. Concise Theory and Problems. *Dover Publications Inc.*

U, V are positive definite symmetric matrices that are termed the right stretch tensor and the left stretch tensor, respectively.

R is a rotation matrix.

The first and the second decompositions are called the right and left polar decompositions, respectively, and will be shown using a series of exercises.

3.2.3.a. Proof of the Polar Decomposition of the Deformation Gradient

Problem 1: Let $F \in \mathbb{M}^3$ such that $\det F > 0$. Show that the matrices $C = F^T F$ and $B = FF^T$ are positive definite symmetric matrices.

Solution: The symmetry of the matrices C and B can be shown, as follows:

$$C^T = (F^T F)^T = F^T F^{T^T} = F^T F = C, \quad B^T = (FF^T)^T = F^{T^T} F^T = FF^T = B$$

Also $\det F > 0 \Longrightarrow \ker F = \{\hat{0}\} \Longrightarrow \forall x \in \mathbb{R}^3, x \neq \hat{0}: \|Fx\|^2 = Fx \cdot Fx = x \cdot F^T Fx = x \cdot Cx > 0 \Longrightarrow C$ is a positive definite symmetric matrix. Similarly, B is a positive definite symmetric matrix.

Problem 2: Let $M \in \mathbb{M}^3$ such that M is a positive definite symmetric matrix. Show that there exists a unique square root positive definite symmetric matrix for M denoted by $M^{1/2}$ such that $M = M^{1/2}M^{1/2}$.

Solution: Since M is a positive definite symmetric matrix, then, using the results of sections 1.2.6.f and 1.2.6.g, it has three positive eigenvalues (note that they are not necessarily distinct) and three orthogonal eigenvectors. Denoting the eigenvalues of M by M_1, M_2 and M_3 and choosing an orthonormal coordinate system aligned with the eigenvectors of M, then M has the following form:

$$M = \begin{pmatrix} M_1 & 0 & 0 \\ 0 & M_2 & 0 \\ 0 & 0 & M_3 \end{pmatrix}$$

Denoting $\lambda_i = M_i^{1/2}$, then, the positive definite symmetric matrix $M^{1/2}$ can be constructed, as follows:

$$M^{1/2} = \begin{pmatrix} \lambda_1 & 0 & 0 \\ 0 & \lambda_2 & 0 \\ 0 & 0 & \lambda_3 \end{pmatrix}$$

Notice that the identity $M = M^{1/2}M^{1/2}$ is maintained in any coordinate system (**Why?**). Clearly, M_1, M_2 and M_3 are the eigenvalues of M and λ_1, λ_2 and λ_3 are the eigenvalues of $M^{1/2}$.

The uniqueness of $M^{1/2}$ can be shown by assuming that two distinct positive definite symmetric matrices are the square roots of M. Denote the two distinct square roots by U_1 and U_2. It will be shown here that if M_1 is an eigenvalue of M associated with the eigenvector v, then, $\lambda_1 = \sqrt{M_1}$ is an eigenvalue of both U_1 and U_2 and v is an eigenvector of both as well. The following result shows that either λ_1 or $-\lambda_1$ is an eigenvalue of U_1 (Why?):

$$Mv = M_1 v \Longrightarrow U_1 U_1 v = \lambda_1^2 v \Longrightarrow (U_1 + \lambda_1 I)(U_1 - \lambda_1 I)v = \hat{0}$$

Since, by assumption, U_1 is positive definite, therefore, λ_1 is an eigenvalue of U_1 and v is an eigen–vector of U_1; otherwise, $-\lambda_1$ is an eigenvalue of U_1 contradicting it being positive definite.

Similarly, λ_1 is an eigenvalue of U_2, and v is an eigenvector of U_2. Since U_1 and U_2 are symmetric positive definite matrices that share the same eigenvalues and the same eigenvectors, then they have the same spectral decomposition (section 1.2.6.f) and, thus, they are identical.

Notice: The uniqueness shown in the above example is very important for the polar decomposition of the deformation gradient. Consider the positive definite symmetric matrix:

$$M = \begin{pmatrix} 4 & 0 & 0 \\ 0 & 4 & 0 \\ 0 & 0 & 1 \end{pmatrix}$$

The unique positive definite symmetric square root of M is:

$$M^{\frac{1}{2}} = \begin{pmatrix} 2 & 0 & 0 \\ 0 & 2 & 0 \\ 0 & 0 & 1 \end{pmatrix}$$

However, there are other square roots of M; for example, consider the matrix:

$$A = \sqrt{2}\begin{pmatrix} 1 & 1 & 0 \\ 1 & -1 & 0 \\ 0 & 0 & 1/\sqrt{2} \end{pmatrix}$$

A is symmetric and $M = AA = A^2$; however, A is not positive definite. In fact, many other matrices are similar to A, but the one and only positive definite symmetric matrix that is the square root of M is the matrix $M^{1/2}$ shown above.

Problem 3: Let $F \in \mathbb{M}^3$ be such that $\det F > 0$. Then, verify that F can be uniquely decomposed into:

$$F = RU = VR$$

where $\mathbb{R} \in \mathbb{M}^3$ is a rotation matrix and $U, V \in M^3$ are positive definite symmetric matrices.

Solution: Since $\det F > 0$ then, from problems 1 and 2 above, the matrices $C = F^T F$ and $B = FF^T$ are positive definite symmetric matrices and thus possess positive definite symmetric square root matrices.

Denote $U = C^{1/2}$ and $V = B^{1/2}$. Notice that U and V are invertible (**Why?**). Let

$$R = FU^{-1}, \quad S = V^{-1}F$$

The matrices R and S are invertible (**Why?**). In addition:

$$RR^T = FU^{-1}U^{-1}F^T = F(F^T F)^{-1}F^T = FF^{-1}F^{-T}F^T = I$$

$$S^T S = F^T V^{-1} V^{-1} F = F^T \left(FF^T\right)^{-1} F = F^T F^{-T} F^{-1} F = I$$

Thus, both R and S are orthogonal matrices. And since $\det R = \det S = 1$ then R and S are rotation matrices.

F can thus be decomposed, as follows:

$$F = RU = VS$$

where R and S are rotation matrices.

The uniqueness of U and V immediately leads to the uniqueness of S and R as follows: Let $F = RU = R'U'$ be two different polar decompositions for F, and then $F^T F = U^2 = U'^2$ and, from the previous problem, U and U' are identical, being the unique positive definite symmetric square root of $F^T F$. Therefore, $R = R'$. Similarly, VS is a unique polar decomposition for F.

It is left to show that R and S are identical:

$$F = RU = VS = IVS = (SS^T)VS = S(S^T VS) = S(S^T V^{1/2} V^{1/2} S)$$

where $V^{1/2}$ is the unique positive definite symmetric square root of the positive definite symmetric matrix V. Let $U'' = S^T V^{1/2} V^{1/2} S$. U'' is a positive definite symmetric matrix (**Why?**). Therefore:

$$F = RU = SU''$$

Then, by the uniqueness of the decomposition $F = RU$: $R = S$ and $U = U''$

The relationship between V and U is easily shown to be:

$$U = R^T VR$$

It is important to notice that a properly defined strain function should be dependent on the matrices U and V and should be independent of the rotation matrix R. For example, the Green strain defined in a later section as $\varepsilon_{green} = \frac{1}{2}(F^T F - I) = \frac{1}{2}(U^2 - I)$ is independent of R. However, the small strain defined in a later section as $\varepsilon_{small} = \frac{1}{2}(Grad u + Grad u^T) = \frac{1}{2}(F + F^T) - I$ is dependent on both R and U and, thus, is not appropriate for large rotations!

3.2.3.b. Singular-Value Decomposition of the Deformation Gradient

Let $F \in \mathbb{M}^3$ such that $\det F > 0$. F admits the decomposition $F = RU$ where R is a rotation matrix and U is a positive definite symmetric matrix. From the results of section 1.2.6.f, U can be represented using the spectral decomposition as $U = QDQ^T$ where D is a diagonal matrix whose diagonal entries λ_1, λ_2 and λ_3 are the eigenvalues of U, while Q is a rotation matrix made out of the three corresponding orthogonal eigenvectors of U. Therefore:

$$F = RU = RQDQ^T = PDQ^T = P\begin{pmatrix} \lambda_1 & 0 & 0 \\ 0 & \lambda_2 & 0 \\ 0 & 0 & \lambda_3 \end{pmatrix} Q^T$$

The matrix $P = RQ$ is a rotation matrix. The entries λ_1, λ_2 and λ_3 are the positive square roots of the eigenvectors of the matrix $C = F^T F = U^2$ and are also denoted the singular values of F. It should be noted that the eigenvalues of the matrices $C = F^T F$ and $B = FF^T$ are identical. Additionally, the singular–value decomposition is not unique since the diagonal entries could be switched around and a corresponding rotation matrix Q' would be used instead of Q.

3.2.3.c. Physical Interpretation and Examples

The unique decomposition of the deformation gradient F into a rotation and a stretch indicates that any smooth deformation can be decomposed at any point inside the continuum body into a unique stretch described by U followed by a unique rotation described by R. As shown in Figure 3–13, a circle representing the directions of all the vectors in $\mathbb{R}^2$ is deformed into an ellipse under the action of $F:\mathbb{R}^2 \to \mathbb{R}^2$ where $\det F > 0$. The decomposition $F = RU$ is schematically shown by first stretching the circle into an ellipse whose major axes are the eigenvectors of U followed by a rotation of the ellipse through the matrix R (Figure 3–13a). The decomposition $F = VR$ is schematically shown by first rotating the circle through the matrix R and then stretching the circle into an ellipse whose major axes are the eigenvectors of V. Notice that the eigenvectors of V and the eigenvectors of U differ by a mere rotation.

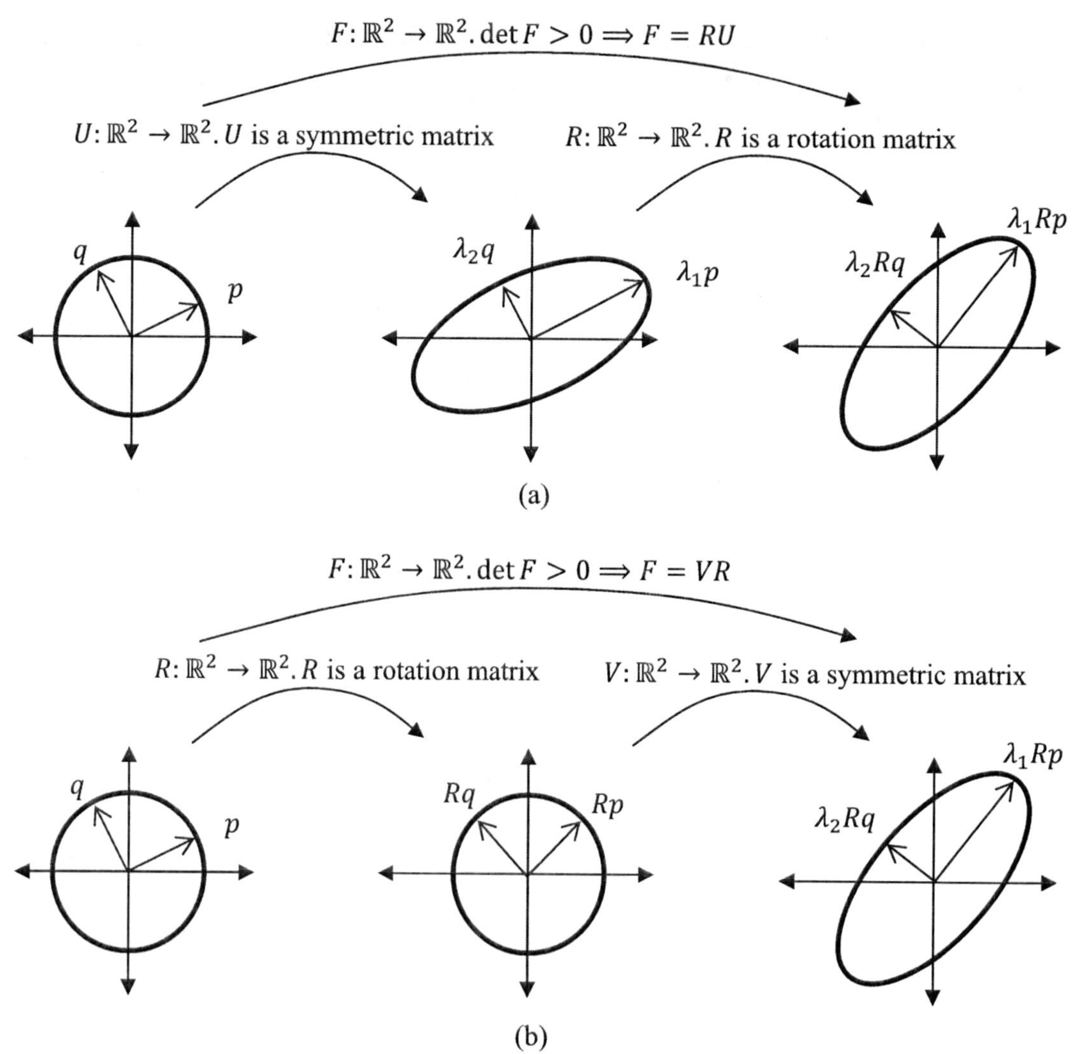

FIGURE 3-13 Schematic representation of the polar decomposition of a matrix $F \in \mathbb{M}^2$. (a) $F = RU$, (b) $F = VR$.

▶ **BONUS EXERCISE:** Let $F \in \mathbb{M}^3 \det F > 0$. Let $F = RU = VR$ be the polar decomposition, as shown above. Show that the eigenvalues of V and U are identical and that the eigenvectors of V and U differ by a rotation. (Hint: show that if x is an eigenvector of U associated with the eigen–value λ, then Rx is an eigenvector of V associated with the same eigenvalue. Then, show if y is an eigenvector of V associated with the eigenvalue β, then $R^T y$ is an eigenvector of U associated with the same eigenvalue.)

Problem: Find the left and right polar decompositions of the matrix F, and find the eigenvectors and eigenvalues of U and V:

$$F = \begin{pmatrix} 1.1 & 0.1 & 0.3 \\ 0.1 & 0.9 & 0.3 \\ 0 & -0.1 & 0.8 \end{pmatrix}$$

Solution: U and V can be constructed using the matrices $F^T F$ and FF^T or using the Mathe–matica function `SingularValueDecomposition`, which returns the matrices P, D and Q in the singular–value decomposition.

$$F = PDQ^T = \begin{pmatrix} -0.842 & 0.502 & -0.197 \\ -0.453 & -0.857 & -0.246 \\ -0.292 & -0.118 & 0.949 \end{pmatrix} \begin{pmatrix} 1.243 & 0 & 0 \\ 0 & 0.872 & 0 \\ 0 & 0 & 0.751 \end{pmatrix} \begin{pmatrix} -0.782 & -0.372 & -0.501 \\ 0.535 & -0.813 & -0.230 \\ -0.321 & -0.448 & 0.835 \end{pmatrix}$$

The right and left polar decompositions can thus be evaluated as:

$$F = PDQ^T = \left(PQ^T\right) QDQ^T = RU = \begin{pmatrix} 0.990 & -0.006 & 0.142 \\ -0.025 & 0.975 & 0.219 \\ -0.140 & -0.220 & 0.965 \end{pmatrix} \begin{pmatrix} 1.086 & 0.090 & 0.178 \\ 0.090 & 0.899 & 0.115 \\ 0.178 & 0.115 & 0.88 \end{pmatrix}$$

$$F = PDQ^T = PDQ^T \left(QP^T PQ^T\right) = PDP^T \left(PQ^T\right) = VR$$

$$= \begin{pmatrix} 1.131 & 0.136 & 0.114 \\ 0.136 & 0.941 & 0.0775 \\ 0.114 & 0.0775 & 0.794 \end{pmatrix} \begin{pmatrix} 0.990 & -0.006 & 0.142 \\ -0.025 & 0.975 & 0.219 \\ -0.140 & -0.220 & 0.965 \end{pmatrix}$$

The eigenvalues of both U and V are the entries of the diagonal matrix D. The eigenvectors of U are the rows of the matrix Q, while the eigenvectors of V are the rows of the matrix P. The following Mathematica code was utilized for the above calculations:

```
F={{1.1,0.1,0.3},{0.1,0.9,0.3},{0,-0.1,0.8}};
{P,Dd,Q}=SingularValueDecomposition[F];
P//MatrixForm
Dd//MatrixForm
Qt=Transpose[Q];
Qt//MatrixForm
```

```
R=P.Qt;R//MatrixForm
U=Q.Dd.Qt; U//MatrixForm
V=P.Dd.Transpose[P]; V//MatrixForm
Chop[F-R.U]
Chop[F-V.R]
```

Illustrative Example: In this example, the right and left polar decompositions are evaluated using the manipulate function in Mathematica. The code below draws a circle of radius 1 unit and then applies the right polar decomposition in two steps, first through the stretch U and then through the rotation R. The left polar decomposition is applied in two steps, first through the rotation R and then through the stretch V. The values of the different components of F can be changed to view their effect on the decomposition (Figure 3–14). The code is not executed when $\det F < 0$.

```
Manipulate[F={{F11,F12},{F21,F22}};If[Det[F]> 0,(
{P,Dd,Q}=SingularValueDecomposition[F];
Qt=Transpose[Q];
R=P.Qt;
U=Q.Dd.Qt;
V=P.Dd.Transpose[P];
(*Rightpolardecomposition*)
evU=Table[Q[[i]],{i,1,2}];
Lines11=Table[Arrow[{{0,0},evU[[i]]}],{i,1,2}];
Lines12=Table[Arrow[{{0,0},Dd[[i,i]]*evU[[i]]}],{i,1,2}];
Lines13=Table[Arrow[{{0,0},Dd[[i,i]]*R.evU[[i]]}],{i,1,2}];
evV=Table[P[[i]],{i,1,2}];
c11=Circle[];
c12=GeometricTransformation[c11,U];
c13=GeometricTransformation[c12,R];
(*Leftpolardecomposition*)
evV=Table[P[[i]],{i,1,2}];
Lines1=Table[Arrow[{{0,0},evU[[i]]}],{i,1,2}];
Lines2=Table[Arrow[{{0,0},evV[[i]]}],{i,1,2}];
Lines3=Table[Arrow[{{0,0},Dd[[i,i]]*evV[[i]]}],{i,1,2}];
evV=Table[P[[i]],{i,1,2}];
c1=Circle[];
c2=GeometricTransformation[c1,R];
c3=GeometricTransformation[c2,V];
{{Graphics[{Lines11,c11}],Graphics[{Lines12,c12}],
Graphics[{Lines13,c13}]},
{Graphics[{Lines1,c1}],Graphics[{Lines2,c2}],
Graphics[{Lines3,c3}]}}//MatrixForm),
Graphics[{Text["=DetF<0"Det[F]]}]],{F11,0.4,1.4},{F12,-1,
1},{F21,-1,1},{F22,0.4,1.4}]
```

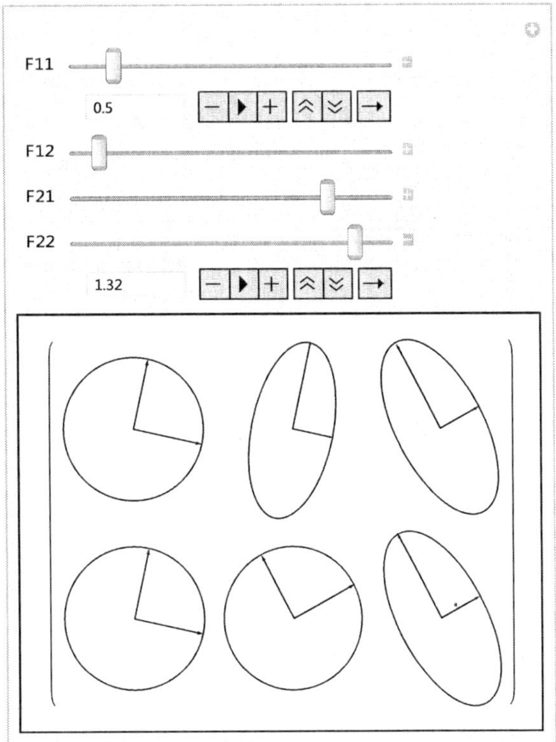

FIGURE 3-14 Illustration of the right (top) and the left (bottom) polar decomposition of a matrix using Mathematica.

3.3. STRAIN MEASURES

3.3.1. Introduction

Strain is a measure of the movement of material points relative to one another and thus is applicable for continuum solid bodies that remain intact throughout any deformation. The term strain involves stretching and/or shear distortions. While in the previous chapter, stress was used as a failure criterion, some materials may be better analyzed using a strain–based failure criteria. Strain also plays an important role in the solution of the equations of motion, as will be shown in subsequent chapters.

To properly define a strain measure, a few important assumptions (continuity, smoothness, and bijectivity) are required in the mathematical description of how a material deforms. The strain

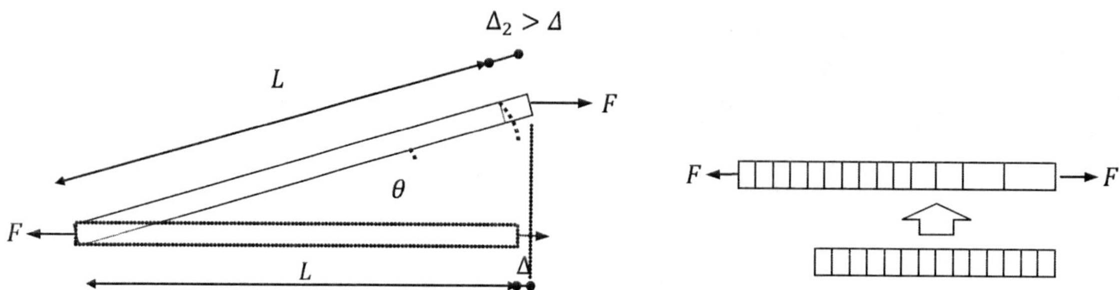

FIGURE 3-15 Failure of the formula $\frac{\Delta}{L}$ in describing the strain of a) a rotating bar, b) nonuniform elongation.

formula $\varepsilon = \frac{\Delta}{L}$ is commonly used to define strain; however, the use of such formula experimentally or numerically without taking proper precautions can produce erroneous results. For example, if a bar is rotating while being stretched, and if the magnitude of the rotation is large enough, then the horizontal displacement might not adequately describe the strain in the material. As shown in Figure 3–15a, if the horizontal displacement Δ is used instead of the inclined displacement Δ_2, then the calculated strain would be smaller than the actual strain. In addition, if the material is elongating nonuniformly, the above formula should not be used. Figure 3–15b shows a bar that has equally spaced vertical stripes. After stretching it by a force of magnitude F, the stripes are no longer equidistant and thus the "strain" calculated from the total extension of the bar is different from the strain calculated for each portion between the stripes.

The existence of a strain measure relies on the existence of a function $f: X \in \Omega_0 \longmapsto x \in \Omega$ such that $x = f(X)$ where f is continuous, smooth (differentiable), and bijective. This mathematical assumption implies that once cracks or holes develop inside the body, we can no longer use the defined strain measure. There are numerous strain measures, and their adequacy depends on the material and the type of deformation being investigated. In the following text, the strain measures in a uniaxial direction will be investigated, and then, their generalization will be presented.

3.3.2. Uniaxial (Normal) Strain Measures

Let a bar of length L_0 be stretched or contracted *uniformly* to a length L. The ratio $\lambda = \frac{L}{L_0}$ is termed the stretch ratio. The following are some of the common strain measures that can be defined for such deformation.

3.3.2.a. Engineering Longitudinal Strain

The engineering strain is widely used for traditional materials where very small deformations are allowed and is defined as:

$$\varepsilon_{eng} = \frac{(L - L_0)}{L_0} = \lambda - 1$$

This strain measure is traditionally applicable for strains of magnitude up to ± 0.25.

3.3.2.b. True (Logarithmic) Strain

The true strain is widely used for metals and is defined as:

$$\varepsilon_{true} = \ln \frac{L}{L_0} = \ln \lambda$$

Notes:

- The advantage of using the true strain in some applications is that the true strain is **additive**. If a bar of length L_0 is extended to a length L_1 and then to a length L_2, then the value of the true strain is the same whether we compare the final length with the initial length or we add the strains developed in the two stages:

$$\varepsilon_{true} = \ln \frac{L_2}{L_0} = \ln \frac{L_1 L_2}{L_0 L_1} = \ln \frac{L_1}{L_0} + \ln \frac{L_2}{L_1}$$

- For small strains ($\lambda \approx 1$), and using the Taylor expansion of the Natural Logarithm function, the two measures of strain coincide:

$$\varepsilon_{true} = \ln \lambda = (\lambda - 1) - \frac{(\lambda - 1)^2}{2} + \cdots \approx \lambda - 1 = \varepsilon_{eng}$$

3.3.2.c. Green (Lagrangian) Strain

The Green strain is widely used for rubber materials that exhibit large deformations and is defined as follows:

$$\varepsilon_{green} = \frac{1}{2} \frac{(L^2 - L_0^2)}{L_0^2} = \frac{1}{2}(\lambda^2 - 1)$$

Notes:

- If a bar is compressed such that the length L is approaching zero, the compressive Green strain approaches the value of -0.5. Thus, such strain measure is not applicable when a body undergoes excessive compressive deformations.
- For small strains ($\lambda \approx 1$) both the Green and the engineering measures of strain coincide, $\varepsilon_{green} = \frac{1}{2}(\lambda - 1)(\lambda + 1) \approx \lambda - 1 = \varepsilon_{eng}$.

3.3.2.d. Example for Comparison

Use Mathematica to draw the relationship between the engineering, logarithmic, and Green strains developed in a bar of unit length and its stretch ratio λ (Figure 3–16). Then, draw the relationship for the following (Figure 3–17):

- The %difference between the logarithmic and engineering strains as a function of the engineering strain.
- The %difference between the Green and engineering strains as a function of the engineering strain.

```
Needs["PlotLegends`"];
eeng=lambda-1;
etrue=Log[lambda];
egreen=1/2*(lambda^2-1);
Plot[{eeng,etrue,egreen},{lambda,0,2},AxesLabel->{Lambda,Strain},PlotStyle-
>{{Black,Thickness[0.02]},{GrayLevel[0.2],
Thickness[0.01]},{GrayLevel[0.5],Thickness[0.005]}},PlotLegend-
>{Eng,True,"Green"},LegendPosition->{1.1,-0.4}]
```

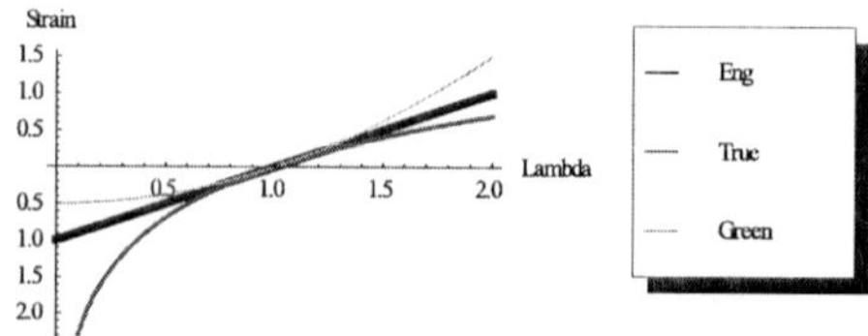

FIGURE 3-16 Different measures of strain as a function of the stretch ratio.

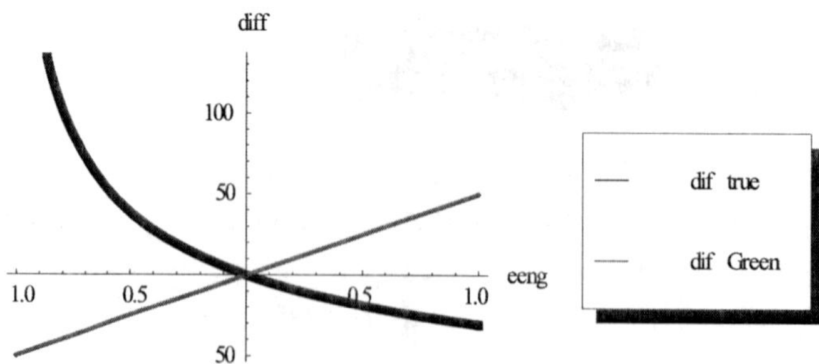

FIGURE 3-17 %Difference between different strain measures as a function of the Engineering strain.

The relationship between the logarithmic, the Green, and the engineering strains is:

$$\varepsilon_{true} = \ln \lambda = \ln(\varepsilon_{eng} + 1)$$

$$\varepsilon_{green} = \frac{1}{2}(\lambda^2 - 1) = \frac{1}{2}((1 + \varepsilon_{eng})^2 - 1) = \varepsilon_{eng} + \frac{\varepsilon_{eng}^2}{2}$$

Thus, the following code can be used to find the %difference as a function of the engineering strain:

```
Needs["PlotLegends'"];
Clear[eeng]
etrue=Log[eeng+1];
egreen=eeng+eeng^2/2;
Plot[{(etrue-eeng)/eeng*100,(egreen-eeng)/eeng*100},{eeng,-1,1},AxesLabel-
>{eeng,"%diff"},PlotStyle->{{Black,Thickness[0.02]},
{GrayLevel[0.2],Thickness[0.01]}},PlotLegend->{ "%dif true","%dif
Green"},LegendPosition->{1.1,-0.4}]
```

3.3.3. Three-Dimensional Strain Tensors

Uniaxial strain measures defined previously can be generalized to three–dimensional strain mea–sures in the continuum. Let $\Omega_0 \in \mathbb{R}^3$ be a set representing a reference configuration. Let $\Omega \in \mathbb{R}^3$ be a set representing a deformed configuration. Let $f{:}X \in \Omega_0 \longmapsto x \in \Omega$ be a bijective smooth function. The displacement vector at each point $u \in \mathbb{R}^3$ is defined as:

$$u = x - X = f(X) - X$$

The following are different definitions of three–dimensional strain measures.

3.3.3.a. Small (Infinitesimal) Strain Tensor

Let $B \subset \mathbb{R}^3$ be the traditional basis. $B = \{e_1, e_2, e_3\}$ where $e_1 = \{1, 0, 0\}$, $e_2 = \{0, 1, 0\}$ and $e_3 = \{0, 0, 1\}$. The small strain tensor, as defined in section 3.2.2.a. is the symmetric part of the dis–placement gradient ∇u and has the following form:

$$\varepsilon = \frac{1}{2}\left(\nabla u + \nabla u^T\right) = \begin{pmatrix} \frac{\partial u_1}{\partial X_1} & \frac{1}{2}\left(\frac{\partial u_1}{\partial X_2} + \frac{\partial u_2}{\partial X_1}\right) & \frac{1}{2}\left(\frac{\partial u_1}{\partial X_3} + \frac{\partial u_3}{\partial X_1}\right) \\ \frac{1}{2}\left(\frac{\partial u_1}{\partial X_2} + \frac{\partial u_2}{\partial X_1}\right) & \frac{\partial u_2}{\partial X_2} & \frac{1}{2}\left(\frac{\partial u_2}{\partial X_3} + \frac{\partial u_3}{\partial X_2}\right) \\ \frac{1}{2}\left(\frac{\partial u_1}{\partial X_3} + \frac{\partial u_3}{\partial X_1}\right) & \frac{1}{2}\left(\frac{\partial u_2}{\partial X_3} + \frac{\partial u_3}{\partial X_2}\right) & \frac{\partial u_3}{\partial X_3} \end{pmatrix}$$

Using the summation convention described in section 1.2.5.a:

$$\varepsilon_{ij} = \frac{1}{2}\left(\frac{\partial u_i}{\partial X_j} + \frac{\partial u_j}{\partial X_i}\right)$$

Remarks:

The Strain Matrix is a Linear Map: If a new orthonormal basis set $B' = \{e'_1, e'_2, e'_3\}$ is chosen such that $Q \in \mathbb{M}^3$ is the rotation matrix associated with the coordinate transformation (Section 1.2.6.c), then the new components u' and X' are related to u and X by $u' = Qu$ and $X' = QX$. These rela–tionships can be used to show the relationship between ε' and ε:

$$\begin{aligned}
\varepsilon' &= \frac{1}{2}\left(\frac{\partial u'}{\partial X'} + \left(\frac{\partial u'}{\partial X'}\right)^T\right) = \frac{1}{2}\left(\frac{\partial u'}{\partial u}\frac{\partial u}{\partial X}\frac{\partial X}{\partial X'} + \left(\frac{\partial u'}{\partial u}\frac{\partial u}{\partial X}\frac{\partial X}{\partial X'}\right)^T\right) = \frac{1}{2}\left(Q\nabla uQ^T + \left(Q\nabla uQ^T\right)^T\right) \\
&= \frac{1}{2}\left(Q\nabla uQ^T + Q\nabla u^TQ^T\right) = \frac{1}{2}\left(Q\left(\nabla u + \nabla u^T\right)Q^T\right) = Q\varepsilon Q^T
\end{aligned}$$

where if $a, b \in \mathbb{R}^3$ then $\frac{\partial a}{\partial b} \in \mathbb{M}^3$ with components $\frac{\partial a_i}{\partial b_j}$.

Engineering Longitudinal Strains along Basis Vectors: As shown in section 3.2.2.b, when the deformations are small, the diagonal components of the infinitesimal strain matrix are equal to the change in length per unit length of the three material vectors aligned with the three basis vectors. If dL_1, dL_2 and dL_3 are the lengths of the material vectors aligned with e_1, e_2 and e_3 respectively, and, if dl_1, dl_2 and dl_3 are their corresponding lengths after deformation, then the engineering strains in the corresponding directions are given by the diagonal components of ε:

$$\frac{dl_1 - dL_1}{dL_1} \approx \varepsilon_{11} = \frac{\partial u_1}{\partial X_1}, \quad \frac{dl_2 - dL_2}{dL_2} \approx \varepsilon_{22} = \frac{\partial u_2}{\partial X_2}, \quad \frac{dl_3 - dL_3}{dL_3} \approx \varepsilon_{33} = \frac{\partial u_3}{\partial X_3}$$

Engineering Longitudinal Strains along General Vectors: Given a vector $dX \in \mathbb{R}^3$ that deforms into the vector $dx \in \mathbb{R}^3$, then the engineering strain along the direction of the vector dX is given by:

$$\varepsilon_{eng\ along\ dX} = \frac{\|dx\| - \|dX\|}{\|dX\|}$$

As per section 3.2.2.a, the relationship between dx and dX is given by:

$$dx = dX + \varepsilon dX + WdX$$

$\|dx\|$ can be calculated, as follows:

$$\begin{aligned}
\|dx\| &= \sqrt{(dX + \varepsilon dX + WdX) \cdot (dX + \varepsilon dX + WdX)} \\
&= \sqrt{(dX \cdot dX + 2dX \cdot \varepsilon dX + 2dX \cdot WdX + 2\varepsilon dX \cdot WdX + \varepsilon dX \cdot \varepsilon dX + WdX \cdot WdX)}
\end{aligned}$$

Since W is a skew–symmetric matrix, then $2dX \cdot WdX = 0$. Also, for small deformations, the terms $2\varepsilon dX \cdot WdX + \varepsilon dX \cdot \varepsilon dX + WdX \cdot WdX$ are very small compared to the other terms; then:

$$\|dx\| \approx \sqrt{(dX \cdot dX + 2dX \cdot \varepsilon dX)} = \|dX\| \left(1 + \frac{2dX \cdot \varepsilon dX}{dX \cdot dX}\right)^{\frac{1}{2}} \approx \|dX\| + \frac{dX \cdot \varepsilon dX}{\|dX\|}$$

Then, the engineering strain along the vector dX is given by the formula:

$$\varepsilon_{eng\ along\ dX} = \frac{\|dX\| + \frac{dX \cdot \varepsilon dX}{\|dX\|} - \|dX\|}{\|dX\|} = \frac{dX \cdot \varepsilon dX}{\|dX\| \|dX\|}$$

Engineering Shear Strains between Basis Vectors: As shown in section 3.2.2.b, when the deformations are small, the off–diagonal components of the infinitesimal strain matrix are equal to half the engineering shear strains between the basis vectors. If γ_{12}, γ_{13} and γ_{23} are the engineering shear strains between the vectors: $e_1 - e_2$, $e_1 - e_3$ and $e_2 - e_3$, respectively, then:

$$\gamma_{12} \approx 2\varepsilon_{12} = \frac{\partial u_2}{\partial X_1} + \frac{\partial u_1}{\partial X_2}, \quad \gamma_{13} \approx 2\varepsilon_{13} = \frac{\partial u_1}{\partial X_3} + \frac{\partial u_3}{\partial X_1}, \quad \gamma_{23} \approx 2\varepsilon_{23} = \frac{\partial u_2}{\partial X_3} + \frac{\partial u_3}{\partial X_2}$$

Angle Change and Shear Strain: Given two vectors $dX, dY \in \mathbb{R}^3$ separated by an angle Θ that deform into the vectors $dx, dy \in \mathbb{R}^3$ separated by an angle θ (Figure 3–18a), **half** the change in the dot product between the two vectors after and before deformation is determined by the formula (Section 1.1.8.b):

$$\frac{1}{2}\frac{dx \cdot dy - dX \cdot dY}{\|dX\| \|dY\|} = \frac{1}{2}\frac{\|dx\| \|dy\| \cos\theta - \|dX\| \|dY\| \cos\Theta}{\|dX\| \|dY\|}$$

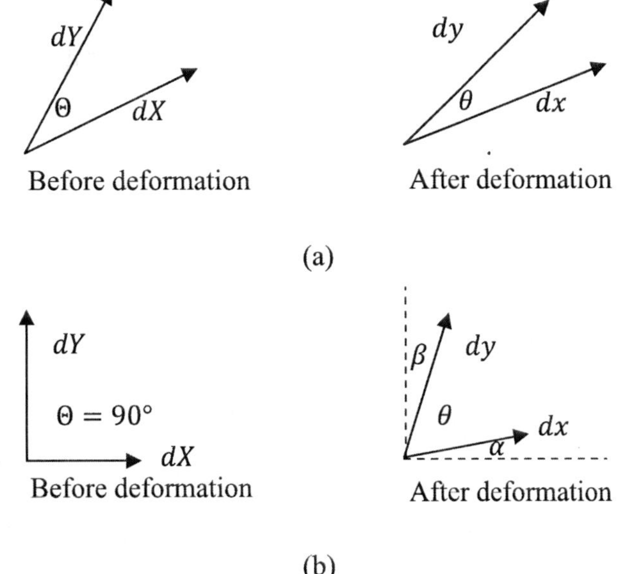

FIGURE 3-18 (a) Angle change between vectors, and (b) shear strain of planes parallel to dX and perpendicular to:

dy

Using section 3.2.2.a, $dx \cdot dy$ can be expanded, as follows:

$$dx \cdot dy = (dX + \varepsilon dX + WdX) \cdot (dY + \varepsilon dY + WdY)$$

Using the symmetry of ε and the skew symmetry of W, and if the deformations are small such that terms of small values can be ignored, then $dx \cdot dy$ can be approximated as:

$$dx \cdot dy \approx dX \cdot dY + 2dX \cdot \varepsilon dY$$

Then, **half** the change in the dot product between the two vectors after and before deformation can be determined by ε, as follows:

$$\frac{1}{2}\frac{dx \cdot dy - dX \cdot dY}{\|dX\|\|dY\|} \approx \frac{dX \cdot \varepsilon dY}{\|dX\|\|dY\|}$$

When the deformations are small such that $\|dx\| \approx \|dX\|$ and $\|dy\| \approx \|dY\|$, then half the difference in the cosines of the angles before and after deformation can be calculated as follows:

$$\frac{\cos\theta - \cos\Theta}{2} \approx \frac{dX \cdot \varepsilon dY}{\|dX\|\|dY\|}$$

If the original vectors dX, dY are orthogonal to each other ($\cos\Theta = 0$), and if the changes in lengths and in angles is very small, then the engineering shear strain of planes parallel to dX and perpendicular to dY is given in terms of the angles α, β (Figure 3–18b) by:

$$\frac{\alpha + \beta}{2} \approx \frac{\sin(\alpha + \beta)}{2} = \frac{\sin(90 - \theta)}{2} = \frac{\cos\theta}{2} \approx \frac{dX \cdot \varepsilon dY}{\|dX\|\|dY\|}$$

Symmetry and Principal Strains: Similar to the symmetric nature of the stress matrix, the symmetry of the infinitesimal strain matrix implies that there is a coordinate system in which the strain matrix is diagonal. The principal strains are the eigenvalues of the strain matrix. If an orthonormal coordinate system aligned with the eigenvectors is chosen for the basis set, then the strain matrix is diagonal (no shear strains). In other words, in this coordinate system, *no* shear strains exist on the planes perpendicular to the basis vectors.

Volumetric Strains: The volumetric strain (change of volume per unit volume) in a continuum mechanics problem characterized by small deformations can be calculated using the infinitesimal strain tensor. The volumetric strain can be computed as $\frac{V - V_0}{V_0} = \varepsilon_{11} + \varepsilon_{22} + \varepsilon_{33}$ (**Why?**).

Behaviour under Rotations: A major disadvantage of the infinitesimal strain measure is that it predicts strains for bodies that undergo large rotations even when the actual strain is zero (See section 3.2.2.c for an example.). Consider a rigid rotation (See section 3.1.1.b.); then the displacement function is given by:

$$u = (Q - I)X$$

In this particular case, the infinitesimal strain tensor is given by:

$$\varepsilon = \frac{1}{2}\left((Q-I)+(Q-I)^T\right) = \frac{1}{2}(Q+Q^T-2I)$$

Let $Q = \begin{pmatrix} \cos\theta & \sin\theta & 0 \\ -\sin\theta & \cos\theta & 0 \\ 0 & 0 & 1 \end{pmatrix}$, then $\varepsilon = \begin{pmatrix} \cos\theta - 1 & 0 & 0 \\ 0 & \cos\theta - 1 & 0 \\ 0 & 0 & 0 \end{pmatrix} \neq \hat{0}$. Thus, this strain measure is not applicable for large values of θ.

Practical Applicability: The infinitesimal strain tensor is applicable to materials with very small deformations (e.g., concrete and steel). To put this in perspective, the average failure strain of normal concrete is 0.003, while the yielding strain of steel is around 0.005. Mild steel can attain strains as high as 0.2 before failure. However, these values are still minute in comparison to pos–sible strains exhibited by other elastic materials, such as rubber.

3.3.3.b. Green (Lagrangian) Strain Tensor

Let $B \subset \mathbb{R}^3$ be the traditional basis. $B = \{e_1, e_2, e_3\}$ where $e_1 = \{1,0,0\}$, $e_2 = \{0,1,0\}$ and $e_3 = \{0,0,1\}$. The Green strain tensor $\varepsilon \in \mathbb{M}^3$ is defined as a function of the deformation gradient F, such that:

$$\varepsilon = \frac{1}{2}(F^TF - I)$$

If the displacement gradient ∇u is used instead of the deformation gradient F, then:

$$\varepsilon = \frac{1}{2}((\nabla u + I)^T(\nabla u + I) - I) = \frac{1}{2}(\nabla u + \nabla u^T + \nabla u^T\nabla u)$$

In component form, ε can be written as:

$$\varepsilon_{ij} = \frac{1}{2}\left(\frac{\partial u_i}{\partial X_j} + \frac{\partial u_j}{\partial X_i} + \sum_{k=1}^{3}\frac{\partial u_k}{\partial X_i}\frac{\partial u_k}{\partial X_j}\right)$$

If the right polar decomposition of the deformation gradient is used (See section 3.2.3.), the Green strain tensor is given by:

$$\varepsilon = \frac{1}{2}(U^2 - I)$$

Thus, the Green strain matrix is function of the right stretch part U of the deformation gradient and independent of the rigid–body rotation R.

Remarks:

The strain matrix is a linear map: Similar to the infinitesimal strain matrix, the Green strain matrix is a symmetric linear map, as defined in section 1.2.6.f. If a new orthonormal basis set $B' = \{e_1', e_2', e_3'\}$ is chosen such that $Q \in \mathbb{M}^3$ is the rotation matrix associated with the coordinate transformation (Section 1.2.6.c), the strain matrix expressed in the new coordinate system has the form:

$$\varepsilon' = Q\varepsilon Q^T$$

Longitudinal Green Strain: The Green strain matrix is the three–dimensional generalization of the uniaxial Green strain measure. Given a material vector $dX \in \mathbb{R}^3$ with length $l_0 = \sqrt{dX \cdot dX}$ that deforms into the vector $dx \in \mathbb{R}^3$ whose length is $l = \sqrt{dx \cdot dx}$, the Green strain along the vector is determined using the Green strain tensor:

$$\begin{aligned}
\varepsilon_{Green\ along\ dX} &= \frac{1}{2}\left(\frac{l^2 - l_0^2}{l_0^2}\right) = \frac{1}{2}\left(\frac{dx \cdot dx - dX \cdot dX}{\|dX\|\,\|dX\|}\right) = \frac{1}{2}\left(\frac{FdX \cdot FdX - dX \cdot dX}{\|dX\|\,\|dX\|}\right)\\
&= \frac{1}{2}\left(\frac{dX \cdot F^T F dX - dX \cdot dX}{\|dX\|\,\|dX\|}\right) = \frac{dX \cdot \frac{1}{2}\left(F^T F - I\right) dX}{\|dX\|\,\|dX\|}\\
&= \frac{dX \cdot \varepsilon dX}{\|dX\|\,\|dX\|}
\end{aligned}$$

Angle Change and Shear Strain: Given two vectors $dX, dY \in \mathbb{R}^3$ separated by an angle Θ that deform into the vectors $dx, dy \in \mathbb{R}^3$ separated by an angle θ (Figure 3–18), **half** the change in the dot product between the vectors before and after deformation is determined using the Green strain tensor, as follows (See section 1.1.8.b):

$$\frac{1}{2}\frac{dx \cdot dy - dX \cdot dY}{\|dX\|\|dY\|} = \frac{dX \cdot \varepsilon dY}{\|dX\|\|dY\|}$$

Symmetry and Principal Strains: Similar to the symmetric nature of the stress matrix, the symmetry of the Green strain matrix implies that there is a coordinate system in which the strain matrix is diagonal. The principal strains are the eigenvalues of the strain matrix. If an orthonormal coordinate system aligned with the eigenvectors is chosen for the basis set, then the strain matrix is diagonal (no shear strains). In other words, in this coordinate system, *no* shear strains exist on the planes perpendicular to the basis vectors.

Behaviour under Rotations: A major advantage of the Green strain measure is that it predicts zero strains for bodies that undergo large rotations. Consider a rigid–body rotation (See section 3.1.1.b.), then the displacement function is given by:

$$u = (Q - I)X$$

In this particular case, the Green strain tensor is equal to zero:

$$\varepsilon = \frac{1}{2}((Q-I)+(Q-I)^T+(Q-I)^T(Q-I)) = \frac{1}{2}(Q+Q^T-2I+Q^TQ-Q-Q^T+I) = \hat{0}$$

Applicability: This strain measure is applicable for materials that exhibit very large strains and rotations (e.g., rubber and some biological tissue).

3.3.3.c. Other Strain Tensors

Many other strain measures can be utilized to describe the relative deformation of material points. For such strain measures to be independent of rigid–body rotations, they are defined in terms of the right (U) [or the left (V)] stretch tensor of the deformation gradient. The following are several strain measures as a function of the right stretch tensor:

$$\begin{aligned}
\textit{Hencky Strain} &= \ln U \\
\textit{Nominal Strain} &= U - I \\
\textit{Green Strain} &= \frac{1}{2}(U^2 - I) \\
\textit{General Strain Measure} &= \frac{1}{m}(U^m - I)
\end{aligned}$$

3.3.3.d. Example Problems

Problem 1: Strain Matrices in Mathematica:

The `Table` function can be used to define both the Infinitesimal and the Green strain tensors. For example, let a motion be described by:

$$\begin{pmatrix} x_1 \\ x_2 \\ x_3 \end{pmatrix} = \begin{pmatrix} 1.2 & 0.2 & 0.2 \\ 0.2 & 1.3 & 0.1 \\ 0.9 & 0.5 & 1 \end{pmatrix} \begin{pmatrix} X_1 \\ X_2 \\ X_3 \end{pmatrix}$$

Find the displacement function, the infinitesimal strain tensor, the Green strain tensor, the infinitesimal uniaxial strain along the direction of the vector $u = \{1, 1, 1\}$, and the uniaxial Green strain along the direction of the vector u.

Solution: The displacement function is:

$$\begin{pmatrix} u_1 \\ u_2 \\ u_3 \end{pmatrix} = \begin{pmatrix} 0.2 & 0.2 & 0.2 \\ 0.2 & 0.3 & 0.1 \\ 0.9 & 0.5 & 0 \end{pmatrix} \begin{pmatrix} X_1 \\ X_2 \\ X_3 \end{pmatrix}$$

The infinitesimal strain tensor is:

$$\varepsilon_{small} = \frac{1}{2}\left(\nabla u + \nabla u^T\right) = \begin{pmatrix} 0.2 & 0.2 & 0.55 \\ 0.2 & 0.3 & 0.3 \\ 0.55 & 0.3 & 0 \end{pmatrix}$$

The Green strain tensor is:

$$\varepsilon_{Green} = \frac{1}{2}\left(\nabla u + \nabla u^T + \nabla u^T \nabla u\right) = \begin{pmatrix} 0.645 & 0.475 & 0.58 \\ 0.475 & 0.49 & 0.335 \\ 0.58 & 0.335 & 0.025 \end{pmatrix}$$

Notice that the strains are dramatically different according to the strain measure. This is because the deformation is very large (Figure 3–19).

The infinitesimal strain and the Green strain along the vector u can be obtained, as follows:

$$n = \frac{u}{\|u\|} = \left\{\frac{1}{\sqrt{3}}, \frac{1}{\sqrt{3}}, \frac{1}{\sqrt{3}}\right\}$$

$$n \cdot \varepsilon_{small} n = 0.87, \quad n \cdot \varepsilon_{Green} n = 1.31$$

The following code is used for the above calculations:

```
F={{1.2,0.2,0.2},{0.2,1.3,0.1},{0.9,0.5,1}};
X={X1,X2,X3};
x=F.X;
u=x-X;
Gradu=Table[D[u[[i]],X[[j]]],{i,1,3},{j,1,3}]
einfinitesimal=1/2*(Gradu+Transpose[Gradu]);
%//MatrixForm
egreen=1/2*(Gradu+Transpose[Gradu]+Transpose[Gradu].Gradu);
%//MatrixForm
u={1,1,1};
n=u/Norm[u];
n.einfinitesimal.n
n.egreen.n
Box1=Cuboid[{0,0,0},{1,1,1}];
Box2=GeometricTransformation[Box1,F];
Graphics3D[{GrayLevel[0.5],EdgeForm[Thickness[0.01]],Box1,
GrayLevel[0.8],EdgeForm[Thickness[0.005]],Box2},
Lighting->"Neutral",Axes->True,AxesOrigin->{0,0,0},
BaseStyle->Directive[Bold,15],AxesLabel->{e1,e2,e3}]
```

Problem 2: A body undergoes a rigid–body translation in $\mathbb{R}^3$. Use Mathematica to show that the strains are equal to zero.

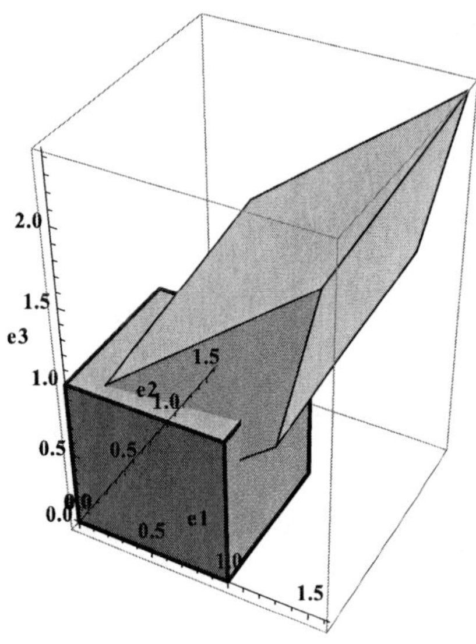

FIGURE 3-19 Large deformation of a cube with a unit side length.

Solution: The Mathematica code:

```
X={X1,X2,X3};
c={c1,c2,c3};
x=X+c;
u=x-X;
Gradu=Table[D[u[[i]],X[[j]]],{i,1,3},{j,1,3}];
einfinitesimal=1/2*(Gradu+Transpose[Gradu]);
%//MatrixForm
egreen=1/2*(Gradu+Transpose[Gradu]+Transpose[Gradu].Gradu);
%//MatrixForm
```

Problem 3: A cube of unit dimension undergoes a rigid–body rotation in $\mathbb{R}^3$ of 20°, then 20°, and then 30° around the basis vectors e_3, e_2 and e_1, respectively. Use Mathematica to visualize the box and to show that, for such large rotations, the infinitesimal strain tensor should not be used as a strain measure.

Solution: The deformation gradient is given by:

$$F = Q_{e_1} Q_{e_2} Q_{e_3} = \begin{pmatrix} 0.883 & -0.3214 & 0.3420 \\ 0.4569 & 0.7553 & -0.4698 \\ -0.1073 & 0.5712 & 0.8138 \end{pmatrix}$$

The Green strain tensor is identically zero while the small strain tensor predicts the following strains:

$$\varepsilon = \frac{1}{2}\left(\nabla u + \nabla u^T\right) = \frac{1}{2}\left(2I - F - F^T\right) = \begin{pmatrix} -0.117 & 0.068 & 0.117 \\ 0.068 & -0.245 & 0.050 \\ 0.117 & 0.051 & -0.186 \end{pmatrix}$$

Mathematica code: (Notice that the decomposition of the linear mappings is done such that $Q_{e_3} = Q_z$ appears on the right since the rotation around e_3 is applied first.)

```
Qx=RotationMatrix[thx,{1,0,0}];
Qy=RotationMatrix[thy,{0,1,0}];
Qz=RotationMatrix[thz,{0,0,1}];
thx=30Degree
thy=20Degree
thz=20Degree
F=Qx.Qy.Qz;
X={X1,X2,X3};
x=F.X;
u=x-X;
Gradu=Table[D[u[[i]],X[[j]]],{i,1,3},{j,1,3}];
einfinitesimal=1/2*(Gradu+Transpose[Gradu]);
einfinitesimal=FullSimplify[einfinitesimal];
%//MatrixForm
egreen=1/2*(Gradu+Transpose[Gradu]+Transpose[Gradu].Gradu);
egreen=FullSimplify[egreen];
%//MatrixForm
Box1=Cuboid[{0,0,0},{1,1,1}];
Box2=GeometricTransformation[Box1,F];
Graphics3D[{Box1,Box2},Lighting->"Neutral",Axes->True,AxesOrigin->{0,0,},
AxesLabel->{e1,e2,e3}]
```

Problem 4: A cube with unit length undergoes uniform shearing motion of an angle θ in the direction of e_1 and perpendicular to e_2. Find the infinitesimal and the Green strain measures that describe this motion. Estimate the value of θ at which the infinitesimal strain should not be used.

Solution: The position function is given by:

$$x = \begin{pmatrix} x_1 \\ x_2 \\ x_3 \end{pmatrix} = \begin{pmatrix} X_1 + \tan\theta \\ X_2 \\ X_3 \end{pmatrix}$$

The deformation gradient is given by:

$$F = \begin{pmatrix} 1 & \tan\theta & 0 \\ 0 & 1 & 0 \\ 0 & 0 & 1 \end{pmatrix}$$

The infintisemal strain tensor is given by:

$$\varepsilon = \begin{pmatrix} 0 & \frac{\tan\theta}{2} & 0 \\ \frac{\tan\theta}{2} & 0 & 0 \\ 0 & 0 & 0 \end{pmatrix}$$

The Green strain tensor is:

$$\varepsilon = \begin{pmatrix} 0 & \frac{\tan\theta}{2} & 0 \\ \frac{\tan\theta}{2} & \frac{\tan^2\theta}{2} & 0 \\ 0 & 0 & 0 \end{pmatrix}$$

The following code calculates the general Green and Infinitesimal strain tensors for a shearing motion, as described in section 3.1.1.e. Then it draws the cube before and after a shear of an angle 30°. Next, it solves for the value of θ at which the value of ε_{22} calculated from the Green strain tensor is equal to 0.01. The solution is $\pm 8^{\circ}$.

```
F={{1,Tan[theta],0},{0,1,0},{0,0,1}};
X={X1,X2,X3};
x=F.X;
u=x-X;
Gradu=Table[D[u[[i]],X[[j]]],{i,1,3},{j,1,3}]
einfinitesimal=1/2*(Gradu+Transpose[Gradu]);
einfinitesimal=FullSimplify[einfinitesimal];
%//MatrixForm
egreen= 1/2*(Gradu+Transpose[Gradu]+Transpose[Gradu].Gradu);
egreen=FullSimplify[egreen];
%//MatrixForm
Box1=Cuboid[{0,0,0},{1,1,1}];
Box2=GeometricTransformation[Box1,F];
theta=30Degree
Graphics3D[{Box1,Box2},Axes->True,AxesOrigin->{0,0,0},AxesLabel->{e1,e2,e3}]
s=Solve[Tan[theta2]^2/2==0.1,theta2]
(theta2/.s)/Pi*180
```

Problem 5: A displacement function produces the following state of strain inside a two–dimensional body:

$$\varepsilon = \begin{pmatrix} 0.01 & 0.012 \\ 0.012 & 0 \end{pmatrix}$$

- Sketch the shape of the deformation on an infinitesimal square inside the body.
- Find the principal strains and their directions.
- Sketch the shape of the deformation on an infinitesimal square in a coordinate system aligned with the principal directions.

Solution: The shape of the deformation of the square:

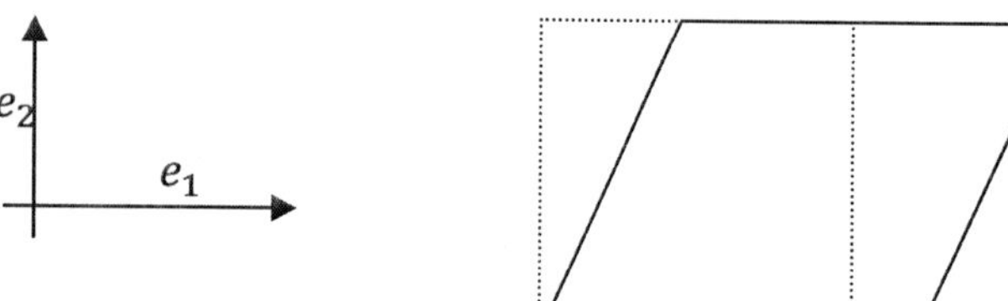

The principal strains and their directions using the Mathematica function `Eigensystem` are:

```
{{0.018,-0.008},{{-0.83205,-0.5547},{0.5547,-0.83205}}}
```

Notice that if $ev = \{-0.8305, -0.5547\}$ is an eigenvector of ε, then $-ev = \{0.8305, 0.5547\}$ is also an eigenvector of ε. Thus, a coordinate system aligned with the following two eigenvectors can be chosen to ensure a proper rotation ($\det Q = 1$):

$$e_1' = \begin{pmatrix} 0.83205 \\ 0.5547 \end{pmatrix}, \quad e_2' = \begin{pmatrix} -0.5547 \\ 0.83205 \end{pmatrix}$$

The transformation matrix is given by:

$$Q = \begin{pmatrix} 0.83205 & 0.5547 \\ -0.5547 & 0.83205 \end{pmatrix}$$

The strain matrix in the new coordinate system has the form:

$$\varepsilon' = Q\varepsilon Q^T = \begin{pmatrix} 0.018 & 0 \\ 0 & -0.008 \end{pmatrix}$$

The shape of the deformation in the new coordinate system:

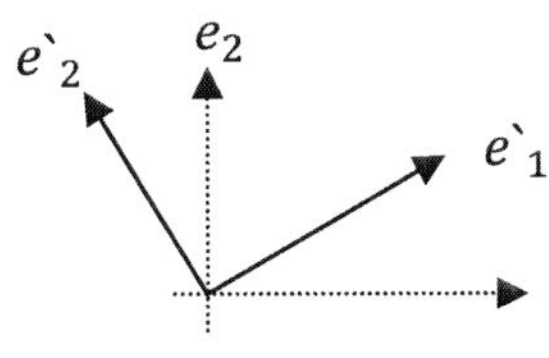

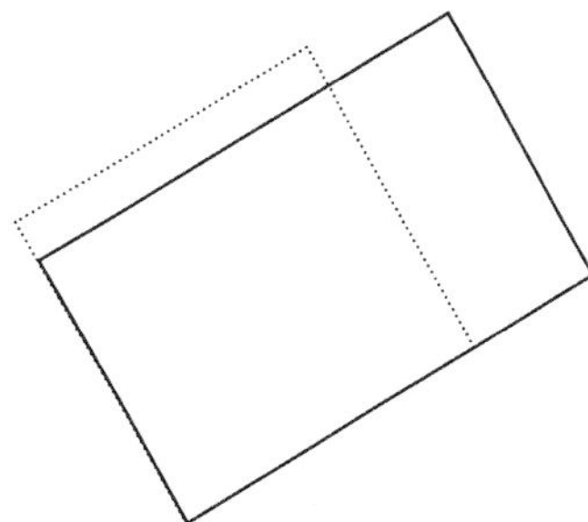

The following Mathematica code was utilized:

```
eps={{0.01,0.012},{0.012,0}};
s=Eigensystem[eps]
Q={-s[[2,1]],-s[[2,2]]}
epsdash=Chop[Q.eps.Transpose[Q]]
```

Problem 6: A displacement function of a 2–units–by–2–units plate has the form:

$$u_1 = 0.2X_1$$
$$u_2 = 0.09X_2X_1$$

- Find the deformed configuration function $x = f(X)$.
- Find the expression for the infinitesimal strain tensor as a function of the original position (X_1, X_2).
- Find the expression for the Green strain tensor as a function of the original position (X_1, X_2).
- Use Mathematica to draw the vector plot of the displacement function to visualize the plate displacements.
- Use Mathematica to draw the contour plot for the following:

$$u_1, u_2, \varepsilon_{11}^{Green}, \varepsilon_{12}^{Green}, \varepsilon_{11}^{Small}, \varepsilon_{12}^{Small}$$

X_2

2 units

X_1

2 units

Solution: The deformed configuration function:

$$x = f(X) = u + X = \begin{pmatrix} 1.2X_1 \\ (1 + 0.09X_1)\,X_2 \end{pmatrix}$$

The infinitesimal strain tensor:

$$\varepsilon = \begin{pmatrix} 0.2 & 0.045X_2 \\ 0.045X_2 & 0.09X_1 \end{pmatrix}$$

The Green strain tensor:

$$\varepsilon = \begin{pmatrix} 0.22 + 0.00405X_2^2 & (0.045 + 0.00405X_1)\,X_2 \\ (0.045 + 0.00405X_1)\,X_2 & (0.09 + 0.00405X_1)\,X_1 \end{pmatrix}$$

The two matrices produce very similar values for the strain components except for the value of ε_{11}, which can also be noticed from the plots. Examples of the required plots are shown after the Mathematica code below:

```
Clear[x1,x2,X1,X2]
X={X1,X2} ;
u={0.2X1,0.09X2*X1} ;
x=u+X;
Gradu=Table[D[u[[i]],X[[j]]],{i,1,2},{j,1,2} ]
einfinitesimal=1/2*(Gradu+Transpose[Gradu]);
egreen=1/2*(Gradu+Transpose[Gradu]+Transpose[Gradu].Gradu);
einfinitesimal//MatrixForm
FullSimplify[egreen]//MatrixForm
VectorPlot[u,{X1,0,2},{X2,0,2},VectorStyle->Black,BaseStyle->Directive[Bold,
15]AspectRatio->Automatic,PlotLabel->"u"]
ContourPlot[u[[1]],{X1,0,2},{X2,0,2},ColorFunction->"GrayTones",BaseStyle-
>Directive[Bold,15],AspectRatio->Automatic,ContourLabels->All,PlotLabel-
>"u1"]
ContourPlot[u[[2]],{X1,0,2},{X2,0,2}, ColorFunction->"GrayTones",BaseStyle-
>Directive[Bold,15],AspectRatio->Automatic,ContourLabels->All,PlotLabel-
>"u2"]
ContourPlot[einfinitesimal[[1,1]],{X1,0,2},{X2,0,2} ,ColorFunction-
>"GrayTones",BaseStyle->Directive[Bold,15],AspectRatio-
>Automatic,ContourLabels->All,PlotLabel->"einf11"]
ContourPlot[egreen[[1,1]],{X1,0,2},{X2,0,2} ,ColorFunction-
>"GrayTones",BaseStyle->Directive[Bold,15],AspectRatio-
>Automatic,ContourLabels->All,PlotLabel->"egreen11"]
ContourPlot[einfinitesimal[[1,2]],{X1,0,2},{X2,0,2} ,ColorFunction-
>"GrayTones",BaseStyle->Directive[Bold,15],AspectRatio-
>Automatic,ContourLabels->All,PlotLabel->"einf12"]
ContourPlot[egreen[[1,2]],{X1,0,2},{X2,0,2} ,ColorFunction-
```

```
>"GrayTones",BaseStyle->Directive[Bold,15],AspectRatio-
>Automatic,ContourLabels->All,PlotLabel->"egreen12"]
ContourPlot[einfinitesimal[[2,2]],{X1,0,2},{X2,0,2} ,ColorFunction-
>"GrayTones",BaseStyle->Directive[Bold,15],AspectRatio-
>Automatic,ContourLabels->All,PlotLabel->"einf22"]
ContourPlot[egreen[[2,2]],{X1,0,2},{X2,0,2} ,ColorFunction-
>"GrayTones",BaseStyle->Directive[Bold,15],AspectRatio-
>Automatic,ContourLabels->All,PlotLabel->"egreen22"]
```

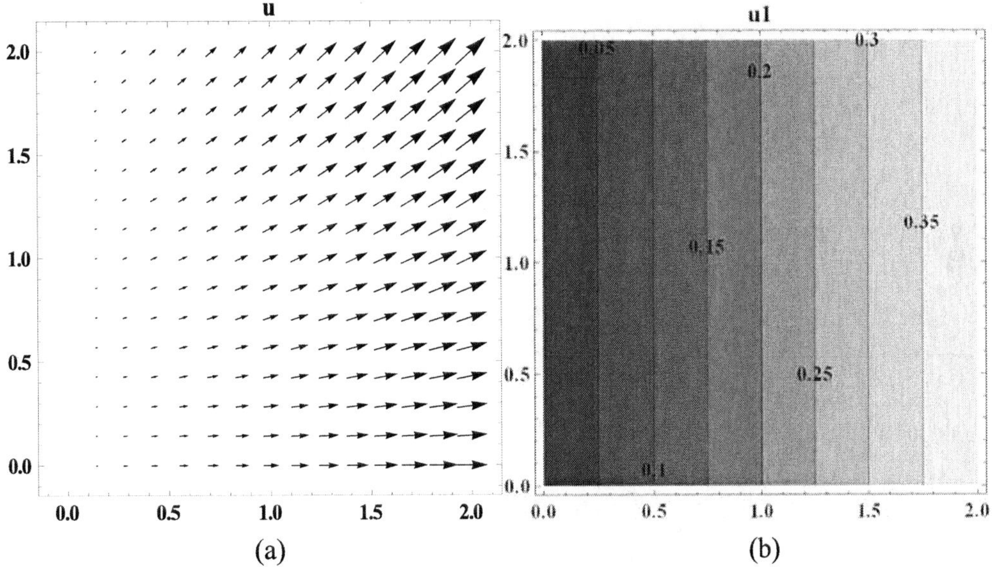

(a) (b)

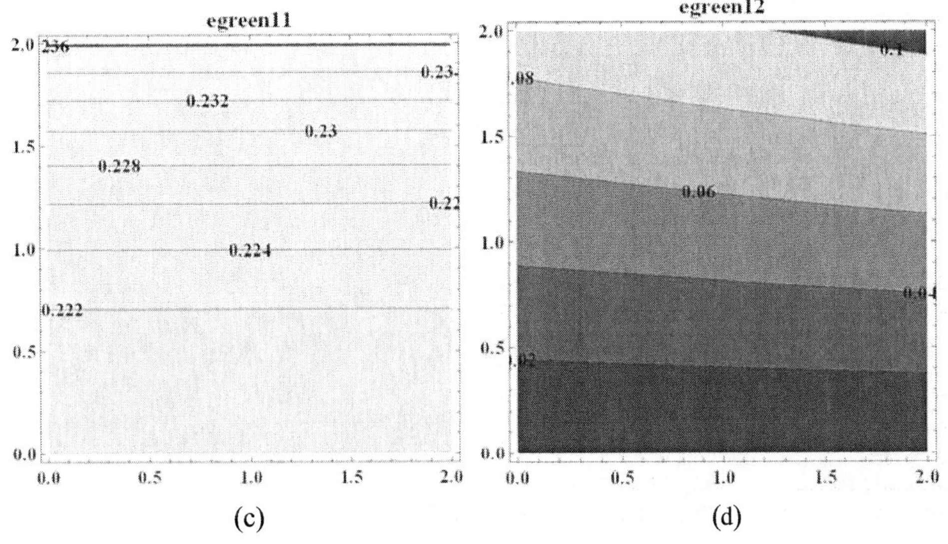

(c) (d)

Problem 7: The small strain matrix given below describes the three–dimensional state of strain in a deformation problem characterized by small strains and rotations:

$$\varepsilon = \begin{pmatrix} -0.01 & 0.002 & -0.023 \\ 0.002 & 0.05 & 0 \\ -0.023 & 0 & 0.02 \end{pmatrix}$$

Determine the following (units of meter):

- The longitudinal strain along the direction of the vector $dX = \{1, 2, 1\}$.
- The change in the cosines of the angles between the vectors $dX = \{1, 0, 1\}$, and $dY = \{1, 1, 1\}$, and comment on whether the angle between the vectors decreased on increased after deformation.
- The shear strain of planes parallel to the vector $dX = \{-1, 1, 0\}$ and perpendicular to the vector $dY = \{1, 1, 0\}$.

Solution: The longitudinal strain along the direction of the vector $dX = \{1, 2, 1\}$ can be obtained, as follows:

$$\frac{\|dx\| - \|dX\|}{\|dX\|} \approx \frac{dX \cdot \varepsilon dX}{\|dX\| \, \|dX\|}$$

The values for $\|dX\|$ and the vector εdX are:

$$\|dX\| = \sqrt{6} \text{ meters}, \quad \varepsilon dX = \begin{pmatrix} -0.01 & 0.002 & -0.023 \\ 0.002 & 0.05 & 0 \\ -0.023 & 0 & 0.02 \end{pmatrix} \begin{pmatrix} 1 \\ 2 \\ 1 \end{pmatrix} = \begin{pmatrix} -0.029 \\ 0.102 \\ -0.003 \end{pmatrix}$$

The strain along the direction of dX is then given by:

$$\frac{dX \cdot \varepsilon dX}{\|dX\| \|dX\|} = 0.0287$$

The change in the cosines of the angles between $dX = \{1, 0, 1\}$ and $dY = \{1, 1, 1\}$ is given by the formula (See section 3.3.3.a.):

$$\frac{\cos\theta - \cos\Theta}{2} \approx \frac{dX \cdot \varepsilon dY}{\|dX\| \|dY\|} = -0.0139$$

The original angle cosine ($\cos\Theta$) can be obtained from the dot product formula between the orig–inal two vectors:

$$\cos\Theta = \frac{dX \cdot dY}{\|dX\| \|dY\|} = \sqrt{\frac{2}{3}} \Longrightarrow \Theta = 0.6155^{\text{rad}} = 35.26^{\circ}$$

Therefore, the new angle between the vectors is increasing:

$$\cos\theta - \sqrt{\frac{2}{3}} = -0.0248 \Longrightarrow \theta = 0.662^{\text{rad}} = 37.93^{\circ}$$

The shear strain of planes parallel to the vector $dX = \{-1, 1, 0\}$ and perpendicular to the vector $dY = \{1, 1, 0\}$:

$$\frac{dX \cdot \varepsilon dY}{\|dX\| \, \|dY\|} = 0.03^{\text{rad}}$$

The Mathematica code utilized is:

```
Eps={{-0.01,0.002,-0.023},{0.002,0.05,0},{-0.023,0,0.02}};
dX={1,2,1};
Norm[dX]
Eps.dX
dX.Eps.dX/Norm[dX]^2
dX={1,0,1};
dY={1,1,1};
anglechange=dX.Eps.dY/Norm[dX]/Norm[dY]
THETA=N[ArcCos[dX.dY/Norm[dX]/Norm[dY]]]
THETA/Degree
theta=ArcCos[2*anglechange+Cos[THETA]]
theta/Degree
dX={-1,1,0};
dY={1,1,0};
ShearSTrain=dX.Eps.dY/Norm[dX]/Norm[dY]
```

Problem 8: The longitudinal engineering strains on the surface of a test specimen were mea– sured using a strain rosette to be 0.005, 0.002, and –0.001 along the three directions: a, b and $c \in \mathbb{R}^2$, respectively, where $a = \{1, 0\}$, $b = \{1, 1\}$, and $c = \{0, 1\}$. If the material is assumed to be in a small strain state, find the principal strains and the principal directions on the surface of the test specimen.

Solution: The state of strain on the surface of the material can be represented by the symmetric small strain matrix:

$$\varepsilon = \begin{pmatrix} \varepsilon_{11} & \varepsilon_{12} \\ \varepsilon_{21} & \varepsilon_{22} \end{pmatrix} = \begin{pmatrix} \varepsilon_{11} & \varepsilon_{12} \\ \varepsilon_{12} & \varepsilon_{22} \end{pmatrix}$$

Since the longitudinal strain is known in three directions, three equations can be written to find the components of the strain matrix:

$$\frac{a \cdot \varepsilon a}{\|a\|^2} = 0.005, \quad \frac{b \cdot \varepsilon b}{\|b\|^2} = 0.002, \quad \frac{c \cdot \varepsilon c}{\|c\|^2} = -0.001$$

The three equations can be solved to find the three unknowns:

$$\|a\|^2 = 1, \quad a \cdot \varepsilon a = \begin{pmatrix} 1 \\ 0 \end{pmatrix} \cdot \begin{pmatrix} \varepsilon_{11} \\ \varepsilon_{21} \end{pmatrix} = \varepsilon_{11} = 0.005$$

$$\|b\|^2 = 2, \quad \frac{1}{2}(b \cdot \varepsilon b) = \begin{pmatrix} 1/2 \\ 1/2 \end{pmatrix} \cdot \begin{pmatrix} \varepsilon_{11} + \varepsilon_{12} \\ \varepsilon_{12} + \varepsilon_{22} \end{pmatrix} = \frac{1}{2}\varepsilon_{11} + \varepsilon_{12} + \frac{1}{2}\varepsilon_{22} = 0.002$$

$$\|c\|^2 = 1, \quad c \cdot \varepsilon c = \varepsilon_{22} = -0.001$$

Therefore, the strain matrix is:

$$\varepsilon = \begin{pmatrix} 0.005 & 0 \\ 0 & -0.001 \end{pmatrix}$$

Thus, the principal strains are 0.005 and -0.001, while the principal directions are a and c. The following Mathematica code is used to solve the above equations.

```
eps=Table[Subscript[e,i,j],{i,1,2},{j,1,2}]
a={1,0};
b={1,1};
c={0,1};
a.eps.a/Norm[a]^2
b.eps.b/Norm[b]^2
c.eps.c/Norm[c]^2
Solve[{a.eps.a/Norm[a]^2==0.005,b.eps.b/Norm[b]^2==0.002,c.eps.c/Norm[c]^2==
-0.001,eps[[1,2]]==eps[[2,1]]},Flatten[eps]]
```

Problem 9: The longitudinal engineering strains on the surface of a test specimen were measured using a strain rosette to be 0.005, 0.002 and –0.001 along the three directions: a, b and $c \in \mathbb{R}^2$, respectively, where $a = \{\cos 30^\circ, \sin 30^\circ\}$, $b = \{0, 1\}$ and $c = \{-\cos 30^\circ, \sin 30^\circ\}$. If the material is assumed to be in a small strain state, find the principal strains and the principal directions on the surface of the test specimen.

Solution: The state of strain on the surface of the material can be represented by the symmetric small strain matrix:

$$\varepsilon = \begin{pmatrix} \varepsilon_{11} & \varepsilon_{12} \\ \varepsilon_{21} & \varepsilon_{22} \end{pmatrix} = \begin{pmatrix} \varepsilon_{11} & \varepsilon_{12} \\ \varepsilon_{12} & \varepsilon_{22} \end{pmatrix}$$

Since the longitudinal strain is known in three directions, three equations can be written to find the components of the strain matrix:

$$\frac{a \cdot \varepsilon a}{\|a\|^2} = 0.005, \quad \frac{b \cdot \varepsilon b}{\|b\|^2} = 0.002, \quad \frac{c \cdot \varepsilon c}{\|c\|^2} = -0.001$$

The three equations can be solved to find the three unknowns:

$$\|a\|^2 = 1, \quad a \cdot \varepsilon a = \frac{1}{4}\left(3\varepsilon_{11} + 2\sqrt{3}\varepsilon_{12} + \varepsilon_{22}\right) = 0.005$$

$$\|b\|^2 = 1, \quad b \cdot \varepsilon b = \begin{pmatrix} 0 \\ 1 \end{pmatrix} \cdot \begin{pmatrix} \varepsilon_{12} \\ \varepsilon_{22} \end{pmatrix} = \varepsilon_{22} = 0.002$$

$$\|c\|^2 = 1, \quad c \cdot \varepsilon c = \frac{1}{4}\left(3\varepsilon_{11} - 2\sqrt{3}\varepsilon_{12} + \varepsilon_{22}\right) = -0.001$$

Solving the above three equations in the three unknown strain entries yields:

$$\varepsilon = \begin{pmatrix} \varepsilon_{11} & \varepsilon_{12} \\ \varepsilon_{12} & \varepsilon_{22} \end{pmatrix} = \begin{pmatrix} 0.002 & 0.0034641 \\ 0.0034641 & 0.002 \end{pmatrix}$$

The eigenvalues of ε are: 0.0054641 and –0.0014641, which are the principal strains of ε. The corresponding eigenvectors (principal directions) are:

$$ev1 = \left\{\frac{1}{\sqrt{2}}, \frac{1}{\sqrt{2}}\right\}, \quad ev2 = \left\{-\frac{1}{\sqrt{2}}, \frac{1}{\sqrt{2}}\right\}$$

The following Mathematica code is used to solve the above equations.

```
eps=Table[Subscript[e,i,j],{i,1,2},{j,1,2}]
a={Cos[30Degree],Sin[30Degree]};
b={0,1};
c={-Cos[30Degree],Sin[30Degree]};
FullSimplify[a.eps.a/Norm[a]^2]
FullSimplify[b.eps.b/Norm[b]^2]
FullSimplify[c.eps.c/Norm[c]^2]
a=Solve[{a.eps.a/Norm[a]^2==0.005,b.eps.b/Norm[b]^2==0.002,c.eps.c/Norm[c]^2
==-0.001,eps[[1,2]]==eps[[2,1]]},Flatten[eps]]
eps=eps/.a[[1]];
Eigensystem[eps]//MatrixForm
```

3.4. INSTANTANEOUS MEASURES OF DEFORMATION: THE VELOCITY GRADIENT

In the previous section, it was shown that the deformation gradient contains all the information regarding rotation and stretching of material vectors when comparing the undeformed configuration with a single deformed configuration. This section, however, will be concerned with deformations that are evolving with time and, thus, the instantaneous measures of deformation will be investigated to identify the speed of spinning and/or stretching of material vectors as time evolves. To study such measures, the velocity of material points, as well as the velocity gradient, will be defined.

3.4.1. Definitions

Let Ω_0 and Ω represent the reference and the deformed configuration of a body embedded in $\mathbb{R}^3$. Let $x(X,t) \in \Omega$ represent the image of $X \in \Omega_0$ at a certain point in time $t \in \mathbb{R}$ under the bijective smooth mapping $f_t:\Omega_0 \to \Omega$. The set Ω, in this case, is a function of time. Then, the velocity vector v at each material point is the rate of change of the position x with respect to time:

$$v = \frac{\partial x}{\partial t} = \begin{pmatrix} \frac{\partial x_1}{\partial t} \\ \frac{\partial x_2}{\partial t} \\ \frac{\partial x_3}{\partial t} \end{pmatrix} = \begin{pmatrix} v_1 \\ v_2 \\ v_3 \end{pmatrix}$$

The gradient of the velocity field with respect to the material coordinates X_1, X_2 and X_3 is denoted by Grad v and is equal to:

$$\text{Grad}\, v = \begin{pmatrix} \frac{\partial v_1}{\partial X_1} & \frac{\partial v_1}{\partial X_2} & \frac{\partial v_1}{\partial X_3} \\ \frac{\partial v_2}{\partial X_1} & \frac{\partial v_2}{\partial X_2} & \frac{\partial v_2}{\partial X_2} \\ \frac{\partial v_3}{\partial X_1} & \frac{\partial v_3}{\partial X_2} & \frac{\partial v_3}{\partial X_3} \end{pmatrix} = \dot{F}$$

The gradient of the velocity field (the velocity gradient) with respect to the spatial coordinates x_1, x_2 and x_3 is denoted by L or grad v:

$$L = \text{grad}\, v = \begin{pmatrix} \frac{\partial v_1}{\partial x_1} & \frac{\partial v_1}{\partial x_2} & \frac{\partial v_1}{\partial x_3} \\ \frac{\partial v_2}{\partial x_1} & \frac{\partial v_2}{\partial x_2} & \frac{\partial v_2}{\partial x_3} \\ \frac{\partial v_3}{\partial x_1} & \frac{\partial v_3}{\partial x_2} & \frac{\partial v_3}{\partial x_3} \end{pmatrix}$$

The velocity gradient is related to the deformation gradient through the relationship:

$$L_{ij} = \frac{\partial v_i}{\partial x_j} = \frac{\partial v_i}{\partial X_k}\frac{\partial X_k}{\partial x_j} \Longrightarrow L = \dot{F}F^{-1}$$

Unlike the displacement gradient, the velocity gradient is instantaneous and is evaluated with respect to the coordinates of the deformed configuration. The **stretching tensor** D and the **spin tensor** W are the symmetric and the skew–symmetric parts of the velocity gradient L:

$$D = \frac{1}{2}\left(L + L^T\right), \quad W = \frac{1}{2}\left(L - L^T\right)$$

3.4.2. Physical Interpretation of the Velocity Gradient and its Decomposition

Let a material curve $dx \in \mathbb{R}^3$ in the deformed configuration be the image of the material curve $dX \in \mathbb{R}^3$ in the reference configuration. Assuming that the deformation is evolving in time, then the rate of change of dx with respect to time is given by:

$$\dot{dx} = \dot{F}dX = \dot{F}F^{-1}dx = Ldx$$

Therefore, the velocity gradient L when applied to a material curve dx gives the vector $\dot{dx}$ representing the rate of change of the norm (magnitude) and the direction of the material curve dx. The stretch tensor D is a symmetric tensor and gives the velocity of the stretching of the material vectors aligned with its principal directions, while the spin tensor W gives the velocity of rotation of material vectors. The magnitude of the stretching and the magnitude of the rotation can be obtained by integrating over time or multiplying by a small increment in time. The physical interpretation of both follow that of the displacement gradient decompositions in section 3.2.2 with the slight difference in the fact that the velocity gradient is a truly instantaneous rate of change of lengths and angles and is evaluated with respect to the spatial (deformed) coordinates.

Problem 1: A body is undergoing a rotation around e_3 with a frequency of $f = 1\text{Hz}$. Show that, at any time during the rotation, the stretch tensor D is zero, while the spin tensor W is not.

Solution: The position function is equal to:

$$x = QX$$

where, Q is the rotation tensor around e_3 with an angle $\theta = 2\pi f t = 2\pi t$.

The deformation gradient is:

$$F = Q = \begin{pmatrix} \cos 2\pi t & -\sin 2\pi t & 0 \\ \sin 2\pi t & \cos 2\pi t & 0 \\ 0 & 0 & 1 \end{pmatrix}$$

The velocity gradient is equal to:

$$L = \dot{F}F^{-1} = \begin{pmatrix} 0 & -2\pi & 0 \\ 2\pi & 0 & 0 \\ 0 & 0 & 0 \end{pmatrix}$$

The symmetric part of the velocity gradient is zero and the skew–symmetric part is equal to L:

$$D = \begin{pmatrix} 0 & 0 & 0 \\ 0 & 0 & 0 \\ 0 & 0 & 0 \end{pmatrix}, \quad W = L$$

By comparing this problem to Problem 1 in section 3.2.2.c, it can be noticed that the velocity gradient is a better measure for large displacements compared to the displacement gradient since the stretching is always zero (rigid–body rotation), and the spin tensor gives the exact value of the speed of rotation.

The following is the Mathematica code used in this problem:

```
Clear[t,x1,x2,x3,X1,X2,X3]
X={X1,X2,X3};
Q=RotationMatrix[t*2*Pi,{0,0,1}];
Q//MatrixForm
x=Q.X;
F=Table[D[x[[i]],X[[j]]],{i,1,3},{j,1,3}];
Finv=FullSimplify[Inverse[F]];
Fdot=FullSimplify[D[F,t]];
L=FullSimplify[Fdot.Finv];
L//MatrixForm
Dd=FullSimplify[1/2*(L+Transpose[L])]
W=FullSimplify[1/2*(L-Transpose[L])]
```

Problem 2: Show that for any rigid–body rotation, the stretch tensor D is zero, while the spin tensor W gives the rate of rotation.

Solution: This example is just a generalization to the previous example. A rigid–body rotation is described by the function:

$$x = QX$$

Therefore, the deformation gradient is equal to the rotation matrix Q, and the following equalities are obtained for $\dot{F}$, L, D and W:

$$F = Q, \quad F^{-1} = Q^T, \quad \dot{F} = \dot{Q}, \quad L = \dot{Q}Q^T, \quad D = \frac{1}{2}\left(\dot{Q}Q^T + Q\dot{Q}^T\right), \quad W = \frac{1}{2}\left(\dot{Q}Q^T - Q\dot{Q}^T\right)$$

However, a rotation matrix always satisfies:

$$QQ^T = I$$

Taking the time derivative of both sides:

$$\frac{d(QQ^T)}{dt} = \hat{0} \Longrightarrow \dot{Q}Q^T + Q\dot{Q}^T = \hat{0}$$

Therefore:

$$D = \hat{0}, \quad W = L$$

Problem 3: Show that the simple shearing motion parallel to e_1 and perpendicular to e_2 with a shear angle θ where $\tan\theta = t/100$ constitutes:

- a clockwise rotation of an angular velocity 1/200 rad/second
- an extension in the directions of $u = \{1, 1, 0\}$ with a speed of 1/200 units of length/second
- a contraction in the directions of $u = \{-1, 1, 0\}$ with a speed of 1/200 units of length/second

Solution: The position function is:

$$x = \begin{pmatrix} x_1 \\ x_2 \\ x_3 \end{pmatrix} = \begin{pmatrix} X_1 + \frac{t}{100} X_2 \\ X_2 \\ X_3 \end{pmatrix}$$

The different measures of deformation are:

$$F = \begin{pmatrix} 1 & \frac{t}{100} & 0 \\ 0 & 1 & 0 \\ 0 & 0 & 1 \end{pmatrix}, \quad F^{-1} = \begin{pmatrix} 1 & \frac{-t}{100} & 0 \\ 0 & 1 & 0 \\ 0 & 0 & 1 \end{pmatrix}, \quad \dot{F} = \begin{pmatrix} 0 & \frac{1}{100} & 0 \\ 0 & 0 & 0 \\ 0 & 0 & 0 \end{pmatrix}, \quad L = \begin{pmatrix} 0 & \frac{1}{100} & 0 \\ 0 & 0 & 0 \\ 0 & 0 & 0 \end{pmatrix}$$

$$D = \begin{pmatrix} 0 & \frac{1}{200} & 0 \\ \frac{1}{200} & 0 & 0 \\ 0 & 0 & 0 \end{pmatrix}, \quad W = \begin{pmatrix} 0 & \frac{1}{200} & 0 \\ \frac{-1}{200} & 0 & 0 \\ 0 & 0 & 0 \end{pmatrix}$$

The axial vector of W is $w = \left\{0, 0, -\frac{1}{200}\right\}$ and, thus, this constitutes a clockwise rotation of angular velocity 1/200 rad/sec. The eigenvalues and the corresponding eigenvectors of the stretch tensor are:

$$\lambda = \left\{\frac{1}{200}, \frac{-1}{200}, 0\right\}, \quad ev1 = \begin{pmatrix} 1 \\ 1 \\ 0 \end{pmatrix}, \quad ev2 = \begin{pmatrix} -1 \\ 1 \\ 0 \end{pmatrix}, \quad ev3 = \begin{pmatrix} 0 \\ 0 \\ 1 \end{pmatrix}$$

This indicates that vectors in the direction of $ev1$ and $ev2$ will have velocities of stretch equal to λ_1 and λ_2 respectively:

$$(\dot{ev1}) = Lev1 = Dev1 + Wev1 = \frac{1}{200}ev1 + Wev1$$

$$(\dot{ev2}) = Lev2 = Dev2 + Wev2 = -\frac{1}{200}ev2 + Wev2$$

Thus, the stretch parts of $(\dot{ev1})$ and $(\dot{ev2})$ are $\frac{1}{200}ev1$ and $-\frac{1}{200}\,ev2$ respectively

The Mathematica code used is:

```
Clear[t,x1,x2,x3,X1,X2,X3]
X={X1,X2,X3};
x={x1,x2,x3};
x1=X1+t/100*X2;
x2=X2;
x3=X3;
F=Table[D[x[[i]],X[[j]]],{i,1,3},{j,1,3}];
Finv=FullSimplify[Inverse[F]];
Fdot=FullSimplify[D[F,t]];
L=FullSimplify[Fdot.Finv];
L//MatrixForm
Dd=FullSimplify[1/2*(L+Transpose[L])];
Eigensystem[Dd]
W=FullSimplify[1/2*(L-Transpose[L])];
W//MatrixForm
```

Problem 4: Denoting the determinant of F by J, show that:

$$\frac{DJ}{Dt} = J \operatorname{trace} L$$

Solution: Using the components of F, J is given by:

$$J = \varepsilon_{ijk}\frac{\partial x_1}{\partial X_i}\frac{\partial x_2}{\partial X_j}\frac{\partial x_3}{\partial X_k}$$

Then, $\frac{DJ}{Dt}$ can be written as:

$$\begin{aligned}
\frac{DJ}{Dt} &= \varepsilon_{ijk}\left(\frac{\partial \dot{x}_1}{\partial X_i}\frac{\partial x_2}{\partial X_j}\frac{\partial x_3}{\partial X_k} + \frac{\partial x_1}{\partial X_i}\frac{\partial \dot{x}_2}{\partial X_j}\frac{\partial x_3}{\partial X_k} + \frac{\partial x_1}{\partial X_i}\frac{\partial x_2}{\partial X_j}\frac{\partial \dot{x}_3}{\partial X_k}\right) \\
&= \varepsilon_{ijk}\left(\left(\frac{\partial \dot{x}_1}{\partial x_1}\frac{\partial x_1}{\partial X_i} + \frac{\partial \dot{x}_1}{\partial x_2}\frac{\partial x_2}{\partial X_i} + \frac{\partial \dot{x}_1}{\partial x_3}\frac{\partial x_3}{\partial X_i}\right)\frac{\partial x_2}{\partial X_j}\frac{\partial x_3}{\partial X_k} + \frac{\partial x_1}{\partial X_i}\left(\frac{\partial \dot{x}_2}{\partial x_1}\frac{\partial x_1}{\partial X_j} + \frac{\partial \dot{x}_2}{\partial x_2}\frac{\partial x_2}{\partial X_j} + \frac{\partial \dot{x}_2}{\partial x_3}\frac{\partial x_3}{\partial X_j}\right)\frac{\partial x_3}{\partial X_k}\right. \\
&\quad \left. + \frac{\partial x_1}{\partial X_i}\frac{\partial x_2}{\partial X_j}\left(\frac{\partial \dot{x}_3}{\partial_1}\frac{\partial x_1}{\partial X_k} + \frac{\partial \dot{x}_3}{\partial x_2}\frac{\partial x_2}{\partial X_k} + \frac{\partial \dot{x}_3}{\partial x_3}\frac{\partial x_3}{\partial X_k}\right)\right)
\end{aligned}$$

However, notice that expressions of the following form are equal to zero from the properties of the determinant:

$$\frac{\partial \dot{x}_1}{\partial x_2}\varepsilon_{ijk}\left(\frac{\partial x_2}{\partial X_i}\frac{\partial x_2}{\partial X_j}\frac{\partial x_3}{\partial X_k}\right)=0$$

Therefore,

$$\begin{aligned}\frac{DJ}{Dt}&=\varepsilon_{ijk}\left(\left(\frac{\partial \dot{x}_1}{\partial x_1}\frac{\partial x_1}{\partial X_i}\right)\frac{\partial x_2}{\partial X_j}\frac{\partial x_3}{\partial X_k}+\frac{\partial x_1}{\partial X_i}\left(\frac{\partial \dot{x}_2}{\partial x_2}\frac{\partial x_2}{\partial X_j}\right)\frac{\partial x_3}{\partial X_k}+\frac{\partial x_1}{\partial X_i}\frac{\partial x_2}{\partial X_j}\left(\frac{\partial \dot{x}_3}{\partial x_3}\frac{\partial x_3}{\partial X_k}\right)\right)\\&=\varepsilon_{ijk}\frac{\partial x_1}{\partial X_i}\frac{\partial x_2}{\partial X_j}\frac{\partial x_3}{\partial X_k}\left(\frac{\partial \dot{x}_1}{\partial x_1}+\frac{\partial \dot{x}_2}{\partial x_2}+\frac{\partial \dot{x}_3}{\partial x_3}\right)\\&=J\operatorname{trace}L\end{aligned}$$

The above proof relied on explicitly using the components of F in a certain coordinate system. However, the results of section 1.2.6.e can be used to obtain the same result. Let a, b and c be three arbitrary, linearly independent vectors in $\mathbb{R}^3$ and then:

$$\begin{aligned}\frac{DJ}{Dt}&=\frac{D\frac{(Fa\cdot(Fb\times Fc))}{a\cdot(b\times c)}}{Dt}=\frac{\dot{F}a\cdot(Fb\times Fc)+Fa\cdot(\dot{F}b\times Fc)+Fa\cdot(Fb\times\dot{F}c)}{a\cdot(b\times c)}\\&=\frac{\dot{F}F^{-1}Fa\cdot(Fb\times Fc)+Fa\cdot(\dot{F}F^{-1}Fb\times Fc)+Fa\cdot(Fb\times\dot{F}F^{-1}Fc)}{a\cdot(b\times c)}\\&=\frac{\dot{F}F^{-1}Fa\cdot(Fb\times Fc)+Fa\cdot(\dot{F}F^{-1}Fb\times Fc)+Fa\cdot(Fb\times\dot{F}F^{-1}Fc)}{Fa\cdot(Fb\times Fc)}\frac{(Fa\cdot(Fb\times Fc))}{a\cdot(b\times c)}\\&=J\operatorname{trace}\dot{F}F^{-1}=J\operatorname{trace}L\end{aligned}$$

Problem 5: Show that an isochoric motion has a velocity gradient whose trace is equal to zero.

Solution: An isochoric motion is a motion that characterizes incompressible materials. Such materials deform so that the local volume in the deformed configuration is always equal to the local volume in the reference configuration. For such motion, the determinant of F is always constant and is equal to 1. Denoting the determinant of F by J, we can write:

$$J=\det F=1\Longrightarrow\frac{DJ}{Dt}=0$$

Let $v=\{v_1,v_2,v_3\}=\{\dot{x}_1,\dot{x}_2,\dot{x}_3\}$ be the velocity vector of a material particle, and then, using the results from the previous problem, the trace of the velocity gradient is equal to zero:

$$\frac{DJ}{Dt}=J\operatorname{trace}L=\operatorname{trace}L=\frac{\partial v_1}{\partial x_1}+\frac{\partial v_2}{\partial x_2}+\frac{\partial v_3}{\partial x_3}=0$$

The above formula can be found in most fluid mechanics books to describe the incompressibility of a fluid.

3.5. MATERIAL-TIME DERIVATIVE

It is often required to quantify the derivatives of scalar–, vector–, and tensor–valued fields on a set representing the configuration of a body embedded in $\mathbb{R}^3$. However, certain difficulties arise when trying to compute the time derivative of different fields when viewing a deformed configuration since the coordinates of each point vary in time as well. There are two different time derivatives of any field, the spatial and the material–time derivative. The spatial–time derivative can be used to track the value of a field at a fixed position in space (at a specific vector in $\mathbb{R}^3$). The material–time derivative, on the other hand, can be used to track the change of a field associated with a material point that could be moving in space. For example, imagine the steady state flow of a fluid flowing through a tube with a varying cross–sectional area. Since the fluid is in a steady state, the velocity at each point inside the tube is constant. However, the fluid particles themselves may exhibit accelerations. So, while the spatial time derivative of the velocity in this particular example is zero, the material–time derivative of the velocity vector is not.

Let Ω_0 and Ω represent the reference and the deformed configuration of a body embedded in $\mathbb{R}^3$. Let $x(X, t) \in \Omega$ represent the image of $X \in \Omega_0$ at a certain point in time $t \in \mathbb{R}$ under the bijective mapping $f_t : \Omega_0 \to \Omega$. Let $\phi : \Omega \times \mathbb{R} \to \mathbb{R}$ be a scalar–valued field that is a function of the position inside Ω and of the time $t \in \mathbb{R}$. The material–time derivative of ϕ, represented by $\dot{\phi}$, is the rate of change of ϕ with respect to t while tracking the material point (while keeping the value of the vector X constant). The material–time derivative of ϕ is given by:

$$\frac{d\phi}{dt} = \dot{\phi} = \frac{\partial \phi(x, t)}{\partial t} + \frac{\partial \phi(x, t)}{\partial x}\frac{\partial x}{\partial t} = \frac{\partial \phi(x, t)}{\partial t} + \operatorname{grad}\phi \cdot v$$

In component form, this can be written as:

$$\frac{d\phi}{dt} = \dot{\phi} = \frac{\partial \phi(x, t)}{\partial t} + \sum_{i=1}^{3} \frac{\partial \phi(x_1, x_2, x_3, t)}{\partial x_i}\frac{\partial x_i}{\partial t}$$

Note:

The material–time derivative is important for quantities that are expressed as functions of the coordinates in the deformed configuration. If a field is a function in the coordinates of the ref–erence configuration, then the material–time derivative is the partial derivative of the quantity in question with respect to time. So, if $\phi : \Omega_0 \times \mathbb{R} \to \mathbb{R}$ is a scalar–valued function of the position in the reference coordinates and in time, then the material–time derivative is simply:

$$\frac{D\phi}{Dt} = \dot{\phi} = \frac{\partial \phi(X, t)}{\partial t} + \sum_{i=1}^{3} \frac{\partial \phi(X_1, X_2, X_3, t)}{\partial X_i}\frac{\partial X_i}{\partial t} = \frac{\partial \phi(X, t)}{\partial t}$$

(The capital D in $\frac{D\phi}{Dt}$ is used to indicate that the quantity ϕ is a function of the coordinates in the reference configuration.)

Problem 1: A tube has a varying cross–sectional area $A = (1 + x)$ in units of square meter situated from $x = 0$ to $x = L$ meter. The horizontal velocity of the fluid entering the tube is constant at $v = 1$ meter/sec. The fluid is incompressible. Show the spatial distribution of the horizontal acceleration is equal to zero, and find the acceleration of a fluid particle situated at a point x.

Solution: The spatial distribution of the velocity in the tube can be obtained by using the fact that the fluid is incompressible, which means that the discharge q is always constant:

$$q = Av = (1 + x)v = constant$$

The discharge at entry ($x = 0$) is:

$$q = 1\text{m}^3/\text{s}$$

The velocity at any point inside the tube can then be obtained, as follows:

$$q = 1 = (1 + x)\, v \Longrightarrow v = \frac{1}{1 + x}\text{meter/sec}$$

The spatial acceleration is obtained by taking the partial derivative of the velocity with respect to time:

$$\frac{\partial v}{\partial t} = 0$$

However, the material acceleration (material–time derivative of the velocity) is obtained by tracking the particles, and thus:

$$a = \frac{dv}{dt} = \dot{v} = \frac{\partial v}{\partial t} + \frac{\partial v}{\partial x}\frac{\partial x}{\partial t} = -\frac{1}{(1 + x)^3}$$

Naturally, there is a negative acceleration since the tube increases in size downstream, and thus, the particles' velocities decrease.

Problem 2: The plate shown below has dimensions 4×2, units and is rotating around the origin through a medium with a speed of $\omega = \frac{2\pi}{1000}$ rad/second. Assume that the spatial distribution of the temperature in the medium is given by $T = 70(1 - \frac{x_1}{5})(1 - \frac{x_2}{10})^{\circ}C$. Assuming that the speed of the plate is slow enough such that the plate assumes the temperature of the medium, draw the contour plot of the material–time derivative of the temperature on the plate at $t = 0$.

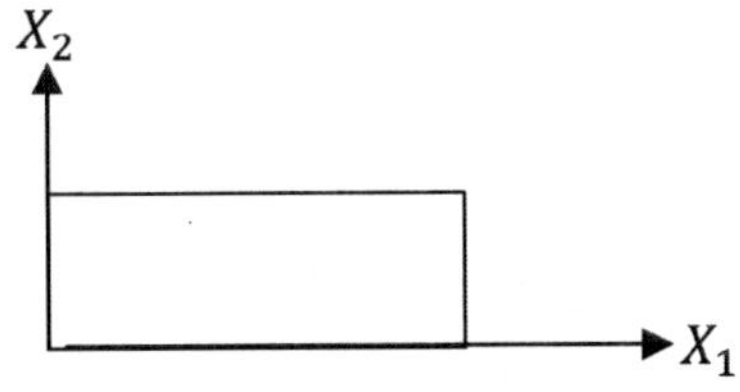

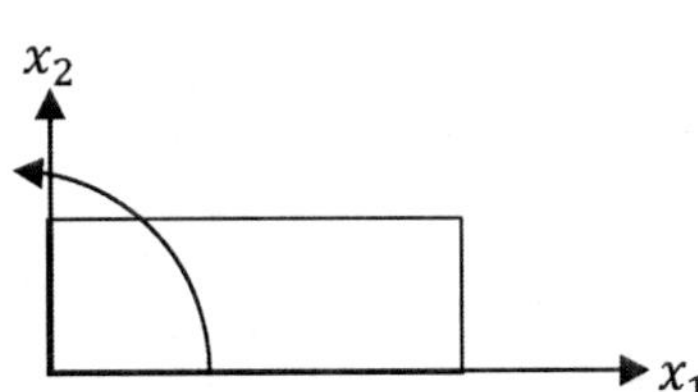

Solution: The position function at time t is:

$$x = \begin{pmatrix} x_1 \\ x_2 \end{pmatrix} = QX = \begin{pmatrix} \cos\omega t & -\sin\omega t \\ \sin\omega t & \cos\omega t \end{pmatrix} \begin{pmatrix} X_1 \\ X_2 \end{pmatrix}$$

where $\omega = 2\pi/1000$

The spatial velocity distribution is given by:

$$v = \dot{Q}X = \dot{Q}Q^{-1}x = \begin{pmatrix} -\dfrac{\pi x_2}{500} \\ -\dfrac{\pi x_1}{500} \end{pmatrix}$$

The spatial gradient of the temperature is given by:

$$\text{grad } T = \begin{pmatrix} -14\left(1 - \dfrac{x_2}{10}\right) \\ -7\left(1 - \dfrac{x_1}{5}\right) \end{pmatrix}$$

The material–time derivative of T is given by:

$$\frac{dT}{dt} = \frac{\partial T}{\partial t} + \frac{\partial T}{\partial x} \cdot \frac{\partial x}{\partial t} = 0 + \text{grad } T \cdot v$$

At $t = 0$:

$$\frac{dT}{dt} = \frac{7\pi\left((x_1 - 5)x_1 - x_2(x_2 - 10)\right)}{2500}\ {}^{\circ}\text{C/sec}$$

Figure 3–20 shows the velocity vector, the spatial gradient of temperature, and the contour plot of the material–time derivative of the temperature at $t = 0$. The following is the Mathematica code used:

```
x={x1,x2};
T=70*(1-x1/5)*(1-x2/10);
gradT={D[T,x1],D[T,x2]};
Q=RotationMatrix[t*2*Pi/1000];
Qn1=Inverse[Q];
Qdot=D[Q,t];
v=Qdot.Qn1.x;
VectorPlot[v/.t->0,{x1,0,4},{x2,0,2},VectorStyle->Black,BaseStyle-
>Directive[Bold, 15]AspectRatio->Automatic,PlotLabel->"v"]
VectorPlot[gradT,{x1,0,4} ,{x2,0,2}, VectorStyle->Black,BaseStyle-
>Directive[Bold, 15]AspectRatio->Automatic,PlotLabel->"gradT"]
ContourPlot[v.gradT/.t->0,{x1,0,4},{x2,0,2},ColorFunction->"GrayTones"
BaseStyle->Directive[Bold,15],ContourLabels->All,AspectRatio-
>Automatic,PlotLabel->"dTemp/dt"]
```

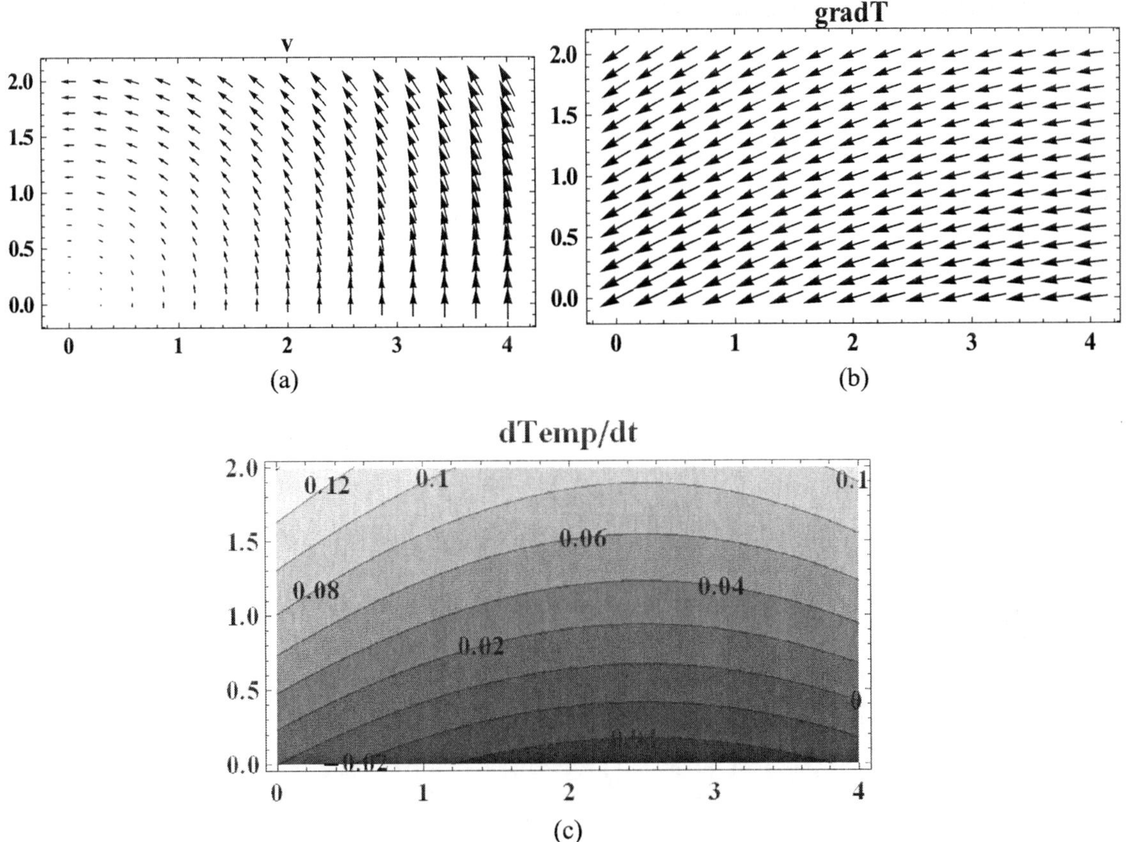

FIGURE 3-20 (a) The velocity vector at $t = 0$, (b) the spatial gradient of the temperature, and (c) the material–time derivative of the temperature at $t = 0$.

New Mathematica Functions Explored in chapter 3

```
Flatten[]
ShearingMatrix[]
Subscript[]
```

▶ EXERCISES

(Answers are given in round brackets for some problems.):

1. Which of the following functions describing motion is (are) physically possible?
   - $x_1 = X_1, \quad x_2 = X_2, \quad x_3 = -X_3$
   - $x_1 = X_1, \quad x_2 = X_2, \quad x_3 = X_2$
   - $x_1 = X_1 + X_2, \quad x_2 = X_2, \quad x_3 = X_3$
2. Let a cube of unit length be represented by the set of vectors $\Omega_0 \subset R^3$ such that three of the cube's sides are aligned with the three orthonormal basis set vectors $e_1 e_2$ and e_3, and

one of the cube vertices lies on the origin of the coordinate system. Let $\Omega \subset \mathbb{R}^3$ be the deformed configuration such that $f{:}X \in \Omega_0 \longmapsto x \in \Omega, x = f(X)$ is smooth, differentiable and bijective, defined as:

$$x_1 = 1.1X_1 + 0.02X_1^2 + 0.01X_2 + 0.03X_3$$
$$x_2 = 0.001X_1 + 0.9X_2 + 0.003X_3$$
$$x_3 = 0.001X_1 + 0.005X_2 + 0.009X_3^2 + 0.9X_3$$

Find the following:

- The displacement function u.
- The new position of the 8 vertices of the cube.
- The deformed curves of three sides of the cube that are aligned with the basis vector e_1, e_2, and e_3.
- The infinitesimal strain matrix and the volumetric strain as a function of the position inside the cube.
- The Green strain matrix as a function of the position inside the cube.
- Evaluate the infinitesimal strain matrix at two of the eight cube vertices.
- Evaluate the Green strain matrix at two of the eight cube vertices.

3. A cube undergoes a deformation such that the infinitesimal strain is described by the matrix:

$$\varepsilon = \begin{pmatrix} 0.015 & 0.005 & 0 \\ 0.005 & -0.002 & 0 \\ 0 & 0 & 0 \end{pmatrix}$$

Find the following:

- The principal strains and the directions of the principal strains.
- The coordinate system transformation in which the strain matrix is diagonal and express the strain matrix in the new coordinate system.
- The longitudinal strain along the direction of the vector $dX = \{1, 1, 1\}$.
- The shear strain of planes perpendicular to dX and parallel to $dY = \{-1, 1, 0\}$, and describe the approximate change in the angles between the vectors dX and $dZ = \{-1, 1, 1\}$.

(**Answer:** $\varepsilon_1 = 0.016, \varepsilon_2 = -0.0034, \varepsilon_3 = 0, Q = \{\{0.965, 0.263, 0\}, \{-0.263, 0.965, 0\}, \{0, 0, 1\}\}$)

4. The longitudinal engineering strains on the surface of a test specimen were measured using a strain rosette to be 0.005, 0.002, and –0.001 along the three directions: a, b and

$c \in \mathbb{R}^2$, respectively (see figure). If the material is assumed to be in a small strain state, find the principal strains and the principal directions on the surface of the test specimen.

(**Answer:** $\varepsilon_1 = 0.006745$, $\varepsilon_2 = -0.005745$, $Q = \{\{0.927517, 0.37378\}, \{0.37378, -0.927517\}\}$)

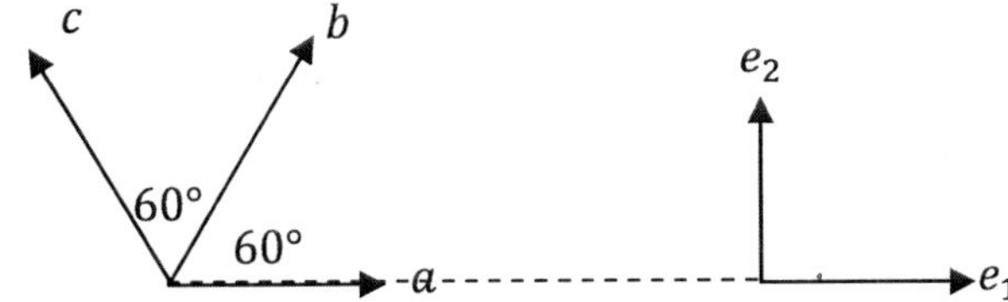

5. The displacement function of the two–dimensional plate shown below is:

$$u_1 = \left(-0.005X_1^2 + 0.004X_1\right) X_2$$

$$u_2 = -0.00016X_1^3 + 0.0012X_1^2$$

Find the following:

- The deformed configuration function $x = f(X)$.
- The expression for the infinitesimal strain tensor as a function of the original position (X_1, X_2).
- The expression for the Green strain tensor as a function of the original position (X_1, X_2).
- Use Mathematica to draw the vector plot of the displacement function to visualize the plate displacements.
- Use Mathematica to draw the contour plot for the following:

$$u_1 u_2, \varepsilon_{11}^{Green}, \varepsilon_{12}^{Green}, \varepsilon_{11}^{Small}, \varepsilon_{12}^{Small}$$

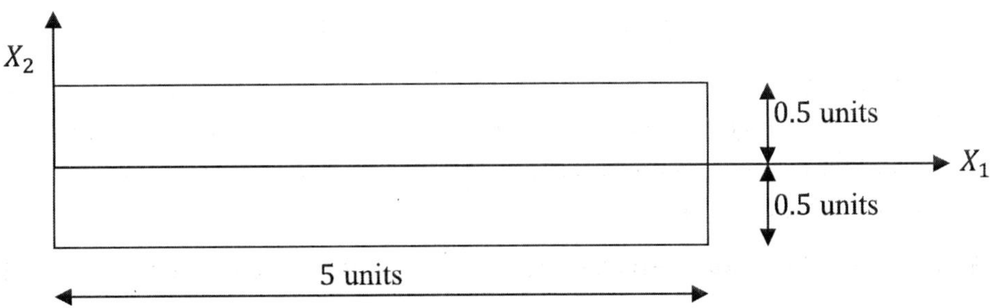

6. Find the infinitesimal rotation matrix as a function of the position for problems 2 and 5 above.

7. The components of the small strain tensor at a point in a body is given by:

$$\varepsilon = \begin{pmatrix} 0.0028 & 0.0004 & 0.0053 \\ 0.0004 & -0.0015 & 0.0008 \\ 0.0053 & 0.0008 & 0.0045 \end{pmatrix}$$

Find the following:

- The normal engineering strain in the direction given by the unit vector $n = \{0.45, -0.28, 0.848\}$. (**Answer:** $\varepsilon_{eng} = 0.00725$)
- The principal strains and the principle directions.

(**Answer:** $\{\{0.00909, -0.00187, -0.00142\}, \{\{-0.64530, -0.08176, -0.75954\}, \{-0.58927, -0.57947, 0.56301\}, \{0.48616, -0.81089, -0.32575\}\}\}$)

8. The measured strains for the three axes of a strain gauge rosette are: $\varepsilon_1 = -0.0004$, $\varepsilon_2 = 0.000085$, $\varepsilon_3 = 0.00025$. If ε_1 coincides with the basis vector e_1 while ε_2 and ε_3 coincide with vectors that are counter–clockwise 60° and 120° respectively from e_1. Find the components of the strain tensor in a Cartesian coordinate system defined by the basis set $B = \{e_1, e_2\}$.

(**Answer:** $\varepsilon_{11} = -0.0004$, $\varepsilon_{12} = -0.0000953$, $\varepsilon_{22} = 0.000357$)

9. The general equation for an ellipsoidal surface with axes in the coordinate directions is:

$$\left(\frac{x_1}{a}\right)^2 + \left(\frac{x_2}{b}\right)^2 + \left(\frac{x_3}{c}\right)^2 = 1$$

where, a, b and c are the three semi–diameters. If $a = b = c = r$ the surface becomes a sphere with radius r. Consider a body in which a homogeneous state of strain exists (constant throughout the body), given by the infinitesimal strain tensor ε. Assume the material to be stretching without rotation $W_{inf} = \hat{0}$. Consider the material particles that, before deformation, were on the spherical surface of radius r given by (using the summation convention described in section 1.2.5.a): $X_i X_i = r^2$.

- By expressing the initial coordinates X_i in terms of the final coordinates x_i, show that the spherical surface becomes an ellipsoid, with the axes of the ellipsoid oriented in the principal strain directions (for convenience, assume that the X_i coordinate axes are in the principal strain directions).
- Discuss the relationship between the semi–diameters of this ellipsoid, the principal components of strain, and the radius of the original sphere.

10. Consider the following in the Euclidian space $\mathbb{R}^2$. Three unit square elements are deformed, as shown in the figures. For each of these:

- Write expressions for the displacement components u_1 and u_2 as functions of X_1 and X_2 and the variables shown for each unit square.
- Determine the components of the infinitesimal strain tensors ε and the infinitesimal rotation tensor W_{inf}.
- Determine the components of the Green strain tensor ε.

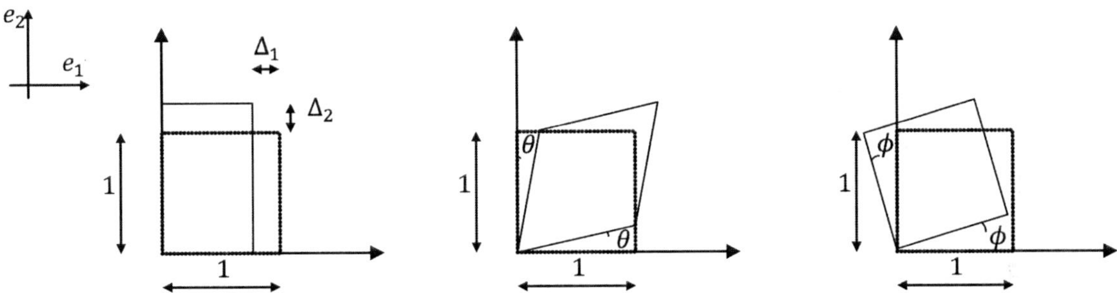

11. Consider the two–dimensional motion:

$$x_1 = X_1 + X_2 t$$

$$x_2 = X_2 + X_1 \frac{t}{2}$$

- Evaluate the following at time $t = 1$:
  a. The deformation gradient F
  b. The velocity gradient L
  c. The rate of deformation tensor D and the rate of spin tensor W
- Calculate $\frac{DJ}{Dt}$. Explain the physical meaning of the value that you get.
- Find the time t at which the above relation will stop being physically possible.

12. Find the stretch and spin tensors of the deformation described by:

$$\begin{pmatrix} x_1 \\ x_2 \\ x_3 \end{pmatrix} = \begin{pmatrix} \lambda_1(t) X_1 \\ \lambda_2(t) X_2 \\ \lambda_3(t) X_3 \end{pmatrix}$$

13. A state of deformation is described by $F = \lambda I$, where λ is constant and I is the unit matrix. The state of stress is hydrostatic, and the Cauchy stress is given by $\sigma = -pI$, where p is the hydrostatic pressure. Find the first and second Piola–Kirchhoff stress tensors.

CHAPTER 4
Mass Balance and Equilibrium Equations

Newton's laws of motion form perhaps the most successful physical model in engineering history. The dimensions and materials of many engineering components are chosen such that they are able to sustain the external loads applied on them; these components include: a rocket travelling to the moon, the wing of an aeroplane, a column of a building, the engine block of a car, cables, ropes, etc. External loads calculated on the different components are calculated by solving Newton's laws of motion.

Newton formulated three laws for the analysis of mechanical systems. The first law states that a body will stay at rest or move with a constant velocity unless acted upon by external forces. The second law states that the resultant force $F \in \mathbb{R}^3$ acting on a body with a constant mass $m \in \mathbb{R}$ will cause the body to have an acceleration $a \in \mathbb{R}^3$ such that $F = ma$. And, finally, the third law states that the action and reaction between two bodies are equal and opposite in direction. In this chapter, the physical laws of mass balance and Newton's equations of equilibrium will be applied to continuum bodies in their different configurations.

4.1. EULERIAN VERSUS LAGRANGIAN FORMULATIONS

The term "Lagrangian Formulation" is usually used to describe an equation written as a function of the coordinates in the reference coordinate system, while the term "Eulerian Formulation" is usually used to describe an equation written as a function of the coordinates in the deformed coordinate system. In a Lagrangian formulation, the coordinates of a vector $X \in \Omega_0$ are directed to a material point and are fixed in time, so differential equations written in terms of $X \in \mathbb{R}^3$ describe the material rates of change of the different variables appearing in the differential equation. In an Eulerian formulation, quantities are written as a function of the spatial coordinates of the vector $x \in \mathbb{R}^3$, thus, the quantities appearing in the differential equation point to a position in space rather than a material point. Traditionally, equations for solids usually use the Lagrangian formulations while those for fluids use the Eulerian formulations.

4.2. MASS BALANCE (CONTINUITY EQUATION)

4.2.1. Relationship between Densities in the Different Configurations

Let Ω_0 and Ω represent the reference and the deformed configurations, respectively, of a body embedded in $\mathbb{R}^3$. Let $x(X, t) \in \Omega$ represent the image of $X \in \Omega_0$ at a certain point in time $t \in \mathbb{R}$ under the bijective smooth mapping $f_t: \Omega_0 \to \Omega$. Notice that the set Ω in this case is a function of time. The velocity vector is described by the vector valued function $v = \dot{x}(X, t)$ with F and L being the deformation and the velocity gradients, respectively, while $J = \det F$. The density distribution of the continuum body in the reference configuration is given by the function $\rho_0: \Omega_0 \to \mathbb{R}$. As the body deforms, the density at each material point changes with time $t \in \mathbb{R}$ and thus can be described by a function of both time and material points $\rho_t: \Omega_0 \times \mathbb{R} \to \mathbb{R}$. The spatial density distribution of the continuum body in the deformed configuration is a function of the position x and of the time t and is described by the function $\rho: \Omega \times \mathbb{R} \to \mathbb{R}$. The difference between ρ_t and ρ is merely a change of coordinates from X to x:

$$\rho_t(X, t) = \rho(f_t(X), t) = \rho(x(X, t), t)$$

In this section, we will use the physical law of mass balance to establish the relationship between the mass densities ρ_0, ρ_t and ρ and their time derivatives. If dv and dV represent an infinitesi–mal volume in the deformed and reference configurations, respectively, then, according to section 3.2.1, the relationship between the volume in the deformed versus the reference configurations is given by:

$$JdV = dv$$

The symbol dv is reserved for the infinitesimal volume in the deformed configuration, and v is reserved for the velocity field. If m is the mass of both infinitesimal volumes then the relationship between the densities is:

$$\rho_0(X) = \frac{m}{dV} = \frac{m}{dv}\frac{dv}{dV} = \rho_t(X, t)J = \rho(x(X, t), t)J$$

This relationship can also be obtained using integrals over the whole continuum body. The total mass M_0 of the reference configuration can be calculated by the integral:

$$M_0 = \int_{\Omega_0} \rho_0 dV$$

The total mass M in the deformed configuration can be calculated by the integral:

$$M = \int_{\Omega} \rho dv$$

The volume integral over Ω can be replaced by the volume integral over Ω_0 by replacing dv with JdV:

$$M = \int_{\Omega} \rho dv = \int_{\Omega_0} \rho J dV$$

Since the total mass is conserved, $M_0 = M$ and since dV is arbitrary, we retrieve the relationship between the densities:

$$\rho_0(X) = \rho(x(X,t),t)J$$

4.2.2. The Continuity Equation

The equation describing the rate of change of mass per unit time is termed the continuity equation. First, recall the results from Problem 4 in section 3.4.2:

$$\frac{DJ}{Dt} = J \text{trace } L$$

Assuming that the rate of change of the density in the reference configuration with respect to time is equal to zero (mass is neither created nor destroyed) leads to the following equivalent forms of the continuity equation.

4.2.2.a. Lagrangian Formulation

In a Lagrangian formulation, the continuity equation can be written in terms of ρ_0 as follows:

$$\frac{D\rho_0(X)}{Dt} = 0$$

Alternatively, in a Lagrangian formulation, the continuity equation can be written in terms of ρ_t, as follows:

$$\frac{D(\rho_t(X,t)J)}{Dt} = \frac{\partial \rho_t(X,t)}{\partial t}(J) + \rho_t(X,t)(J)\text{trace } L = 0$$

Since $J = \det F > 0$, the continuity equation becomes:

$$\frac{\partial \rho_t(X,t)}{\partial t} + \rho_t(X,t)\text{trace } L = 0$$

4.2.2.b. Eulerian Formulation

The continuity equation in an Eulerian formulation is given in terms of $\rho(x,t)$ as follows:

$$\frac{d(\rho(x,t)J)}{dt} = \left(\frac{\partial \rho(x,t)}{\partial t} + \frac{\partial \rho(x,t)}{\partial x} \cdot v\right)(J) + \rho(x,t)(J)\text{trace } L = 0$$

Since $J = \det F > 0$, the continuity equation becomes:

$$\frac{\partial \rho(x,t)}{\partial t} + \operatorname{div}(\rho(x,t)v) = 0$$

▶ **BONUS EXERCISE:** Verify the following equality:

$$\operatorname{div}(\rho(x,t)v) = \frac{\partial \rho(x,t)}{\partial x} \cdot v + \rho(x,t)\,\text{trace}\,L$$

Remarks:

- The capital D in $\frac{D}{Dt}$ is used to indicate that the quantity being differentiated is a function of the coordinates in the reference configuration.
  If the motion is isochoric ($J = 1$) at every material point, the above relations reduce to:

$$\text{trace}\,L = \operatorname{div} v = \frac{\partial v_1}{\partial x_1} + \frac{\partial v_2}{\partial x_2} + \frac{\partial v_3}{\partial x_3} = 0$$

- The above formula is most commonly used to describe the continuity equation of incompressible fluids.

For theories of growth or mass transfer, the density in the reference configuration is assumed to vary in time and its rate of change is equal to the source or the sink of the local mass change $c(X,t)$:

$$\frac{D\rho_0(X,t)}{Dt} = \dot{\rho}_0(X,t) = c(X,t)$$

Problem 1: A displacement function of a 2–units–by–2–units plate has the form:

$$u_1 = 0.2X_1\left(1 + \frac{t}{100}\right)$$

$$u_2 = 0.09X_2X_1$$

$$u_3 = 0$$

The density of the material in the reference configuration before deformation is equal to 1 unit. Find the material density $\rho_t(X,t)$ as time evolves. Draw the contour plot of the density after deformation at $t = 5$ units of time using the original coordinates.

Solution: The position function is:

$$x = X + u = \begin{Bmatrix} x_1 \\ x_2 \\ x_3 \end{Bmatrix} = \begin{Bmatrix} (1.2 + 0.002t)X_1 \\ X_2 + 0.09X_2X_1 \\ X_3 \end{Bmatrix}$$

The deformation gradient and its determinant are:

$$F = \begin{pmatrix} 1.2 + 0.002t & 0 & 0 \\ 0.09X_2 & 1 + 0.09X_1 & 0 \\ 0 & 0 & 1 \end{pmatrix}, \quad J = 1.2 + 0.002t + 0.108X_1 + 0.00018tX_1$$

The density as time evolves expressed in the reference coordinate system is:

$$\rho_t(X, t) = \frac{\rho_0}{J} = \frac{1}{1.2 + 0.02t + 0.108X_1 + 0.00018tX_1}$$

The Mathematica code to produce the above equations and the required plot is:

```
X={X1,X2,X3};
u={0.2*X1,0.09*X2*X1,0};
x=X+u;
F=Table[D[x[[i]],X[[j]]],{i,1,3},{j,1,3}];
F//MatrixForm
Det[F]
Rhot=1/Det[F]
ContourPlot[Rhot,{X1,0,2},{X2,0,2},ContourLabels->True]
```

Problem 2: The position function of a 2–units–by–2–units plate has the form:

$$x_1 = X_1 + tX_2^2 \cos 80^\circ$$
$$x_2 = X_2 + tX_1^2 \cos 80^\circ$$
$$x_3 = X_3$$

This position function is applicable for values of the time $0 \leq t \leq 1$.

The density of the material in the reference configuration at $t = 0$ is described by the function:

$$\rho_0(X) = 1 + X_1 X_2 \text{ units of density}$$

Find the material density $\rho_t(X, t)$ as time evolves. Draw the contour plots of the determinant of the deformation gradient, $\rho_0(X)$, $\rho_t(X, t)$ and $\rho(x, t)$ at $t = 0, 1/2$ and 1 unit of time.

Solution: The deformation gradient is:

$$F = \begin{pmatrix} 1 & 2tX_2 \cos 80^\circ & 0 \\ 2tX_1 \cos 80^\circ & 1 & 0 \\ 0 & 0 & 1 \end{pmatrix}$$

The determinant of the deformation gradient is:

$$\det F = 1 - 4t^2 X_1 X_2 \left(\cos 10^\circ\right)^2$$

The density $\rho_t(X, t)$ as time evolves is equal to:

$$\rho_t(X, t) = \frac{\rho_0(X)}{\det F} = \frac{1 + X_1 X_2}{1 - 4t^2 X_1 X_2 \left(\cos 10^\circ\right)^2}$$

Drawing the contour plot of $\rho(x, t)$ requires expressing $\rho_t(X, t)$ in terms of x rather than X. It also requires finding the set $\Omega \in \mathbb{R}^2$ that represents the position of the deformed plate in the plane defined by x_1 and x_2. To find Ω, the position functions of the four sides of the plates at the specified times is obtained using the `Solve[]`. Since $X_1 = 0$ and $0 \leq X_2 \leq 2$ on side 1, the deformed shape in the spatial coordinates of side 1 can be obtained at $t = 1$, as follows:

$$x_1 = X_2^2 \cos 80^\circ, \quad x_2 = X_2 \Longrightarrow x_1 = x_2^2 \cos 80^\circ$$

Similarly for side 2, $X_1 = 2$ and $0 \leq X_2 \leq 2$ and, therefore, the relationship for side 2 is:

$$x_1 = 2 + \left(x_2 - 4\cos 80^\circ\right)^2 \cos 80^\circ$$

Notice that Mathematica outputs the equivalent form:

$$x_1 = -4x_2 + 4x_2 \cos 20^\circ + 12 \sin 10^\circ + x_2^2 \sin 10^\circ$$

Similar equations can be obtained for sides 3 and 4. The following sketch illustrates the deforma– tion of the different sides at $t = 1$:

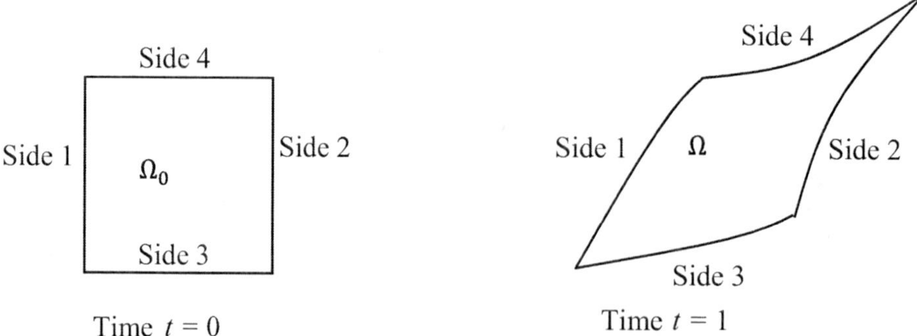

Expressing the density in spatial coordinates is a complicated task since the position function is highly nonlinear and it is not possible to simply invert it. In the Mathematica code presented below,

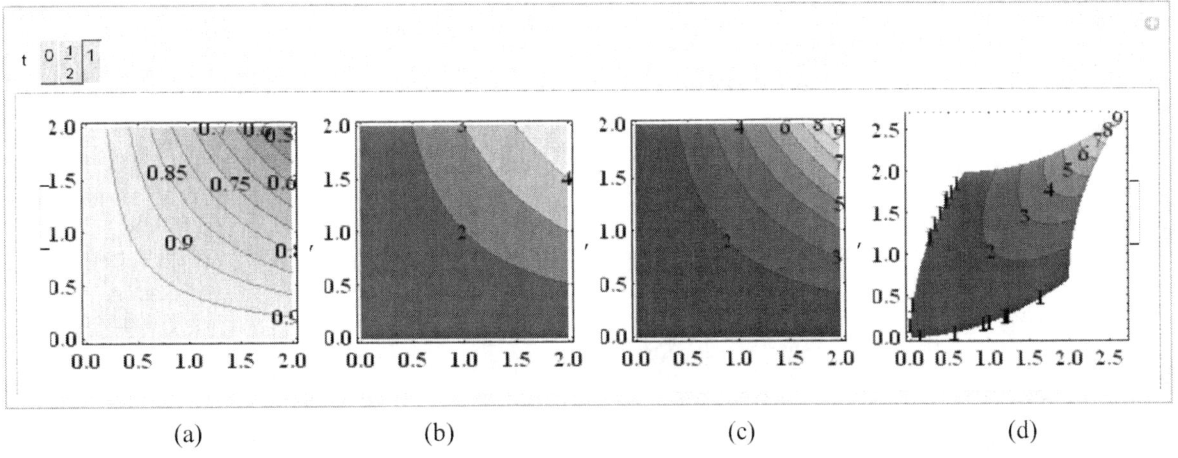

(a) (b) (c) (d)

FIGURE 4-1 Contour plots of (a) $\det F(X, 1)$, (b) $\rho_0(X)$, (c) $\rho_t(X, 1)$ and (d) $\rho(x, 1)$.

the Mathematica `Solve[]` function is used to find an expression for $\rho(x, t)$ by solving a set of nonlinear equations that considers the limiting conditions $0 \le X_1 \le 2$ and $0 \le X_2 \le 2$. The contour plot of $\rho(x, t)$ can be drawn by using the region function that ensures that $\rho(x, t)$ is drawn in the region specified by Ω. The difference between the functions $\rho(x, t)$ and $\rho_t(X, t)$ can be visualized in Figure 4–1. The contour plot of the function $\rho(x, t)$ is drawn on the plate after deformation Figure 4–1d, while the contour plot of the function $\rho_t(X, t)$ is drawn on the plate before deformation Figure 4–1c. The two functions essentially produce the same contour plot for the same material points at a specified t. The following Mathematica code performs the above calculations for the required time slots. Copy and paste the code into your Mathematica software and you can visualize the deformation of the plate at $t = 0, 1/2$ and 1 time units.

```
Manipulate[Clear[X1,X2,X3,x1,x2,x3,y1,y2,x,X,theta,beta];
X={X1,X2,X3};
x1=X1+X2^2*Cos[theta]*t;
x2=X2+X1^2*Cos[beta]*t;
x3=X3;
x={x1,x2,x3};
F=Table[D[x[[i]],X[[j]]],{i,1,3},{j,1,3}];
rhoreference=1+X1*X2;
theta=80Degree;
beta=80Degree;


Rhot=rhoreference/Det[F];


a=Simplify[
```

```
Solve[{r==Rhot,y1==x1,y2==x2,0<=X1<=2,
0<=X2<=2},{r,X1,X2},Reals]];
Rho=r/.a[[1,1]];
Rho=FullSimplify[N[Rho[[1]]]];

con=Table[i-1,{i,1,11}];
side1=Solve[{y1==x1/.X1->0,y2==x2/.X1->0},{y1,X2}];
side2=Solve[{y1==x1/.X1->2,y2==x2/.X1->2},{y1,X2}];
side3=Solve[{y1==x1/.X2->0,y2==x2/.X2->0},{y2,X1}];
side4=Solve[{y1==x1/.X2->2,y2==x2/.X2->2},{y2,X1}];

{ContourPlot[Det[F],{X1,0,2},{X2,0,2},
BaseStyle->Directive[Bold,15],ColorFunction->"GrayTones",
ContourLabels->True,PlotRange->All],
ContourPlot[rhoreference,{X1,0,2},{X2,0,2},
BaseStyle->Directive[Bold,15],ColorFunction->"GrayTones",
ContourLabels->True,Contours->con,PlotRange->All],
ContourPlot[Rhot,{X1,0,2},{X2,0,2},
BaseStyle->Directive[Bold,15],ColorFunction->"GrayTones",
ContourLabels->True,Contours->con,PlotRange->All],
ContourPlot[Rho,{y1,0,3},{y2,0,3},
RegionFunction->
Function[{y1,y2,z},
y1>(y1/.side1[[1,1]])&&y1<(y1/.side2[[1,1]])&&
y2>(y2/.side3[[1,1]])&&y2<(y2/.side4[[1,1]])],
BaseStyle->Directive[Bold,15],ColorFunction->"GrayTones",
ContourLabels->True,Contours->con,PlotRange->All]},{t,{0,
1/2,1}}]
```

4.3. DIFFERENTIAL EQUATIONS OF EQUILIBRIUM

4.3.1. Derivation Using a Differential Volume

While Newton's laws of motion were originally formulated for a rigid body or a system of rigid particles mechanically interacting with each other, the introduction of the concept of stress inside a material body allows the expression of the "differential" form of Newton's laws of motion.

Let $\Omega \in \mathbb{R}^3$ be a set representing a deformed configuration, and let $F = \{F_1, F_2, F_3\} \in \mathbb{R}^3$ be the resultant force acting on a differential cube of mass $dm \in \mathbb{R}$ inside the body whose dimensions are dx_1, dx_2 and dx_3 and whose mass density is $\rho = \frac{dm}{dv}$. The volume dv of the cube is $dv = dx_1 dx_2 dx_3$. Let $b = \{b_1, b_2, b_3\} \in \mathbb{R}^3$ be a body forces vector per unit mass acting on the cube, and assume $a = \{a_1, a_2, a_3\} \in \mathbb{R}^3$ to be the acceleration vector of the differential cube. Then, the statement of

Newton's second law is:

$$F = ma = \rho dv\, a$$

To write the differential form of the equilibrium equations, the stresses are assumed to be **continuous functions**. Such assumption allows the use of the first component in the Taylor expansion for the stresses (Figure 4–2). Thus, the forces in the direction of the first basis vector can be evaluated, as follows:

$$\begin{aligned} F_1 &= -\sigma_{11}dx_2dx_3 + \left(\sigma_{11} + \frac{\partial \sigma_{11}}{\partial x_1}dx_1\right)dx_2dx_3 - \sigma_{21}dx_1dx_3 + \left(\sigma_{21} + \frac{\partial \sigma_{21}}{\partial x_2}dx_2\right)dx_1dx_3 - \sigma_{31}dx_2dx_3 \\ &+ \left(\sigma_{31} + \frac{\partial \sigma_{31}}{\partial x_3}dx_3\right)dx_1dx_2 + \rho b_1 dx_1dx_2dx_3 \\ &= \frac{\partial \sigma_{11}}{\partial x_1}dx_1dx_2dx_3 + \frac{\partial \sigma_{21}}{\partial x_2}dx_2dx_1dx_3 + \frac{\partial \sigma_{31}}{\partial x_3}dx_3dx_1dx_2 + \rho b_1 dx_1dx_2dx_3 \end{aligned}$$

Using the statement of Newton's second law in the first basis vector:

$$\frac{\partial \sigma_{11}}{\partial x_1} + \frac{\partial \sigma_{21}}{\partial x_2} + \frac{\partial \sigma_{31}}{\partial x_3} + \rho b_1 = \rho a_1$$

Similarly, in the other two directions:

$$\begin{aligned} \frac{\partial \sigma_{12}}{\partial x_1} + \frac{\partial \sigma_{22}}{\partial x_2} + \frac{\partial \sigma_{32}}{\partial x_3} + \rho b_2 = \rho a_2 \\ \frac{\partial \sigma_{13}}{\partial x_1} + \frac{\partial \sigma_{23}}{\partial x_2} + \frac{\partial \sigma_{33}}{\partial x_3} + \rho b_3 = \rho a_3 \end{aligned}$$

Remarks:

- The above three equilibrium equations can be written in the following compact form:

$$\sum_{j=1}^{3} \frac{\partial \sigma_{ji}}{\partial x_j} + \rho b_i = \rho a_i$$

where $i = 1, 2$ and 3 for the sum of forces in the direction of the first, second, and third basis vectors, respectively. By adopting the summation convention in section 1.2.5.a, the summation symbol can be dropped and the three equilibrium equations can be written as:

$$\frac{\partial \sigma_{ji}}{\partial x_j} + \rho b_i = \rho a_i$$

If the divergence operator defined in section 1.3.1.c is used, the equations of equilibrium can be written in the following vector form:

$$\text{div}\sigma + \rho b = \rho a$$

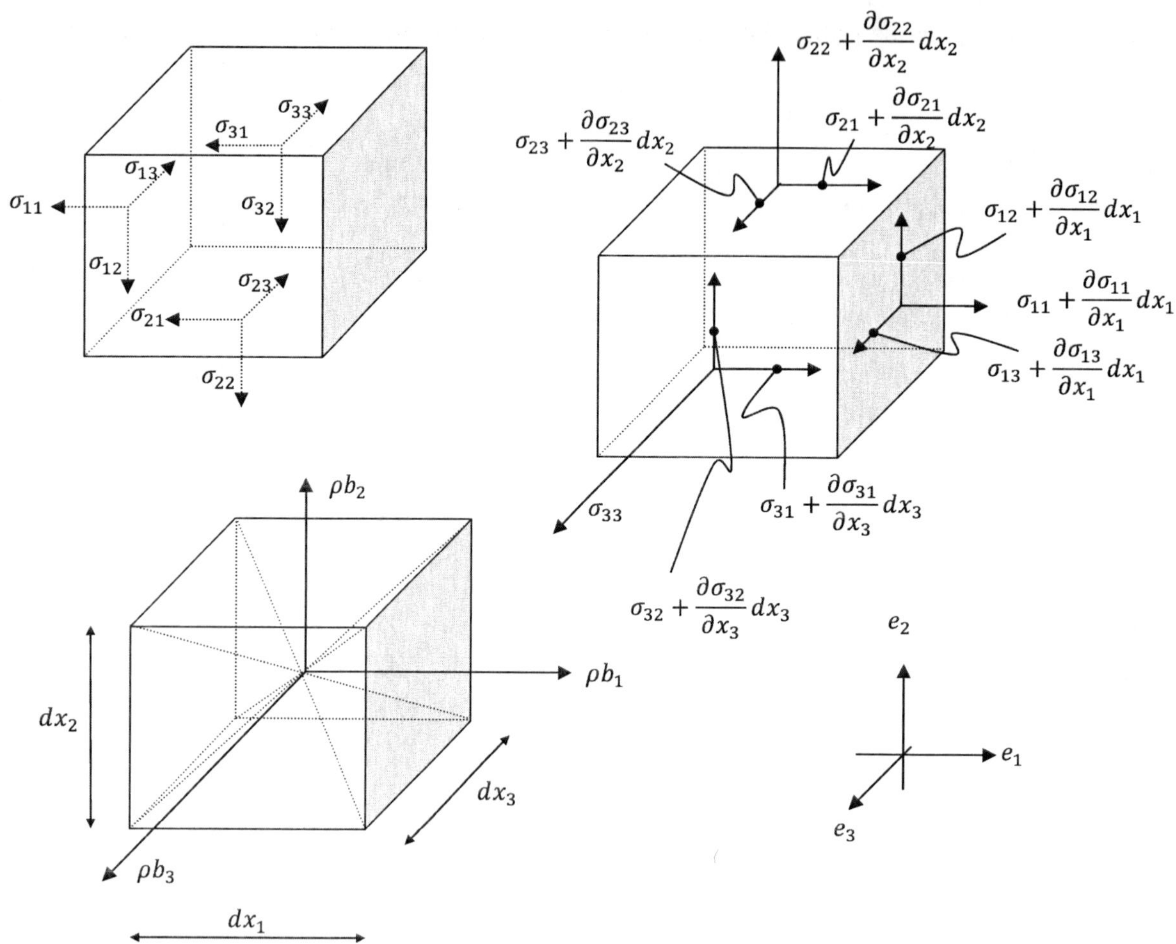

FIGURE 4-2 Forces acting on an infinitesimal cube of dimensions $dx_1 \times dx_2 \times dx_3$.

- The equilibrium equations of moment have already been used in section 2.1.3 in setting $\sigma_{ij} = \sigma_{ji}$.

4.3.2. Derivation Using Integration over the Continuum Body

4.3.2.a. Conservation of Linear Momentum

The same equations of equilibrium can be obtained by integrating over the deformed configura– tion. Let the assumptions in sections 4.2 and 4.3.1 hold. The rate of change of the linear momentum of the continuum body is caused by the applied external traction forces vectors integrated over the surface of the continuum, in addition to the applied body forces integrated over the continuum body. The rate of change of the linear momentum of the continuum body can be evaluated using the following integral:

$$\text{Rate of change of the linear momentum} = \frac{d}{dt}\int_{\Omega} (\rho, t)\, v\, dv$$

As the integration domain Ω is function in time, the integration operator and the differential operator cannot be interchanged. To interchange the integration and the differential operators, the integration domain Ω is replaced by Ω_0 and dv with JdV. Thus, rate of change of the momentum is written as:

$$\text{Rate of change of the linear momentum} = \int_{\Omega_0} \frac{d\rho v J}{dt} dV = \int_{\Omega_0} \left(\frac{d\rho J}{dt} v + \rho \frac{dv}{dt} J\right) dV$$

Recalling the continuity equation in section 4.2.2, the rate of change of the momentum can be written as:

$$\text{Rate of change of the linear momentum} = \int_{\Omega_0} \left(\frac{d\rho J}{dt} v + \rho \dot{v} J\right) dV = \int_{\Omega} \rho \dot{v} dv$$

The rate of change in linear momentum is caused by the action of the external traction vectors ($t_n = \sigma^T n$) and the body forces vectors per unit mass b:

$$\int_{\Omega} \rho \dot{v}\, dv = \int_{\Omega} \rho b\, dv + \int_{\partial\Omega} \sigma^T n dS = \int_{\Omega} \rho b\, dv + \int_{\Omega} \text{div}\sigma\, dv$$

where the divergence theorem in section 1.3.1.c is utilized.

The above equation is in terms of the deformed configuration coordinate system and, thus, it is considered an **Eulerian Formulation** and is equivalent to:

$$\int_{\Omega} (\rho \dot{v} - \rho b - \text{div}\sigma)\, dv = \hat{0}$$

The above equation is a vector equation, and, hence, $\hat{0}$ is used to denote the zero vector. Since the integration volume is arbitrary and denoting the acceleration $\dot{v}$ by a the Eulerian form of the differential equation of equilibrium is obtained:

$$\text{div}\sigma + \rho b = \rho a$$

The **Lagrangian Formulation** can be similarly obtained in terms of the first Piola Kirchhoff stress:

$$\frac{D}{Dt}\int_{\Omega} \rho_0(X) v(X, t)\, dV = \int_{\Omega_0} \rho_0(X) b(X, t)\, dV + \int_{\partial\Omega_0} PNdS = \int_{\Omega_0} (\rho_0(X) b(X, t) + \text{div}P^T)\, dV$$

Similarly, since the integration volume is arbitrary and denoting the acceleration $\dot{v}$ by a the differential equation of equilibrium is obtained:

$$\text{div}P^T + \rho_0 b(X, t) = \rho_0 a(X, t)$$

Note that the equations of motion can be written in the reference coordinate system because the Piola Kirchhoff stress, as shown in section 2.5.2, preserves the force vectors in the reference configuration.

4.3.2.b. Conservation of Angular Momentum

In section 2.1.3, the conservation of angular momentum was used to establish the symmetry of the Cauchy stress tensor using a differential volume. The same result can also be obtained using integral equations. According to Euler's equations of motion, the rate of change of angular momentum around a reference point is equal to the moment of the applied external and body forces vectors around the same reference point. Let $r \in \mathbb{R}^3$ be the position vector with respect to a reference point. The equation of conservation of angular momentum can be written in a Eulerian formulation, as follows:

$$\frac{d}{dt}\int_{\Omega} r \times \rho(x, t)v\, dv = \int_{\Omega} r \times \rho(x, t)b\, dv + \int_{\partial\Omega} r \times \sigma^T n\, dS$$

The reference configuration can be used as the integration domain for the left–hand side of the above equation and the continuity equation in section 4.2.2 can be used to reach to the following equality:

$$\frac{d}{dt}\int_{\Omega} r \times \rho(x, t)v\, dv = \frac{d}{dt}\int_{\Omega_0} r \times \rho(x, t)vJ\, dV = \int_{\Omega_0} r \times \frac{d\rho(x, t)vJ}{dt} dV = \int_{\Omega} r \times \rho(x, t)\dot{v}\, dv$$

Therefore, the equation of conservation of angular momentum can be written as:

$$\int_{\Omega} r \times (\rho(x, t)\dot{v} - \rho(x, t)b) dv = \int_{\partial\Omega} r \times \sigma^T n dS$$

Using the balance of mass and linear momentum equations, the left–hand side of the above equations can be replaced by:

$$\int_{\Omega} r \times (\rho(x, t)\dot{v} - \rho(x, t)b) dv = \int_{\Omega} r \times \text{div}\sigma\, dv$$

Therefore, the balance of angular momentum can be rewritten as:

$$\int_{\partial\Omega} r \times \sigma^T n dS = \int_{\Omega} r \times \text{div}\sigma \, dv$$

The symmetry of the Cauchy stress tensor can be obtained by investigating the component form of the left–hand and the right–hand sides of the above equation:

$$\text{Left hand side} = \left(\int_{\partial\Omega} r \times \sigma^T n dS \right)_i = \int_{\partial\Omega} \varepsilon_{ijk} r_j \sigma_{lk} n_l dS$$

The divergence theorem can be used to replace the surface integral with a volume integral:

$$\begin{aligned} \text{Left–hand side} &= \int_{\partial\Omega} \varepsilon_{ijk} r_j \sigma_{lk} n_l dS = \int_{\Omega} \frac{\partial \varepsilon_{ijk} r_j \sigma_{lk}}{\partial x_l} dv \\ &= \int_{\Omega} \varepsilon_{ijk} \frac{\partial r_j}{\partial x_l} \sigma_{lk} + \varepsilon_{ijk} r_j \frac{\partial \sigma_{lk}}{\partial x_l} dv = \int_{\Omega} \varepsilon_{ijk} \delta_{jl} \sigma_{lk} + \varepsilon_{ijk} r_j \frac{\partial \sigma_{lk}}{\partial x_l} dv \\ &= \int_{\Omega} \varepsilon_{ijk} \sigma_{jk} + \varepsilon_{ijk} r_j \frac{\partial \sigma_{lk}}{\partial x_l} dv \end{aligned}$$

$$\text{Right–hand side} = \left(\int_{\Omega} r \times \text{div}\sigma \, dv \right)_i = \int_{\Omega} \varepsilon_{ijk} r_j \frac{\partial \sigma_{lk}}{\partial x_l} dv$$

Equating both sides leads to the symmetry of the Cauchy stress tensor, as follows:

$$\int_{\Omega} \varepsilon_{ijk} \sigma_{jk} dv = 0 \Longrightarrow \sigma_{jk} = \sigma_{kj}$$

The symmetry of the Cauchy stress tensor can be obtained using the properties of the cross product without using any coordinate system, as demonstrated by Chadiwck[1].

4.3.3. Solution of the Equilibrium Equations

A continuum problem is solved if a stress field (the stress at every point inside the body) and an acceleration field are found such that the three equations of force equilibrium (conservation of linear momentum equations) are satisfied at every point. Recall that the three equations of

[1]Chadwick, P. (1999). Continuum Mechanics. Concise Theory and Problems. *Dover Publications Inc.*

moment equilibrium (or conservation of angular momentum equations) were exhausted in show– ing the stress matrix is in fact a symmetric matrix; that is, by assuming that the stress matrix is symmetric, no additional information can be obtained from considering the three equations of moment equilibrium. The unknown stress fields are required to also satisfy certain boundary con– ditions, which are traditionally given as either displacements or external traction forces on the exterior of the body and initial conditions for velocities. In static equilibrium, which is the pur– pose of many engineering applications, the acceleration vector is assumed to be zero and, thus, the equilibrium equation is simplified and the stress field becomes the only unknown. The problem is mathematically formulated as follows:

Let $\Omega_0 \in \mathbb{R}^3$ be a set representing a reference configuration. Given a constant (or variable) body forces vector $b \in \mathbb{R}^3$, find the distribution of stresses inside the body that satisfies the following equations:

$$\frac{\partial \sigma_{11}}{\partial x_1} + \frac{\partial \sigma_{21}}{\partial x_2} + \frac{\partial \sigma_{31}}{\partial x_3} + \rho b_1 = 0$$

$$\frac{\partial \sigma_{12}}{\partial x_1} + \frac{\partial \sigma_{22}}{\partial x_2} + \frac{\partial \sigma_{32}}{\partial x_3} + \rho b_2 = 0$$

$$\frac{\partial \sigma_{13}}{\partial x_1} + \frac{\partial \sigma_{23}}{\partial x_2} + \frac{\partial \sigma_{33}}{\partial x_3} + \rho b_3 = 0$$

with the following boundary conditions specified on part of the boundary of Ω_0:

The traction vectors $t_n \in \mathbb{R}^3$ on Ω_n are known, so the stresses on the boundary can be given as $\sigma^T n = t_n$.

The displacement vectors $u \in \mathbb{R}^3$ on Ω_u are known where the boundary of $\Omega_0 = \partial\Omega_0 = \Omega_n \cup \Omega_u$ and $\Omega_n \cap \Omega_u = \emptyset$.

Difficulties Associated with Obtaining a Solution to the Equilibrium Equations:

Upon careful investigation of the equilibrium equations, the following two issues arise. First, there are three equations of static equilibrium. However, there are six unknowns (six stress variables). Thus, in their current form, many solutions could possibly satisfy the equilibrium equations! The second issue is that some of the boundary conditions contain expressions of displacements, while the equations themselves in this form do not have displacements as variables! These two major issues impel replacing the six unknown stress variables in the above equations with three unknown variables (usually the three displacements u_1, u_2 and u_3). This can be performed by using a "con– stitutive equation" that describes the relationship between the stress and the strains inside the material. By replacing the stresses with the strains, and by using the relationship between the strains and the displacements described in Chapter 3, the problem becomes well posed, i.e., a solution can be obtained.

Problem 1: A stress field over a body that is in static equilibrium and embedded in $\mathbb{R}^3$ has the fol–lowing form:

$$\sigma = \begin{pmatrix} 5x_1^2 + 3x_2 + x_3 & -x_2x_1 & 0 \\ -x_2x_1 & 5x_2 & 0 \\ 0 & 0 & 9x_3^2 \end{pmatrix}$$

where, x_1, x_2 and x_3 are the coordinates inside the body. Find the body forces vector field that is in equilibrium with this stress field.

Solution: The stress matrix is symmetric so it satisfies the equations of moment equilibrium.

The equations of force equilibrium are:

$$\frac{\partial\sigma_{11}}{\partial x_1} + \frac{\partial\sigma_{21}}{\partial x_2} + \frac{\partial\sigma_{31}}{\partial x_3} + \rho b_1 = 0 \Longrightarrow \rho b_1 = -9x_1$$

$$\frac{\partial\sigma_{12}}{\partial x_1} + \frac{\partial\sigma_{22}}{\partial x_2} + \frac{\partial\sigma_{32}}{\partial x_3} + \rho b_2 = 0 \Longrightarrow \rho b_2 = x_2 - 5$$

$$\frac{\partial\sigma_{13}}{\partial x_1} + \frac{\partial\sigma_{23}}{\partial x_2} + \frac{\partial\sigma_{33}}{\partial x_3} + \rho b_3 = 0 \Longrightarrow \rho b_3 = -18x_3$$

The following Mathematica code can be utilized:

```
X={X1,X2,X3};
s={{5*X1^2+3X2+X3,-X2*X1,0},{-X2*X1,5X2,0},{0,0,9X3^2}};
-Sum[D[s[[i,1]],X[[i]]],{i,1,3}]
-Sum[D[s[[i,2]],X[[i]]],{i,1,3}]
-Sum[D[s[[i,3]],X[[i]]],{i,1,3}]
```

Problem 2: The shown vertical beam has a varying circular cross–sectional area with a radius $r = 2$ meters at the top varying linearly to $r = 1$ meters at the bottom (Figure 4–3). The beam is used to carry a concentrated load of $P = 50N$, has a density $\rho = 20\,\text{kg/m}^3$, and is subjected to a gravity field with $g = 10\,\text{m/sec}^2$. Assume that the only nonzero stress component is σ_{22}, find the distribution of the stress component σ_{22} along the length of the beam.

Solution: The shown beam has a varying cross–sectional area and a varying nonuniform vertical load due to its own weight.

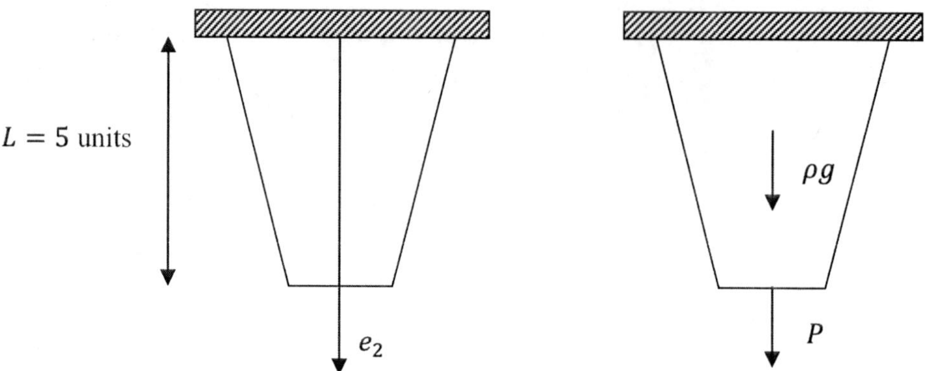

FIGURE 4-3 Beam with a varying cross–sectional area.

The radius r and the area A vary with X_2 according to the equations:

$$r = \left(2 - \frac{X_2}{L}\right)$$

$$A = \pi\left(2 - \frac{X_2}{L}\right)^2$$

A horizontal slice of the beam can be analyzed to find the equilibrium equation, as follows:

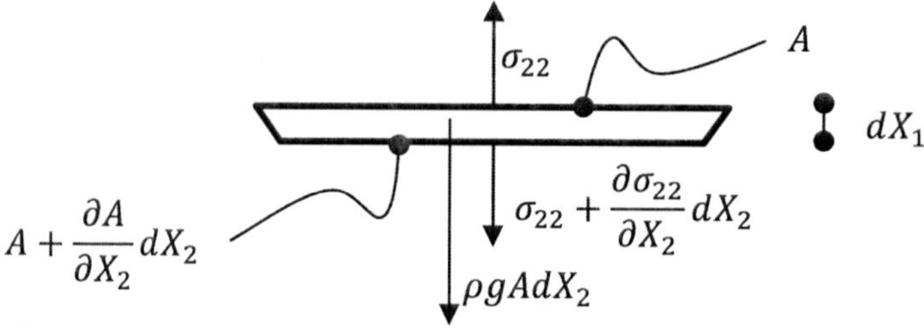

The equilibrium equation can be written as follows:

$$-\sigma_{22}A + \rho g A dX_2 + \left(\sigma_{22} + \frac{\partial \sigma_{22}}{\partial X_2}dX_2\right)\left(A + \frac{\partial A}{\partial X_2}dX_2\right) = 0$$

The final equilibrium equation has following form:

$$\frac{\partial \sigma_{22}}{\partial X_2}A + \frac{\partial A}{\partial X_2}\sigma_{22} + \rho g A = 0$$

The boundary conditions are:

$$@X_2 = L \rightarrow \sigma_{22} = \frac{P}{A} = \frac{P}{4\pi}$$

Even with the very simplified assumption that σ_{22} is the only nonzero component of the stress matrix, the differential equation is relatively complicated, and the Mathematica software will be used to find the distribution of the stress:

$$\sigma_{22} = \frac{1}{3}\left(2\rho gL - \rho gX_2 + \left(\frac{L^2(3P - \rho gL\pi)}{\pi(X_2 - 2L)^2}\right)\right) = \frac{1}{3}\left(2000 - \frac{23806.3}{(X_2 - 10)^2} - 200X_2\right) N/m^2$$

The Mathematica code utilized is shown below. Notice the use of the function `DSolve` to solve the differential equation obtained.

```
Clear[s,x,L,g]
A=Pi*(2-x/L)^2;
a=FullSimplify[DSolve[{s'[x]*A+D[A,x]*s[x]+ro*g*A==0,(s[L]==P/A/.x-
>L)},s[x],x]]
ss=s[x]/.a[[1]]
N[ss/.{L->5,ro->20,g->10,P->50}]
```

Problem 3: The shown horizontal beam has a density $\rho = 1$ kg/m^3 and is under a constant body force $b_1 = 10$ N/Kg in the direction of e_1 (Figure 4–4). Assuming that σ_{11} is the only nonzero stress component, find the distribution of σ_{11} along the beam.

Solution: The assumption that five of the stress matrix components have zero values is a very strong assumption. For example, closer to the fixed boundary, there could be other nonzero stress com–ponents. However, to easily achieve a closed form solution to the equilibrium equations, such assumption is required. Otherwise, a numerical solution (for example, using finite element anal–ysis) would be required to find all the components of the stress.

The equation of equilibrium in the direction of e_1 reads:

$$\frac{\partial\sigma_{11}}{\partial x_1} + \rho b_1 = 0 \Longrightarrow \sigma_{11} = -10x_1 + C$$

where C is a constant that can be obtained from the boundary condition given for the stress:

$$\sigma_{11}|_{x_1=5} = 20 \Longrightarrow -10(5) + C = 20 \Longrightarrow C = 70\text{N/m}^2$$

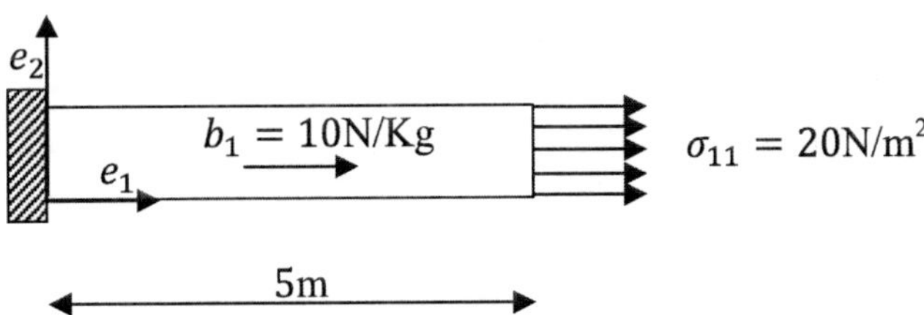

FIGURE 4-4 Beam under a constant body force.

Therefore, the stress distribution is:

$$\sigma_{11} = -10x_1 + 70\text{N/m}^2$$

If the end of the beam $x_1 = 5$m was also fixed, then the given boundary conditions would not be sufficient to solve the differential equation of equilibrium and a constitutive law would be required to replace the stresses with expressions of the displacement in the differential equation of equilibrium.

▶ EXERCISES:

(Answer key is given in round brackets for some problems.):

1. The two–dimensional displacement field of a 2–units–by–3–units plate is given by:

$$x_1 = X_1 + 0.01tX_2$$
$$x_2 = X_2 + 0.02tX_1$$

   If the density in the reference configuration is given by the function $\rho_0 = 1 + X_1X_2$ units of density, find expressions for $\rho_t(X, t)$, $\rho(x, t)$. For $t = 1$, draw the contour plots of $\det F$, $\rho_t(X, 1)$ and $\rho(x, 1)$.

2. A stress field on a body has the following form:

$$\sigma = \begin{pmatrix} \alpha x_1^2 + \delta x_1 + 3x_2 + x_3 & -\beta x_2 x_1 & 0 \\ -\beta x_2 x_1 & 5x_2^2 & 0 \\ 0 & 0 & \gamma x_3^2 \end{pmatrix}$$

   The body forces vector applied is given by:

$$\rho b = \begin{pmatrix} 10 \\ 0 \\ 0 \end{pmatrix}$$

   Find the values of the constants α, β, δ and γ such that the stress field is in equilibrium with the body forces vector field. (**Answer:** 5, 10, –10, 0)

3. A stress field on a body has the following form:

$$\sigma = \begin{pmatrix} \alpha x_1(x_1 - 5) + \gamma & \beta x_1 x_2(x_1 - 5)(x_2 - 5) & 0 \\ \beta x_1 x_2(x_1 - 5)(x_2 - 5) & 0 & 0 \\ 0 & 0 & 0 \end{pmatrix}$$

Find the values of the constants α, β and γ such that the stress field is in equilibrium with a zero body forces vector field and satisfies the boundary condition: $@x_1 = 0, \sigma_{11} = 5$. (**Answer:** 0, 0, 5)

4. A two–dimensional stress field on a plate of dimensions 1 and 5 units and 0.01 units thickness has the following form:

$$\sigma = \begin{pmatrix} 5x_1x_2 + B & A & 0 \\ A & 0 & 0 \\ 0 & 0 & 0 \end{pmatrix}$$

Find the equilibrium body forces vector and the constants A, B given that for $x_1 = 0$, $\sigma_{11} = 2$ units and $\sigma_{12} = 2$ units. Then, draw the contour plot of the von Mises stress on the plate, and identify the critical location (maximum von Mises stress).

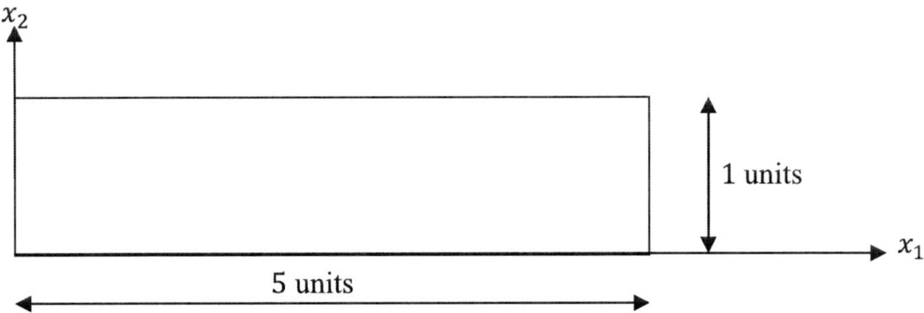

CHAPTER 5
Linear Elastic Materials Constitutive Equations

Experiments on different materials to find their response to mechanical loading is the first step before any design process can be employed. The process of finding the appropriate constitutive model for a material involves different steps. In the first step, preliminary experiments are conducted to identify the mathematical model that would best describe the behavior of the material. Once a model is chosen to best describe the material behavior, different constants in that model are identified as variables to be obtained from experiments. An experimental design is then set up to find the designated material constants by curve fitting the experimental data to the chosen mathematical model.

Different mathematical models are available to describe the behavior of materials. Traditional material–constitutive models describe two stages of material loading. In the first stage, the material shows no sign of damage during loading or unloading, and the original shape can be restored upon removal of the loading. This stage is termed the *elastic* response stage. Once a specified limiting load is reached, the material exhibits signs of damage and behaves *plastically*, usually by exhibiting some degradation in stiffness or some permanent deformation. In this chapter, some classifications of materials according to their mechanical constitutive behavior are listed. The behavior of the materials in the elastic response stage is fully described for linear elastic materials. The nonlinear elastic materials are described in a later chapter.

5.1. CLASSIFICATIONS OF MATERIALS MECHANICAL RESPONSE

5.1.1. According to Linearity

The mechanical response of a material can be classified as either *linear* or *nonlinear*. For a **linear** material, the load is directly proportional to the response. In such materials, the plot of a measure of the stress versus a measure of the strain exhibits a linear behavior during both loading and unloading. A **nonlinear** material is a material that exhibits any other response.

5.1.2. According to Energy Dissipation

The mechanical response of a material can be classified as either *elastic* or *inelastic*. An **elastic** material will not dissipate energy through a loading cycle (i.e., the amount of energy stored or released is path independent, and a specimen will return to its initial position once the load is removed). Such materials are also called **Green Elastic** or **Hyper–elastic**. An inelastic material, on the other hand, will dissipate energy through loading cycles.

Notice that elasticity and linearity are two different concepts. Traditional construction materials like concrete and steel have a linear elastic response where the load is proportional to the defor–mation and energy is not dissipated through loading cycles in the elastic loading stage. Rubber is an example of a nonlinear elastic material. The energy stored inside the material during loading is fully recoverable upon unloading, provided the load is within a certain limit. Thus, rubber is con–sidered elastic. However, the relationship between the load and the deformation is not necessarily linear.

5.1.3. According to Direction or Material Symmetries

The mechanical response of a material can be classified as either *isotropic* or *anisotropic*. Anisotropic materials could be *othrotropic* or *transversely isotropic*. The description of each is as follows:

- An **isotropic** material is a material whose response is independent of the loading orientation; i.e., if a specimen is loaded in any direction, the response would be the same. Examples of materials that behave in this manner are traditional concrete and steel materials.
- An **anisotropic** material possesses material properties that are dependent on the orientation of the specimen. There are different levels of anisotropy:
    - **General anisotropy**: A generally anisotropic material is one that exhibits no symmetric response along any specified axes.
    - **Orthotropic material**: Such material has a symmetric response along three perpendicular axes (e.g., a material with different fibers oriented in two or three orthogonal directions).
    - **Transversely isotropic material**: The material has a symmetric response along an axis that is normal to a transverse plane of isotropy. If the material is loaded in the transverse plane of isotropy, then the behavior is isotropic, i.e., the material properties will be independent of the orientation within that plane. Wood materials with fibers oriented in a certain direction are perfect examples of transversely

isotropic materials. Many biological tissues possessing internal fibers, such as ligaments, can also be considered transversely isotropic.

5.1.4. According to Homogeneity

The mechanical response of a material can be classified as either *homogenous* or *non–homogenous*. A **homogeneous** material has a response that is independent of the specimen used during a mechanical loading experiment. Most traditional engineering materials (e.g., steel) can be considered homogeneous. A **non–homogeneous** material has material properties that vary according to the location of the specimen used in the experiment.

5.1.5. According to Time Dependence

The mechanical response of a material can be classified as either *viscous* or *non–viscous*. A **viscous** material is a material whose response is time dependent (also called rate dependent). The mechanical response of a viscous material is dependent on the rate of application of the load or on whether the response varies with time for a static load. Creep and stress relaxation are examples of phenomena exhibited by viscous materials. A viscous material can exhibit creep under a constant load; i.e., under a constant load, the deformation will increase. A viscous material can also exhibit stress relaxation when subjected to a specified deformation; i.e., the load required to hold a specified deformation will decrease with time when that deformation is held constant. Many materials exhibit creep and stress relaxation, but the speed by which such materials creep or relax will dictate whether such viscous response is required to be studied for a certain application of the material. A **non–viscous** material is a material whose response is time independent (also called rate independent). The response of non–viscous materials is independent of the rate of the application of the load. Most engineering materials are time independent when the rate of load application is relatively low.

5.2. LINEAR ELASTIC MATERIALS

From a mathematical sense, a linear elastic material is fully represented by a linear map C_S between the stress matrix $\sigma \in \mathbb{M}^3$ and the infinitesimal strain matrix $\varepsilon \in \mathbb{M}^3$ such that $C_S: \sigma \longmapsto \varepsilon$. Generally speaking, such a linear map can be represented by a fourth–order tensor or a matrix with $3 \times 3 \times 3 \times 3 = 81$ components (See section 1.2.7.). These components are unique for a particular linear elastic material, and thus, they are material constants. By adopting the summation convention described in section 1.2.5.a, the linear map between the stress and the strain tensors can be written in component form as $\sigma_{ij} = C_{ijkl}\varepsilon_{kl}$. The components of the fourth–order tensor C_s can be visualized in the following form:

$$C_s = \begin{pmatrix} \begin{pmatrix} C_{1111} & C_{1112} & C_{1113} \\ C_{1121} & C_{1122} & C_{1123} \\ C_{1131} & C_{1132} & C_{1133} \end{pmatrix} & \begin{pmatrix} C_{1211} & C_{1212} & C_{1213} \\ C_{1221} & C_{1222} & C_{1223} \\ C_{1231} & C_{1232} & C_{1233} \end{pmatrix} & \begin{pmatrix} C_{1311} & C_{1312} & C_{1313} \\ C_{1321} & C_{1322} & C_{1323} \\ C_{1331} & C_{1332} & C_{1333} \end{pmatrix} \\ \begin{pmatrix} C_{2111} & C_{2112} & C_{2113} \\ C_{2121} & C_{2122} & C_{2123} \\ C_{2131} & C_{2132} & C_{2133} \end{pmatrix} & \begin{pmatrix} C_{2211} & C_{2212} & C_{2213} \\ C_{2221} & C_{2222} & C_{2223} \\ C_{2231} & C_{2232} & C_{2233} \end{pmatrix} & \begin{pmatrix} C_{2311} & C_{2312} & C_{2313} \\ C_{2321} & C_{2322} & C_{2323} \\ C_{2331} & C_{2332} & C_{2333} \end{pmatrix} \\ \begin{pmatrix} C_{3111} & C_{3112} & C_{3113} \\ C_{3121} & C_{3122} & C_{3123} \\ C_{3131} & C_{3132} & C_{3133} \end{pmatrix} & \begin{pmatrix} C_{3211} & C_{3212} & C_{3213} \\ C_{3221} & C_{3222} & C_{3223} \\ C_{3231} & C_{3232} & C_{3233} \end{pmatrix} & \begin{pmatrix} C_{3311} & C_{3312} & C_{3313} \\ C_{3321} & C_{3322} & C_{3323} \\ C_{3331} & C_{3332} & C_{3333} \end{pmatrix} \end{pmatrix}$$

5.2.1. Considerations for Symmetries of the Stress and Strain Tensors

The symmetry of the stress tensor, $\sigma_{ij} = \sigma_{ji}$, leads to the following relationship among the components of C_s:

$$C_{ijkl} = C_{jikl}$$

The symmetry of the strain tensor, $\varepsilon_{ij} = \varepsilon_{ji}$, leads to the following relationship among the components of C_s (**Why?**):

$$C_{ijkl} = C_{ijlk}$$

Thus, adopting these relationships, the number of independent material constants or components in C_s is reduced from 81 to 36 material constants, relating 6 components of stress to 6 components of strain. This reduced relation can be represented in matrix form by assuming that the stress and the infinitesimal strain matrices are vectors in $\mathbb{R}^6$ with the following components:

$$\sigma = \{\sigma_{11}, \sigma_{22}, \sigma_{33}, \sigma_{12}, \sigma_{13}, \sigma_{23}\}, \quad \varepsilon = \{\varepsilon_{11}, \varepsilon_{22}, \varepsilon_{33}, 2\varepsilon_{12}, 2\varepsilon_{13}, 2\varepsilon_{23}\}$$

Traditionally, the engineering shear strains: $\gamma_{12}, \gamma_{13}, \gamma_{23}$ are used in the constitutive relations. The engineering shear strains are equal to double the shear strains ε_{12}, ε_{13} and ε_{23} defined in the strain tensor, and hence, the constant 2 is introduced. The linear map between the stress and the strain can be represented by a matrix $C \in \mathbb{M}^6$. The subscript s in C_s is used to differentiate the fourth–order tensor C_s from the matrix representation and C. The relationship between the stress vector and the strain vector has the following form:

$$\sigma = C\varepsilon$$

Or, in component form:

$$\begin{pmatrix} \sigma_{11} \\ \sigma_{22} \\ \sigma_{33} \\ \sigma_{12} \\ \sigma_{13} \\ \sigma_{23} \end{pmatrix} = \begin{pmatrix} C_{1111} & C_{1122} & C_{1133} & C_{1112} & C_{1113} & C_{1123} \\ C_{2211} & C_{2222} & C_{2233} & C_{2212} & C_{2213} & C_{2223} \\ C_{3311} & C_{3322} & C_{3333} & C_{3312} & C_{3312} & C_{3323} \\ C_{1211} & C_{1222} & C_{1233} & C_{1212} & C_{1213} & C_{1223} \\ C_{1311} & C_{1322} & C_{1333} & C_{1312} & C_{1313} & C_{1323} \\ C_{2311} & C_{2322} & C_{2333} & C_{2312} & C_{2313} & C_{2323} \end{pmatrix} \begin{pmatrix} \varepsilon_{11} \\ \varepsilon_{22} \\ \varepsilon_{33} \\ 2\varepsilon_{12} \\ 2\varepsilon_{13} \\ 2\varepsilon_{23} \end{pmatrix}$$

While the stress and the small strain matrices are symmetric, so far, there is no restriction on C to be a symmetric matrix. However, in the next section it will be shown that, indeed, C is symmetric for energy considerations.

5.2.2. Energy Considerations

Since the material is elastic, then the energy stored during loading and/or the energy released during unloading have to be independent of the path of loading and/or unloading. Thus, for such energy considerations[1], a potential energy function $\bar{U}$ has to exist such that the stress is derived from this potential energy function (See sections 7.2 and 7.3):

$$\sigma_{ij} = \frac{\partial \bar{U}}{\partial \varepsilon_{ij}}$$

Considering C_s as the fourth–order tensor $C_s\colon \mathbb{M}^3 \longrightarrow \mathbb{M}^3$, then the components of C_s can be related to the derivatives of $\bar{U}$, as follows:

$$\sigma_{ij} = C_{ijkl}\varepsilon_{kl} \Longrightarrow C_{ijkl} = \frac{\partial \sigma_{ij}}{\partial \varepsilon_{kl}} = \frac{\partial^2 \bar{U}}{\partial \varepsilon_{ij}\partial \varepsilon_{kl}}$$

The equality of the second partial derivatives of $\bar{U}$ lead to the symmetry of C, as follows:

$$\frac{\partial^2 \bar{U}}{\partial \varepsilon_{ij}\partial \varepsilon_{kl}} = \frac{\partial^2 \bar{U}}{\partial \varepsilon_{kl}\partial \varepsilon_{ij}} \Longrightarrow C_{ijkl} = C_{klij}$$

[1]See Section 6 in Martin H. Sadd (2005). *Elasticity: Theory, Applications, and Numerics.* Elsevier.

Thus, the resulting fourth–order tensor C_s, after adopting the stress and strain symmetries and the symmetry due to the existence of a strain energy function, can be visualized as follows:

$$C_s = \begin{pmatrix} \begin{pmatrix} C_{1111} & C_{1112} & C_{1113} \\ C_{1112} & C_{1122} & C_{1123} \\ C_{1113} & C_{1123} & C_{1133} \end{pmatrix} & \begin{pmatrix} C_{1112} & C_{1212} & C_{1213} \\ C_{1212} & C_{1222} & C_{1223} \\ C_{1213} & C_{1223} & C_{1233} \end{pmatrix} & \begin{pmatrix} C_{1113} & C_{1213} & C_{1313} \\ C_{1213} & C_{1322} & C_{1323} \\ C_{1313} & C_{1323} & C_{1333} \end{pmatrix} \\ \begin{pmatrix} C_{1112} & C_{1212} & C_{1213} \\ C_{1212} & C_{1222} & C_{1223} \\ C_{1213} & C_{1223} & C_{1233} \end{pmatrix} & \begin{pmatrix} C_{1122} & C_{1222} & C_{1322} \\ C_{1222} & C_{2222} & C_{2223} \\ C_{1322} & C_{2223} & C_{2233} \end{pmatrix} & \begin{pmatrix} C_{1123} & C_{1223} & C_{1323} \\ C_{1223} & C_{2223} & C_{2323} \\ C_{1323} & C_{2323} & C_{2333} \end{pmatrix} \\ \begin{pmatrix} C_{1113} & C_{1213} & C_{1313} \\ C_{1213} & C_{1322} & C_{1323} \\ C_{1313} & C_{1323} & C_{1333} \end{pmatrix} & \begin{pmatrix} C_{1123} & C_{1223} & C_{1323} \\ C_{1223} & C_{2223} & C_{2323} \\ C_{1323} & C_{2323} & C_{2333} \end{pmatrix} & \begin{pmatrix} C_{1133} & C_{1233} & C_{1333} \\ C_{1233} & C_{2233} & C_{2333} \\ C_{1333} & C_{2333} & C_{3333} \end{pmatrix} \end{pmatrix}$$

The number of independent constants in the above tensor is indeed 21. If the vector representation of the stresses and the strains is considered, then the resulting relationship between the stress and the small strain is symmetric (The matrix C is symmetric). The matrix C thus has 21 independent constants, and its components can be visualized in the following relationship between the stress and strain vectors:

$$\begin{pmatrix} \sigma_{11} \\ \sigma_{22} \\ \sigma_{33} \\ \sigma_{12} \\ \sigma_{13} \\ \sigma_{23} \end{pmatrix} = \begin{pmatrix} C_{1111} & C_{1122} & C_{1133} & C_{1112} & C_{1113} & C_{1123} \\ & C_{2222} & C_{2233} & C_{2212} & C_{2213} & C_{2223} \\ & & C_{3333} & C_{3312} & C_{3312} & C_{3323} \\ & & & C_{1212} & C_{1213} & C_{1223} \\ & Symm & & & C_{1313} & C_{1323} \\ & & & & & C_{2323} \end{pmatrix} \begin{pmatrix} \varepsilon_{11} \\ \varepsilon_{22} \\ \varepsilon_{33} \\ 2\varepsilon_{12} \\ 2\varepsilon_{13} \\ 2\varepsilon_{23} \end{pmatrix}$$

The number of constants can be further reduced, depending on the symmetries of the material (See section 5.1.3.). If a material possesses a certain symmetry, this implies that the components of the matrix C_s do not change upon coordinate transformations within the specified symmetry. Since C_s is a fourth–order tensor, upon changing the coordinate system, the new material constants have the following form (See section 1.2.7.):

$$C'_{ijkl} = Q_{im}Q_{jn}Q_{ko}Q_{lp}C_{mnop}$$

If a material's behavior is independent of a certain change of the coordinate system described by an orthogonal matrix Q, then the components of C'_s should be equal to those of C_s. Such orthogonal matrix Q is termed a material symmetry. For example, if a material response is symmetric around a plane, then a reflection matrix Q across this plane would produce the same components of C_s' as those of C_s. If a material response is the same within any rotation inside a plane, then a rotation Q within that plane would produce the same components for C_s' as those for C_s. The number of the independent constants of C_s can be further reduced from 21, depending on the level of anisotropy of the material. In the following section, the different levels of anisotropy are described, and the physical interpretation of some of the material constants is given.

5.2.3. Orthotropic Linear Elastic Materials

In an orthotropic linear elastic material, the 21 material constants are reduced to 9 material con–stants fully describing the relationship between the stress and the small strain. Assuming that the basis vectors e_1, e_2 and e_3 are the axis of orthotropy, then the relationship between the stress and the strain is given by:

$$\begin{pmatrix} \sigma_{11} \\ \sigma_{22} \\ \sigma_{33} \\ \sigma_{12} \\ \sigma_{13} \\ \sigma_{23} \end{pmatrix} = \begin{pmatrix} C_{1111} & C_{1122} & C_{1133} & 0 & 0 & 0 \\ & C_{2222} & C_{1122} & 0 & 0 & 0 \\ & & C_{3333} & 0 & 0 & 0 \\ & & & C_{1212} & 0 & 0 \\ & Symm & & & C_{1313} & 0 \\ & & & & & C_{2323} \end{pmatrix} \begin{pmatrix} \varepsilon_{11} \\ \varepsilon_{22} \\ \varepsilon_{33} \\ 2\varepsilon_{12} \\ 2\varepsilon_{13} \\ 2\varepsilon_{23} \end{pmatrix}$$

For orthotropic materials, the relationship between the small strains and the stresses is tradition–ally given as function of nine physical material constants E_{11}, E_{22}, E_{33}, ν_{12}, ν_{13}, ν_{23}, G_{12}, G_{13}, and G_{23} where:

E_{ii}: The ratio of the uniaxial stress to uniaxial strain in the direction of the basis vector e_i

ν_{ij}: Negative the ratio between the axial strain in the direction of the basis vector e_j to the axial strain in the direction of the basis vector e_i when the material is uniaxially stressed in the direction of the basis vector e_i

G_{ij}: The ratio between the shear stress σ_{ij} to the engineering shear strain $\gamma_{ij} = 2\varepsilon_{ij}$ in the plane defined by the directions of e_i and e_j.

The relationship is written as follows:

$$\begin{aligned} \varepsilon_{11} &= \frac{\sigma_{11}}{E_{11}} - \nu_{21}\frac{\sigma_{22}}{E_{22}} - \nu_{31}\frac{\sigma_{33}}{E_{33}} \\ \varepsilon_{22} &= -\nu_{12}\frac{\sigma_{11}}{E_{11}} + \frac{\sigma_{22}}{E_{22}} - \nu_{32}\frac{\sigma_{33}}{E_{33}} \end{aligned}$$

$$\varepsilon_{33} = -\nu_{13}\frac{\sigma_{11}}{E_{11}} - \nu_{23}\frac{\sigma_{22}}{E_{22}} + \frac{\sigma_{33}}{E_{33}}$$

$$2\varepsilon_{12} = \gamma_{12} = \frac{\sigma_{12}}{G}, \quad 2\varepsilon_{13} = \gamma_{13} = \frac{\sigma_{13}}{G}, \quad 2\varepsilon_{23} = \gamma_{23} = \frac{\sigma_{23}}{G}$$

In this case, the relationship between the strains and the stresses can be written in matrix form as:

$$\begin{pmatrix} \varepsilon_{11} \\ \varepsilon_{22} \\ \varepsilon_{33} \\ \gamma_{12} \\ \gamma_{13} \\ \gamma_{23} \end{pmatrix} = \begin{pmatrix} \varepsilon_{11} \\ \varepsilon_{22} \\ \varepsilon_{33} \\ 2\varepsilon_{12} \\ 2\varepsilon_{13} \\ 2\varepsilon_{23} \end{pmatrix} = \begin{pmatrix} \frac{1}{E_{11}} & \frac{-\nu_{21}}{E_{22}} & \frac{-\nu_{31}}{E_{33}} & 0 & 0 & 0 \\ \frac{-\nu_{12}}{E_{11}} & \frac{1}{E_{22}} & \frac{-\nu_{32}}{E_{33}} & 0 & 0 & 0 \\ \frac{-\nu_{13}}{E_{11}} & \frac{-\nu_{23}}{E_{22}} & \frac{1}{E_{33}} & 0 & 0 & 0 \\ 0 & 0 & 0 & \frac{1}{G_{12}} & 0 & 0 \\ 0 & 0 & 0 & 0 & \frac{1}{G_{13}} & 0 \\ 0 & 0 & 0 & 0 & 0 & \frac{1}{G_{23}} \end{pmatrix} \begin{pmatrix} \sigma_{11} \\ \sigma_{22} \\ \sigma_{33} \\ \sigma_{12} \\ \sigma_{13} \\ \sigma_{23} \end{pmatrix}$$

Notice that symmetry of the matrix C implies that $\frac{\nu_{ij}}{E_{ii}} = \frac{\nu_{ji}}{E_{jj}}$.

5.2.4. Transversely Isotropic Elastic Materials

In a transversely isotropic linear elastic material, five material constants describe the relationship between the stress and the small strain. Assuming full isotropy in the plane formed by the basis vectors e_1 and e_2 (the material response is identical given any rotation around e_3) and that it is also a plane of symmetry, then, the relationship between the stress and the strain is given by five material constants:

$$\begin{pmatrix} \sigma_{11} \\ \sigma_{22} \\ \sigma_{33} \\ \sigma_{12} \\ \sigma_{13} \\ \sigma_{23} \end{pmatrix} = \begin{pmatrix} C_{1111} & C_{1122} & C_{1133} & 0 & 0 & 0 \\ & C_{1111} & C_{1133} & 0 & 0 & 0 \\ & & C_{3333} & 0 & 0 & 0 \\ & & & \frac{C_{1111} - C_{1122}}{2} & 0 & 0 \\ & Symm & & & C_{1313} & 0 \\ & & & & & C_{1313} \end{pmatrix} \begin{pmatrix} \varepsilon_{11} \\ \varepsilon_{22} \\ \varepsilon_{33} \\ 2\varepsilon_{12} \\ 2\varepsilon_{13} \\ 2\varepsilon_{23} \end{pmatrix}$$

For transversely isotropic materials, the relationship between the small strains and the stresses are traditionally given as function of five physical material constants E, ν, E_{33}, ν_{13}, and G_{13} where:

E: The ratio of the uniaxial stress to uniaxial strain in any direction in the isotropy plane

ν: Negative the ratio between the transverse strain to the axial strain when the material is uniaxially stressed in any direction in the isotropy plane

E_{33}: The ratio of the uniaxial stress to uniaxial strain in the direction perpendicular to the isotropy plane

ν_{13}: Negative the ratio between the axial strain ε_{33} to the axial strain ε_{11} when the material is stressed uniaxially in the direction of e_1

G_{13}: The ratio between the shear stress σ_{13} to the engineering shear strain $\gamma_{31} = 2\varepsilon_{13}$ in the plane defined by the directions of e_1 and e_3.

In this case, the relationship between the strains and the stresses can be written as:

$$\varepsilon_{11} = \frac{\sigma_{11}}{E} - \nu\frac{\sigma_{22}}{E} - \nu_{31}\frac{\sigma_{33}}{E_{33}}$$

$$\varepsilon_{22} = -\nu\frac{\sigma_{11}}{E} + \frac{\sigma_{22}}{E} - \nu\frac{\sigma_{33}}{E}$$

$$\varepsilon_{33} = -\nu_{13}\frac{\sigma_{11}}{E} - \nu_{13}\frac{\sigma_{22}}{E} + \frac{\sigma_{33}}{G_{33}}$$

$$2\varepsilon_{12} = \gamma_{12} = \frac{\sigma_{12}}{G}, \quad 2\varepsilon_{13} = \gamma_{13} = \frac{\sigma_{13}}{G_{13}}, \quad 2\varepsilon_{23} = \gamma_{23} = \frac{\sigma_{23}}{G_{13}}$$

Thus, the relationship between the stresses and the small strains has the following form:

$$\begin{pmatrix} \varepsilon_{11} \\ \varepsilon_{22} \\ \varepsilon_{33} \\ \gamma_{12} \\ \gamma_{13} \\ \gamma_{23} \end{pmatrix} = \begin{pmatrix} \varepsilon_{11} \\ \varepsilon_{22} \\ \varepsilon_{33} \\ 2\varepsilon_{12} \\ 2\varepsilon_{13} \\ 2\varepsilon_{23} \end{pmatrix} = \begin{pmatrix} \frac{1}{E} & \frac{-\nu}{E} & \frac{-\nu_{31}}{E_{33}} & 0 & 0 & 0 \\ \frac{-\nu}{E} & \frac{1}{E} & \frac{-\upsilon_{31}}{E_{33}} & 0 & 0 & 0 \\ \frac{-\nu_{13}}{E} & \frac{-\nu_{13}}{E} & \frac{1}{E_{33}} & 0 & 0 & 0 \\ 0 & 0 & 0 & \frac{1}{G} & 0 & 0 \\ 0 & 0 & 0 & 0 & \frac{1}{G_{13}} & 0 \\ 0 & 0 & 0 & 0 & 0 & \frac{1}{G_{13}} \end{pmatrix} \begin{pmatrix} \sigma_{11} \\ \sigma_{22} \\ \sigma_{33} \\ \sigma_{12} \\ \sigma_{13} \\ \sigma_{23} \end{pmatrix}$$

Notice that $G = \frac{E}{2(1+\nu)}$, while the symmetry of the matrix C implies that $\frac{\nu_{31}}{E_{33}} = \frac{\nu_{13}}{E}$.

5.2.5. Isotropic Linear Elastic Materials

In an isotropic linear elastic material, the response is independent of the coordinate system chosen. This assumption reduces the number of material constants that describe the behavior of the material in any direction to only two. The two constants that are generally used are the Young's modulus E and Poisson's ratio ν, with the following physical definitions:

Young's modulus E: The slope of the stress–strain curve during uniaxial tensile tests of a sample of a *linear elastic* material.

Poisson's ratio ν: Negative the ratio of the transverse strain (perpendicular to the applied load) to the axial strain (in the direction of the applied load) during uniaxial tensile tests of a samples of a *linear elastic* material. For such materials, the strains are given by the following equations:

$$\varepsilon_{11} = \frac{\sigma_{11}}{E} - \nu\frac{\sigma_{22}}{E} - \nu\frac{\sigma_{33}}{E}$$

$$\varepsilon_{22} = -\nu\frac{\sigma_{11}}{E} + \frac{\sigma_{22}}{E} - \nu\frac{\sigma_{33}}{E}$$

$$\varepsilon_{33} = -\nu\frac{\sigma_{11}}{E} - \nu\frac{\sigma_{22}}{E} + \frac{\sigma_{33}}{E}$$

$$2\varepsilon_{12} = \gamma_{12} = \frac{\sigma_{12}}{G}, \quad 2\varepsilon_{13} = \gamma_{13} = \frac{\sigma_{13}}{G}, \quad 2\varepsilon_{23} = \gamma_{23} = \frac{\sigma_{23}}{G}$$

and, thus, in matrix form:

$$\begin{pmatrix} \varepsilon_{11} \\ \varepsilon_{22} \\ \varepsilon_{33} \\ \gamma_{12} \\ \gamma_{13} \\ \gamma_{23} \end{pmatrix} = \begin{pmatrix} \varepsilon_{11} \\ \varepsilon_{22} \\ \varepsilon_{33} \\ 2\varepsilon_{12} \\ 2\varepsilon_{13} \\ 2\varepsilon_{23} \end{pmatrix} = \begin{pmatrix} \frac{1}{E} & \frac{-\nu}{E} & \frac{-\nu}{E} & 0 & 0 & 0 \\ \frac{-\nu}{E} & \frac{1}{E} & \frac{-\nu}{E} & 0 & 0 & 0 \\ \frac{-\nu}{E} & \frac{-\nu}{E} & \frac{1}{E} & 0 & 0 & 0 \\ 0 & 0 & 0 & \frac{1}{G} & 0 & 0 \\ 0 & 0 & 0 & 0 & \frac{1}{G} & 0 \\ 0 & 0 & 0 & 0 & 0 & \frac{1}{G} \end{pmatrix} \begin{pmatrix} \sigma_{11} \\ \sigma_{22} \\ \sigma_{33} \\ \sigma_{12} \\ \sigma_{13} \\ \sigma_{23} \end{pmatrix}$$

The inverse of the above relationship can be written as:

$$\begin{pmatrix} \sigma_{11} \\ \sigma_{22} \\ \sigma_{33} \\ \sigma_{12} \\ \sigma_{13} \\ \sigma_{23} \end{pmatrix} = \frac{E}{(1-2\nu)(1+\nu)} \begin{pmatrix} 1-\nu & \nu & \nu & 0 & 0 & 0 \\ \nu & 1-\nu & \nu & 0 & 0 & 0 \\ \nu & \nu & 1-\nu & 0 & 0 & 0 \\ 0 & 0 & 0 & \frac{1-2\nu}{2} & 0 & 0 \\ 0 & 0 & 0 & 0 & \frac{1-2\nu}{2} & 0 \\ 0 & 0 & 0 & 0 & 0 & \frac{1-2\nu}{2} \end{pmatrix} \begin{pmatrix} \varepsilon_{11} \\ \varepsilon_{22} \\ \varepsilon_{33} \\ 2\varepsilon_{12} \\ 2\varepsilon_{13} \\ 2\varepsilon_{23} \end{pmatrix}$$

The following Mathematica code was used to invert the strain–stress relationship:

```
Clear[Ee,Nu,G]
Cc={{1/Ee,-Nu/Ee,-Nu/Ee,0,0,0},{-Nu/Ee,1/Ee,-Nu/Ee,0,0,0},{-Nu/Ee,-
Nu/Ee,1/Ee,0,0,0},{0,0,0,1/G,0,0},{0,0,0,0,1/G,0},{0.,0,0,0,0,1/G}};
Dd=FullSimplify[Inverse[Cc]];
G=Ee/2/(1+Nu)
FullSimplify[Dd]//MatrixForm
FullSimplify[1/(Ee/(1-2Nu)/(1+Nu))*Dd]//MatrixForm
```

Problem: Starting from a state of pure shear stress σ_{12}, show that $G = \frac{E}{2(1+\nu)}$ for a linear elastic isotropic material.

Solution: To prove such relationship, we first assume a pure state of stress with σ_{12} being the only non–zero stress component. Using the constitutive relationships for linear elastic isotropic mate–rial, the only nonzero strain component is $\varepsilon_{12} = \frac{\gamma}{2} = \frac{\sigma_{12}}{2G}$. The stress and strain matrices are:

$$\sigma = \begin{pmatrix} 0 & \sigma_{12} & 0 \\ \sigma_{12} & 0 & 0 \\ 0 & 0 & 0 \end{pmatrix}, \quad \varepsilon = \begin{pmatrix} 0 & \varepsilon_{12} & 0 \\ \varepsilon_{12} & 0 & 0 \\ 0 & 0 & 0 \end{pmatrix}$$

Using the procedures described in the examples following section 2.1.4, the stress matrix can be diagonalized in a new basis system using the matrix of eigenvectors:

$$Q = \begin{pmatrix} 1/\sqrt{2} & 1/\sqrt{2} & 0 \\ -1/\sqrt{2} & 1/\sqrt{2} & 0 \\ 0 & 0 & 1 \end{pmatrix}, \quad \sigma' = Q\sigma Q^T = \begin{pmatrix} \sigma_{12} & 0 & 0 \\ 0 & -\sigma_{12} & 0 \\ 0 & 0 & 0 \end{pmatrix} = \begin{pmatrix} \sigma'_{11} & 0 & 0 \\ 0 & \sigma'_{22} & 0 \\ 0 & 0 & 0 \end{pmatrix}$$

The strain matrix can also be written in the new basis system:

$$\varepsilon' = Q\varepsilon Q^T = \begin{pmatrix} \varepsilon_{12} & 0 & 0 \\ 0 & -\varepsilon_{12} & 0 \\ 0 & 0 & 0 \end{pmatrix} = \begin{pmatrix} \varepsilon'_{11} & 0 & 0 \\ 0 & \varepsilon'_{22} & 0 \\ 0 & 0 & 0 \end{pmatrix}$$

Since the material is linear isotropic, the relationship between the stress and the strain is the same in any coordinate system; therefore, in the new coordinate system, we have:

$$\varepsilon'_{11} = \frac{\sigma'_{11}}{E} - \frac{\nu\sigma'_{22}}{E} \Longrightarrow \varepsilon_{12} = \frac{\sigma_{12}}{E} + \frac{\nu\sigma_{12}}{E}$$

Using the relationship between ε_{12} and σ_{12} in terms of G:

$$\frac{\sigma_{12}}{2G} = \frac{\sigma_{12}}{E} + \frac{\nu\sigma_{12}}{E} \Longrightarrow G = \frac{E}{2(1+\nu)}$$

The following Mathematica code shows how to solve the eigenvalue problem described above.

```
sigma={{0,s12,0},{s12,0,0},{0,0,0}}
epsilon={{0,e12,0},{e12,0,0},{0,0,0}};
s=Eigensystem[sigma]
Q={s[[2,3]]/Norm[s[[2,3]]],s[[2,2]]/Norm[s[[2,2]]],s[[2,1]]/Norm[s[[2,1]]]};
Q//MatrixForm
sdash=Q.sigma.Transpose[Q];
epsilondash=Q.epsilon.Transpose[Q];
sdash//MatrixForm
epsilondash//MatrixForm
```

5.2.5.a. Alternative Material Constants

The stress–strain relationships described above rely on the definition of two material constants: Young's modulus and Poisson's ratio. However, many other material constants can also be used; in particular, the shear modulus G, the bulk modulus K, and the Lamé's constants λ and μ are often used instead. The following are the physical definitions:

Shear modulus: It is the slope of the shear stress versus engineering shear strain curve of a sample of a *linear elastic* material. It was shown in the previous problem that:

$$G = \frac{E}{2(1+\nu)}$$

Bulk modulus: If a sample of a *linear elastic* material is subjected to a hydrostatic state of stress $\sigma = pI$, then the ratio between p and the volumetric strain defined as $(\varepsilon_v = \varepsilon_{11} + \varepsilon_{22} + \varepsilon_{33})$ is termed the bulk modulus. It is left to the reader **(Bonus Exercise)** to show that:

$$K = \frac{E}{3(1-2\nu)}$$

Lamé's constants: These are two constants that provide a simplified way of writing the constitutive equations of linear elastic isotropic materials. They have the following form:

$$\lambda = \frac{E\nu}{(1+\nu)(1-2\nu)}, \quad \mu = G = \frac{E}{2(1+\nu)}$$

5.2.5.b. Derivations Using Mathematica

The following Mathematica code is used to reduce the number of independent coefficients in the fourth–order tensor. In the first part, the symmetry of the stress tensor and the symmetry of the strain tensor are utilized to reduce the number of independent coefficients from 81 to 36. Then, the energy considerations concept is utilized to reduce the number of independent coefficients from 36 to 21. Next, reflections along the directions of e_1, e_2 and e_3 are applied to find the number of coefficients required for an orthotropic material, with the planes of symmetry being the planes perpendicular to the three basis vectors. The number of independent constants is shown to be 9. Then, the number of independent constants in a transversely isotropic material is found by utilizing the orthotropic material fourth–order tensor and assuming that the behavior of the material is independent of a 45° rotation in the plane of e_1 and e_2. The resulting number of independent material constants is shown to be 5. Finally, another 45° rotation around e_1 is used to show that the number of independent material constants for an isotropic material is just 2. Copy and paste the code into your Mathematica browser, and follow it step by step.

▶ **BONUS EXERCISE:** Show that the results for transversely isotropic and isotropic materials are independent of the angle of rotation used in the derivation, as long as the angle is not a multiple of $\frac{\pi}{2}$.

```
(*General Ctensor (fourth order tensor)*)
Ctensor=Table[Row[{c,i,j,k,l}],{i,1,3},{j,1,3},{k,1,3},{l,1,3}];
Print["A general C matrix relating 9 stress tensor components to 9 strain
tensor components:"]
Ctensor//MatrixForm

(*Applying stress symmetry*)
RuleStressSymmetry={};
Do[AppendTo[RuleStressSymmetry,Ctensor[[i,j,k,l]]-
>Ctensor[[j,i,k,l]]],{i,2,3},{j,1,i-1},{k,1,3},{l,1,3}];
Print["Stress tensor symmetry leads to the following
relations:",RuleStressSymmetry]
CtensorSt=Ctensor/.RuleStressSymmetry;
Print["The resulting C matrix from stress tensor symmetry is:"]
CtensorSt//MatrixForm

(*Applying Strain symmetry*)
RuleStrainSymmetry={};
Do[AppendTo[RuleStrainSymmetry,CtensorSt[[i,j,k,l]]-
>CtensorSt[[i,j,l,k]]],{i,1,3},{j,1,3},{k,2,3},{l,1,k-1}];
Print["Strain tensor symmetry leads to the following
relations:",RuleStrainSymmetry]
CtensorStStr=CtensorSt/.RuleStrainSymmetry;
Print["The resulting C matrix from both stress and strain tensors symmetry
is:"]
CtensorStStr//MatrixForm

(*Using Energy symmetry*)
RuleEnergySymmetry={};
Do[AppendTo[RuleEnergySymmetry,CtensorStStr[[i,j,k,l]]-
>CtensorStStr[[k,l,i,j]]],{i,1,3},{j,i,3},{k,1,i},{l,k,If[k<i,3,j]}];
Print["Symmetry due to energy function leads to the following
relations:",RuleEnergySymmetry]
CtensorStStrEn=CtensorStStr/.RuleEnergySymmetry;
```

```
Print["The resulting C matrix after using stress, strain and energy
symmetries is:"]
CtensorStStrEn//MatrixForm

Print["The original C tensor and the fully symmetric Ctensor:"]
{Ctensor//MatrixForm,CtensorStStrEn//MatrixForm}
Print["The difference (minus) between C tensor and the fully symmetric
Ctensor:"]
Ctensor-CtensorStStrEn//MatrixForm

(*Adopting the new Ctensor after symmetries*)
Ctensor=CtensorStStrEn;

(*Order of the stress vector*)
order={{1,1},{2,2},{3,3},{1,2},{1,3},{2,3}};
(*Converting Ctensor to Cmatrix*)
Cmatrix=Table[ToExpression[ToString[Row[{"Ctensor","[[",Row[Table[If[OddQ[i]
,Flatten[{order[[j]],order[[k]]}][[(i+1)/2]],","],{i,1,7}]],"]]"}]]],{j,1,6}
,{k,1,6}];
Print["The Following is the resulting C matrix after adopting all the
symmetries:"]
Cmatrix//MatrixForm


(*Orthotropic material*)
Print["Deriving C tensor and C matrix for Orthotropic Materials"]
(*The following are three reflections around the three planes of symmetry
perpendicular to e1=ex and e2=ey and e3=ez*)
Qx=ReflectionMatrix[{1,0,0}];
Qy=ReflectionMatrix[{0,1,0}];
Qz=ReflectionMatrix[{0,0,1}];
(*Using plane perpendicular to x as a symmetry plane*)
Cdashtensor=Table[Sum[Qx[[i,m]]*Qx[[j,n]]*Qx[[k,o]]*Qx[[l,p]]*Ctensor[[m,n,o,
p]],{m,1,3},{n,1,3},{o,1,3},{p,1,3}],{i,1,3},{j,1,3},{k,1,3},{l,1,3}];
Print["The following is the C tensor after reflection around the plane
perpendicular to e1=ex:"]

Cdashtensor//MatrixForm
Print["Following is the difference between the C tensor after reflection
around the plane perpendicular to e1=ex and the original symmetric C
tensor:"]
(Ctensor-Cdashtensor)//MatrixForm
zero=Table[0,{i,1,3},{j,1,3},{k,1,3},{l,1,3}];
Print["Following are the relationships after equating C tensor
after reflection to the Ctensor before reflection around the plane
perpendicular to e1=ex:"]
ax=Solve[(Ctensor-Cdashtensor)==zero,Flatten[Ctensor]]
Ctensorx=Ctensor/.ax[[1]];
Print["Following is the resulting C tensor after adopting the previous
relations:"]
```

```
Ctensorx//MatrixForm

(*Using plane perpendicular to y as a symmetry plane*)
Cdashtensor=Table[Sum[Qy[[i,m]]*Qy[[j,n]]*Qy[[k,o]]*Qy[[l,p]]*Ctensor[[m,n,o,
p]],{m,1,3},{n,1,3},{o,1,3},{p,1,3}],{i,1,3},{j,1,3},{k,1,3},{l,1,3}];
Print["Following is the C tensor after reflection around the plane
perpendicular to e2=ey:"]
Cdashtensor//MatrixForm
Print["Following is the difference between the C tensor after reflection
around the plane perpendicular to e2=ey and the original symmetric C
tensor:"]
(Ctensor-Cdashtensor)//MatrixForm
Print["Following are the relationships after equating C tensor after
reflection to the Ctensor before reflection around the plane perpendicular
to e2=ey:"]
ay=Solve[(Ctensor-Cdashtensor)==zero,Flatten[Ctensor]]
Ctensory=Ctensor/.ay[[1]];
Print["Following is the resulting C tensor after adopting the previous
relations:"]
Ctensory//MatrixForm

(*Using plane perpendicular to z as a symmetry plane*)
Cdashtensor=Table[Sum[Qz[[i,m]]*Qz[[j,n]]*Qz[[k,o]]*Qz[[l,p]]*Ctensor[[m,n,o
,p]],{m,1,3},{n,1,3},{o,1,3},{p,1,3}],{i,1,3},{j,1,3},{k,1,3},{l,1,3}];
Print["Following is the C tensor after reflection around the plane
perpendicular to e3=ez:"]
Cdashtensor//MatrixForm
Print["Following is the difference between the C tensor after reflection
around the plane perpendicular to e3=ez and the original symmetric C
tensor:"]
(Ctensor-Cdashtensor)//MatrixForm
Print["Following are the relationships after equating C tensor after
reflection to the Ctensor before reflection around the plane perpendicular
to e3=ez:"]
az=Solve[(Ctensor-Cdashtensor)==zero,Flatten[Ctensor]]
Ctensorz=Ctensor/.az[[1]];
Print["Following is the resulting C tensor after adopting the previous
relations:"]
Ctensorz//MatrixForm

CtensorOrthotropic=Ctensor/.Flatten[{az[[1]],ay[[1]],ax[[1]]}];
Print["This is the resulting C tensor after adopting the orthotropic
material symmetries:"]
CtensorOrthotropic//MatrixForm


(*Order of the stress vector*)
order={{1,1},{2,2},{3,3},{1,2},{1,3},{2,3}};
(*Converting Ctensor to Cmatrix*)
CmatrixOrthotropic=Table[ToExpression[ToString[Row[{"CtensorOrthotropic","[["
```

```
,Row[Table[If[OddQ[i],Flatten[{order[[j]],order[[k]]}][[(i+1)/2]],","],{i,1,
7}]],"]]"}]]],{j,1,6},{k,1,6}];
Print["Following is the resulting C matrix after adopting the orthotropic
material symmetries:"]
CmatrixOrthotropic//MatrixForm


(*Transverse Isotropy*)
Print["Deriving C tensor of a Transversely Isotropic material from C matrix
of Orthotropic Materials assumine that e1=ex and e2=ey form the plane of
isotropy"]
(*Rotation around z with 45 Degrees*)
Q=RotationMatrix[45Degree,{0,0,1}];
Cdashtensor=Table[Sum[Q[[i,m]]*Q[[j,n]]*Q[[k,o]]*Q[[l,p]]*CtensorOrthotropic
[[m,n,o,p]],{m,1,3},{n,1,3},{o,1,3},{p,1,3}],{i,1,3},{j,1,3},{k,1,3},{l,1,3}
];
Print["Following is the resulting C tensor after rotating the orthotropic
material C matrix with some angle in the plane of isotropy:"]
Cdashtensor=FullSimplify[Cdashtensor];
Cdashtensor//MatrixForm
Print["Following are the differences between the previous C tensor and the
Orthotropic material C tensor:"]
(CtensorOrthotropic-Cdashtensor)//MatrixForm

zero=Table[0,{i,1,3},{j,1,3},{k,1,3},{l,1,3}];
Vars={};
Do[If[CtensorOrthotropic[[i,j,k,l]]==0,,,AppendTo[Vars,CtensorOrthotropic[[i
,j,k,l]]]],{i,3,1,-1},{j,3,1,-1},{k,3,1,-1},{l,3,1,-1}];

Vars;
Print["Following are the relationships after equating C tensor after
rotating in the plane of isotropy to the Ctensor of orthotropic material:"]
a=Solve[(CtensorOrthotropic-Cdashtensor)==zero,Vars]
CtensorTransverseIsotropic=CtensorOrthotropic/.a[[1]];
Print["Following is the C tensor for a Transversely Isotropic Material:"]
CtensorTransverseIsotropic=FullSimplify[CtensorTransverseIsotropic];
CtensorTransverseIsotropic//MatrixForm

(*Order of the stress vector*)
order={{1,1},{2,2},{3,3},{1,2},{1,3},{2,3}};
(*Converting Ctensor to Cmatrix*)
CmatrixTransverseIsotropic=Table[ToExpression[ToString[Row[{"CtensorTransver
seIsotropic","[[",Row[Table[If[OddQ[i],Flatten[{order[[j]],order[[k]]}][[(i
+1)/2]],","],{i,1,7}]],"]]"}]]],{j,1,6},{k,1,6}];
CmatrixTransverseIsotropic=FullSimplify[CmatrixTransverseIsotropic];
Print["Following is the C matrix for a Transversely Isotropic Material:"]
CmatrixTransverseIsotropic//MatrixForm

(*Full Isotropy*)
Print["Deriving C tensor of an Isotropic material from the previous
```

```
Transversely Isotropic Materials C tensor"]
(*Rotation around x with 45 Degrees*)
Q=RotationMatrix[45Degree,{1,0,0}];
Cdashtensor=Table[Sum[Q[[i,m]]*Q[[j,n]]*Q[[k,o]]*Q[[l,p]]*CtensorTransverseI
sotropic[[m,n,o,p]],{m,1,3},{n,1,3},{o,1,3},{p,1,3}],{i,1,3},{j,1,3},{k,1,3}
,{l,1,3}];
Print["Following is the resulting C tensor after rotating the Transversely
Isotropic material C matrix with some angle around x:"]
Cdashtensor=FullSimplify[Cdashtensor];
Cdashtensor//MatrixForm
Print["Following are the differences between the previous C tensor and the
Transversely Isotropic material C tensor:"]
(CtensorTransverseIsotropic-Cdashtensor)//MatrixForm

zero=Table[0,{i,1,3},{j,1,3},{k,1,3},{l,1,3}];
Print["Following are the relationships after equating C tensor after
rotating to the Ctensor of transversely isotropic material:"]
a=Solve[(CtensorTransverseIsotropic-Cdashtensor)==zero,Vars]
CtensorIsotropic=CtensorTransverseIsotropic/.a[[1]];
Print["Following is the C tensor for an Isotropic Material:"]
CtensorIsotropic=FullSimplify[CtensorIsotropic];
CtensorIsotropic//MatrixForm

(*Order of the stress vector*)
order={{1,1},{2,2},{3,3},{1,2},{1,3},{2,3}};
(*Converting Ctensor to Cmatrix*)
CmatrixIsotropic=Table[ToExpression[ToString[Row[{"CtensorIsotropic","[[",Ro
w[Table[If[OddQ[i],Flatten[{order[[j]],order[[k]]}][[(i+1)/2]],","],{i,1,7}]
],"]]"}]]],{j,1,6},{k,1,6}];
CmatrixIsotropic=FullSimplify[CmatrixIsotropic];
Print["Following is the C matrix for an Isotropic Material:"]
CmatrixIsotropic//MatrixForm
```

5.2.6. Restrictions on the Elastic Moduli

The strain energy density $\bar{U}$ (strain energy per unit volume) stored in any linear elastic material under a general state of stress can be shown to be equal to (See section 7.2.):

$$\bar{U} = \frac{1}{2}\sigma_{ij}\varepsilon_{ij} = \frac{1}{2}(\sigma_{11}\varepsilon_{11} + \sigma_{22}\varepsilon_{22} + \sigma_{33}\varepsilon_{33} + 2\sigma_{12}\varepsilon_{12} + 2\sigma_{13}\varepsilon_{13} + 2\sigma_{23}\varepsilon_{23})$$

As shown in section 5.2, we can consider the stress σ and the strain to be vectors in $\mathbb{R}^6$. We can then use the dot product to find a simplified expression for the strain–energy density function:

$$\bar{U} = \frac{1}{2}(\sigma \cdot \varepsilon) = \frac{1}{2}(\varepsilon \cdot \sigma) = \frac{1}{2}(\varepsilon \cdot C\varepsilon)$$

If a material is stable and deformable, then the material would deform in the direction of the applied stress and would do that in a stable manner such that the strain energy stored is greater than zero. In such case, C is restricted to satisfy $\forall\varepsilon \neq \hat{0}: \bar{U} > 0$. Since C is a symmetric matrix, it also happens to be positive definite (See section 1.2.6.g.). Mathematically, this invokes restrictions on the different material constants.

5.2.6.a. Restrictions on the Numerical Values of the Linear Elastic Isotropic Material Constants

The process of finding restrictions on the different moduli requires ensuring that $\forall\varepsilon \neq \hat{0}$: $\bar{U} = \frac{1}{2}(\varepsilon \cdot C\varepsilon) > 0$.

The first restriction is that Young's modulus has to be higher than 0. This can be shown by assuming a state of uniaxial stress. In this case, the nonzero strains are ε_{11}, ε_{22}, and ε_{33}. Half the dot product between the strains and the stresses gives the strain energy stored per unit volume of the material:

$$\frac{1}{2}\varepsilon \cdot C\varepsilon = \begin{pmatrix} \frac{\sigma_{11}}{E} \\ -\frac{\nu\sigma_{11}}{E} \\ -\frac{\nu\sigma_{11}}{E} \\ 0 \\ 0 \\ 0 \end{pmatrix} \cdot \frac{E}{2(1-2\nu)(1+\nu)} \begin{pmatrix} 1-\nu & \nu & \nu & 0 & 0 & 0 \\ \nu & 1-\nu & \nu & 0 & 0 & 0 \\ \nu & \nu & 1-\nu & 0 & 0 & 0 \\ 0 & 0 & 0 & \frac{1-2\nu}{2} & 0 & 0 \\ 0 & 0 & 0 & 0 & \frac{1-2\nu}{2} & 0 \\ 0 & 0 & 0 & 0 & 0 & \frac{1-2\nu}{2} \end{pmatrix}$$

$$\cdot \begin{pmatrix} \frac{\sigma_{11}}{E} \\ -\frac{\nu\sigma_{11}}{E} \\ -\frac{\nu\sigma_{11}}{E} \\ 0 \\ 0 \\ 0 \end{pmatrix} = \frac{\sigma_{11}^2}{2E}$$

Thus, for the material to be deformable and stable under uniaxial loading, the following is a restriction on Young's modulus:

$$\frac{\sigma_{11}^2}{E} > 0 \Longrightarrow \infty > E > 0$$

Another restriction can be obtained by assuming a state of pure shear stress σ_{12}. Half the dot product between the stresses and the strains in this case is:

$$\frac{1}{2}\varepsilon\cdot C\varepsilon=\begin{pmatrix}0\\0\\0\\\sigma_{12}/G\\0\\0\end{pmatrix}\cdot\frac{E}{2\,(1-2\nu)\,(1+\nu)}\begin{pmatrix}1-\nu & \nu & \nu & 0 & 0 & 0\\ \nu & 1-\nu & \nu & 0 & 0 & 0\\ \nu & \nu & 1-\nu & 0 & 0 & 0\\ 0 & 0 & 0 & \frac{1-2\nu}{2} & 0 & 0\\ 0 & 0 & 0 & 0 & \frac{1-2\nu}{2} & 0\\ 0 & 0 & 0 & 0 & 0 & \frac{1-2\nu}{2}\end{pmatrix}\begin{pmatrix}0\\0\\0\\\sigma_{12}/G\\0\\0\end{pmatrix}$$

$$=\frac{\sigma_{12}^2}{2G}=\frac{2\,(1+\nu)\,\sigma_{12}^2}{2E}$$

Thus, for the material to be deformable and stable under shear loading, the following is a restriction on Poisson's ratio:

$$\frac{2(1+\nu)\sigma_{12}^2}{2E}>0\Longrightarrow(1+\nu)>0\Longrightarrow\nu>-1$$

The third and final restriction can be obtained by assuming a state of hydrostatic stress $\sigma=pI$. The dot product between the stresses and the strains in this case is:

$$\frac{\varepsilon\cdot C\varepsilon}{2}=\begin{pmatrix}\frac{p(1-2\nu)}{E}\\\frac{p(1-2\nu)}{E}\\\frac{p(1-2\nu)}{E}\\0\\0\\0\end{pmatrix}\cdot\frac{E}{2(1-2\nu)\,(1+\nu)}\begin{pmatrix}1-\nu & \nu & \nu & 0 & 0 & 0\\ \nu & 1-\nu & \nu & 0 & 0 & 0\\ \nu & \nu & 1-\nu & 0 & 0 & 0\\ 0 & 0 & 0 & \frac{1-2\nu}{2} & 0 & 0\\ 0 & 0 & 0 & 0 & \frac{1-2\nu}{2} & 0\\ 0 & 0 & 0 & 0 & 0 & \frac{1-2\nu}{2}\end{pmatrix}$$

$$\cdot\begin{pmatrix}\frac{p\,(1-2\nu)}{E}\\\frac{p\,(1-2\nu)}{E}\\\frac{p\,(1-2\nu)}{E}\\0\\0\\0\end{pmatrix}=\frac{p^2}{2K}=\frac{3\,(1-2\nu)\,p^2}{2E}$$

Thus, for the material to be deformable and stable under hydrostatic loading, the following is a restriction on Poisson's ratio:

$$\frac{3(1-2\nu)p^2}{2E} > 0 \Longrightarrow (1-2\nu) > 0 \Longrightarrow \nu < 1/2$$

It can be easily shown that the above restrictions ensure that, under any other state of stress or strain, the material is stable and deformable. Thus, the restrictions on the elastic moduli for a linear–elastic isotropic material are:

$$E > 0, \quad -1 < \nu < 0.5$$

The above restrictions on Young's modulus and Poisson's ratio lead to the following restrictions on the other material constants described before:

$$G, K, \mu > 0, \quad \lambda + \frac{2}{3}\mu > 0$$

5.2.6.b. Restrictions on the Numerical Values of the Linear-Elastic Orthotropic Material Constants

Following similar arguments, the following restrictions are obtained for the linear–elastic ortho–tropic material constants:

$$E_{11}, E_{22}, E_{33}, G_{12}, G_{13}, G_{23} > 0$$

$$|\nu_{ij}| < \left(\frac{E_i}{E_j}\right)^{1/2}$$

$$1-\nu_{12}\nu_{21}-\nu_{13}\nu_{31}-\nu_{23}\nu_{32}-2\nu_{21}\nu_{13}\nu_{32} > 0$$

The above equations show that values for the different orthotropic Poisson's ratios are not bounded by –1 and 0.5. In fact, orthotropic materials (typically, fibrous biological materials like ligaments) can exhibit values beyond those bounds. For example, ligaments can exhibit a Poisson's ratio that is higher than 0.5.

▶ **BONUS EXERCISES:**

- Deduce the different restrictions on the orthotropic linear–elastic material constants using physical and/or mathematical arguments.
- What does a Poisson's ratio higher than 0.5 indicate in terms of volume changes upon uniaxial loading?

Important Note: The above restrictions ensure that, physically, a material is stable (possessing a rising stress strain curve) and that any state of stress will cause nonzero strain values inside the material. It is important to realize that it cannot be argued that $E < 0$ and/or that $\nu = 0.5$ are physically impossible. They are simply unacceptable constants for *linear–elastic materials* with a positive definite constitutive model. Materials that become unstable with loading (e.g., decrease in stress accompanying an increase in strain) or incompressible materials (materials exhibiting no strain for a stress state $\sigma = pI, \sigma \in \mathbb{M}^3, p \in \mathbb{R}$) cannot be modeled using the linear–elastic constitutive model.

5.2.7. Plane Isotropic Linear Elastic Materials Constitutive Laws

In many solid mechanics problems, it is possible, due to loading and geometry symmetries, to simplify particular problems to be solved in $\mathbb{R}^2$ rather than $\mathbb{R}^3$. In such cases, the constitutive laws can be reduced to account for such symmetries. In the following, the material is assumed to be isotropic linear elastic and the measure of strain is the infinitesimal strain tensor.

5.2.7.a. Plane Stress

As the title implies, plane stress is the condition that the material is allowed to freely expand or contract in a certain direction (called the thickness direction). Loads and stresses are only applied in a plane perpendicular to thickness direction (Figure 5–1a). A thin sheet of material under in–plane loading without including any out–of plane relative deformations is usually the prime example for plane stress. Assuming that the thickness direction is the direction of the basis vector e_3, then the stress components σ_{23}, σ_{32}, and σ_{33} have zero values. Additionally, the engineering strain components γ_{23} and γ_{32} also have zero values. The strain component ε_{33} is not restricted to be zero but does not appear in the constitutive equation. The constitutive relationship has the form:

$$\begin{pmatrix} \varepsilon_{11} \\ \varepsilon_{22} \\ \gamma_{12} = 2\varepsilon_{12} \end{pmatrix} = \frac{1}{E}\begin{pmatrix} 1 & -\nu & 0 \\ -\nu & 1 & 0 \\ 0 & 0 & 2(1+\nu) \end{pmatrix}\begin{pmatrix} \sigma_{11} \\ \sigma_{22} \\ \sigma_{12} \end{pmatrix}$$

The axial strain ε_{33}, measuring the extension or contraction in the thickness direction can be calculated using the equation:

$$\varepsilon_{33} = -\frac{\nu}{E}(\sigma_{11} + \sigma_{22})$$

The relationship between the strains and the stresses can be inverted to have the following form:

$$\sigma = C\varepsilon$$

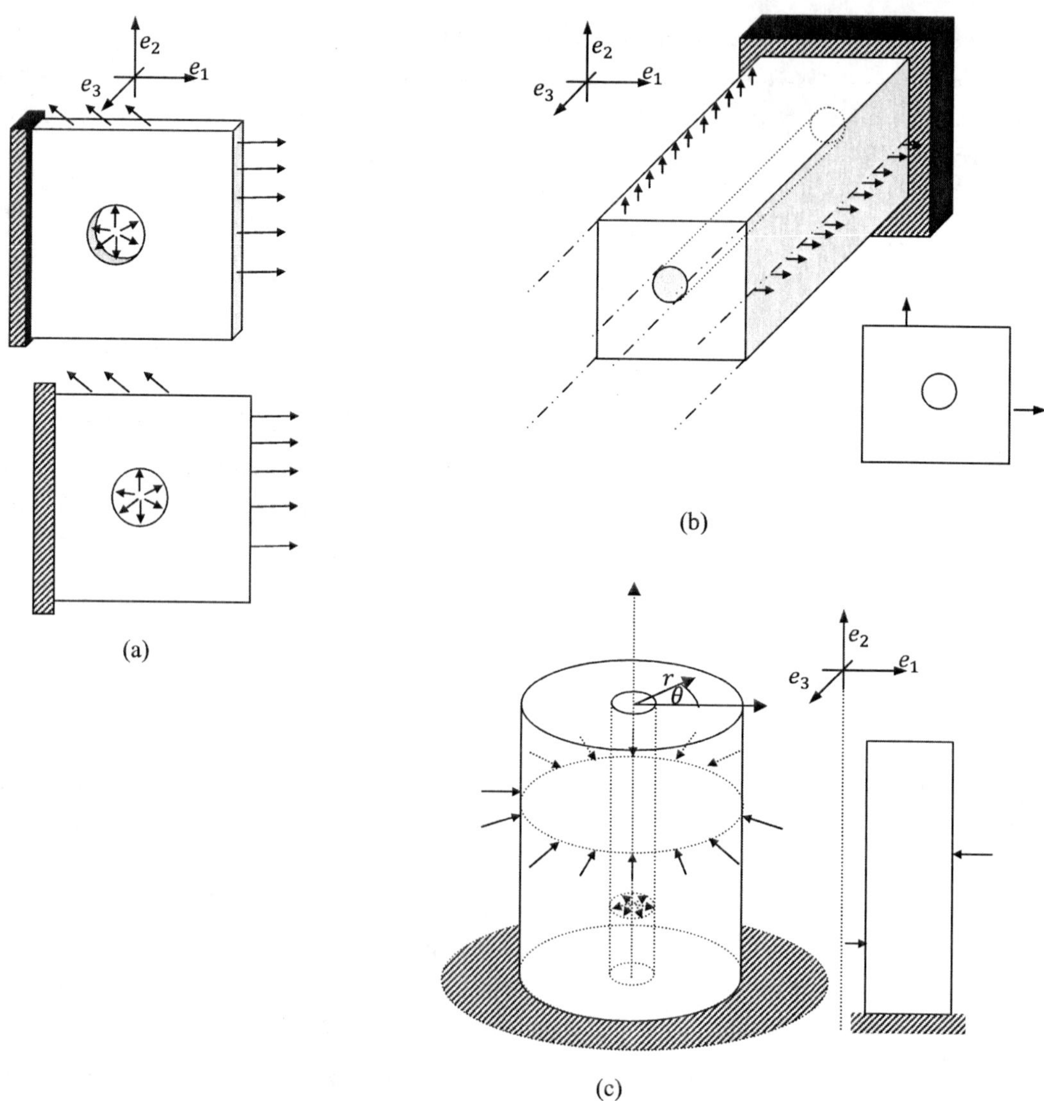

FIGURE 5-1 Examples illustrating a) Plane stress: 3D view and section to be modeled, b) Plane strain: 3D view of part of the infinitely long model and the cross section to be modeled, and c) Axisymmetry: 3D view and the cross section to be modelled.

where, C is a 3×3 material constitutive relationship matrix and has the following component form:

$$\begin{pmatrix} \sigma_{11} \\ \sigma_{22} \\ \sigma_{12} \end{pmatrix} = \frac{E}{(1+\nu)} \begin{pmatrix} \dfrac{1}{1-\nu} & \dfrac{\nu}{1-\nu} & 0 \\ \dfrac{\nu}{1-\nu} & \dfrac{1}{1-\nu} & 0 \\ 0 & 0 & \dfrac{1}{2} \end{pmatrix} \begin{pmatrix} \varepsilon_{11} \\ \varepsilon_{22} \\ \gamma_{12} = 2\varepsilon_{12} \end{pmatrix}$$

The following Mathematica code was utilized to obtain C:

```
Dd=1/Ee*{{1,-nu,0},{-nu,1,0},{0,0,2(1+nu)}};
Cc=FullSimplify[Inverse[Dd]]
Cc//MatrixForm
```

5.2.7.b. Plane Strain

As the title implies, in plane strain, the material is constrained from deformation in a certain direction (called the thickness direction). As a result, stresses are induced in the thickness direction. Any loading in the plane perpendicular to the thickness direction is assumed to exist on every cross section perpendicular to the thickness direction (Figure 5–1b). Assuming that the thickness direction is the direction of the basis vector e_3, then the infinitesimal strain components ε_{23}, ε_{32}, and ε_{33} have zero values. Additionally, the stress components σ_{23} and σ_{32} also have zero values. The stress component σ_{33} is not restricted to be zero, but it does not appear in the constitutive equation.

Equating ε_{33} to zero in the general three–dimensional constitutive relationship the following relationship is obtained:

$$\varepsilon_{33} = \frac{\sigma_{33}}{E} - \frac{\nu\left(\sigma_{11} + \sigma_{22}\right)}{E} = 0 \Longrightarrow \sigma_{33} = \nu\left(\sigma_{11} + \sigma_{22}\right).$$

Substituting the above relationship for σ_{33} in the general three–dimensional constitutive relationship for the linear–elastic isotropic material, the following plane–strain constitutive relationship is obtained:

$$\begin{pmatrix} \varepsilon_{11} \\ \varepsilon_{22} \\ \gamma_{12} = 2\varepsilon_{12} \end{pmatrix} = \frac{1}{E}\begin{pmatrix} 1-\nu^2 & -\nu(1+\nu) & 0 \\ -\nu(1+\nu) & 1-\nu^2 & 0 \\ 0 & 0 & 2(1+\nu) \end{pmatrix}\begin{pmatrix} \sigma_{11} \\ \sigma_{22} \\ \sigma_{12} \end{pmatrix}$$

The stress component σ_{33} can be calculated from the following equation, which forms part of the constitutive law:

$$\sigma_{33} = \nu\left(\sigma_{11} + \sigma_{22}\right)$$

The inverse of the stress strain relationship of materials under plane–strain conditions has the following form:

$$\sigma = C\varepsilon \Longrightarrow \qquad \begin{pmatrix} \sigma_{11} \\ \sigma_{22} \\ \sigma_{12} \end{pmatrix} = \frac{E}{1+\nu}\begin{pmatrix} \frac{1-\nu}{1-2\nu} & \frac{\nu}{1-2\nu} & 0 \\ \frac{\nu}{1-2\nu} & \frac{1-\nu}{1-2\nu} & 0 \\ 0 & 0 & \frac{1}{2} \end{pmatrix}\begin{pmatrix} \varepsilon_{11} \\ \varepsilon_{22} \\ \gamma_{12} = 2\varepsilon_{12} \end{pmatrix}$$

The following Mathematica code was utilized to obtain C:

```
Dd=1/Ee*{{1-nu^2,-nu(1+nu),0},{-nu(1+nu),1-nu^2,0},{0,0,2(1+nu)}};
Cc=FullSimplify[Inverse[Dd]]
Cc//MatrixForm
```

5.2.7.c. Axisymmetry

Axisymmetry is used for modeling cylindrical shapes where the load and boundary conditions are also cylindrical. Assuming a coordinate system as shown in Figure 5–1c, it is possible to analyze a cross section of the model at $\theta = 0$. The strain–displacement relationships are:

$$\varepsilon_{11} = \frac{\partial u_1}{\partial X_1}, \quad \varepsilon_{22} = \frac{\partial u_2}{\partial X_2}, \quad \varepsilon_{33} = \frac{\partial u_3}{\partial X_3} = \frac{u_1}{r}, \quad \gamma_{12} = 2\varepsilon_{12} = \frac{\partial u_1}{\partial X_2} + \frac{\partial u_2}{\partial X_1}, \quad \gamma_{13} = \gamma_{23} = 0$$

▶ **BONUS EXERCISE:** Show that $\varepsilon_{33} = \dfrac{u_1}{r}$.

The relationship between the nonzero strain values and the stress components has the following form:

$$\begin{pmatrix} \varepsilon_{11} \\ \varepsilon_{22} \\ \varepsilon_{33} \\ \gamma_{12} = 2\varepsilon_{12} \end{pmatrix} = \frac{1}{E}\begin{pmatrix} 1 & -\nu & -\nu & 0 \\ -\nu & 1 & -\nu & 0 \\ -\nu & -\nu & 1 & 0 \\ 0 & 0 & 0 & 2(1+\nu) \end{pmatrix}\begin{pmatrix} \sigma_{11} \\ \sigma_{22} \\ \sigma_{33} \\ \sigma_{12} \end{pmatrix}$$

The inverse of the above relationship is:

$$\sigma = C\varepsilon \Longrightarrow \quad \begin{pmatrix} \sigma_{11} \\ \sigma_{22} \\ \sigma_{33} \\ \sigma_{12} \end{pmatrix} = \frac{E}{1+\nu}\begin{pmatrix} \frac{1-\nu}{1-2\nu} & \frac{\nu}{1-2\nu} & \frac{\nu}{1-2\nu} & 0 \\ \frac{\nu}{1-2\nu} & \frac{1-\nu}{1-2\nu} & \frac{\nu}{1-2\nu} & 0 \\ \frac{\nu}{1-2\nu} & \frac{\nu}{1-2\nu} & \frac{1-\nu}{1-2\nu} & 0 \\ 0 & 0 & 0 & \frac{1}{2} \end{pmatrix}\begin{pmatrix} \varepsilon_{11} \\ \varepsilon_{22} \\ \varepsilon_{33} \\ \gamma_{12} \end{pmatrix}$$

The following Mathematica code was utilized to obtain C:

```
Dd=1/Ee*{{1,-nu,-nu,0},{-nu,1,-nu,0},{-nu,-nu,1,0},{0,0,0,2(1+nu)}};
Cc=FullSimplify[Inverse[Dd]]
Cc//MatrixForm
```

5.2.8. Problems

Problem 1: In this problem, the symmetry of the coefficient matrix will be shown to be crucial so that the material model would not predict energy creation or dissipation during loading cycles. Assume that a linear elastic material specimen is loaded through the following three steps:

- Step 1 σ_{11} is increased from 0 to 100 MPa.
- Step 2 σ_{11} is kept constant at 100 MPa, while σ_{22} is increased from 0 to 100 MPa.
- Step 3 σ_{11} and σ_{22} are decreased from 100 MPa to 0 MPa such that $\sigma_{11} = \sigma_{22}$.

Assume that the material has two the different Poisson's ratios of 0.3 and 0.4 such that the relationships between the applied stresses and the strains ε_{11} and ε_{22} are:

$$\begin{pmatrix} \varepsilon_{11} \\ \varepsilon_{22} \end{pmatrix} = \begin{pmatrix} \frac{1}{E} & -\frac{0.3}{E} \\ -\frac{0.4}{E} & \frac{1}{E} \end{pmatrix} \begin{pmatrix} \sigma_{11} \\ \sigma_{22} \end{pmatrix}$$

Assume that all the other stress and strain components are zero. Find the energy created or dissipated through the entire loading process.

Solution: The variation of strain increments due to any stress increments is according to the formula:

$$\begin{pmatrix} d\varepsilon_{11} \\ d\varepsilon_{22} \end{pmatrix} = \begin{pmatrix} \frac{1}{E} & -\frac{0.3}{E} \\ -\frac{0.4}{E} & \frac{1}{E} \end{pmatrix} \begin{pmatrix} d\sigma_{11} \\ d\sigma_{22} \end{pmatrix}$$

In step 1, the changes in the stress and strain components are as follows:

$$\sigma_{11} = 0 \to 100, \quad \sigma_{22} = 0 \to 0$$

$$\varepsilon_{11} = 0 \to \frac{100}{E}, \quad \varepsilon_{22} = 0 \to -\frac{0.4\,(100)}{E}$$

The strain energy per unit volume stored inside the material in step 1 can be calculated as follows:

$$\begin{aligned} Energy &= \int_0^{\frac{100}{E}} \sigma_{11} d\varepsilon_{11} + \int_0^{-\frac{0.4(100)}{E}} \sigma_{22} d\varepsilon_{22} \\ &= \int_0^{100} \frac{\sigma_{11}}{E} d\sigma_{11} + \int_0^{0} \frac{-0.3\sigma_{11}}{E} d\sigma_{22} + \int_0^{-\frac{0.4(100)}{E}} 0 d\varepsilon_{22} \\ &= \frac{(100)^2}{2E} \end{aligned}$$

In step 2, the changes in the stress and strain components are:

$$\sigma_{11} = 100 \to 100, \quad \sigma_{22} = 0 \to 100$$

$$\varepsilon_{11} = \frac{100}{E} \to \frac{100}{E} - \frac{0.3\,(100)}{E}, \quad \varepsilon_{22} = -\frac{0.4\,(100)}{E} \to \frac{100}{E} - \frac{0.4\,(100)}{E}$$

The strain energy per unit volume stored inside the material in step 2 can be calculated as follows:

$$Energy = \int_{\frac{100}{E}}^{\frac{100}{E}-\frac{0.3(100)}{E}} \sigma_{11} d\varepsilon_{11} + \int_{-\frac{0.4(100)}{E}}^{\frac{100}{E}-\frac{0.4(100)}{E}} \sigma_{22} d\varepsilon_{22}$$

$$= \int_{100}^{100} \frac{\sigma_{11}}{E} d\sigma_{11} + \int_{0}^{100} \frac{-0.3\sigma_{11}}{E} d\sigma_{22} + \int_{100}^{100} -\frac{0.4\sigma_{22}}{E} d\sigma_{11} + \int_{0}^{100} \frac{\sigma_{22}}{E} d\sigma_{22}$$

$$= \frac{(100)^2}{2E} - \frac{0.3(100)^2}{E}$$

During loading step 3, the changes in the stress and strain components are as follows:

$$\sigma_{11} = 100 \to 0, \quad \sigma_{22} = 100 \to 0$$

$$\varepsilon_{11} = \frac{100}{E} - \frac{0.3\,(100)}{E} \to 0, \quad \varepsilon_{22} = \frac{100}{E} - \frac{0.3\,(100)}{E} \to 0$$

The strain energy per unit volume in step 3 can be calculated using the following integrals:

$$Energy = \int_{\frac{100}{E}-\frac{0.3(100)}{E}}^{0} \sigma_{11} d\varepsilon_{11} + \int_{\frac{100}{E}-\frac{0.4(100)}{E}}^{0} \sigma_{22} d\varepsilon_{22}$$

$$= \int_{100}^{0} \frac{\sigma_{11}}{E} d\sigma_{11} + \int_{100}^{0} \frac{-0.3\sigma_{11}}{E} d\sigma_{22} + \int_{100}^{0} -\frac{0.4\sigma_{22}}{E} d\sigma_{11} + \int_{100}^{0} \frac{\sigma_{22}}{E} d\sigma_{22}$$

$$= \int_{100}^{0} \frac{\sigma_{11}}{E} d\sigma_{11} + \int_{100}^{0} \frac{-0.3\sigma_{22}}{E} d\sigma_{22} + \int_{100}^{0} -\frac{0.4\sigma_{11}}{E} d\sigma_{11} + \int_{100}^{0} \frac{\sigma_{22}}{E} d\sigma_{22}$$

$$= -\frac{(100)^2}{2E} + \frac{0.3\,(100)^2}{2E} + \frac{0.4\,(100)^2}{2E} - \frac{(100)^2}{2E}$$

The total energy stored inside the material during the loading cycle is:

$$
\begin{aligned}
Total\ Energy &= \frac{(100)^2}{2E} + \frac{(100)^2}{2E} - \frac{0.3\,(100)^2}{E} - \frac{(100)^2}{2E} + \frac{0.3\,(100)^2}{2E} + \frac{0.4\,(100)^2}{2E} - \frac{(100)^2}{2E} \\
&= \frac{(0.4 - 0.3)\,(100)^2}{2E}
\end{aligned}
$$

Since the material is elastic, the total energy stored inside the material should be equal to zero since the load was removed in the final step. However, the calculation of the energy stored inside the material reveals a different scenario. The energy is dissipated by cycling the material through the above loading sequence. This energy dissipation is due to the difference between the off–diagonal entries in the matrix of material coefficients C. Had the matrix C been symmetric, the total energy stored or generated during the loading cycle would have been zero. This problem shows that, if the coefficients matrix is not symmetric, the material does not conserve energy.

Problem 2: The state of strain inside a continuum is represented by the small strain matrix:

$$
\varepsilon = \begin{pmatrix} 0.002 & 0.001 & 0.002 \\ 0.001 & -0.003 & 0.001 \\ 0.002 & 0.001 & 0.0015 \end{pmatrix}
$$

Assume that the material is steel with Young's modulus of 210 GPa and Poisson's ratio of 0.3 and follows the von Mises yield criterion with $\sigma_{max} = 750$ MPa. Find the principal strains and the prin–cipal strains directions, the principal stresses and their directions, and comment on whether this state of strain is possible before yielding of the material.

Solution: The principal strains (eigenvalues) are 0.004, –0.0033, –0.00026 and the corresponding principal directions (eigenvectors) of the strains are:

$$\{0.7324, 0.1964, 0.6519\},\ \{0.1282, -0.9801, 0.1512\}\text{ and }\{-0.6687, 0.02718, 0.7431\}.$$

To apply the linear–elastic constitutive relations, the strain matrix will be converted into a vector of strains, as follows:

$$
\varepsilon = \begin{pmatrix} \varepsilon_{11} \\ \varepsilon_{22} \\ \varepsilon_{33} \\ 2\varepsilon_{12} \\ 2\varepsilon_{13} \\ 2\varepsilon_{23} \end{pmatrix} = \begin{pmatrix} 0.002 \\ -0.003 \\ 0.0015 \\ 0.002 \\ 0.004 \\ 0.002 \end{pmatrix}
$$

The stresses can be obtained using the relationship shown in section 5.2.5:

$$\begin{pmatrix} \sigma_{11} \\ \sigma_{22} \\ \sigma_{33} \\ \sigma_{12} \\ \sigma_{13} \\ \sigma_{23} \end{pmatrix} = \frac{E}{(1-2\nu)(1+\nu)} \begin{pmatrix} 1-\nu & \nu & \nu & 0 & 0 & 0 \\ \nu & 1-\nu & \nu & 0 & 0 & 0 \\ \nu & \nu & 1-\nu & 0 & 0 & 0 \\ 0 & 0 & 0 & \frac{1-2\nu}{2} & 0 & 0 \\ 0 & 0 & 0 & 0 & \frac{1-2\nu}{2} & 0 \\ 0 & 0 & 0 & 0 & 0 & \frac{1-2\nu}{2} \end{pmatrix} \begin{pmatrix} \varepsilon_{11} \\ \varepsilon_{22} \\ \varepsilon_{33} \\ 2\varepsilon_{12} \\ 2\varepsilon_{13} \\ 2\varepsilon_{23} \end{pmatrix}$$

Therefore, the stresses are:

$$\begin{pmatrix} \sigma_{11} \\ \sigma_{22} \\ \sigma_{33} \\ \sigma_{12} \\ \sigma_{13} \\ \sigma_{23} \end{pmatrix} = \begin{pmatrix} 383.654 \\ -424.04 \\ 302.885 \\ 161.538 \\ 323.077 \\ 161.538 \end{pmatrix} \text{MPa}$$

The stress matrix is:

$$\sigma = \begin{pmatrix} 383.654 & 161.538 & 323.077 \\ 161.538 & -424.038 & 161.538 \\ 323.077 & 161.538 & 302.885 \end{pmatrix} \text{MPa}$$

The principal stresses (eigenvalues) of the stress matrix are: 714.53, –470.1 and 18.064 MPa, and the principal directions are:

{0.7324,0.1964,0.6519}, {0.1282,–0.9801,0.1512} and {–0.6687,0.02718,0.7431}.

Since the material is an isotropic linear–elastic material, the eigenvectors of the stress and the strain matrix coincide.

The maximum principal stress is 714.53 MPa; however, the von Mises stress is equal to 1031.19 MPa, which is higher than $\sigma_{max} = 750$ MPa. This indicates that the material cannot sustain the stress matrix above, and in fact, the strain state above is impossible to occur before yielding or failure of the steel metal.

The following is the Mathematica code utilized:

```
str={{0.002,0.001,0.002},{0.001,-0.003,0.001},{0.002,0.001,0.0015}};
a=Eigensystem[str];
a//MatrixForm
str//MatrixForm
Ee=210000;
Nu=0.3;
G=Ee/2/(1+Nu);
Strainvector={str[[1,1]],str[[2,2]],str[[3,3]],2*str[[1,2]],2*str[[1,3]],2*s
tr[[2,3]]};
Cc={{1/Ee,-Nu/Ee,-Nu/Ee,0,0,0},{-Nu/Ee,1/Ee,-Nu/Ee,0,0,0},{-Nu/Ee,-
Nu/Ee,1/Ee,0,0,0},{0,0,0,1/G,0,0},{0,0,0,0,1/G,0},{0,0,0,0,0,1/G}};
Dd=FullSimplify[Inverse[Cc]];
stressvector=Dd.Strainvector
Stressmatrix={{stressvector[[1]],stressvector[[4]],stressvector[[5]]},{stres
svector[[4]],stressvector[[2]],stressvector[[6]]},{stressvector[[5]],stressv
ector[[6]],stressvector[[3]]}};
Stressmatrix//MatrixForm
b=Eigensystem[Stressmatrix];
b//MatrixForm
VonMises[sigma_]:=Sqrt[1/2*((sigma[[1,1]]-sigma[[2,2]])^2+(sigma[[2,2]]-
sigma[[3,3]])^2+(sigma[[3,3]]-
sigma[[1,1]])^2+6*(sigma[[1,2]]^2+sigma[[1,3]]^2+sigma[[2,3]]^2))];
VonMises[Stressmatrix]
```

Problem 3: The position function in dimensions of m of a plate adheres to the following form:

$$x_1 = X_1 + 0.001X_2 + 0.0002X_1^2$$

$$x_2 = 1.001X_2 - 0.0002X_2^2$$

$$x_3 = X_3 \quad .$$

If the plate has dimensions 2 m in the direction of e_1, and 1 m in the direction of e_2, with the bottom left corner coinciding with the origin of the coordinate system, then:

- Show that this is a state of plane strain.
- Find the engineering small–strain matrix, and the Green–strain matrix and show that the engineering small strain is adequate to describe the motion.
- Plot the vector plot of the displacement function.
- If Young's modulus is equal to 210 GPa and Poisson'a ratio = 0.3, then find the equilibrium body forces to which this plate is subjected.
  Draw the contour plot of the von Mises stress and the out–of–plane stress component.

Solution: The displacement function is:

$$u = x - X = \begin{pmatrix} 0.0002X_1^2 + 0.001X_2 \\ 0.001X_2 - 0.0002X_2^2 \\ 0 \end{pmatrix}$$

The vector plot of the displacement function is

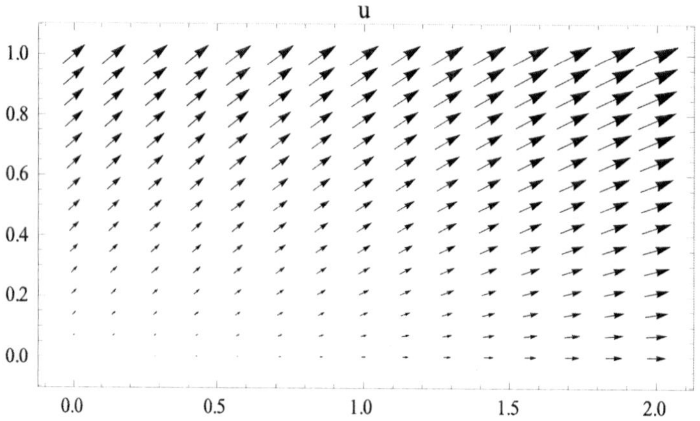

The gradient of the displacement tensor is:

$$M = \text{Grad}\, u = \begin{pmatrix} 0.0004X_1 & 0.001 & 0 \\ 0 & 0.001 - 0.0004X_2 & 0 \\ 0 & 0 & 0 \end{pmatrix}$$

The engineering small–strain matrix and the Green–strain matrix are:

$$\varepsilon_{eng} = \frac{1}{2}\left(M + M^T\right) = \begin{pmatrix} 0.0004X_1 & 0.0005 & 0 \\ 0.0005 & 0.001 - 0.0004X_2 & 0 \\ 0 & 0 & 0 \end{pmatrix}$$

$$\varepsilon_{Green} = \frac{1}{2}\left(M + M^T + M^TM\right) = \begin{pmatrix} 0.0004X_1 + 8 \times 10^{-8}X_1^2 & 0.005 + 2 \times 10^{-7}X_1 & 0 \\ 0.005 + +2 \times 10^{-7}X_1 & 0.001 - 0.0004X_2 + 8 \times 10^{-8}X_2^2 & 0 \\ 0 & 0 & 0 \end{pmatrix}$$

The difference between the engineering and the Green strain is very small, and thus, the engineer–ing strain can be used to describe the motion. In addition, the displacement component $u_3 = 0$, and the displacement function is independent of the coordinate X_3; therefore, the strains ε_{33}, ε_{13}, and ε_{23} are equal to zero, indicating a state of plane strain. The strain vector is:

$$\varepsilon = \begin{pmatrix} \varepsilon_{11} \\ \varepsilon_{22} \\ \varepsilon_{33} \\ 2\varepsilon_{12} \\ 2\varepsilon_{13} \\ 2\varepsilon_{23} \end{pmatrix} = \begin{pmatrix} 0.0004X_1 \\ 0.001 - 0.0004X_2 \\ 0 \\ 0.001 \\ 0 \\ 0 \end{pmatrix}$$

Using the constitutive equation for the linear–elastic isotropic material, the stress vector is:

$$\sigma = \begin{pmatrix} 121.154 + 113.077X_1 - 48.4615X_2 \\ 282.692 + 48.4615X_1 - 113.077X_2 \\ 121.154 + 48.4615X_1 + 48.4615X_2 \\ 80.7692 \\ 0 \\ 0 \end{pmatrix} \text{MPa}$$

Since the displacement is small, the coordinates x_1 can be used instead of the coordinates X_1 to find the equilibrium body forces:

$$\rho b_1 = -\frac{\partial \sigma_{11}}{\partial x_1} - \frac{\partial \sigma_{21}}{\partial x_2} - \frac{\partial \sigma_{31}}{\partial x_3} = -113.077\ \text{MN/m}^3$$

$$\rho b_2 = -\frac{\partial \sigma_{12}}{\partial x_1} - \frac{\partial \sigma_{22}}{\partial x_2} - \frac{\partial \sigma_{32}}{\partial x_3} = 113.077\ \text{MN/m}^3$$

$$\rho b_3 = -\frac{\partial \sigma_{13}}{\partial x_1} - \frac{\partial \sigma_{23}}{\partial x_2} - \frac{\partial \sigma_{33}}{\partial x_3} = 0\ \text{MN/m}^3$$

The von Mises stress is:

$$\textit{von Mises stress} = \sqrt{45665.7 + 4175.15\,(-2.5 + X_1)\,X_1 - 20875.7X_2 + 4175.15X_1X_2 + 4175.15X_2^2}$$

The contour plot of the von Mises stress shows that the areas with the maximum von Mises stresses are the right and left bottom corners of the plate. Since the plate is under a state of plane strain, the component σ_{33} exists, and its contour plot is shown below (units are in MPa).

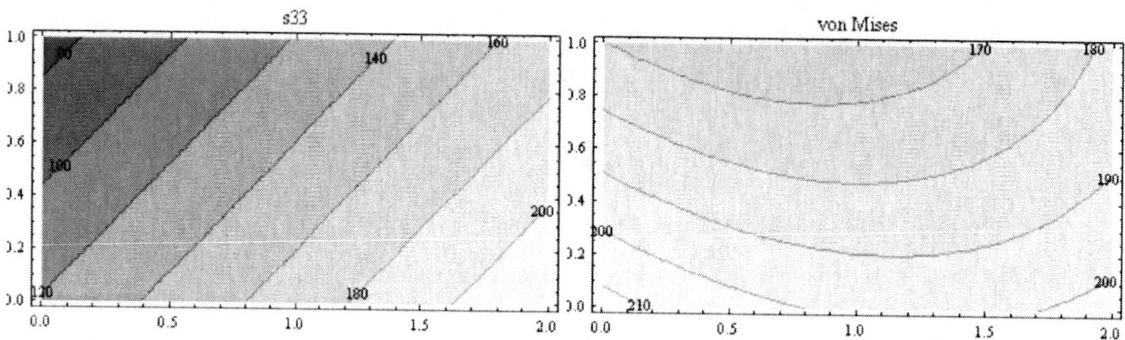

The Mathematica code utilized is:

```
Clear[Ee,Nu]
x1=X1+0.001*X2+0.0002*X1^2;
x2=1.001*X2-0.0002*X2^2;
x3=X3;
X={X1,X2,X3};
x={x1,x2,x3};
u=x-X;
u//MatrixForm
uplane=Table[u[[i]],{i,1,2}]
VectorPlot[uplane,{X1,0,2},{X2,0,1},AspectRatio->Automatic,PlotLabel-
>"u",VectorStyle->Black]
gradu=Table[D[u[[i]],X[[j]]],{i,1,3},{j,1,3}];
gradu//MatrixForm
strGreen=FullSimplify[1/2*(gradu+Transpose[gradu]+Transpose[gradu].gradu)];
Expand[strGreen]//MatrixForm
str=FullSimplify[1/2*(gradu+Transpose[gradu])];
str//MatrixForm
Strainvector={str[[1,1]],str[[2,2]],str[[3,3]],2*str[[1,2]],2*str[[1,3]],2*s
tr[[2,3]]};
Strainvector//MatrixForm
Ee=210000;
Nu=0.3;
G=Ee/2/(1+Nu);
Cc={{1/Ee,-Nu/Ee,-Nu/Ee,0,0,0},{-Nu/Ee,1/Ee,-Nu/Ee,0,0,0},{-Nu/Ee,-
Nu/Ee,1/Ee,0,0,0},{0,0,0,1/G,0,0},{0,0,0,0,1/G,0},{0,0,0,0,0,1/G}};
Dd=FullSimplify[Inverse[Cc]];
stressvector=FullSimplify[Chop[Dd.Strainvector]];
stressvector//MatrixForm
Stressmatrix={{stressvector[[1]],stressvector[[4]],stressvector[[5]]},{stres
svector[[4]],stressvector[[2]],stressvector[[6]]},{stressvector[[5]],stressv
ector[[6]],stressvector[[3]]}};
Stressmatrix=FullSimplify[Chop[Stressmatrix]];
Stressmatrix//MatrixForm
X={X1,X2,X3};
s=Stressmatrix;
```

```
-Sum[D[s[[i,1]],X[[i]]],{i,1,3}]
-Sum[D[s[[i,2]],X[[i]]],{i,1,3}]
-Sum[D[s[[i,3]],X[[i]]],{i,1,3}]
VonMises[sigma_]:=Sqrt[1/2*((sigma[[1,1]]-sigma[[2,2]])^2+(sigma[[2,2]]-
sigma[[3,3]])^2+(sigma[[3,3]]-
sigma[[1,1]])^2+6*(sigma[[1,2]]^2+sigma[[1,3]]^2+sigma[[2,3]]^2))];
a=FullSimplify[Chop[VonMises[Stressmatrix]]]
ContourPlot[a,{X1,0,2},{X2,0,1},AspectRatio->Automatic,ContourLabels-
>All,PlotLabel->"vonMises",ColorFunction->"GrayTones"]
ContourPlot[Stressmatrix[[3,3]],{X1,0,2},{X2,0,1},AspectRatio-
>Automatic,ContourLabels->All,PlotLabel->"s33",ColorFunction->"GrayTones"]
```

Problem 4: In a confined compression test of a linear–elastic isotropic material, the confining stress was −500 MPa while the vertical stress was −100 MPa. The axial strains measured along the directions of the vectors e_1, e_2 and e_3 were −0.003, 0.001, and −0.003, respectively. Find Young's modulus and Poisson's ratio.

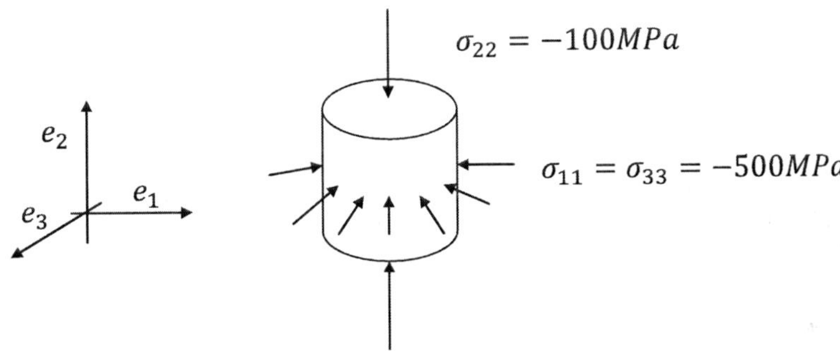

Solution: Using the constitutive relationships for linear–elastic isotropic materials:

$$\varepsilon_{11} = -\frac{500}{E} + \frac{\nu\,(100)}{E} + \frac{\nu\,(500)}{E} = -0.003 \Longrightarrow -500 + 600\nu = -0.003E$$

$$\varepsilon_{22} = \frac{\nu\,(500)}{E} - \frac{100}{E} + \frac{\nu\,(500)}{E} = 0.001 \Longrightarrow -100 + 1000\nu = 0.001E$$

Thus, Poisson's ratio and Young's modulus values are:

$$\nu = 0.222, \quad E = 122,222.00\ \text{MPa}$$

Problem 5: Consider a transversely isotropic linear–elastic piece of material with the isotropy plane being the plane of e_1 and e_2. The values of Young's modulus and Poisson's ratio in the plane of isotropy are equal to 100 MPa and 0.2, respectively. In a uniaxial tension test, when the material was stretched in the direction of e_1 by a stress of a value 1 MPa the strain measured in the direction of the vector e_3 was recorded to be –0.0005. In another uniaxial tension test, when the material was stretched in the direction of e_3 by a stress of a value 1 MPa, the strain measured in the direction of e_3 was 0.002. Find an estimate for the material constants E_{33}, ν_{13}, and ν_{31}.

Solution: The Young's modulus in the direction of e_3 can be obtained from the second experiment:

$$\varepsilon_{33} = \frac{\sigma_{33}}{E_{33}} \Longrightarrow E_{33} = \frac{1}{0.002} = 500\text{ MPa}$$

From the first experiment, the value of Poisson's ratio ν_{13} can be obtained as follows:

$$\varepsilon_{33} = -\frac{\nu_{13}\sigma_{11}}{E_{11}} \Longrightarrow \nu_{13} = \frac{0.0005\,(100)}{1} = 0.05$$

The value of Poisson's ratio ν_{31} can be obtained from the symmetry of the coefficients matrix:

$$\frac{\nu_{13}}{E_{11}} = \frac{\nu_{31}}{E_{33}} \Longrightarrow \nu_{31} = \frac{0.05\,(500)}{100} = 0.25$$

Problem 6: Consider an orthotropic material whose major axes of orthotropy lie in the directions of the vectors $e'_1 = \{\frac{1}{\sqrt{2}}, \frac{1}{\sqrt{2}}, 0\}$, $e'_2 = \{\frac{-1}{\sqrt{2}}, \frac{1}{\sqrt{2}}, 0\}$ and $e'_3 = \{0, 0, 1\}$. The material properties in the coordinate system defined by the orthotropy axes are: $E_{11} = 100$ MPa, $E_{22} = 200$ MPa, $E_{33} = 350$ MPa, $\nu_{31} = 0.2$, $\nu_{32} = 0.1$, $\nu_{21} = 0.05$, $G_{12} = 50$ MPa, $G_{13} = 70$ MPa and $G_{23} = 90$ MPa. Consider the coordinate system defined by the basis set $B = \{e_1, e_2, e_3\}$, where $e_1 = \{1, 0, 0\}$, $e_2 = \{0, 0, 1\}$ and $e_3 = \{0, 0, 1\}$. If the material is in a state of uniaxial stress such that the only nonzero stress component is $\sigma_{11} = 1$ MPa, find the engineering strain matrix in the coordinate system defined by the basis sets $B = \{e_1, e_2, e_3\}$ and $B' = \{e'_1, e'_2, e'_3\}$.

Solution: Since the material is orthotropic, the coefficients matrix C depends on the coordinate system involved. The values of the material coefficients are given in the coordinate system defined by B'. Therefore, it is best to apply the constitutive relationship in the coordinate system of B'.

The stress matrix in the coordinate system of B is:

$$\sigma = \begin{pmatrix} 1 & 0 & 0 \\ 0 & 0 & 0 \\ 0 & 0 & 0 \end{pmatrix} \text{MPa}$$

The rotation matrix involved with the change of coordinates is:

$$Q = \begin{pmatrix} 1/\sqrt{2} & 1/\sqrt{2} & 0 \\ -1/\sqrt{2} & 1/\sqrt{2} & 0 \\ 0 & 0 & 1 \end{pmatrix}$$

The stress matrix in the coordinate system of B' is:

$$\sigma' = Q\sigma Q^T = \begin{pmatrix} 1/2 & -1/2 & 0 \\ -1/2 & 1/2 & 0 \\ 0 & 0 & 0 \end{pmatrix}$$

The relationship between the stress and the strain in the coordinate system of B' can be used to find the strains in the coordinate system of B', as follows (notice that the symmetry of C is utilized):

$$\begin{pmatrix} \varepsilon'_{11} \\ \varepsilon'_{22} \\ \varepsilon'_{33} \\ 2\varepsilon'_{12} \\ 2\varepsilon'_{13} \\ 2\varepsilon'_{23} \end{pmatrix} = \begin{pmatrix} \frac{1}{E_{11}} & \frac{-\nu_{21}}{E_{22}} & \frac{-\nu_{31}}{E_{33}} & 0 & 0 & 0 \\ \frac{-\nu_{12}}{E_{11}} & \frac{1}{E_{22}} & \frac{-\nu_{32}}{E_{33}} & 0 & 0 & 0 \\ \frac{-\nu_{13}}{E_{11}} & \frac{-\nu_{23}}{E_{22}} & \frac{1}{E_{33}} & 0 & 0 & 0 \\ 0 & 0 & 0 & \frac{1}{G_{12}} & 0 & 0 \\ 0 & 0 & 0 & 0 & \frac{1}{G_{13}} & 0 \\ 0 & 0 & 0 & 0 & 0 & \frac{1}{G_{23}} \end{pmatrix} \begin{pmatrix} \sigma'_{11} \\ \sigma'_{22} \\ \sigma'_{33} \\ \sigma'_{12} \\ \sigma'_{13} \\ \sigma'_{23} \end{pmatrix}$$

$$= \begin{pmatrix} \frac{1}{100} & \frac{-0.05}{200} & \frac{-0.2}{350} & 0 & 0 & 0 \\ \frac{-0.05}{200} & \frac{1}{200} & \frac{-0.1}{350} & 0 & 0 & 0 \\ \frac{-0.2}{350} & \frac{-0.1}{350} & \frac{1}{350} & 0 & 0 & 0 \\ 0 & 0 & 0 & \frac{1}{50} & 0 & 0 \\ 0 & 0 & 0 & 0 & \frac{1}{70} & 0 \\ 0 & 0 & 0 & 0 & 0 & \frac{1}{90} \end{pmatrix} \begin{pmatrix} \frac{1}{2} \\ \frac{1}{2} \\ 0 \\ -\frac{1}{2} \\ 0 \\ 0 \end{pmatrix} = \begin{pmatrix} 0.004875 \\ 0.002375 \\ -0.000428571 \\ -0.01 \\ 0 \\ 0 \end{pmatrix}$$

The strain matrix in the coordinate system of B' is:

$$\varepsilon' = \begin{pmatrix} 0.004875 & -0.005 & 0 \\ -0.005 & 0.002375 & 0 \\ 0 & 0 & -0.000428571 \end{pmatrix}$$

The strain matrix in the coordinate system of B is:

$$\varepsilon' = Q\,\varepsilon\,Q^T \Rightarrow \varepsilon = Q^T\varepsilon' Q = \begin{pmatrix} 0.008625 & 0.00125 & 0 \\ 0.00125 & -0.001375 & 0 \\ 0 & 0 & -0.000428571 \end{pmatrix}$$

It is very important to notice that, while a uniaxial state of stress was applied, shear strain was developed in the same coordinate system. This is because the uniaxial state of stress was applied in a coordinate system different from the orthotropy axis coordinate system!

The following is the Mathematica code utilized.

```
E11=100;E22=200;E33=350;G12=50;G13=70;G23=90;nu31=0.2;nu32=0.1;nu21=0.05;nu1
3=nu31/E33*E11;nu12=nu21/E22*E11;nu23=nu32/E33*E22;
Ss={{1/E11,-nu21/E22,-nu31/E33,0,0,0} ,{-nu12/E11,1/E22,-nu32/E33,0,0,0} ,{-
nu13/E11,-nu23/E22,1/E22,0,0,0},{0,0,0,1/G12,0,0},{0,0,0,0,1/G13,0},
{0,0,0,0,0,1/G23}};
s={{1,0,0},{0,0,0},{0,0,0}};
Q=RotationMatrix[-45Degree,{0,0,1}]
sdash=Q.s.Transpose[Q];
sdash//MatrixForm
mat=sdash
sigmavector={mat[[1,1]],mat[[2,2]],mat[[3,3]],mat[[1,2]],mat[[1,3]],mat[[2,3
]]}
sigmavector//MatrixForm
strvector=Ss.sigmavector;
strvector//MatrixForm
strdash={{strvector[[1]],strvector[[4]]/2,strvector[[5]]/2},{strvector[[4]]/
2,strvector[[2]],strvector[[6]]/2},{strvector[[5]]/2,strvector[[6]]/2,strvec
tor[[3]]}} ;
strdash//MatrixForm
str=Transpose[Q].strdash.Q;
str//MatrixForm
```

Problem 6: The in–plane strain components ε_{11}, ε_{22}, and γ_{12} of a specimen are 0.001, −0.002, and 0.003, respectively. If Young's modulus is 10,000 MPa and Poisson's ratio is 0.2, find the stress and strain matrices if the material is in the state of plane strain and if the material is in the state of plane stress.

Solution: In a plane strain condition, the strain matrix is:

$$\varepsilon = \begin{pmatrix} 0.001 & 0.0015 & 0 \\ 0.0015 & -0.002 & 0 \\ 0 & 0 & 0 \end{pmatrix}$$

The relationship between the stresses and the strains in a plane–strain condition can be utilized to find the stresses, as follows:

$$\begin{pmatrix}\sigma_{11}\\ \sigma_{22}\\ \sigma_{12}\end{pmatrix} = \frac{E}{1+\nu}\begin{pmatrix}\frac{1-\nu}{1-2\nu} & \frac{\nu}{1-2\nu} & 0\\ \frac{\nu}{1-2\nu} & \frac{1-\nu}{1-2\nu} & 0\\ 0 & 0 & \frac{1}{2}\end{pmatrix}\begin{pmatrix}\varepsilon_{11}\\ \varepsilon_{22}\\ \gamma_{12}=2\varepsilon_{12}\end{pmatrix}$$

$$= \begin{pmatrix}11111.1 & 2777.78 & 0\\ 2777.78 & 11111.1 & 0\\ 0 & 0 & 4166.67\end{pmatrix}\begin{pmatrix}0.001\\ -0.002\\ 0.003\end{pmatrix} = \begin{pmatrix}5.56\\ -19.44\\ 12.5\end{pmatrix}\text{MPa}$$

In a plane–strain condition, the movement in the direction of e_3 is restrained, and thus, a stress component σ_{33} develops according to the relationship:

$$\sigma_{33} = \nu\,(\sigma_{11} + \sigma_{22}) = 0.2\,(5.56 - 19.44) = -2.78\,\text{MPa}$$

The stress matrix is thus:

$$\sigma = \begin{pmatrix}5.56 & 12.5 & 0\\ 12.5 & -19.4 & 0\\ 0 & 0 & -2.78\end{pmatrix}$$

The following is the Mathematica code for the above calculations:

```
Ee=10000;Nu=0.2;
Cc=Ee/(1+Nu)*{{(1-Nu)/(1-2Nu),Nu/(1-2Nu),0} ,{Nu/(1-2Nu),(1-Nu)/(1-
2Nu),0},{0,0,1/2}};
Cc//MatrixForm
strvector={0.001,-0.002,0.003};
stressvector=Cc.strvector;
stressvector//MatrixForm
s=Table[0,{i,1,3},{j,1,3}];
s[[1,1]]=stressvector[[1]];
s[[2,2]]=stressvector[[2]];
s[[1,2]]=stressvector[[3]];
s[[2,1]]=s[[1,2]];
s[[3,3]]=Nu (s[[1,1]]+s[[2,2]]);
s//MatrixForm
```

In a plane–stress condition, a strain component ε_{33} develops, and thus, the strain matrix is:

$$\varepsilon = \begin{pmatrix} 0.001 & 0.0015 & 0 \\ 0.0015 & -0.002 & 0 \\ 0 & 0 & \varepsilon_{33} \end{pmatrix}$$

The relationship between the stresses and the strains in a plane–stress condition can be utilized to find the values of the stress components, as follows:

$$\begin{pmatrix} \sigma_{11} \\ \sigma_{22} \\ \sigma_{12} \end{pmatrix} = \frac{E}{(1+\nu)} \begin{pmatrix} \frac{1}{1-\nu} & \frac{\nu}{1-\nu} & 0 \\ \frac{\nu}{1-\nu} & \frac{1}{1-\nu} & 0 \\ 0 & 0 & \frac{1}{2} \end{pmatrix} \begin{pmatrix} \varepsilon_{11} \\ \varepsilon_{22} \\ \gamma_{12} = 2\varepsilon_{12} \end{pmatrix}$$

$$= \begin{pmatrix} 10416.7 & 2083.33 & 0 \\ 2083.33 & 10416.7 & 0 \\ 0 & 0 & 4166.67 \end{pmatrix} \begin{pmatrix} 0.001 \\ -0.002 \\ 0.003 \end{pmatrix} = \begin{pmatrix} 6.25 \\ -18.75 \\ 12.5 \end{pmatrix} \text{MPa}$$

The stress matrix is thus:

$$\sigma = \begin{pmatrix} 6.25 & 12.5 & 0 \\ 12.5 & -18.75 & 0 \\ 0 & 0 & 0 \end{pmatrix}$$

In a plane–stress condition, the strain component ε_{33} develops according to the relationship:

$$\varepsilon_{33} = -\frac{\nu}{E}(\sigma_{11} + \sigma_{22}) = -\frac{0.2}{10000}(6.25 - 18.75) = 0.00025$$

The strain matrix is thus:

$$\varepsilon = \begin{pmatrix} 0.001 & 0.0015 & 0 \\ 0.0015 & -0.002 & 0 \\ 0 & 0 & 0.00025 \end{pmatrix}$$

The following is the Mathematica code for the above calculations:

```
Ee=10000;Nu=0.2;
Cc=Ee/(1+Nu)*{{1/(1-Nu),Nu/(1-Nu),0},{Nu/(1-Nu),1/(1-Nu),0},{0,0,1/2}};
Cc//MatrixForm
strvector={0.001,-0.002,0.003};
```

```
stressvector=Cc.strvector;
stressvector//MatrixForm
s=Table[0,{i,1,3},{j,1,3}];
s[[1,1]]=stressvector[[1]];
s[[2,2]]=stressvector[[2]];
s[[1,2]]=stressvector[[3]];
s[[2,1]]=s[[1,2]];
s[[3,3]]=0;
s//MatrixForm
str=Table[0,{i,1,3} ,{j,1,3}];
str[[1,1]]=strvector[[1]];
str[[2,2]]=strvector[[2]];
str[[1,2]]=str[[2,1]]=strvector[[3]]/2;
str[[3,3]]=-Nu/Ee*(s[[1,1]]+s[[2,2]]);
str//MatrixForm
```

Problem 7: The in –plane displacements of a plate of dimensions of 5 m and 1 m in the directions of e_1 and e_2, respectively, with the bottom left corner situated at the origin are according to the relationship:

$$u_1 = 0.02X_1 + 0.02X_1X_2$$
$$u_2 = 0.0015X_2X_1^2 + 0.02X_2$$

If the plate is in the state of plane stress, draw the contour plot of the change of thickness of the plate, assuming that Young's modulus is 1000 MPa, Poisson's ratio is 0.4, and the thickness of the plate is 1 mm.

Solution: Since the plate is in a state of plane stress, $\sigma_{33} = 0$ while a strain compoenent ε_{33} devel– ops, according to the equation:

$$\varepsilon_{33} = -\frac{\nu}{E}(\sigma_{11} + \sigma_{22})$$

The in–plane strain components can be obtained from the in–plane gradient of the displacement function, as follows:

$$M = \text{Grad}u = \begin{pmatrix} \frac{\partial u_1}{\partial X_1} & \frac{\partial u_1}{\partial X_2} \\ \frac{\partial u_2}{\partial X_1} & \frac{\partial u_2}{\partial X_2} \end{pmatrix} = \begin{pmatrix} 0.02 + 0.02X_2 & 0.02X_1 \\ 0.003X_1X_2 & 0.02 + 0.0015X_1^2 \end{pmatrix}$$

$$\begin{pmatrix} \varepsilon_{11} & \varepsilon_{12} \\ \varepsilon_{21} & \varepsilon_{22} \end{pmatrix} = \frac{1}{2}\left(M + M^T\right) = \begin{pmatrix} 0.02 + 0.02X_2 & (0.01 + 0.0015X_2)X_1 \\ (0.01 + 0.0015X_2)X_1 & 0.02 + 0.0015X_1^2 \end{pmatrix}$$

The stresses can obtained through the constitutive relation of plane stress:

$$\begin{pmatrix}\sigma_{11}\\ \sigma_{22}\\ \sigma_{12}\end{pmatrix} = \frac{E}{(1+\nu)}\begin{pmatrix}\frac{1}{1-\nu} & \frac{\nu}{1-\nu} & 0\\ \frac{\nu}{1-\nu} & \frac{1}{1-\nu} & 0\\ 0 & 0 & \frac{1}{2}\end{pmatrix}\begin{pmatrix}\varepsilon_{11}\\ \varepsilon_{22}\\ \gamma_{12} = 2\varepsilon_{12}\end{pmatrix}$$

$$= \begin{pmatrix}33.33 + 0.714X_1^2 + 23.81X_2\\ 33.33 + 1.786X_1^2 + 9.524X_2\\ X_1\,(7.143 + 1.07X_2)\end{pmatrix} \text{MPa}$$

Thus, the strain in the third direction is:

$$\varepsilon_{33} = -\frac{\nu}{E}(\sigma_{11} + \sigma_{22}) = -0.027 - 0.001X_1^2 - 0.013X_2$$

The change in thickness in mm can be obtained using the formula:

$$\Delta = \varepsilon_{33}t$$

The required contour plot is:

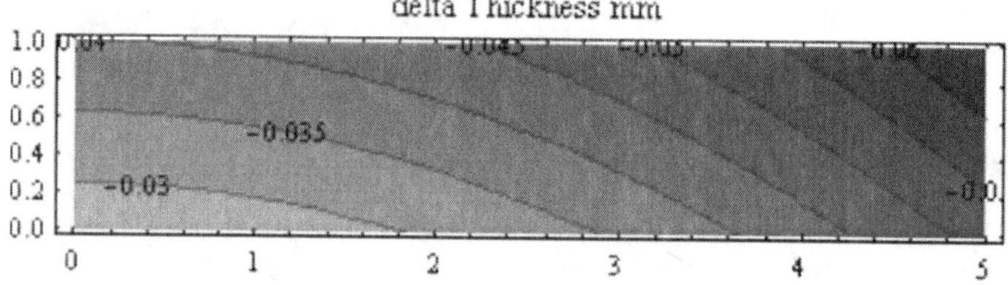

The following is the Mathematica code used for the above calculations:

```
X={X1,X2,X3};
u={0.02X1+0.02X1*X2,0.0015X2*X1^2+0.02X2};
VectorPlot[u,{X1,0,5},{X2,0,1},AspectRatio->Automatic]
gradu=Table[D[u[[i]],X[[j]]],{i,1,2},{j,1,2}];
gradu//MatrixForm
esmall=1/2*(gradu+Transpose[gradu]);
esmall//MatrixForm
strvector={esmall[[1,1]],esmall[[2,2]],2*esmall[[1,2]]};
strvector//MatrixForm
Ee=1000;Nu=0.4;
```

```
Cc=Ee/(1+Nu)*{{1/(1-Nu),Nu/(1-Nu),0},{Nu/(1-Nu),1/(1-Nu),0},{0,0,1/2}};
stressvector=Cc.strvector;
stressvector//MatrixForm
strain=Table[0,{i,1,3},{j,1,3}];
strain[[1;;2,1;;2]]=esmall;
strain[[3,3]]=-Nu/Ee*(stressvector[[1]]+stressvector[[2]]);
strain//MatrixForm
thickness=1;
delta=FullSimplify[strain[[3,3]]*thickness]
ContourPlot[delta,{X1,0,5},{X2,0,1},ContourLabels->All,AspectRatio->Automati
c,PlotLabel->"delta Thickness mm",ColorFunction->"GrayTones"]
```

Problem 8: A cylindrical pressure vessel of Young's modulus of 210 GPa and Poisson's ratio of 0.3 has an internal pressure P of 3 MPa, an average diameter of 2 m, and a thickness of 10 mm. Find the von Mises stress, the change in diameter, and the change in thickness of the cylinder in the following two conditions:

- Condition 1: Capped ends
- Condition 2: Restrained ends

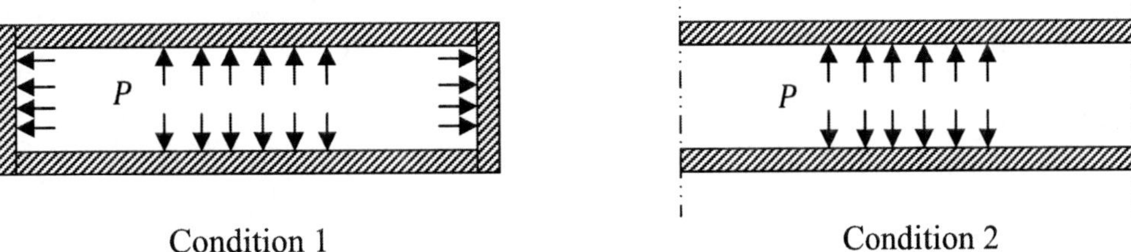

Solution: The orthonormal basis vectors are chosen in this example such that e_1 is aligned with the longitudinal axis of the pipe and e_2 is vertical. The analyzed part of the pressure wall is taken to be perpendicular to e_3, as shown in the following sketch:

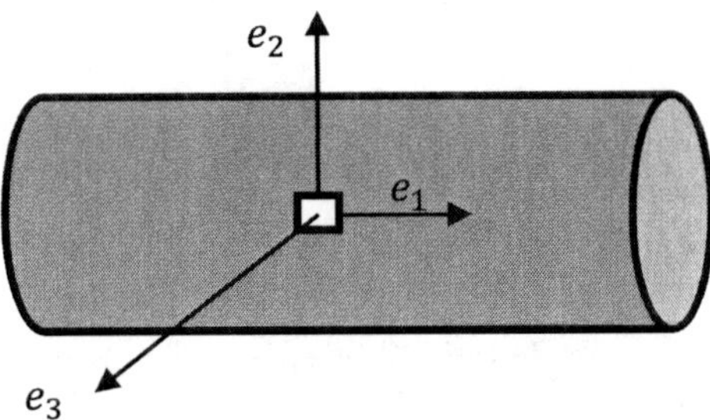

For condition 1:

The normal stress component in the longitudinal direction is:

$$\sigma_{11} = \frac{P\left(\pi r^2\right)}{2\pi rt} = \frac{Pr}{2t} = 150\text{ MPa}$$

The normal stress component in the direction of the vector e_2 is:

$$\sigma_{22} = \frac{Pr}{t} = 300\,\text{MPa}$$

Thus, the stress matrix is:

$$\sigma = \begin{pmatrix} 150 & 0 & 0 \\ 0 & 300 & 0 \\ 0 & 0 & 0 \end{pmatrix} \text{MPa}$$

The von Mises stress is:

$$Von\ Mises\ stress = \sqrt{\frac{(\sigma_{11} - \sigma_{22})^2 + \sigma_{22}^2 + \sigma_{11}^2}{2}} = 150\sqrt{3} = 259.81\,\text{MPa}$$

The strain stress relationship can be used to find the strain vector:

$$\begin{pmatrix} \varepsilon_{11} \\ \varepsilon_{22} \\ \varepsilon_{33} \\ 2\varepsilon_{12} \\ 2\varepsilon_{13} \\ 2\varepsilon_{23} \end{pmatrix} = \begin{pmatrix} \frac{1}{E} & \frac{-\nu}{E} & \frac{-\nu}{E} & 0 & 0 & 0 \\ \frac{-\nu}{E} & \frac{1}{E} & \frac{-\nu}{E} & 0 & 0 & 0 \\ \frac{-\nu}{E} & \frac{-\nu}{E} & \frac{1}{E} & 0 & 0 & 0 \\ 0 & 0 & 0 & \frac{1}{G} & 0 & 0 \\ 0 & 0 & 0 & 0 & \frac{1}{G} & 0 \\ 0 & 0 & 0 & 0 & 0 & \frac{1}{G} \end{pmatrix} \begin{pmatrix} \sigma_{11} \\ \sigma_{22} \\ \sigma_{33} \\ \sigma_{12} \\ \sigma_{13} \\ \sigma_{23} \end{pmatrix} = \begin{pmatrix} 0.000286 \\ 0.001214 \\ -0.00064 \\ 0 \\ 0 \\ 0 \end{pmatrix}$$

Thus, the change in diameter can be obtained, as follows:

$$\varepsilon_{22} = \frac{2\pi\,(r + \Delta r) - 2\pi r}{2\pi r} = \frac{\Delta r}{r} = \frac{2\Delta r}{2r} = 0.001214$$

Therefore, the change in diameter is equal to:

$$2\Delta r = 2\,(1000)\,0.001214 = 2.4\,\text{mm}.$$

The change in the thickness of the pressure vessel is:

$$\Delta t = \varepsilon_{33} t = -0.00064\,(10) = -0.0064\,\text{mm}$$

The Mathematica code is:

```
P=3;r=1000;t=10;Ee=210000;Nu=0.3;
s=Table[0,{i,1,3},{j,1,3}];
s[[1,1]]=P*r/2/t
s[[2,2]]=P*r/t
VonMises[sigma_]:=Sqrt[1/2*((sigma[[1,1]]-sigma[[2,2]])^2+(sigma[[2,2]]-
sigma[[3,3]])^2+(sigma[[3,3]]-
sigma[[1,1]])^2+6*(sigma[[1,2]]^2+sigma[[1,3]]^2+sigma[[2,3]]^2))];
a=VonMises[s]
G=Ee/2/(1+Nu);
Ss={{1/Ee,-Nu/Ee,-Nu/Ee,0,0,0},{-Nu/Ee,1/Ee,-Nu/Ee,0,0,0},{-Nu/Ee,-
Nu/Ee,1/Ee,0,0,0},{0,0,0,1/G,0,0},{0,0,0,0,1/G,0},{0,0,0,0,0,1/G}};
stressvector={s[[1,1]],s[[2,2]],s[[3,3]],s[[1,2]],s[[1,3]],s[[2,3]]}
strainvector=Ss.stressvector;
strainvector//MatrixForm
deltadiameter=strainvector[[2]]*2*r
deltathickness=strainvector[[3]]*t
```

For condition 2:

The stress in the longitudinal direction is not known since the vessel is restrained in the longitudinal direction.

The stress in the direction of the vector e_2 is:

$$\sigma_{22} = \frac{Pr}{t} = 300\,\text{MPa}$$

Thus, the stress matrix is:

$$\sigma = \begin{pmatrix} \sigma_{11} & 0 & 0 \\ 0 & 300 & 0 \\ 0 & 0 & 0 \end{pmatrix} \text{MPa}$$

The strain in the longitudinal direction is zero, and thus, the following formula can be solved to find σ_{11}:

$$\varepsilon_{11} = \frac{\sigma_{11}}{E} - \frac{\nu}{E}(\sigma_{22} + \sigma_{33}) = 0 \Longrightarrow \sigma_{11} = 0.3\,(300) = 90\,\text{MPa}$$

The von Mises stress is:

$$Von\ Mises\ stress = \sqrt{\frac{(\sigma_{11} - \sigma_{22})^2 + \sigma_{22}^2 + \sigma_{11}^2}{2}} = 353.7\,\text{MPa}$$

The strain stress relationship can be used to find the strain vector:

$$\begin{pmatrix} \varepsilon_{11} \\ \varepsilon_{22} \\ \varepsilon_{33} \\ 2\varepsilon_{12} \\ 2\varepsilon_{13} \\ 2\varepsilon_{23} \end{pmatrix} = \begin{pmatrix} \frac{1}{E} & \frac{-\nu}{E} & \frac{-\nu}{E} & 0 & 0 & 0 \\ \frac{-\nu}{E} & \frac{1}{E} & \frac{-\nu}{E} & 0 & 0 & 0 \\ \frac{-\nu}{E} & \frac{-\nu}{E} & \frac{1}{E} & 0 & 0 & 0 \\ 0 & 0 & 0 & \frac{1}{G} & 0 & 0 \\ 0 & 0 & 0 & 0 & \frac{1}{G} & 0 \\ 0 & 0 & 0 & 0 & 0 & \frac{1}{G} \end{pmatrix} \begin{pmatrix} \sigma_{11} \\ \sigma_{22} \\ \sigma_{33} \\ \sigma_{12} \\ \sigma_{13} \\ \sigma_{23} \end{pmatrix} = \begin{pmatrix} 0 \\ 0.0013 \\ -0.00056 \\ 0 \\ 0 \\ 0 \end{pmatrix}$$

Thus, the change in diameter can be obtained, as follows:

$$\varepsilon_{22} = \frac{2\pi\,(r + \Delta r) - 2\pi r}{2\pi r} = \frac{\Delta r}{r} = \frac{2\Delta r}{2r} = 0.0013$$

The change in diameter is equal to:

$$2\Delta r = 2\,(1000)\,0.0013 = 2.6\,\text{mm}.$$

The change in the thickness of the pressure vessel is:

$$\Delta t = \varepsilon_{33} t = -0.00056\,(10) = -0.0056\,\text{mm}$$

The Mathematica code utilized is:

```
P=3;r=1000;t=10;Ee=210000;Nu=0.3;
s=Table[0,{i,1,3},{j,1,3}];
s[[2,2]]=P*r/t
s[[1,1]]=Nu*s[[2,2]]
VonMises[sigma_]:=Sqrt[1/2*((sigma[[1,1]]-sigma[[2,2]])^2+(sigma[[2,2]]-
sigma[[3,3]])^2+(sigma[[3,3]]-
sigma[[1,1]])^2+6*(sigma[[1,2]]^2+sigma[[1,3]]^2+sigma[[2,3]]^2))];
a=VonMises[s]
G=Ee/2/(1+Nu);
Ss={{1/Ee,-Nu/Ee,-Nu/Ee,0,0,0},{-Nu/Ee,1/Ee,-Nu/Ee,0,0,0},{-Nu/Ee,-
Nu/Ee,1/Ee,0,0,0},{0,0,0,1/G,0,0},{0,0,0,0,1/G,0},{0,0,0,0,0,1/G}};
stressvector={s[[1,1]],s[[2,2]],s[[3,3]],s[[1,2]],s[[1,3]],s[[2,3]]}
strainvector=Ss.stressvector;
strainvector//MatrixForm
deltadiameter=strainvector[[2]]*2*r
deltathickness=strainvector[[3]]*t
```

5.3. HYPER-ELASTIC MATERIALS (LARGE ELASTIC DEFORMATIONS)

A hyper–elastic material is a material whose potential elastic deformation energy stored is independent of the path of deformation. In such materials, the stress strain relationships are derived from a potential energy function. Examples for large deformation elastic materials are rubber and some biological materials. More details about the constitutive laws for these materials will be given in section 7.3.5.

EXERCISES

(Answer key is given in round brackets for some problems):

1. Provide an alternative proof that $G = \frac{E}{2(1+\nu)}$ starting with a uniaxial state of stress.
2. What does a negative Poisson's ratio mean? Conduct a literature search and write a small paragraph supporting or opposing that materials with negative Poisson's ratio exist (Cite any references used.).
3. Conduct a literature search to find examples of different materials that are orthotropic and/or non–homogeneous and/or time dependent (Cite any references used.).
4. The constitutive laws for plane isotropic linear elastic materials are given as the linear function $\varepsilon = D\sigma$. Use Mathematica to rewrite the relationship as $\sigma = C\varepsilon$. State your opinion on whether this constitutive relationship is bijective or not and the importance of such notion.
5. Show the following relationship between the stress and the small strains in a linear–elastic isotropic material:

$$\begin{pmatrix} \sigma_{11} \\ \sigma_{22} \\ \sigma_{33} \\ \sigma_{12} \\ \sigma_{13} \\ \sigma_{23} \end{pmatrix} = \begin{pmatrix} 2\mu+\lambda & \lambda & \lambda & 0 & 0 & 0 \\ \lambda & 2\mu+\lambda & \lambda & 0 & 0 & 0 \\ \lambda & \lambda & 2\mu+\lambda & 0 & 0 & 0 \\ 0 & 0 & 0 & 2\mu & 0 & 0 \\ 0 & 0 & 0 & 0 & 2\mu & 0 \\ 0 & 0 & 0 & 0 & 0 & 2\mu \end{pmatrix} \begin{pmatrix} \varepsilon_{11} \\ \varepsilon_{22} \\ \varepsilon_{33} \\ \varepsilon_{12} \\ \varepsilon_{13} \\ \varepsilon_{23} \end{pmatrix}$$

   a. **Note:** Lamé's constants allow the relationship between the stress and the strain to be written in the following compact form (using the summation convention adopted in section 1.2.5.a):

$$\sigma_{ij} = 2\mu\varepsilon_{ij} + \lambda\varepsilon_{kk}\delta_{ij}$$

b. Also show that the corresponding relation between the deviatoric stress and the deviatoric strain, and the relation between the hydrostatic stress and volumetric strain are:

$$S_{ij} = 2\mu\varepsilon'_{ij}$$

$$p = \frac{\sigma_{kk}}{3} = \left(\frac{2}{3}\mu + \lambda\right)\varepsilon_{mm}$$

Where the deviatoric strain tensor is defined as $\varepsilon'_{ij} := \varepsilon_{ij} - \frac{\varepsilon_{kk}}{3}\delta_{ij}$

6. Show that for a linear isotropic elastic material, the components of strain can be expressed in terms of the components of stress using Lamé's constants as:

$$\varepsilon_{ij} = \frac{1}{2\mu}\sigma_{ij} - \frac{\lambda}{2\mu\left(2\mu + 3\lambda\right)}\sigma_{kk}\delta_{ij}$$

7. A linear–elastic material with Young's modulus of 200 GPa and Poisson's ratio 0.3 is in the state of strain described by the infinitesimal strain matrix:

$$\varepsilon = \begin{pmatrix} 0.015 & 0.001 & 0 \\ 0.001 & -0.002 & 0 \\ 0 & 0 & 0 \end{pmatrix}$$

- Find the principal strains and their directions.
(**Answer:** {{0.01506, −0.00206, 0.}, {{0.99829, 0.05852, 0.}, {−0.05852, 0.99829, 0.}, {0., 0., 1.}}})
- Find the principal stresses and their directions.
(**Answer:** {{3.81671, 1.5, 1.18329}, {{0.998286, 0.0585209, 0.}, {0., 0., 1.}, {−0.0585209, 0.998286, 0.}}})
- Are the principal stresses and strains aligned? (**Answer:** Yes)
- Consider the same strain matrix but for an orthotropic linear–elastic material with the following material properties: $E_{11} = 200$ GPa, $E_{22} = 100$ GPa, $E_{33} = 50$ GPa, $\nu_{12} = \nu_{13} = \nu_{23} = 0.3$, $G_{12} = G_{13} = G_{23} = 50$ GPa. Are the principal stresses and strains aligned?
(**Answer:** Principal stresses and strains are not aligned!)

8. The displacement function of the two–dimensional plate shown below is:

$$u_1 = \left(-0.005X_1^2 + 0.004X_1\right) X_2$$

$$u_2 = -0.00016X_1^3 + 0.0012X_1^2$$

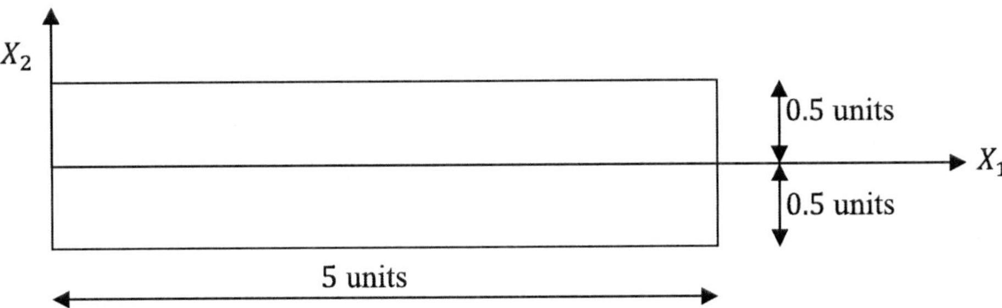

Assume that the plate is made with a linear–elastic material with Young's modulus of 200 GPa and Poisson's ratio 0.3. Draw a contour plot of the stress components σ_{11}, σ_{12}, σ_{22}, the von Mises stress, and the hydrostatic stress for the following two cases:

- The plate is in a state of plane stress.
- The plate is in a state of plane strain.

9. The shown 1m average diameter pressure vessel is made of a material with Young's modulus of 2.1 GPa and Poisson's ratio 0.3. The wall thickness is 10 mm. A constant pressure P of 1.3 MPa is maintained inside the vessel

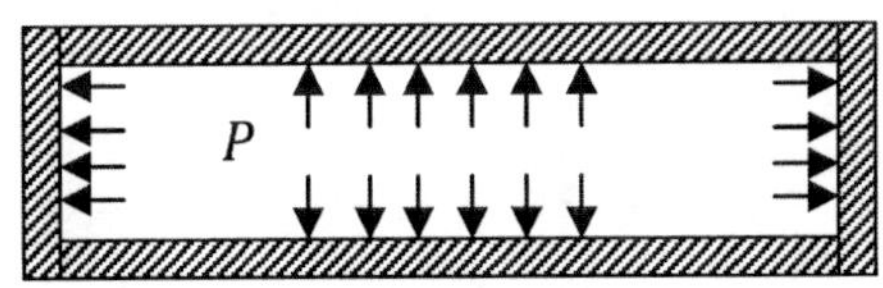

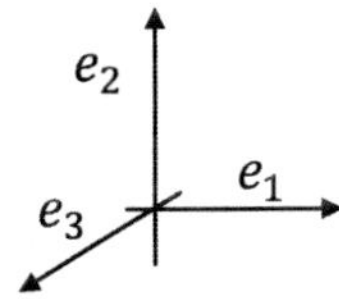

- Assuming that the length of the pressure vessel is 5m, find the increase in length. (**Answer:** 31 mm)
- Find the increase in the diameter of the pressure vessel. (**Answer:** 26.3 mm)
- Find the change in thickness of the pressure vessel. (**Answer:** –0.14 mm)
- Find the volumetric strain throughout the pressure vessel wall. (**Answer:** 1.9%)

10. The shown cylinder is subjected to a confined compression test in which the horizontal stress is kept constant at –200 MPa, while the magnitude of the compressive vertical stress is increased to reach the value of –600 MPa. Find the volumetric strain if the shear and bulk moduli are 22.72 GPa and 20.833 GPa, respectively. (**Answer:** –1.6%)

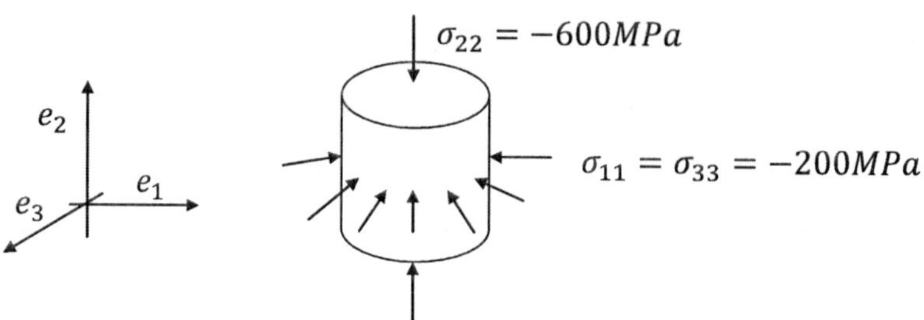

11. An *elastic* cylindrical pressure vessel with average diameter d, average radius r, thickness t, Young's modulus E, Poisson's ratio ν is subjected to an internal pressure P. Find the circumferential stress, the longitudinal stress, the circumferential strain, and the longitudinal strain in the following two conditions:

    Condition 1: The pressure vessel has two caps on its ends.

    Condition 2: The pressure vessel is infinitely long and, thus, cannot expand or contract in the longitudinal direction.

    For both conditions, assume that the thickness is very small compared to the diameter.

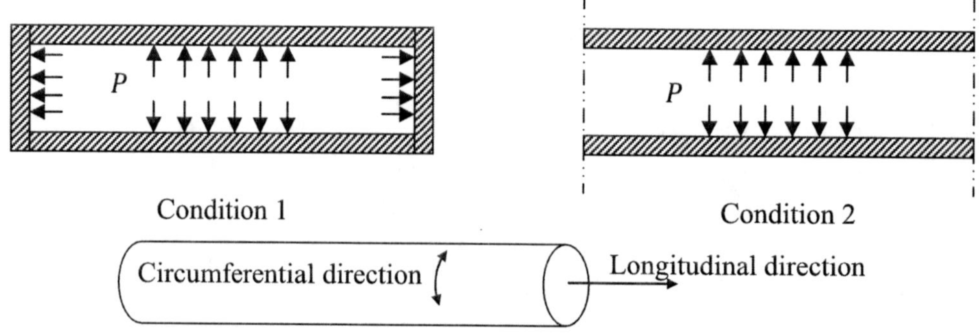

12. Show that if e_1, e_2 and e_3 are the axes of orthotropy of a linear–elastic orthotropic material, then the stress–strain relationship of the material is given by that shown in section 5.2.3. Then, show that if the plane of e_1 and e_2 is a plane of isotropy and symmetry, then the stress–strain relationship of this transversely isotropic material is given by that shown in section 5.2.4.

CHAPTER 6
The Beam Approximations

Many of the large standing towers, bridges, and buildings owe their existence to some simplified versions of the equilibrium equations applied to slender, line–like structures. The first beam approximation is represented in the Euler Bernoulli's beam theory, which originated sometime in the eighteenth century. This approximation has been so successful that it is still the basic approximation used in modern structural analysis taught in current structural analysis courses. In this chapter, the Euler Bernoulli's beam approximation and the Timoshenko's beam approximation will be represented as examples of the equilibrium equations described before.

6.1. EULER BERNOULLI'S BEAM UNDER BENDING

A beam is generally a slender structure with one dimension that is much larger than the other two dimensions. There are three basic physical assumptions behind the Euler Bernoulli's beam equations that will be listed here, but are going to be explained in more details in the following derivation section. These three basic assumptions are:

- **Plane sections perpendicular** to the neutral axis before deformations **remain plane** and **perpendicular** to the neutral axis after deformation (Figure 6–3)
- Small deformations
- Linear elastic isotropic material

As will be shown later, the above basic assumptions lead to the following results:

- The **strain field** is linear along the cross section perpendicular to the longer dimension.
- The **stress field** is linear along the cross section perpendicular to the longer dimension.
- Let E be the Young's modulus of the material, $I = \int X_2^2 dA$ be the moment of inertia of the beam's cross section, q be the loading per unit length of the beam. The relationship

between the deflection y in the direction of the basis vector e_2 and the applied loading is:

$$\frac{d^2\left(EI\frac{d^2y}{dX_1^2}\right)}{dX_1^2} = q$$

If the quantity EI is constant along the longitudinal position X_1, then the relationship is simplified to:

$$EI\frac{d^4y}{dX_1^4} = q$$

When q is equal to zero, y is a third–degree polynomial (cubic function) of the position X_1 and hence, the Euler Bernoulli beam equation is referred to as the **cubic formulation**.

6.1.1. Derivation

6.1.1.a. Normal Force, Shearing Force, and Bending Moment

The beam equations are traditionally written in terms of the internal forces rather than the components of the stress tensor; thus, the first step in beam analysis is to define the internal forces on any cross section. Assume that the right portion of the beam shown in Figure 6–1 has been removed, and its action is represented by the stress distribution on the gray section perpendicular to the X_1 axis. Then, the integration of the stress field over the area would yield the internal forces acting on this cross section. The internal forces acting on the area A are namely the normal (axial) force N, shearing forces V_{12} and V_{13}, the bending moments M_3 and M_2, and the torque T. These are shown in Figure 6–1, and their classical positive sign convention is indicated by the arrows in the figure. The internal forces can be evaluated from the stress distribution, as follows:

$$N = \int \sigma_{11} dA, \quad V_{12} = -\int \sigma_{12} dA, \quad V_{13} = -\int \sigma_{13} dA$$

$$M_3 = -\int X_2\sigma_{11} dA, \quad M_2 = -\int X_3\sigma_{11} dA, \quad T = \int -X_3\sigma_{12} + X_2\sigma_{13} dA$$

Notice that the negative signs in the above equations are introduced to reconcile the difference in the sign convention of the stress tensor components and the internal forces.

6.1.1.b. Equilibrium Equations in Terms of Internal Forces

It is possible to derive the equilibrium equations for the Euler Bernoulli beam from the general equilibrium equations described in Chapter 4 by integration. However, we will follow the

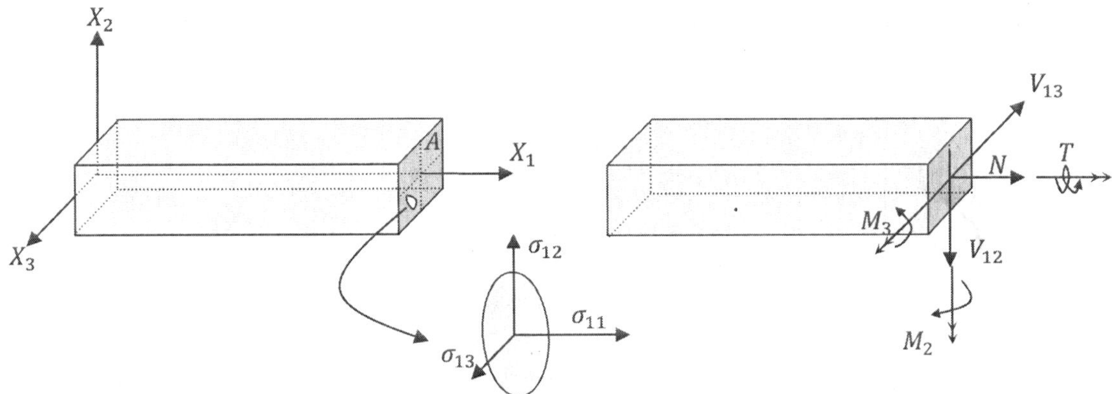

FIGURE 6-1 Internal forces on a beam cross sectional area aligned perpendicular to the coordinate axis X_1. Arrows indicate classical positive directions.

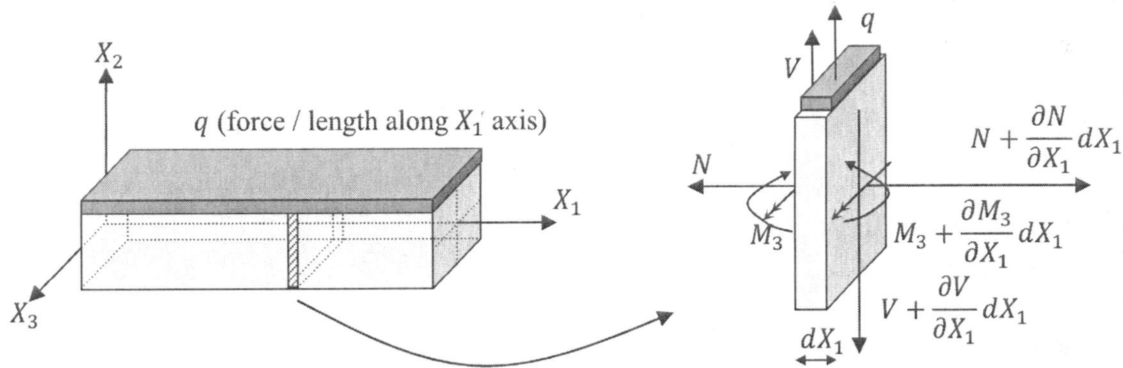

FIGURE 6-2 Equilibrium Equations for Euler Bernoulli and Timoshenko beams subjected to transverse loading.

classical method as described in many structural analysis books following the sign convention shown in Figure 6–2. First, we consider two planes that are very close to each other (separated by a small distance dX_1) and perpendicular to the axis X_1. Then, we can use the first approximation in a Taylor's series to describe the changes in the internal forces between those two planes (Figure 6–2). Assuming loading to be transverse and only in the direction of X_2, and considering the sum of forces in the vertical direction and the sum of moments to be equal to zero and denoting $V = V_{12}$:

$$\frac{\partial V}{\partial X_1} = q, \quad \frac{\partial M_3}{\partial X_1} = V$$

Notice that q, V and M_3 are strictly functions of X_1, and hence, the equilibrium equations can be written as:

$$\frac{dV}{dX_1} = q, \quad \frac{dM_3}{dX_1} = V$$

6.1.1.c. Deformation Assumptions

The approximation in beam theory is inherent in the assumptions of the deformation functions. In Euler Bernoulli formulation, the beam is assumed to deform such that plane sections perpen–dicular to the longitudinal and neutral axis before deformation remain plane and perpendicular to the neutral axis after deformation. Poisson's effects are usually neglected.

Let $\Omega_0 \in \mathbb{R}^3$ represent the reference configuration of a beam in which the neutral axis is aligned with the coordinate axis X_1. Let $\Omega \in \mathbb{R}^3$ represent the deformed configuration such that plane sections remain plane and perpendicular to the neutral axis after deformation and such that the shape and dimensions of the cross sections do not change. Let $y = f(X_1)$ be a smooth function whose slope is termed θ that describes the motion of the neutral axis in the direction of the basis vector e_2 (Figure 6–3). The position functions are given in terms of y and θ as follows:

$$x_1 = X_1 - X_2 \sin\theta$$

$$x_2 = X_2 \cos\theta + y$$

$$x_3 = X_3$$

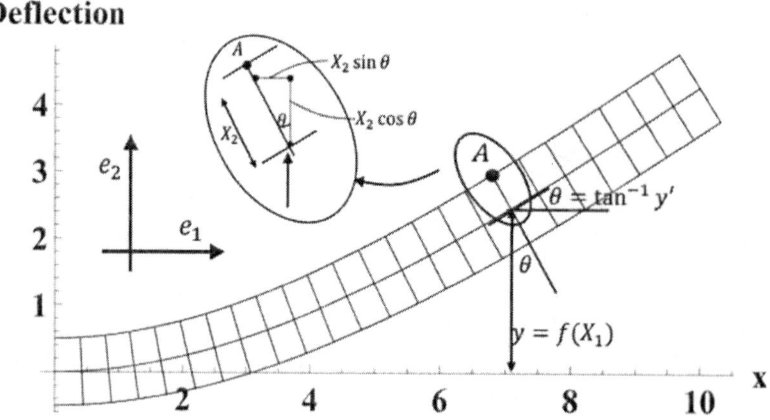

FIGURE 6-3 A beam deforming according to the assumptions behind the Euler Bernoulli's equation. (Plane sections perpendicular to the neutral axis remain plane and perpendicular to the neutral axis after deformation.)

Let y be small enough (small deformations) such that the following approximations are valid:

$$\sin\theta = \tan\theta = \theta = \frac{dy}{dX_1}$$

$$\cos\theta = 1$$

Therefore, the displacement functions are:

$$u_1 = -\theta X_2$$

$$u_2 = y$$

$$u_3 = 0$$

Along the entire beam, ε_{11} is the only nonzero component in the infinitesimal strain tensor:

$$\varepsilon_{11} = -X_2\frac{\partial\theta}{\partial X_1} = -X_2\frac{d^2y}{dX_1^2}, \quad \varepsilon_{22} = \frac{\partial y}{\partial X_2} = 0,$$

$$\varepsilon_{33} = 0, \quad \gamma_{12} = -\theta + \theta = 0, \quad \gamma_{13} = \gamma_{23} = 0$$

Notes:

- On any cross section where X_1 = constant, the value of y and its derivatives are constant since y is a strict function of X_1; therefore, ε_{11} is a linear function of X_2 at X_1 = constant.
- The assumed displacement function results in zero values for the axial strain components ε_{22} and ε_{33}. It is possible to add correction factors in the position functions x_2 and x_3 so that $\varepsilon_{22} = \varepsilon_{33} = -\nu\varepsilon_{11}$. However, this would add a complication to the derivation with no practical significance and might introduce shear strains as well.

6.1.1.d. Constitutive Law

Assuming that the material is isotropic linear elastic and ignoring Poisson's ratio effect, then the component σ_{11} of the stress tensor at X_1 = constant is a linear function of X_2:

$$\sigma_{11} = E\varepsilon_{11} = -EX_2\frac{\partial\theta}{\partial X_1} = -EX_2\frac{d^2y}{dX_1^2}$$

In Euler Bernoulli's beam, Poisson's ratio effects are ignored and therefore, σ_{11} and ε_{11} are the only non–zero components of the **normal** stresses and strains.

Using the formulas in section 6.1.1.a. The bending moment M_3 can be written as a function of the deformation variable $\frac{d^2y}{dX_1^2}$:

$$M_3 = -\int X_2\sigma_{11}dA = E\frac{d^2y}{dX_1^2}\int X_2^2 dA = EI\frac{d^2y}{dX_1^2}$$

Using the linear–elastic isotropic constitutive equation and the assumption of plane sections remaining plane, the famous bending moment formula can be deduced:

$$\sigma_{11} = -\frac{M_3 X_2}{I}$$

It is important to note that the assumption of "plane and perpendicular sections remain plane and perpendicular after deformation" leads to zero shear strains across the entire beam. The zero–shear strains approximation is equivalent to assuming that the shear modulus G is infinite. The shear stresses σ_{12} and σ_{21}, in this case, are not necessarily zero. If b is the thickness of the cross section, then the shear stress component σ_{12} at a point on the crosssectional area at the position X_2 can be calculated using the classical formula:

$$\sigma_{12} = -\frac{VQ(X_2)}{Ib}$$

where $Q(X_2)$ is the moment of area above or below the point under consideration.

6.1.1.e. Final Equations To Be Solved

Assuming that the term EI is constant across the whole beam (i.e., homogeneous material and constant cross section), we can combine the equilibrium equations with the constitutive law for the bending moment to reach the following equations to be solved:

$$M_3 = EI\frac{d^2y}{dX_1^2}, \quad \frac{dM_3}{dX_1} = V = EI\frac{d^3y}{dX_1^3}, \quad \frac{dV}{dX_1} = EI\frac{d^4y}{dX_1^4} = q$$

These equations are solved to find the unknown function $y = f(X_1)$, which can then be used to obtain the distribution of the internal forces along the entire beam.

6.1.2. Solution

The Euler Bernoulli beam equation is a fourth–order ordinary differential equation. Thus, four boundary conditions are required to solve it. From a practical standpoint, the boundary conditions at each end of the beam are usually given as one of the following combinations: shear and moment, shear and rotation, moment and deflection. The following Mathematica code utilizes the `DSolve` function to solve the ordinary differential equation with $q = -a$ (constant) and the boundary conditions of a simple beam. The code also outputs the deflection at the midpoint of the beam. Finally, unit values are assumed for the different constants to be able to plot the deflection, bending moment, shearing force, and rotation (Figure 6–4).

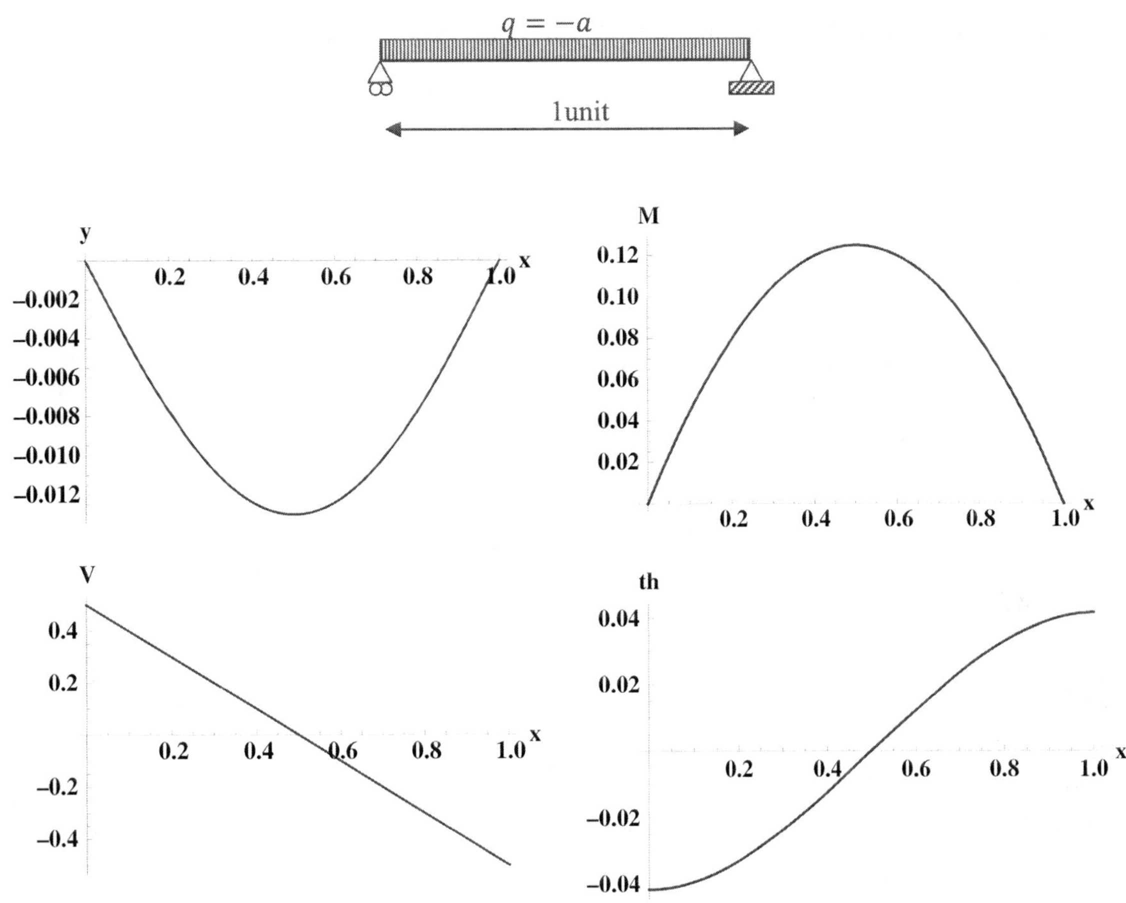

FIGURE 6-4 Plot of the simple beam solution of the Euler Bernoulli beam equation (y, M, V and θ) using Mathematica assuming $L = 1$ units, $a = 1$ units and $EI = 1$ units.

```
Clear[a,x,V,M,y,L,EI]
M=EI*D[y[x],{x,2}];
V=EI*D[y[x],{x,3}];
th=D[y[x],x];
q=-a;
M1=M/.x->0;
M2=M/.x->L;
V1=V/.x->0;
V2=V/.x->L;
y1=y[x]/.x->0;
y2=y[x]/.x->L;
th1=th/.x->0;
th2=th/.x->L;
s=DSolve[{EI*y''''[x]==q,M1==0,M2==0,y1==0,y2==0},y,x];
```

```
y=y/.s[[1]];
y[L/2]
EI=1;
a=1;
L=1;
Plot[y[x],{x,0,L},BaseStyle->Directive[Bold,15],PlotRange->All,AxesLabel-
>{"x","y"},PlotStyle->{Black,Thickness[0.005]}]
Plot[M,{x,0,L},PlotRange->All,BaseStyle->Directive[Bold,15],AxesLabel-
>{"x","M"},PlotStyle->{Black,Thickness[0.005]}]
Plot[V,{x,0,L},BaseStyle->Directive[Bold,15],PlotRange->All,AxesLabel-
>{"x","V"},PlotStyle->{Black,Thickness[0.005]}]
Plot[th,{x,0,L},PlotRange->All,BaseStyle->Directive[Bold,15],AxesLabel-
>{"x","th"},PlotStyle->{Black,Thickness[0.005]}]
```

6.2. TIMOSHENKO BEAM UNDER BENDING

The Timoshenko beam formulation is intentionally derived to better describe beams whose shear deformations cannot be ignored. Short beams are a prime example for which shear deformations cannot be ignored, and thus, the Timoshenko beam approximation is better suited to describe their behavior. The basic physical assumptions behind the Timoshenko beam are similar to those described for the Euler Benroulli beam, except that the shear deformations are allowed. The following are the three basic assumptions behind the Timoshenko beam theory. (Compare with those described in section 6.1.)

- **Plane sections perpendicular** to the neutral axis before deformations **remain plane**, but not necessarily **perpendicular** to the neutral axis after deformation (Figure 6–5)
- Small deformation
- Linear elastic isotropic material

6.2.1. Derivation

6.2.1.a. Normal Force, Shearing Force, and Bending Moment

Similar to the Euler Bernoulli beam in section 6.1.1.a, the internal forces in terms of the stresses are:

$$N = \int \sigma_{11} dA, \quad V_{12} = -\int \sigma_{12} dA, \quad V_{13} = -\int \sigma_{13} dA$$

$$M_3 = -\int X_2 \sigma_{11} dA, \quad M_2 = -\int X_3 \sigma_{11} dA, \quad T = \int -X_3 \sigma_{12} + X_2 \sigma_{13} dA$$

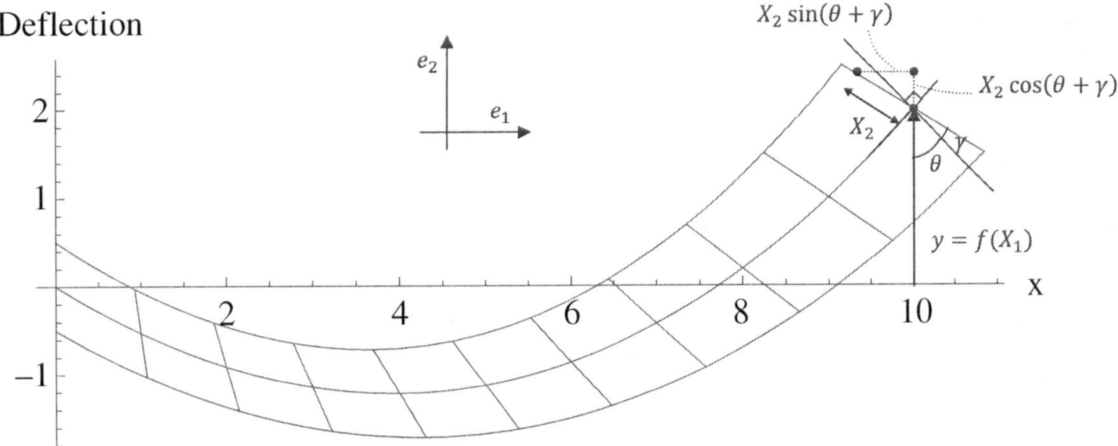

FIGURE 6-5 Beam deformation according to the Timoshenko beam assumptions. (Plane sections perpendicular to the neutral axis remain plane, but not necessarily perpendicular to the neutral axis after deformation.)

6.2.1.b. Equilibrium Equations

Similar to the Euler Bernoulli beam in section 6.1.1.b, the equilibrium equations can be written as:

$$\frac{dV}{dX_1} = q, \quad \frac{dM_3}{dX_1} = V$$

6.2.1.c. Deformation Assumptions

In Timoshenko beam formulation, the beam is assumed to deform such that plane sections perpendicular to the longitudinal and neutral axis before deformation remain plane, but not necessarily perpendicular to the neutral axis after deformation (Figure 6–5). Poisson's ratio effects are usually neglected.

Let $\Omega_0 \in \mathbb{R}^3$ represent the reference configuration of a beam in which the neutral axis is aligned with the coordinate axis X_1. Let $\Omega \in \mathbb{R}^3$ represent the deformed configuration such that plane sections remain plane, but not necessarily perpendicular to the neutral axis after deformation, and such that the shape and dimensions of the cross sections do not change. Let $y = f(X_1)$ be a smooth function (whose slope is termed θ) that describes the smooth translation of the neutral axis in the direction of the basis vector e_2. Let $\gamma = g(X_1)$ be a smooth function describing the shear rotation of the cross section (Figure 6–5), then the position functions are:

$$x_1 = X_1 - X_2 \sin(\theta + \gamma)$$

$$x_2 = X_2 \cos(\theta + \gamma) + y$$

$$x_3 = X_3$$

The full cross–sectional rotation is denoted by $\psi = \theta + \gamma$.

Let y be small enough (small deformations) such that the following approximations are valid:

$$\sin\psi = \psi = \sin(\theta + \gamma) = \tan(\theta + \gamma) = \theta + \gamma = \frac{dy}{dX_1} + \gamma$$

$$\cos\psi = \cos(\theta + \gamma) = 1$$

Therefore, the displacement functions are:

$$u_1 = -\psi X_2$$

$$u_2 = y$$

$$u_3 = 0$$

Throughout the entire beam, ε_{11} and γ_{12} are the only nonzero components in the infinitesimal strain tensor:

$$\varepsilon_{11} = -X_2\frac{\partial\psi}{\partial X_1}, \quad \varepsilon_{22} = \frac{\partial y}{\partial X_2} = 0, \quad \varepsilon_{33} = 0, \quad \gamma_{12} = -\gamma, \quad \gamma_{13} = \gamma_{23} = 0$$

On any cross section where X_1 = constant, the values of y, ψ and their derivatives are constant since y and ψ are strict functions of X_1; therefore, ε_{11} is a linear function of X_2 at X_1 = constant.

6.2.1.d. Constitutive Law

Assuming that the material is isotropic linear elastic, then the component σ_{11} of the stress tensor at X_1 = constant is a linear function of X_2:

$$\sigma_{11} = E\varepsilon_{11} = -EX_2\frac{\partial\psi}{\partial X_1} = -EX_2\frac{d\psi}{dX_1}$$

The component σ_{12} of the stress tensor at X_1 = constant can be **empirically** written as an average value by introducing a cross–sectional area parameter k:

$$\sigma_{12} = kG\gamma_{12} = -kG\gamma$$

The bending moment M_3 and the shearing force V can be now written as a function of the defor–mation variables ψ and γ, as follows:

$$M_3 = -\int X_2\sigma_{11}dA = E\frac{d\psi}{dX_1}\int X_2^2 dA = EI\frac{d\psi}{dX_1}$$

$$V = V_{12} = -\int \sigma_{12}dA = kGA\gamma$$

6.2.1.e. Final Equations To Be Solved

Assuming that the terms *EI* and *kGA* are constant across the whole beam (i.e., homogeneous material and constant cross section), we can combine the equilibrium equations with the constitutive laws for the bending moment and shearing force to reach the following equations to be solved for the variables y and ψ (or y and γ):

$$M_3 = EI\frac{d\psi}{dX_1}, \quad \frac{dM_3}{dX_1} = V = EI\frac{d^2\psi}{dX_1^2}, \quad \frac{dV}{dX_1} = EI\frac{d^3\psi}{dX_1^3} = q,$$

$$\psi = \frac{dy}{dX_1} + \gamma = \frac{dy}{dX_1} + \frac{EI}{kAG}\frac{d^2\psi}{dX_1^2}$$

6.2.2. Solution

The Timoshenko beam equation is a system of differential equations in the unknown functions ψ and y. To find a solution, four boundary conditions are required to solve it. From a practical standpoint, the boundary conditions at each end of the beam are usually given as one of the following combinations: shear and moment, shear and full–section rotation, moment and deflection, deflection, and full–section rotation. The following Mathematica code utilizes the `DSolve` function to solve the two ordinary differential equations for y and ψ with $q = -a$ (constant) and the boundary conditions of a simple beam. The code also outputs the deflection at the midpoint of the beam. Finally, unit values are assumed for the different constants to be able to plot the deflection, bending moment, shearing force, and rotation (Figure 6–6).

```
Clear[a,x,V,M,y,L,psi,EI,kAG]
M=EI*D[psi[x],x];
V=EI*D[psi[x],{x,2}];
q=-a;
M1=M/.x->0;
M2=M/.x->L;
V1=V/.x->0;
V2=V/.x->L;
y1=y[x]/.x->0;
y2=y[x]/.x->L;
psi1=psi[x]/.x->0;
psi2=psi[x]/.x->L;
s=DSolve[{EI*psi'''[x]==q,psi[x]==y'[x]+EI/kAG*psi''[x],M1==0,M2==0,y1==0,y2
==0},{y,psi},x];
y=y/.s[[1,2]];
psi=psi/.s[[1,1]];
y[L/2]
EI=1;
a=1;
L=1;
kAG=1;
```

```
Plot[y[x],{x,0,L},BaseStyle->Directive[Bold,15],PlotRange->All,AxesLabel-
>{"x","y"},PlotStyle->{Black,Thickness[0.005]}]
Plot[psi[x],{x,0,L},PlotRange->All,BaseStyle->Directive[Bold,15],AxesLabel-
>{"x","Psi"},PlotStyle->{Black,Thickness[0.005]}]
Plot[M,{x,0,L},PlotRange->All,AxesLabel->{"x","M"},PlotStyle-
>{Black,Thickness[0.005]}BaseStyle->Directive[Bold,15]]
Plot[V,{x,0,L},PlotRange->All,AxesLabel->{"x","V"},PlotStyle-
>{Black,Thickness[0.005]}BaseStyle->Directive[Bold,15]]
```

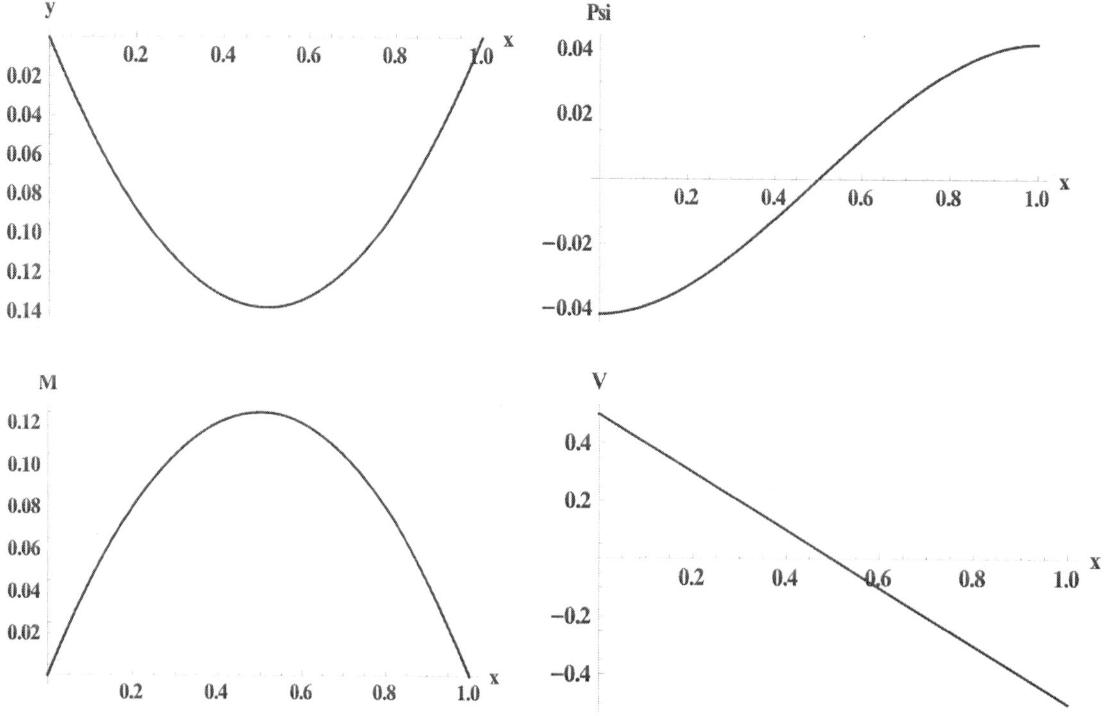

FIGURE 6-6 Plot of the simple beam solution of the Timoshenko beam equation using Mathematica (y, M, V and ψ) assuming $L = 1$ units, $a = 1$ units, $EI = 1$ units and $kAG = 1$ units.

6.3. LINEAR ELASTIC BEAM UNDER AXIAL LOADING

The Euler Bernoulli and the Timoshenko beam formulations described account only for the defor–mations due to bending, without considering any axial deformation in the neutral axis of the beam. It is possible to augment the Euler Bernoulli and the Timoshenko beams so that they also account for axial deformations of the neutral axis, but because of the assumption of small deformations, the resulting equations of axial–loading deformations and bending deformations are uncoupled.

It is worth noting before going through any derivations that this uncoupling is only valid for small deformations and naturally does not allow for the modeling of large deformations phenomena, such as buckling. This is because the axial forces N did not appear in the equilibrium equations derived for the Euler Bernoulli and the Timoshenko beam equations (**Why?**). If a linear–elastic isotropic (Young's modulus = E) beam of length L and a constant cross–section area A is subjected to a constant normal force P, then the elongation Δ can be obtained using the formula:

$$\Delta = \frac{PL}{EA}$$

In this section, the derivation of this formula will be presented using the elements of solid mechanics described in the previous chapters.

6.3.1. Derivation

6.3.1.a. Normal Force and Equilibrium Equation

Similar to the Euler Bernoulli beam in section 6.1.1.a, the normal force in terms of the stresses is:

$$N = \int \sigma_{11} dA$$

Similar to the Euler Bernoulli beam in section 6.1.1.b, the equilibrium equations can be written as:

$$\frac{dN}{dX_1} = 0$$

6.3.1.b. Deformation Assumptions

Under an axial load, a beam is assumed to deform such that the horizontal displacement $u = h(X_1)$ is a smooth function of X_1 and that cross sections remain plane during the deformation. Poisson's ratio effect is usually neglected. Thus, the position functions are:

$$x_1 = X_1 + u$$

$$x_2 = X_2$$

$$x_3 = X_3$$

And, the displacement functions are:

$$u_1 = u$$

$$u_2 = 0$$

$$u_3 = 0$$

Throughout the entire beam, ε_{11} is the only nonzero component in the infinitesimal strain tensor:

$$\varepsilon_{11} = \frac{du}{dX_1}$$

On any cross section where $X_1 = \text{constant}$, the values of u and its derivatives are constant since u is a strict function of X_1; therefore, ε_{11} is constant on any cross section perpendicular to X_1.

6.3.1.c. Constitutive Law

Assuming that the material is isotropic linear elastic, then the component σ_{11} of the stress tensor at $X_1 = \text{constant}$ is:

$$\sigma_{11} = E\varepsilon_{11} = E\frac{du}{dX_1}$$

The normal force can be written in terms of the strains:

$$N = \int \sigma_{11} dA = EA\frac{du}{dX_1}$$

6.3.1.d. Final Equations To Be Solved

Assuming that the term EA is constant across the whole beam (i.e., homogeneous material and constant cross section), we can combine the equilibrium equations with the constitutive laws for the normal force to reach the following equations to be solved:

$$N = EA\frac{du}{dX_1}, \quad \frac{dN}{dX_1} = EA\frac{d^2u}{dX_1^2} = 0$$

6.3.1.e. Solution

The solution to the above equations given the boundary conditions of $N = P$ when $X_1 = L$ and $u = 0$ when $X_1 = 0$ is simply:

$$u = \frac{P}{EA}X_1$$

6.4. EXAMPLES

6.4.1. Visualizing the Difference between Euler Bernoulli and Timoshenko Beam Approximations

The following code analyzes a cantilever beam of length 10 units and height of 1 unit under an upward unit load. The displaced shape of the beam is drawn for various values of EI and kAG. The

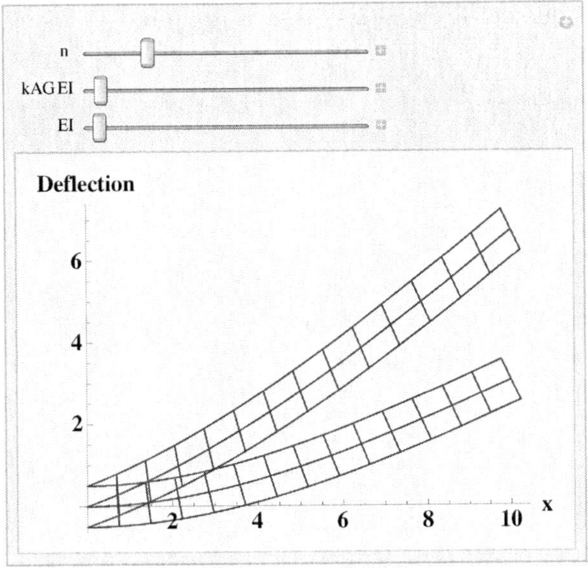

FIGURE 6-7 Timoshenko (more flexible) versus Euler Bernoulli (less deflection) cantilever beam under an upward unit load at the right end using Mathematica.

Euler Bernoulli beam has a deflection less than the Timoshenko beam for a small value of kAG/EI. Notice that plane sections remain plane for both beams. For Euler Bernoulli beam, the sections perpendicular to the neutral axis remain perpendicular after deformation, while this is not the case for the Timoshenko beam (Figure 6–7). At higher values of kAG (higher shear stiffness) the displacement of both beams are almost the same. Copy and paste the code into Mathematica and notice the effect of changing the values of kAG and EI.

```
Manipulate[
 Lx=10;
 Ly=1;
 z=t^2*Lx/2/EI-t^3/6/EI;
 zd=D[z,t];
 ztable=Table[{z,zd}/.t->i*Lx/n,{i,1,n}];
 curvtable=Table[{i/n*Lx+(5-t)/5*Ly/2*ztable[[i,2]],ztable[[i,1]]+(t-
5)/5*Ly/2},{i,1,n}];
kAG=kAGoEI*EI;
 z2=z+t/kAG;
 psi=(2t*Lx-t^2)/2/EI;
 ztable2=Table[{z2,psi}/.t->i*Lx/n,{i,1,n}];
 curvtable2=Table[{i/n*Lx+(5-t)/5*Ly/2*ztable2[[i,2]],ztable2[[i,1]]+(t-
5)/5*Ly/2},{i,1,n}];
 plot1={{t-Ly/2*zd,Ly/2+z},{t,z},{t+Ly/2*zd,-Ly/2+z},curvtable};
 ParametricPlot[{plot1,{t-Ly/2*psi,Ly/2+z2},{t,z2},{t+Ly/2*psi,-
Ly/2+z2},curvtable2},{t,0,10},BaseStyle->Directive[Bold,15],PlotStyle-
```

```
>{Black},AxesLabel-
>{"x","Deflection"}],{n,10,30,2},{{kAGoEI,0.01,"kAG/EI"},0.01,1},{EI,100,100
0}]
```

6.4.2. Euler Bernoulli Beam

6.4.2.a. Cantilever Beam

The following Mathematica code obtains the equation of a cantilever beam loaded at the left end with a shearing force P and a bending moment M (Figure 6–8). The deflection of the centerline follows the following equation:

$$y = \frac{3Mx^2 - 3LPx^2 + Px^3}{6EI}$$

```
Clear[a,x,V,M,y,L,EI]
M=EI*D[y[x],{x,2}];
V=EI*D[y[x],{x,3}];
th=D[y[x],x];
q=-a;
M1=M/.x->0;
M2=M/.x->L;
V1=V/.x->0;
V2=V/.x->L;
y1=y[x]/.x->0;
y2=y[x]/.x->L;
th1=th/.x->0;
th2=th/.x->L;
s=DSolve[{EI*y''''[x]==0,M2==Mom,V2==PP,y1==0,th1==0},y,x]
```

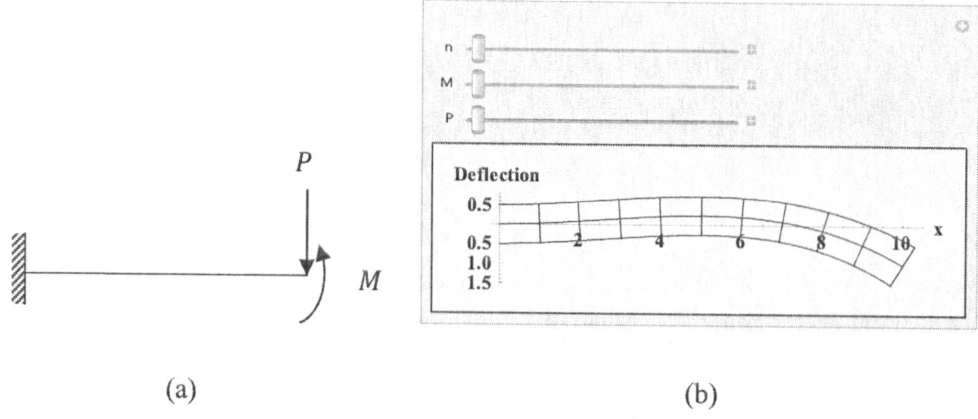

FIGURE 6-8 Euler Bernoulli cantilever beam loaded with an end shear and moment. a) Positive direction of loading, b) beam deflected shape using Mathematica.

The `Manipulate` function is used to visualize the deflected shape of the beam using $EI = 500$ units, $L = 10$ units and beam height = 1 unit.

```
Manipulate[EI=500;
 Lx=10;
 Ly=1;
 z=(3*M*t^2-3*Lx*P*t^2+P*t^3)/EI;
 zd=D[z,t];
 ztable=Table[{z,zd}/.t->i*Lx/n,{i,1,n}];
 curvtable=Table[{i/n*Lx+(5-t)/5*Ly/2*ztable[[i,2]],ztable[[i,1]]+(t-
5)/5*Ly/2},{i,1,n}];
 plot1={{t-Ly/2*zd,Ly/2+z},{t,z},{t+Ly/2*zd,-Ly/2+z},curvtable};
ParametricPlot[{plot1},{t,0,10},BaseStyle->Directive[Bold,15],PlotStyle-
>{Black},AxesLabel->{"x","Deflection"}],{n,10,30,2},{{M,-1,"M"},-
15,15},{{P,-1,"P"},-2,2}]
```

6.4.2.b. Bending and Shear Stresses in an Euler Bernoulli Cantilever Beam

Consider a cantilever Euler Bernoulli beam with Young's modulus $E = 20GPa$. The beam is 5 m long and has a rectangular cross section. The height of the beam is 0.5 m, while the width is 0.25 m. If a load of 125 kN is applied at the free end, draw the contour plot of σ_{11}, σ_{12} and the von Mises stress along the beam. Also, sketch the principal stress directions at the top fibers, the bottom fibers, the neutral axis and at quarter of the height of the beam at the support ($X_1 = 0$). Also draw the vector plot of the displacement of the beam.

Solution: To find the stress distribution in the beam, the bending moment and the shearing force are first obtained by solving the beam differential equation of equilibrium:

$$EI\frac{d^4y}{dX_1^4} = q = 0 \Longrightarrow EIy = \frac{C_1X_1^3}{6} + \frac{C_2X_1^2}{2} + C_3X_1 + C_4$$

where $I = bt^3/12$ is the moment of inertia and $b = 0.25\text{m}$ and $t = 0.5\text{m}$. The constants C_1, C_2, C_3 and C_4 are integration constants and can be obtained from the following boundary conditions:

$$@X_1 = 0 \Longrightarrow y = 0$$

$$@X_1 = 0 \Longrightarrow \theta = \frac{dy}{dX_1} = 0$$

$$@X_1 = 5 \Longrightarrow V = EI\frac{d^3y}{dX_1^3} = 125$$

$$@X_1 = 5 \Longrightarrow M = EI\frac{d^2y}{dX_1^2} = 0$$

Thus, the displacement function y is:

$$y = -\frac{125}{6EI}(3LX_1^2 - X_1^3)$$

The displacement function of the beam along the entire length and height is:

$$u = x - X = \begin{pmatrix} X_1 - X_2 \frac{dy}{dX_1} \\ y + X_2 \\ X_3 \end{pmatrix} - \begin{pmatrix} X_1 \\ X_2 \\ X_3 \end{pmatrix} = \begin{pmatrix} -X_2 \frac{dy}{dx_1} \\ y \\ 0 \end{pmatrix}$$

The vector plot of the displacement shows that the downward vertical displacement is dominant:

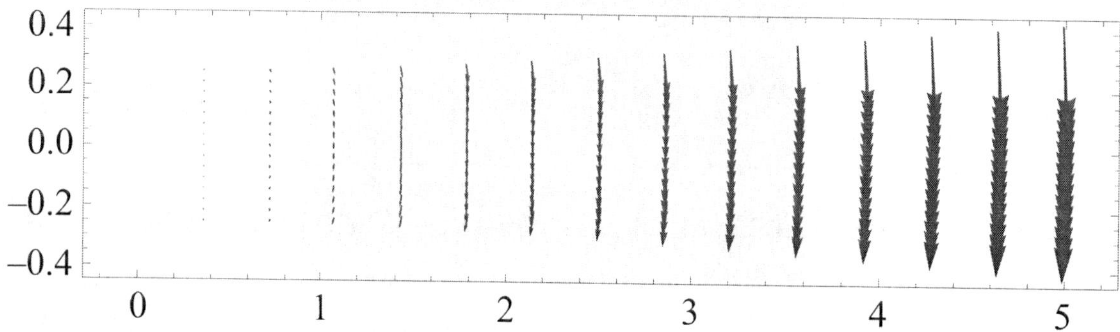

The bending moment and the shearing force functions are:

$$M_3 = EI \frac{d^2y}{dX_1^2} = 125(X_1 - 5)\, kNm., \quad V = EI \frac{d^3y}{dX_1^3} = 125kN$$

The stresses along the length of the beam can be obtained as a function of the beam height and of the beam length:

$$\sigma_{11} = -\frac{M_3(X_1)X_2}{I} = -\frac{125(X_1 - 5)X_2}{I}, \quad \sigma_{12} = \frac{-V(X_1)Q(X_2)}{Ib} = -\frac{125\left(\frac{t}{2} - X_2\right) b \left(\frac{X_2}{2} + \frac{t}{4}\right)}{Ib}$$

The contour plot of σ_{11}, σ_{12} and von Mises stress are shown in Figure 6–9. The bending stress σ_{11} is positive at the top (tension) and negative at the bottom, and increases in magnitude toward the support. The shearing stress component σ_{12} is constant at the neutral axis (**Why?**) throughout the length of the beam and decreases toward zero at the top and bottom fibers of the beam. The von Mises stress distribution follows the bending stress component, but gives a positive value in tension and in compression!

The eigenvalues and the eigenvectors of the stress matrix at the specified locations are obtained using Mathematica and are sketched in Figure 6–10. The shear stresses at the top and bottom fibers are zero, and thus, the principal stresses are aligned with the chosen orthonormal coordinate system. At the neutral axis of the beam, the bending stress component σ_{11} is zero, and the only nonzero component is σ_{12}, thus, the principal stresses at the neutral axis are aligned at 45 degrees.

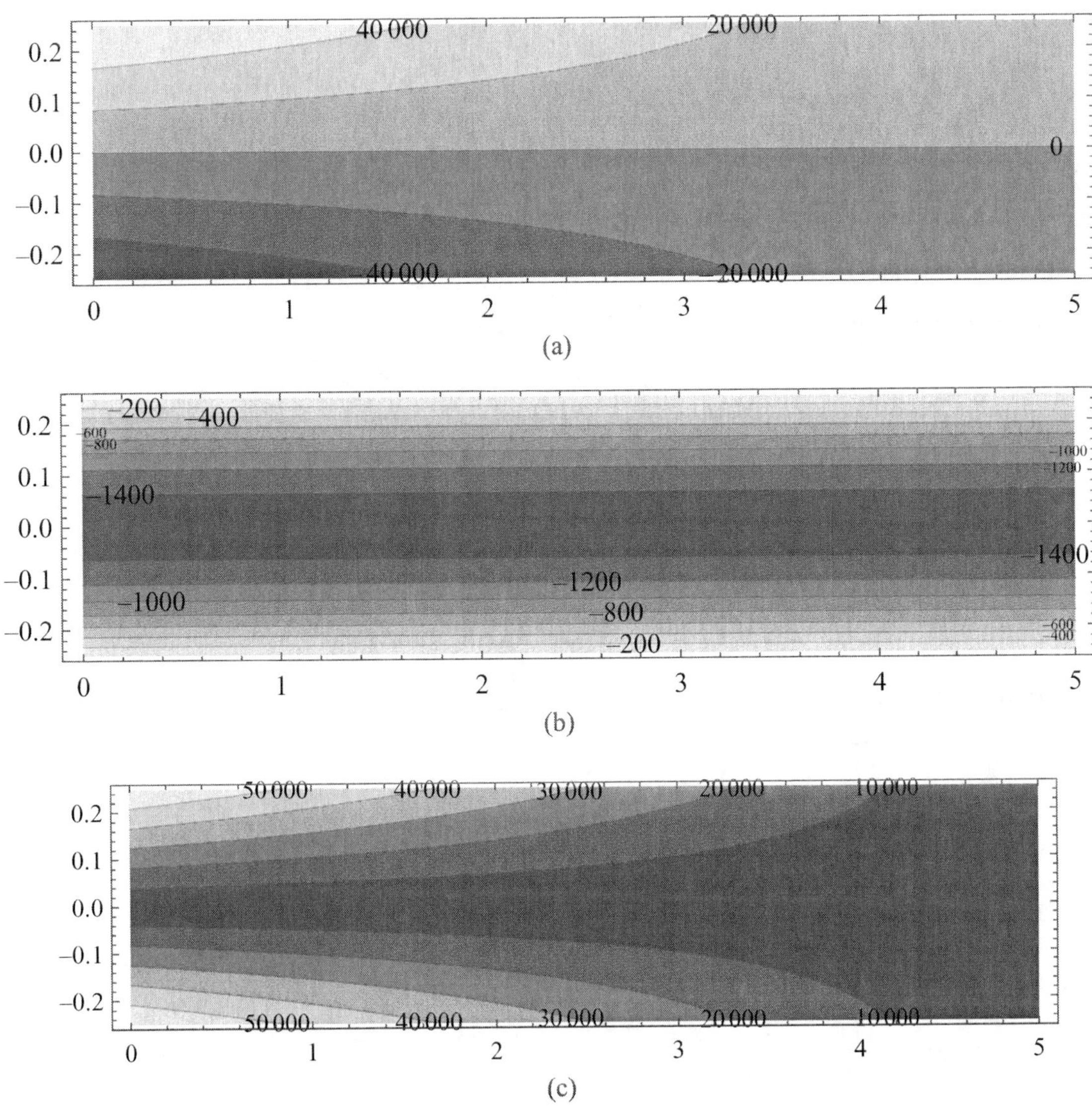

FIGURE 6-9 Stresses in kPa units in an Euler Bernoulli cantilever beam under a downward load at the right end: (a) σ_{11}, (b) σ_{12} and (c) von Mises stresses.

The direction of the principal stress with the highest magnitude at $X_2 = 0.125$ m is along the vec– tor {0.9993, –0.0374212,0.}. At $X_2 = -0.125$ m, the principal stress with magnitude –30.042 MPa is aligned with the vector {0.9993,0.0374212,0.}

▶ **BONUS EXERCISE:** Sketch the vector plot of the directions of the principal stresses and the contour plot of the in plane maximum and minimum principal stresses.

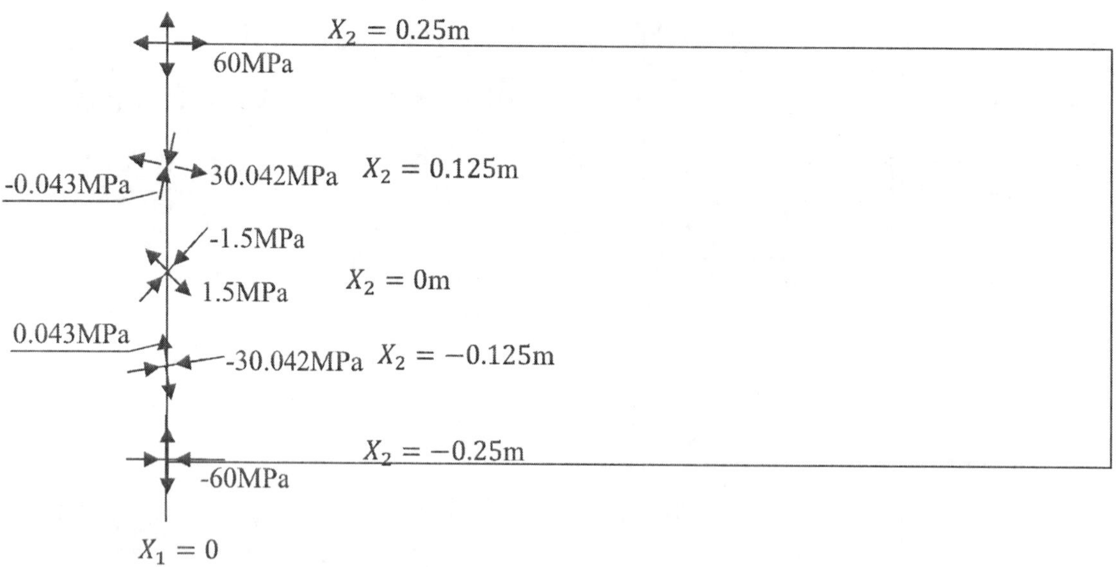

FIGURE 6-10 Directions and magnitude of principal stresses at the support of the cantilever beam.

The following Mathematica code was utilized for the above calculations:

```
Clear[a,x,V,M,y,L,EI]
M=EI*D[y[x],{x,2}];
V=EI*D[y[x],{x,3}];
th=D[y[x],x];
q=-a;
M1=M/.x->0;
M2=M/.x->L;
V1=V/.x->0;
V2=V/.x->L;
y1=y[x]/.x->0;
y2=y[x]/.x->L;
th1=th/.x->0;
th2=th/.x->L;
s=DSolve[{EI*y''''[x]==0,M2==0,V2==125,y1==0,th1==0},y,x]
displacement=y/.s[[1]]
EI=Ee*Ii;
u = {X2*-D[displacement[x], x], displacement[x]}
L=5;
displacement[L]
M=FullSimplify[M/.s[[1]]]
V=V/.s[[1]]
Ee=20000000;
b=0.25;
t=0.5;
Ii=b*t^3/12;
s11=-M*X2/Ii;
```

```
Q=(t/2-X2)*b*(t/4+X2/2)
s12=-V*Q/Ii/b;
smatrix={{s11,s12,0},{s12,0,0},{0,0,0}};
FullSimplify[Chop[smatrix]]//MatrixForm
VonMises[sigma_]:=Sqrt[1/2*((sigma[[1,1]]-sigma[[2,2]])^2+(sigma[[2,2]]-
sigma[[3,3]])^2+(sigma[[3,3]]-
sigma[[1,1]])^2+6*(sigma[[1,2]]^2+sigma[[1,3]]^2+sigma[[2,3]]^2))];
vonmisesstress=VonMises[smatrix]
ContourPlot[s11,{x,0,5},{X2,-0.25,0.25},AspectRatio->0.25,ContourLabels-
>True,ColorFunction->"GrayTones"]
ContourPlot[s12,{x,0,5},{X2,-0.25,0.25},AspectRatio->0.25,ContourLabels-
>True,ColorFunction->"GrayTones"]
ContourPlot[vonmisesstress,{x,0,5},{X2,-0.25,0.25},AspectRatio-
>0.25,ContourLabels->True,ColorFunction->"GrayTones"]
VectorPlot[u, {x, 0, 5}, {X2,-0.25, 0.25}, AspectRatio -> 0.25,
 VectorStyle -> Black]
stressmat1=smatrix/.{x->0,X2->-0.25}
Eigensystem[stressmat1]
stressmat2=smatrix/.{x->0,X2->-0.25/2}
Eigensystem[stressmat2]
stressmat3=smatrix/.{x->0,X2->}
Eigensystem[stressmat3]
stressmat4=smatrix/.{x->0,X2->0.25/2}
Eigensystem[stressmat4]
stressmat5=smatrix/.{x->0,X2->0.25}
Eigensystem[stressmat5]
stressmat1=smatrix/.{x->1,X2->-0.25}
Eigensystem[stressmat1]
stressmat2=smatrix/.{x->1,X2->-0.25/2}
Eigensystem[stressmat2]
stressmat3=smatrix/.{x->1,X2->}
Eigensystem[stressmat3]
stressmat4=smatrix/.{x->1,X2->0.25/2}
Eigensystem[stressmat4]
stressmat5=smatrix/.{x->1,X2->0.25}
Eigensystem[stressmat5]
```

6.4.2.c. Bending and Shear Stresses in an Euler Bernoulli Simple Beam

Consider an Euler Bernoulli simple beam with Young's modulus $E = 20$ GPa. The beam is 8 m long and has a rectangular cross section. The height of the beam is 0.5 m, while the width is 0.25 m. If a load of 125 kN/m is distributed over the beam length, draw the contour plot of σ_{11}, σ_{12} and the von Mises stress along the beam. Also draw the vector plot of the displacement of the beam.

Solution: To find the stress distribution in the beam, the bending moments and the shearing force are first obtained by solving the beam differential equation of equilibrium:

$$EI\frac{d^4y}{dX_1^4} = q = 125 \Longrightarrow EIy = \frac{C_1X_1^3}{6} + \frac{C_2X_1^2}{2} + C_3X_1 + C_4$$

where $I = bt^3/12$ is the moment of inertia and $b = 0.25\,\text{m}$ and $t = 0.5\,\text{m}$. The constants C_1, C_2, C_3 and C_4 are integration constants and can be obtained from the following boundary conditions:

$$@X_1 = 0 \Longrightarrow y = 0$$

$$@X_1 = L \Longrightarrow y = 0$$

$$@X_1 = 0 \Longrightarrow M = EI\frac{d^2y}{dX_1^2} = 0$$

$$@X_1 = L \Longrightarrow M = EI\frac{d^2y}{dX_1^2} = 0$$

Thus, the displacement function y is:

$$y = -\frac{125}{24EI}(L^3X_1 - 2LX_1^3 + X_1^4)$$

The displacement function of the beam along the entire length and height is:

$$u = x - X = \begin{pmatrix} X_1 - X_2\frac{dy}{dX_1} \\ y + X_2 \\ X_3 \end{pmatrix} - \begin{pmatrix} X_1 \\ X_2 \\ X_3 \end{pmatrix} = \begin{pmatrix} -X_2\frac{dy}{dx_1} \\ y \\ 0 \end{pmatrix}$$

The vector plot of the displacement shows that the downward vertical displacement is dominant:

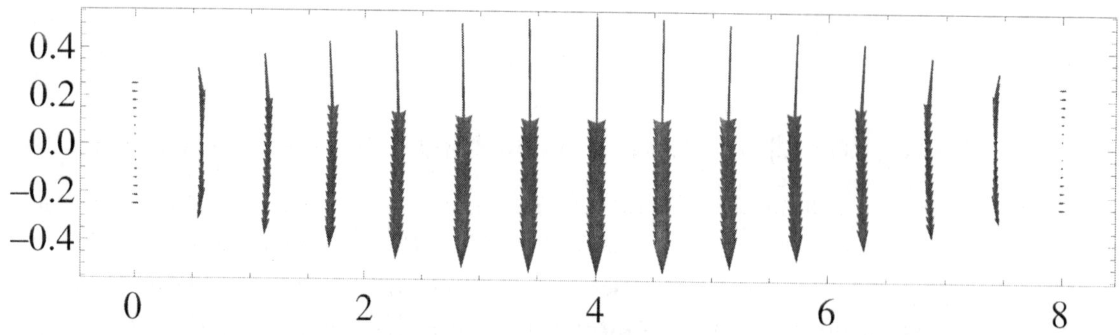

The bending moment and the shearing force functions are:

$$M_3 = EI\frac{d^2y}{dX_1^2} = -\frac{125X}{2}(-8 + X_1)kNm., \quad V = EI\frac{d^3y}{dX_1^3} = -125(X_1 - 4)kN$$

The stresses along the length of the beam can be obtained as a function of the beam height and length:

$$\sigma_{11} = -\frac{M_3(X_1)X_2}{I}, \quad \sigma_{12} = \frac{-V(X_1)Q(X_2)}{Ib} = -\frac{V(\frac{t}{2} - X_2)b(\frac{X_2}{2} + \frac{t}{4})}{Ib}$$

The contour plot of σ_{11}, σ_{12} and von Mises stress are shown in Figure 6–11. Note that the contribution of the shear stresses σ_{12} in the von Mises stress in this beam is very small compared to

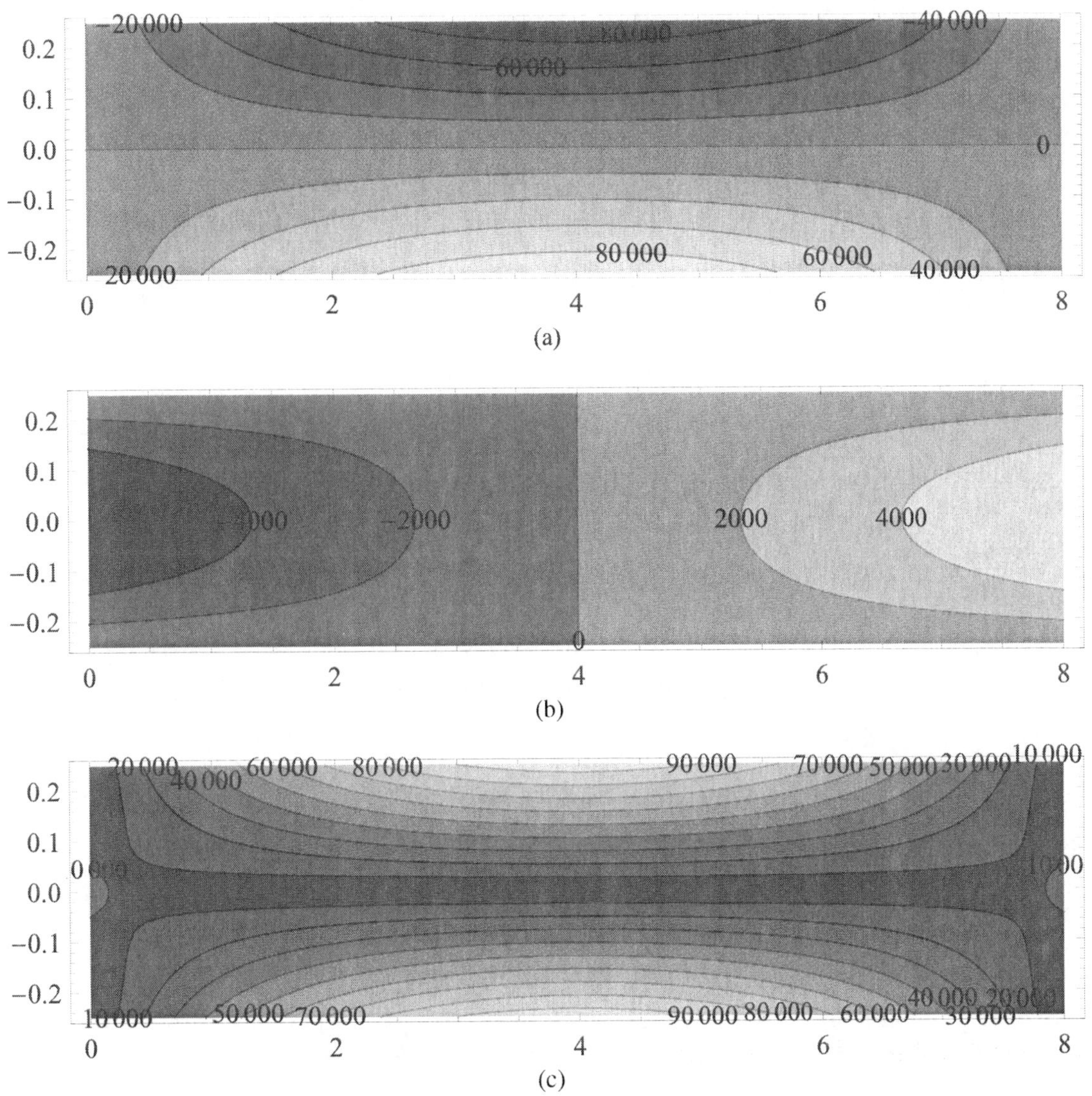

FIGURE 6-11 Stresses in KPa units in an Euler Bernoulli simple beam under a downward distributed load: (a) σ_{11}, (b) σ_{12} and (c) von Mises stresses.

the bending stress σ_{11}. This is natural due to the small ratio between the depth of the beam and its length. The longer the beam, the less significant the shear stresses are.

The following is the Mathematica code utilized for the above calculations:

```
Clear[a,x,V,M,y,L,EI,Ee,Ii]
M=EI*D[y[x],{x,2}];
V=EI*D[y[x],{x,3}];
th=D[y[x],x];
q=-125;
M1=M/.x->0;
M2=M/.x->L;
V1=V/.x->0;
V2=V/.x->L;
y1=y[x]/.x->0;
y2=y[x]/.x->L;
th1=th/.x->0;
th2=th/.x->L;
s=DSolve[{EI*y''''[x] ==q,M2==0,y2==0,y1==0,M1==0},y,x]
displacement=y/.s[[1]]
Middisplacement=displacement[L/2]
EI=Ee*Ii;
u={X2*-D[displacement[x],x],displacement[x]}
L=8;
M=FullSimplify[M/.s[[1]]]
V=FullSimplify[V/.s[[1]]]
Ee=20000000;
b=0.25;
t=0.5;
Ii=b*t^3/12;
s11=-M*X2/Ii
Q=(t/2-X2)*b*(t/4+X2/2)
s12=-V*Q/Ii/b
smatrix={{s11,s12,0},{s12,0,0},{0,0,0}};
FullSimplify[Chop[smatrix]]//MatrixForm

VonMises[sigma_]:=Sqrt[1/2*((sigma[[1,1]]-sigma[[2,2]])^2+(sigma[[2,2]]-
sigma[[3,3]])^2+(sigma[[3,3]]-
sigma[[1,1]])^2+6*(sigma[[1,2]]^2+sigma[[1,3]]^2+sigma[[2,3]]^2))];
vonmisesstress=VonMises[smatrix]
ContourPlot[s11,{x,0,L},{X2,-0.25,0.25},AspectRatio->0.25,ContourLabels-
>True,ColorFunction->"GrayTones"]
ContourPlot[s12,{x,0,L},{X2,-0.25,0.25},AspectRatio->0.25,ContourLabels-
>True,ColorFunction->"GrayTones"]
ContourPlot[vonmisesstress,{x,0,L},{X2,-0.25,0.25},AspectRatio-
>0.25,ContourLabels->True,ColorFunction->"GrayTones"]
VectorPlot[u,{x,0,L},{X2,-0.25,0.25},AspectRatio->0.25,VectorStyle->Black]
```

6.4.3. Beams under Axial Load

6.4.3.a. Displacement Boundary Conditions

A beam fixed at both ends is under a triangular axial load with a maximum value of 125 kN at $X_1 = L$ (Figure 6–12). The beam's length, width, and height are 5, 0.25, and 0.5 m, respectively. Young's modulus is 20 GPa. Ignore Poisson's ratio effect. Find the displacement function of this beam along with the stress distribution along the length of the beam. Draw the contour plot of σ_{11} and the vector plot of the displacement along the beam length.

Solution: The displacement function of the beam is:

$$u = \begin{pmatrix} u_1(X_1) \\ 0 \\ 0 \end{pmatrix}$$

The only nonzero strain component is ε_{11}:

$$\varepsilon_{11} = \frac{du_1}{dX_1}$$

Ignoring Poisson's ratio effect, the only nonzero stress component is σ_{11}:

$$\sigma_{11} = E\varepsilon_{11} = E\frac{du_1}{dX_1}$$

The distribution of the load is according to the equation:

$$q = \frac{125X_1}{L}$$

The equilibrium equation is:

$$\frac{dN}{dX_1} = -q$$

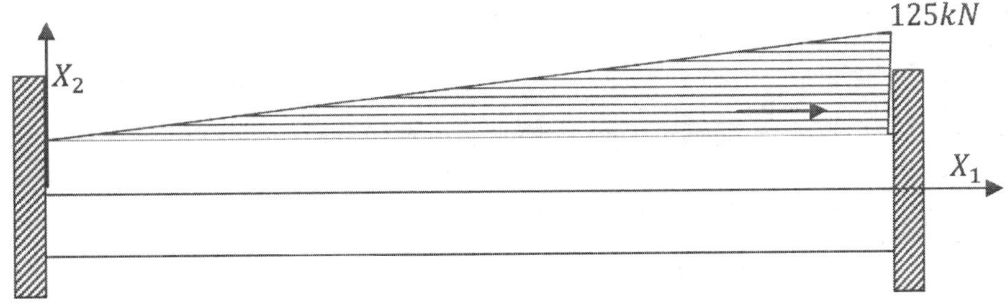

FIGURE 6-12 Beam under a triangular axial load.

The normal force can be replaced with the stress, and the equilibrium equation can now be written in terms of the displacement as follows:

$$\frac{dN}{dX_1} = \frac{d(\sigma_{11}A)}{dX_1} = A\frac{d\sigma_{11}}{dX_1} = EA\frac{d^2u_1}{dX_1^2} = -\frac{125X_1}{L}$$

The boundary conditions are:

$$@X_1 = 0 \Longrightarrow u_1 = 0$$

$$@X_1 = L \Longrightarrow u_1 = 0$$

Thus, the solution to the above differential equation is:

$$u_1 = \frac{125(X_1L^2 - X_1^3)}{6EAL}, \quad \sigma_{11} = \frac{125(L^2 - 3X_1^2)}{6AL}, \quad N = \frac{125(L^2 - 3X_1^2)}{6L}$$

The required contour and vector plots are shown below.

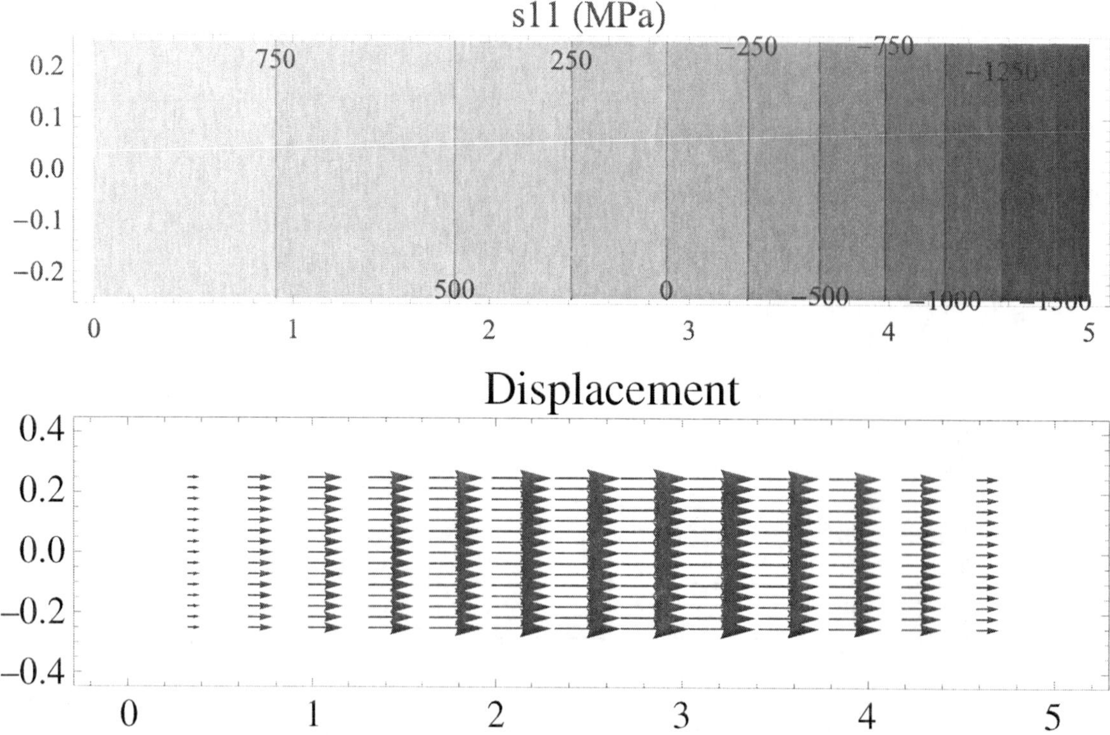

The Mathematica code utilized is:

```
Clear[u,x,L,EA,Ee,A]
EA=Ee*A;
u1=u[x]/.x->0;
u2=u[x]/.x->L;
Normalf=EA*D[u[x]];
qnormal=125*x/L;
s=DSolve[{EA*u''[x]==-qnormal,u1==0,u2==0},u,x]
u=u[x]/.s[[1]]
s11=Ee*D[u,x]
Normalf=Normalf/.s[[1]];
Ee=20000;
A=0.25*0.5;
L=5;
disp={u,0}
ContourPlot[s11,{x,0,5},{X2,-0.25,0.25},AspectRatio->0.25,ContourLabels->True,ColorFunction->"GrayTones",PlotLabel->"s11 (MPa)"]
VectorPlot[disp,{x,0,5},{X2,-0.25,0.25},AspectRatio->0.25,PlotLabel->"Displacement",VectorStyle->Black]
```

6.4.3.b. Displacement and Stress Boundary Conditions

The shown beam has one end fixed and the other free. The beam is under a triangular axial load with a maximum value of 125 kN at $X_1 = L$ (Figure 6–13). The beam's length, width, and height are 5, 0.25, and 0.5 m, respectively. Young's modulus is 20 GPa. Ignore Poisson's ratio effect. Find the displacement function of this beam along with the stress distribution along the length of the beam. Draw the contour plot of σ_{11} and the vector plot of the displacement along the beam length.

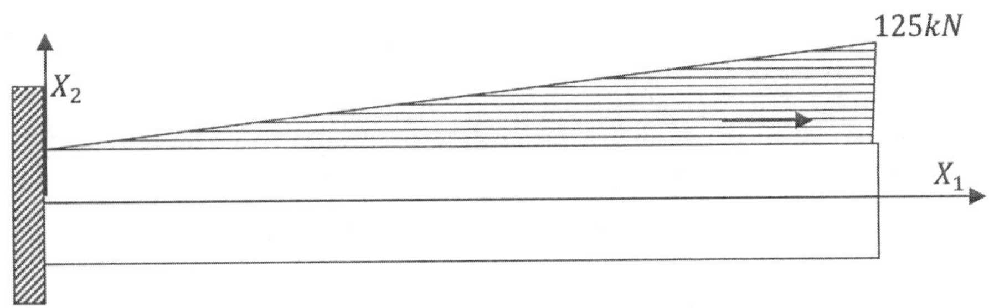

FIGURE 6-13 Beam under a triangular axial load.

Solution: The displacement function of the beam is:

$$u = \begin{pmatrix} u_1(X_1) \\ 0 \\ 0 \end{pmatrix}$$

The only nonzero strain component is ε_{11}:

$$\varepsilon_{11} = \frac{du_1}{dX_1}$$

Ignoring Poisson's ratio effect, the only nonzero stress component is σ_{11}:

$$\sigma_{11} = E\varepsilon_{11} = E\frac{du_1}{dX_1}$$

The distribution of the load is according to the equation:

$$q = \frac{125X_1}{L}$$

The equilibrium equation is:

$$\frac{dN}{dX_1} = -q$$

The normal force can be replaced with the stress, and the equilibrium equation can now be written in terms of the displacement, as follows:

$$\frac{dN}{dX_1} = \frac{d(\sigma_{11}A)}{dX_1} = A\frac{d\sigma_{11}}{dX_1} = EA\frac{d^2u_1}{dX_1^2} = -\frac{125X_1}{L}$$

The boundary conditions are:

$$@X_1 = 0 \Longrightarrow u_1 = 0$$

$$@X_1 = L \Longrightarrow \sigma_{11} = \frac{du_1}{dX_1} = 0$$

Thus, the solution to the above differential equation is:

$$u_1 = \frac{125(3X_1L^2 - X_1^3)}{6EAL}, \quad \sigma_{11} = \frac{125(3L^2 - 3X_1^2)}{6AL}, \quad N = \frac{125(3L^2 - 3X_1^2)}{6L}$$

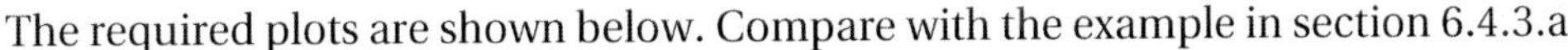

The required plots are shown below. Compare with the example in section 6.4.3.a

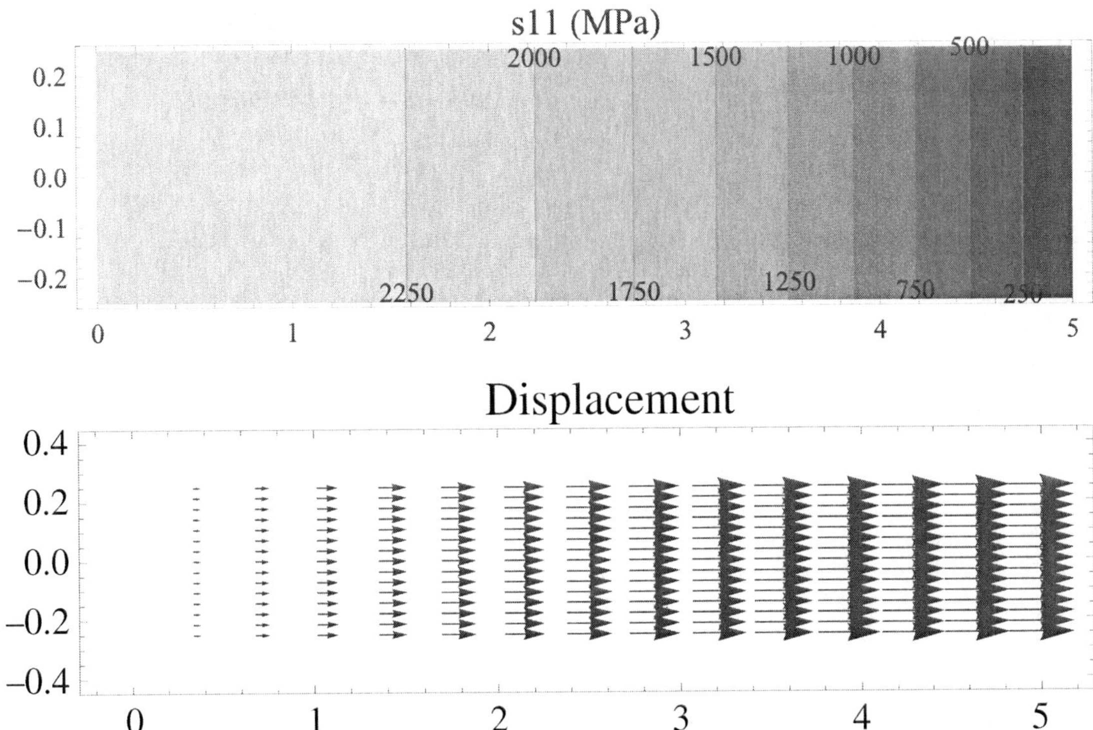

The Mathematica code utilized is:

```
Clear[u,x,L,EA,Ee,A]
EA=Ee*A;
u1=u[x]/.x->0;
u2=u[x]/.x->L;
s2=D[u[x],x]/.x->L;
Normalf=EA*D[u[x]];
qnormal=125*x/L;
s=DSolve[{EA*u'' [x]==qnormal,u1==0,s2==0},u,x]
u=u[x]/.s[[1]]
s11=Ee*D[u,x]
Normalf=Normalf/.s[[1]];
Ee=20000;
A=0.25*0.5;
L=5;
disp={u,0}
ContourPlot[s11,{x,0,5},{X2,-0.25,0.25},AspectRatio->0.25,ContourLabels-
>True,ColorFunction->"GrayTones",PlotLabel->"s11 (MPa)"]
VectorPlot[disp,{x,0,5},{X2,-0.25,0.25},AspectRatio->0.25,PlotLabel-
>"Displacement",VectorStyle->Black]
```

6.4.3.c. Beam with Varying Cross-Sectional Area under Nonuniform Axial Load

The shown beam (Figure 6–14) has a varying circular cross–sectional area with a radius $r = 2$ units at the top varying linearly to $r = 1$ unit at the bottom. The beam is used to carry a concentrated load P and has a weight density ρ. Young's modulus $E = 1000$ units. Find the vertical displacement field of the beam.

Solution: The shown beam has a varying cross–sectional area and a varying nonuniform vertical load due to its own weight.

The radius r and the area A vary with X_1 according to the equations:

$$r = \left(2 - \frac{X_1}{L}\right)$$

$$A = \pi\left(2 - \frac{X_1}{L}\right)^2$$

Following the derivation in section 6.3.1.a, we first derive the equilibrium equation:

$$\frac{dN}{dX_1} + \rho A = 0$$

Then, the position function is assumed to be:

$$x_1 = X_1 + u, \quad x_2 = X_2, \quad x_3 = X_3$$

and the displacement function is:

$$u_1 = u, \quad u_2 = 0, \quad u_3 = 0$$

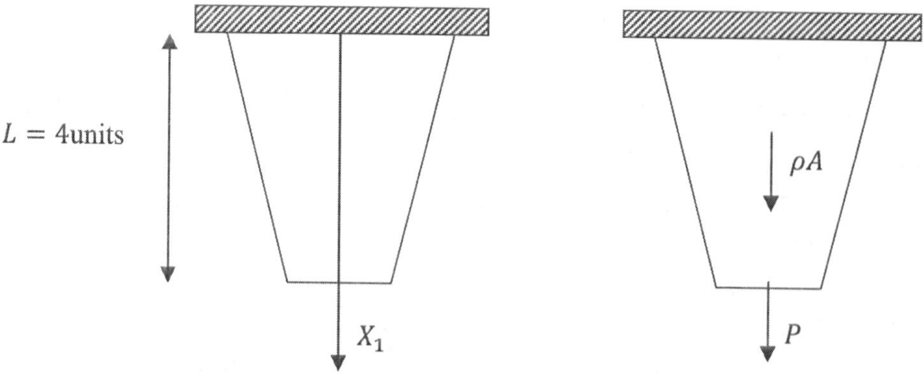

FIGURE 6-14 Beam under varying axial load.

Throughout the entire beam, ε_{11} is the only nonzero component in the infinitesimal strain tensor:

$$\varepsilon_{11} = \frac{du}{dX_1}$$

Then, using the constitutive law:

$$\sigma_{11} = E\frac{du}{dX_1}, \quad N = EA\frac{du}{dX_1}$$

Finally, combining the constitutive law with the equilibrium equation, we reach the equation to be solved (Notice that unlike section 6.3.1.d *EA* is not constant.):

$$\frac{dEA\frac{du}{dX_1}}{dX_1} + \rho A = 0 \Longrightarrow E\frac{dA}{dX_1}\frac{du}{dX_1} + EA\frac{d^2u}{dX_1^2} + \rho A = 0$$

$$\frac{E\pi}{2}\left(\frac{X_1}{4} - 2\right)\frac{du}{dX_1} + E\pi\left(2 - \frac{X_1}{4}\right)^2\frac{d^2u}{dX_1^2} + \pi\rho\left(2 - \frac{X_1}{4}\right)^2 = 0$$

The boundary conditions are $u = 0$ when $X_1 = 0$ and $N = EA\frac{du}{dX_1} = P$ when $X_1 = L$.

We can now use Mathematica to solve for the displacement function:

```
Clear[L,A,A1,ro,Ee,u]
L=4;
A=Pi*(2-x/L)^2;
A1=A/.x->L;
s=DSolve[{D[Ee*A*u'[x],x]+ro*A==0,u[0]==0,u'[L]==P/Ee/A1},u,x]
```

The code above produces the result:

$$u = \frac{-12PX_1 - 112\pi\rho X_1 + 24\pi\rho X_1^2 - \pi\rho X_1^3}{6E\pi\,(x - 8)}$$

You can now copy and paste the following code into Mathematica to see the effect of changing the density and the external load on the displacement function produced:

```
Clear[u,x,Ee,P,L,A]
Manipulate[Ee=1000;
 L=4;
 A=Pi*(2-x/L)^2;
 A1=A/.x->L;
 s=DSolve[{D[Ee*A*u'[x],x]+ro*A==0,u[0] ==0,u'[L] ==P/Ee/A1},u,x];
 s=u/.s[[1]];
 Plot[s[x],{x,0,L}],{P,0,500},{ro,0,100}]
```

6.4.4. Timoshenko Cantilever Beam

The following Mathematica code obtains the equation of a cantilever beam loaded at the left end with a shearing force P and a bending moment M (Figure 6–15). The deflection of the centerline follows the following equation:

$$y = \frac{-6EIPx + 3kAGMx^2 - 3kAGLPx^2 + kAGPx^3}{6EIkAG}, \quad \psi = \frac{2Mx - 2LPx + Px^2}{2EI}$$

```
Clear[a,x,V,M,y,L,psi,EI,kAG]
M=EI*D[psi[x],x];
V=EI*D[psi[x],{x,2}];
q=-a;
M1=M/.x->0;
M2=M/.x->L;
V1=V/.x->0;
V2=V/.x->L;
y1=y[x]/.x->0;
y2=y[x]/.x->L;
psi1=psi[x]/.x->0;
psi2=psi[x]/.x->L;
s=DSolve[{EI*psi'''[x]==q,psi[x]==y'[x]+EI/kAG*psi''[x],y1==0,psi1==0,M2==Mo
m,V2==P},{y,psi},x]
```

The `Manipulate` function is used to visualize the deflected shape of the beam using EI = 500 units, $kAG = 0.01EI$, $L = 10$ units and beam height = 1 unit.

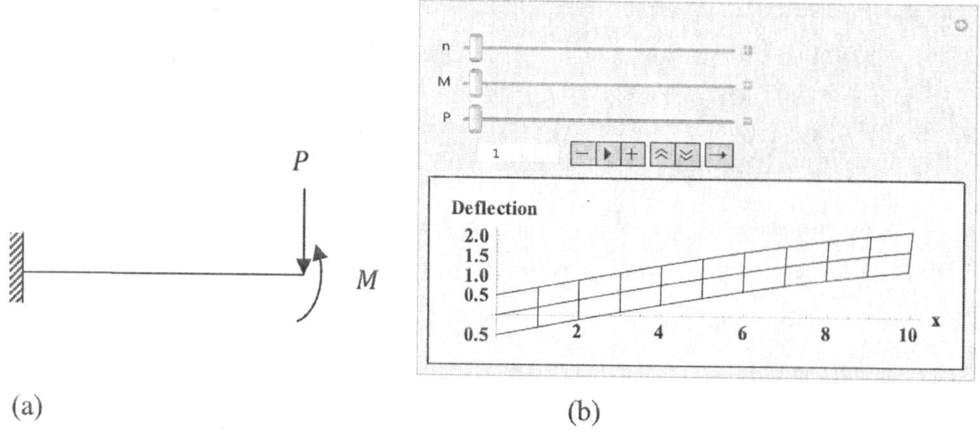

FIGURE 6-15 Timoshenko cantilever beam loaded with an end shear and moment. a) Positive direction of loading, b) beam deflected shape using Mathematica.

```
Manipulate[
EI=500;
kAG=0.01*EI;
Lx=10;
Ly=1;
z2=(-6EI*P*t+3kAG*Mom*t^2-3kAG*Lx*P*t^2+kAG*P*t^3)/6/EI/kAG;
psi=(2Mom*t-2Lx*P*t+P*t^2)/2/EI;
ztable2=Table[{z2,psi}/.t->i*Lx/n,{i,1,n}];
curvtable2=Table[{i/n*Lx+(5-t)/5*Ly/2*ztable2[[i,2]],ztable2[[i,1]]+(t-
5)/5*Ly/2},{i,1,n}];
ParametricPlot[{{t-Ly/2*psi,Ly/2+z2},{t,z2},{t+Ly/2*psi,-
Ly/2+z2},curvtable2},{t,0,10},BaseStyle->Directive[Bold,15],PlotStyle-
>{Black},AxesLabel->{"x","Deflection"}],{n,10,30,2},{{Mom,-5,"M"},-
10,10},{{P,-1,"P"},-1,1}]
```

6.5. FINAL NOTES

The beam approximation is used to analyze and design beams and frames under the effect of com–bined lateral loads, axial loads, and bending moments. In a coordinate system that includes the beam longitudinal axis, lateral loads lead to the development of "shear" stresses in the beam, and thus, the design codes include formulas for the design of beams under the effect of "bending" and "shear" stresses. When more detailed analysis of the stresses state inside a beam is required, finite element analysis of a beam could be more beneficial since it does not include the "plane sections remain plane" assumption and is able to more accurately characterize the stress state inside the beam. Figure 6–16 shows the arrows pointing in the direction of the maximum principal stresses obtained from a finite element analysis of a simple beam under lateral loading. The direction of the principal stresses shows that in the middle of the beam span, the maximum principal stresses are primarily horizontal and located at the bottom fibers of the beam. These stresses develop due to the higher bending loads at mid–span of beams under lateral loading. The direction of the prin–cipal stresses closer to the support, however, are influenced by the "shear" components of the supports and thus tend to be oriented at an angle that is dependent on the interaction between the bending and the shear loads. In concrete beams, for instance, cracks develop in locations closer to the support that are oriented perpendicular to the directions of the maximum principal stresses (direction of the maximum tension), and thus, "shear" reinforcement are required closer to the supports.

This chapter has presented the derivations for the classical small **plane** deformation beam theories. The extension to three–dimensional deformations is straightforward. Note that for large defor–mations, the equations presented here are no longer valid. When the deformations are large, the

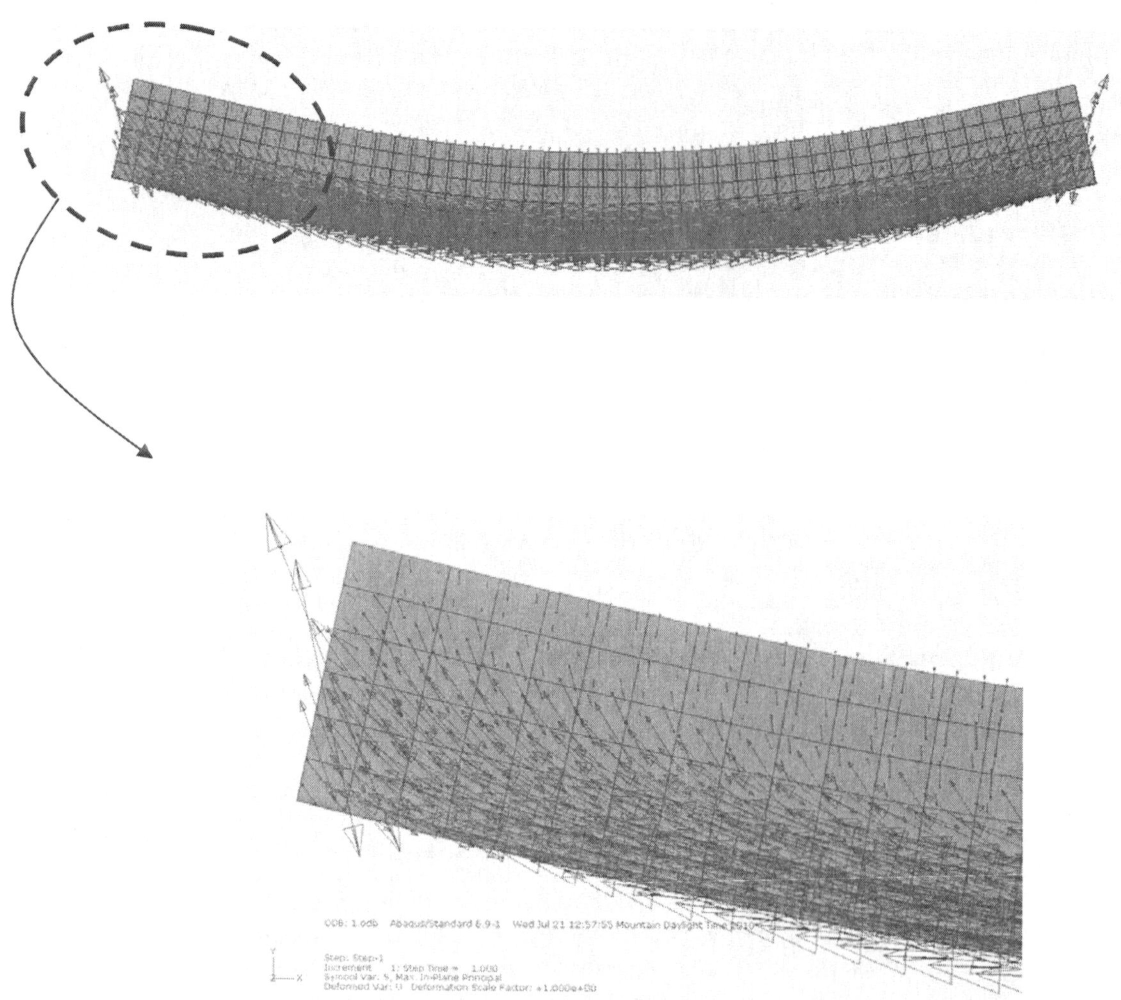

FIGURE 6-16 The direction of the maximum principal stresses in a beam under lateral loading obtained using a commercial finite element analysis (ABAQUS 6.9).

use of the small strains tensor could be erroneous, as shown in section 3.3. In addition, when the equilibrium equations were derived, the horizontal forces and the moment equation were uncoupled. However, for large deformations, the horizontal forces can have large effects on the bending moment induced in a beam. The Euler Bernoulli beam and the Timoshenko beam, as described in the previous sections, do not include the effect of axial forces on the bending moment and thus cannot predict the phenomenon of buckling. In addition, no reference was given to any of the traditional or modern methods (the stiffness method in particular) for solution of frame structures composed of different beams connected together. Interested readers should consult a modern structural analysis book.

▶ EXERCISES:

1. Use Mathematica to solve the Euler Bernoulli beam and the Timoshenko beam equations (shear, moment, rotation (slope), and deflection) for the beams shown in the following figure (assume values for the loads and material constants). Also, find the maximum deflection associated with the loads shown.

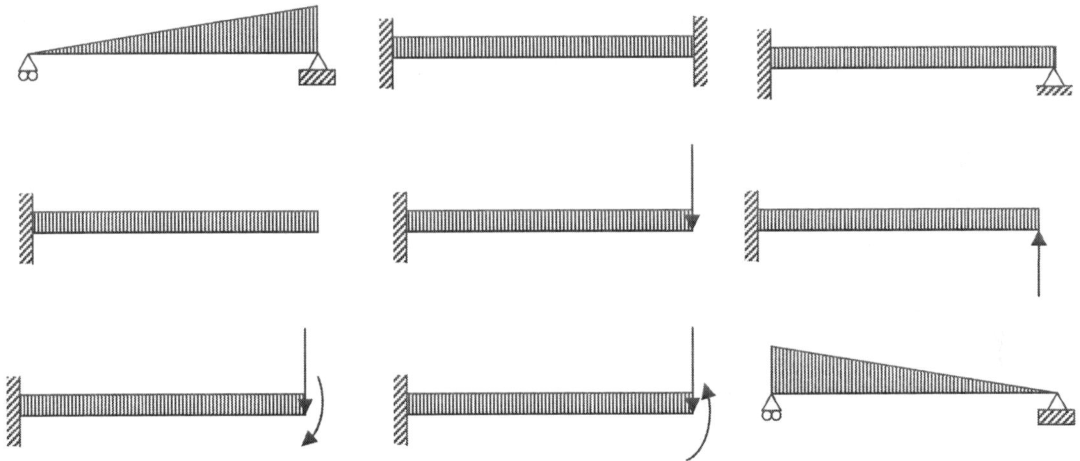

2. From the code given in section 6.4.1, state your opinion about the appropriate ratio of kAG/EI for which the Euler Bernoulli beam approximation is no longer appropriate to model a shear flexible beam.
3. Rewrite the final equations presented in sections 6.1.1.e, 6.2.1.e, and 6.3.1.d, assuming that the cross–sectional area parameters (I and A) are not constant.
4. The empirical shear coefficient k in the Timoshenko beam theory appears in the equation: $V_{12} = -\int \sigma_{12} dA = kGA\gamma$. Conduct a literature search to find the values of k for different cross–sectional shapes. (Square, circle, I beam etc.)
5. A beam is 5 m long with a rectangular cross–sectional area whose width and height are 0.25 m and 0.5 m, respectively. Young's modulus is 20 GPa. Ignore Poisson's ratio effect. A distributed load of a value 125 kN/m is applied vertically downward. Compare the following between a simple beam and a fixed–ends beam:
   a. Deflection, bending moment, and shearing force distribution along the beam
   b. σ_{11}, σ_{12}, von Mises stress, hydrostatic stress (Draw the contour plots and comment on the difference).
6. The shown beam has a rectangular cross–sectional area with a width of 0.25 m. The height of the beam varies linearly along its length as shown in the figure below. Young's modulus is taken as 20 GPa while Poisson's ratio effect is ignored. Assume that the only

nonzero component of the stress is σ_{11} and that plane sections perpendicular to the X_1 axis remain plane and perpendicular to the X_1 axis after deformation. Find the following:

a. A generic form for the position and the displacement functions.

b. The expression for the infinitesimal strain tensor components.

c. The differential equation of equilibrium and its boundary conditions.

d. Solve the differential equation of equilibrium to find the displacement of the beam at every point.

e. Draw the contour plots of σ_{11}.

f. Draw the vector plot of the displacement vector $u = \{u_1, u_2\}$.

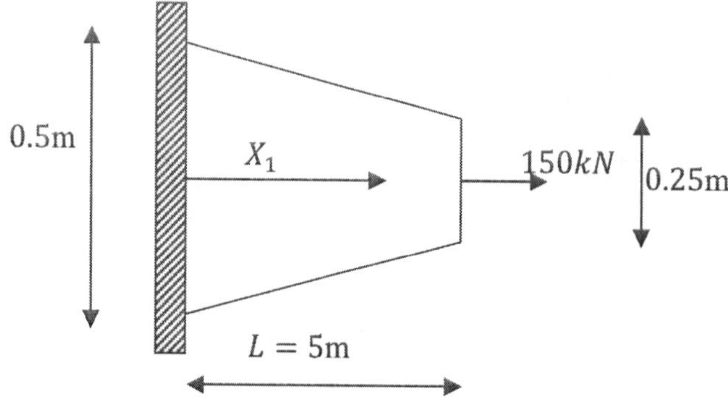

CHAPTER 7

Energy, Elastic Energy, and Virtual Work

Nature has taught us that an energy source is required to get any useful work done. Classical physics have realized experimentally that energy is neither created nor destroyed, but in fact, there is always a trade–off between forms of energy. Physicists have also realized that there are "useful" forms of energy that humans can use for their well–being, and there are other sources of energy that, with our current knowledge, cannot be used to our benefit. For example, chemical energy stored in hydrocarbons can be extracted to produce useful motion and heat. Kinetic energy of rivers can be used to produce electricity. Electricity can then be used to, reciprocally, produce kinetic energy or light energy. Light and heat are examples of forms of energy that, once produced, are harder to recapture. During any man–made energy transformation process, it is almost impossible to extract the full amount of useful energy from its source; instead, there are usually energy losses in forms that cannot be fully used. For example, when the hydrocarbons in a vehicle engine are burned, part of the energy content of the hydrocarbons is wasted as heat energy, which then dissipates in the environment without the ability to be recaptured.

Mechanical energy is perhaps one of the early forms of energy that humans have utilized. When a spring is loaded, an "elastic" potential energy is stored inside the spring and can, in the ideal case, be fully recovered once the loading is removed. In fact, the term "elastic" potential energy indicates that the energy stored is fully recoverable. If, on the other hand, a spring is overloaded, causing some internal damage, some of the energy is lost in the damaging process; this is usually called "plastic" energy, and so, when the spring is unloaded, only part of the energy stored is recovered and the rest is dissipated. In this chapter, we will study the very basic form of mechanical energy: "elastic" potential energy stored in deformable bodies. Solution methods for the equilibrium equations that are based on energy assumptions are also introduced.

7.1. POTENTIAL (DEFORMATION) ENERGY IN SINGLE DEGREES OF FREEDOM SPRINGS

7.1.1. Static Equilibrium Linear-Elastic Spring

A linear–elastic spring has a linear relationship between the spring force $f(x)$ and the extension x. When a weight of value mg is gradually hung from a linear–elastic spring whose spring constant is equal to k (force/unit length), the spring is gradually stretched until equilibrium is achieved (Figure 7–1). At this stage, the spring force is equal to the applied weight and thus:

$$\Delta = \frac{mg}{k}$$

The work done by the external forces during the process of loading is stored as an elastic potential energy inside the spring and can be recovered once the load is removed (Figure 7–1). The spring force acts to reduce the potential energy of the spring by always acting to reduce the spring length to its original unstretched length. At an arbitrary distance x, the potential energy stored in the spring is equal to $\frac{1}{2}kx^2$. In the process of moving a distance x downward, the weight mg lost some of its gravitational potential energy equal to the value mgx. If we consider the combined system of the spring and the mass, then the potential energy Π of the system is equal to:

$$\Pi = \frac{1}{2}kx^2 - mgx$$

Differentiating the potential energy Π with respect to x:

$$\frac{d\Pi}{dx} = kx - mg$$

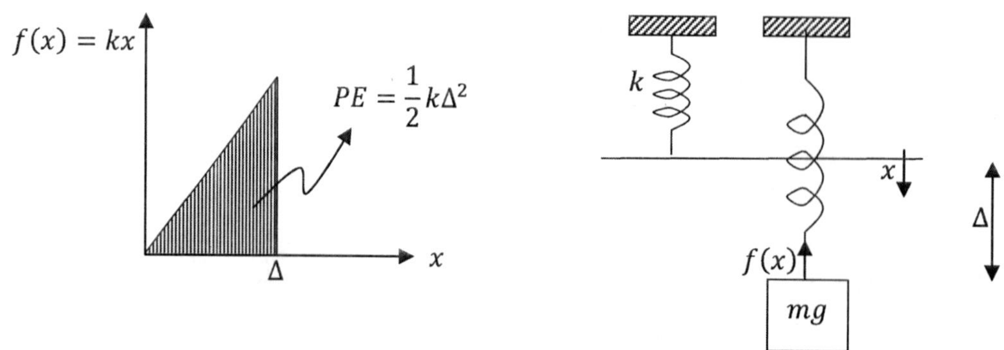

FIGURE 7-1 A linear–elastic spring has a linear relationship between the spring force $f(x)$ and the extension x. The potential energy stored in the spring when it undergoes a displacement Δ is equal to $\frac{1}{2}k\Delta^2$.

Substituting the equilibrium position Δ for x in the above equation shows that the derivative of the potential energy, with respect to the position at equilibrium, is equal to zero!

$$\left.\frac{d\Pi}{dx}\right|_{x=\Delta} = 0$$

The above statement shows that the potential energy is stationary (either minimum or maximum) at the position of equilibrium. Whether the potential energy is maximum or minimum depends on its second derivative:

$$\frac{d^2\Pi}{dx^2} = k$$

Thus, whether the potential energy is maximum or minimum at equilibrium depends on the constant k. For a spring with a positive k, the static equilibrium position renders the potential energy minimum.

7.1.2. Static Equilibrium, Nonlinear-Elastic Spring

The arguments presented in section 7.1.1 hold for elastic springs with nonlinear–force deformation behavior. In this case, if the spring force is equal to $f(x)$, then the elastic potential energy stored inside the spring is equal to:

$$\Pi_{\text{spring}} = \int f(x)dx$$

The assumption that Π_{spring} is elastic is equivalent to it being independent of the integration path. Mathematically, this leads to the force in the spring being the derivative of a "potential energy" function of the spring:

$$f(x) = \frac{d\Pi_{\text{spring}}}{dx}$$

The total potential energy of a system composed of the nonlinear spring in addition to a weight mg is:

$$\Pi = \int \frac{d\Pi_{\text{spring}}}{dx}dx - mgx = \Pi_{\text{spring}} - mgx$$

Taking the derivative with respect to the position, we get:

$$\frac{d\Pi}{dx} = \frac{d\Pi_{\text{spring}}}{dx} - mg = f(x) - mg$$

Similar to section 7.1.1, the potential energy of the system is stationary at equilibrium. The second derivative of the potential energy of the system is:

$$\frac{d^2\Pi}{dx^2} = f'(x)$$

Thus, as long as the spring force has a positive slope ($f'(x) > 0$) with respect to the extension, the potential energy is minimum at the equilibrium position.

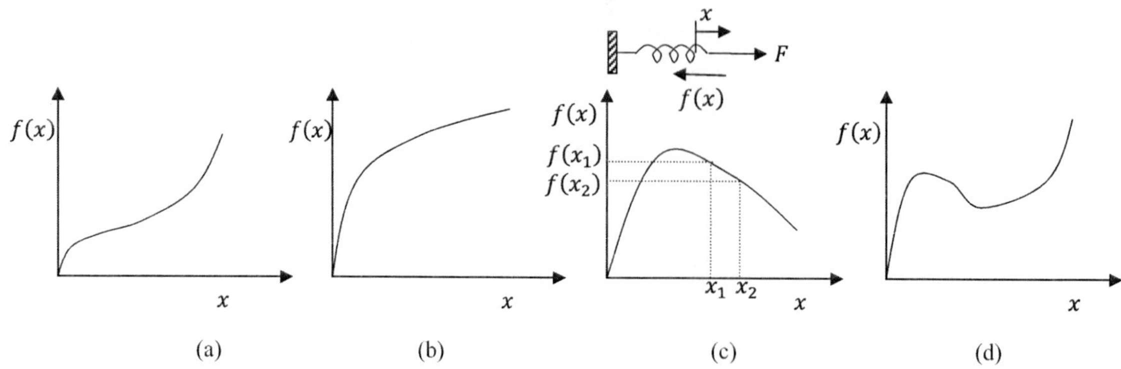

FIGURE 7-2 Force versus extension in nonlinear–elastic springs. (a) and (b) Slope is always positive, and the potential energy is minimum at the equilibrium position. (c) and (d) have negative slopes in some portions, and thus, at these locations, any perturbation in the displacement would lead to instabilities and loss of equilibrium.

Instabilities, load control, and displacement control:

If the slope of the spring force is negative with respect to the extension, the equilibrium position is deemed unstable. In this case, under the effect of an external force F, any small perturbation in the displacement would lead to instabilities in the system. To understand this, let there be an equilibrium position x_1 where the slope is negative (Figure 7–2c & d). At this instance, the external force F is in equilibrium with the internal spring force $f(x_1)$. Imagine then, that a small perturbation in the displacement increases the position to x_2 corresponding to an internal force $f(x_2)$. Since the slope is negative, $f(x_2) < f(x_1) = F$. This, in turn, will lead to further extension in the direction of the higher force F. Thus, the system will become unstable with the external force F acting to increase the extension x in an unstable fashion. However, if the extension of the spring is applied using a displacement–controlling mechanism, this unstable mechanism will not be observed.

Problem: The force versus displacement of a nonlinear–elastic spring follows the following relationship:

$$f = 1000x + 50x^2 - 20x^3$$

where the force and the displacement are given in N and mm, respectively. If the range of applicability of this formula is when $x \in [-5, 5]$, then:

- Draw the relationship between the force versus displacement in the full range of applicability of the relationship.
- Find the strain energy stored inside the spring at full extension and at full contraction.
- Find the displacement in the spring that corresponds to a force of 3kN and a force of –1kN.
- Comment on the behavior of the spring at $x = -3.333$ and $x = 5$ mm.

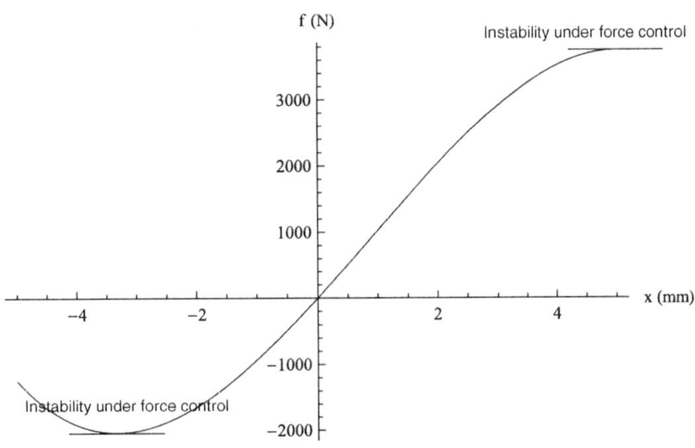

FIGURE 7-3 Nonlinear spring.

Solution: The required plot is shown in Figure 7–3:

The strain energy stored inside the spring at full extension and full contraction can be obtained by the formula:

$$E_{\text{full extension}} = \int_0^5 f dx = 11458.3 \text{ N.mm}, \quad E_{\text{full contraction}} = \int_0^{-5} f dx = 7291.67 \text{ N.mm}$$

The displacements corresponding to 3kN and –1kN can be obtained using the Newton Raphson method, which is implemented in Mathematica using the function `FindRoot`, and they are:

$$1000x + 50x^2 - 20x^3 = 3000 \Longrightarrow x = 3.121 \text{ mm},$$
$$1000x + 50x^2 - 20x^3 = -1000 \Longrightarrow x = -1.084 \text{ mm}$$

The graph of the force versus the displacement of the spring predicts an unstable behavior at full extension $x = 5$ mm and at $x = -3.333$ mm. The forces that can be carried by the spring at those two displacement values are equal to 3.75 kN and -2.037 kN, respectively. At both displacements or at both spring forces values, the slope of the spring force versus extension is equal to zero, indi–cating that a small increase in the external force applied on the spring will lead to an unstable behavior for the spring. Notice that the slope of the spring force versus extension is equal to the second derivative of the strain–energy function:

$$f(x) = \frac{d\Pi_{\text{spring}}}{dx} \Longrightarrow \frac{df(x)}{dx} = \frac{d^2\Pi_{\text{spring}}}{dx^2}$$

Thus, for the sake of stability of the spring, it should be used at force values that are sufficiently less than 3.75 kN in tension and −2.037 kN in compression.

The following Mathematica code was used for the above calculations:

```
Clear[x];
f=1000x+50x^2-20x^3;
Plot[f,{x,-5,5},AxesLabel->{"x (mm)","f (N)"},PlotStyle->Black]
Integrate[f,{x,0.,5}]
Integrate[f,{x,0.,-5}]
FindRoot[f==3000,{x,0}]
FindRoot[f==-1000,{x,0}]
a=D[f,x];
sol=Solve[a==0.,x]
f/.sol[[1,1]]
f/.sol[[2,1]]
```

Note: A spring that is stable under any loading condition requires that the force is always increasing with respect to the displacement, i.e.:

$$\frac{df(x)}{dx} = \frac{d^2\Pi_{\text{spring}}}{dx^2} > 0$$

A continuously smooth function whose second derivative is always higher than zero belongs to the family of "convex" function. If the elastic strain energy density of continuum bodies is convex in its variables, this implies that the material is stable.

7.1.3. Static Equilibrium, Nonlinear Nonelastic Spring

The term nonelastic spring indicates that during a cycle of loading, energy supplied by exter–nal sources dissipates in the system or turns into a form that is not useful. Figure 7–4 shows an elasto–plastic spring during a cycle of gradual loading and unloading. When the mass is gradu–ally removed, the spring does not return back to its original position, but has a residual extension $\Delta_{plastic}$. The energy required to reach from state A to state B is the area under the curve shown in Figure 7–4b. During unloading, the load displacement curve follows a different path and energy is lost. As can be seen in state C shown in Figure 7–4, the energy lost during the loading cycle is manifested in some change in the internal structure of the spring, which now has a longer length than that before loading. In such springs, a potential energy function cannot be easily defined, and the behavior of the spring is history (path) dependent.

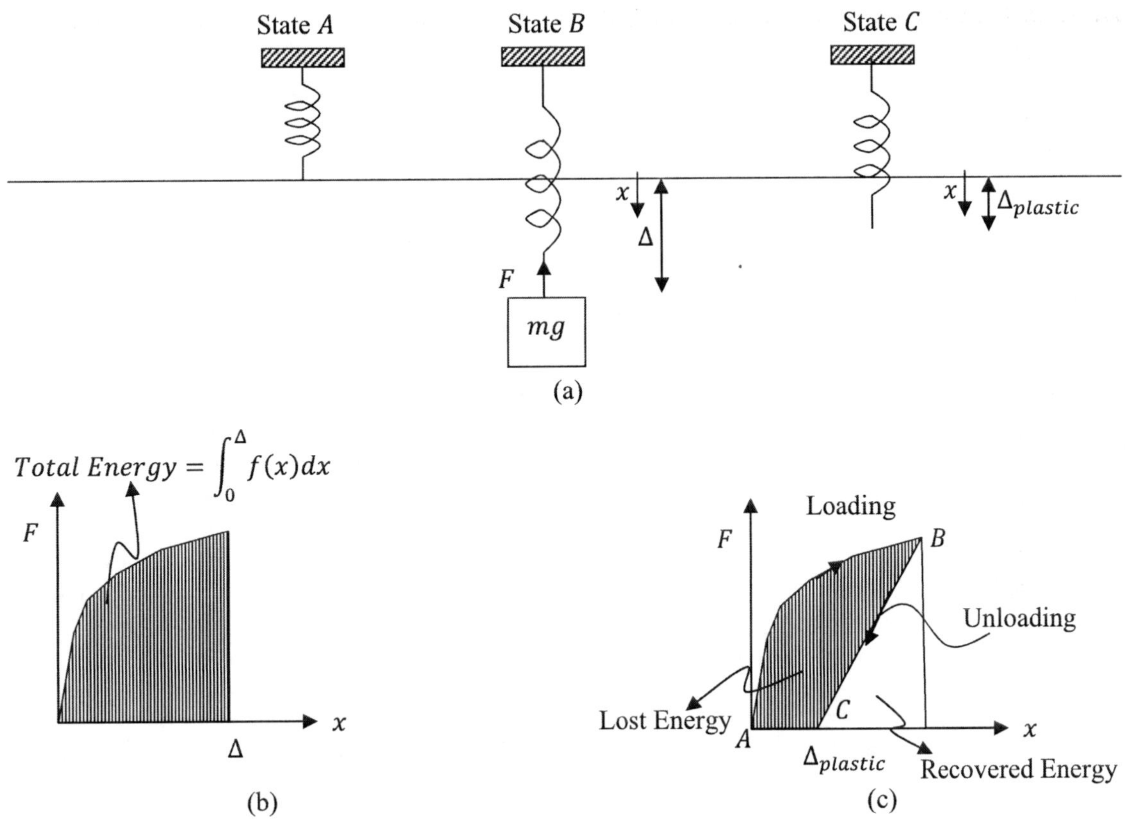

FIGURE 7-4 A cycle of loading and unloading in an elasto–plastic spring. (a) The process of loading and unloading produces a residual plastic extension, (b) energy supplied to the spring during the loading process, (c) part of the energy supplied is recovered and another part is lost in the damaging or plastification effect.

7.1.4. Sudden Application of Loading, Linear-Elastic Spring

In sections 7.1.1 through 7.1.3, the load was applied gradually such that at any point during the process of load application, the velocity, and the acceleration of the applied mass were equal to zero. If, on the other hand, the weight is fully applied to the spring when the extension is equal to zero, then the weight will accelerate downward. To understand the expected motion, the differential equation of motion has to be solved to find the position, velocity, and acceleration as functions of time. Assuming that the spring is linear elastic, then the differential equation of equilibrium (assuming no damping in the system) is:

$$mg - kx = m\ddot{x}$$

The initial conditions at $t = 0$ are: the velocity $\dot{x} = 0$ and the position $x = 0$. The following Mathematica code will be used to solve the above ordinary differential equation of motion:

```
Clear[m,g,x,k]
DSolve[{m*g-k*x[t]==m*x''[t],x[0]==0,x'[0]==0},x,t]
```

The solution is:

$$x = \frac{mg}{k}\left(1 - \cos\sqrt{\frac{k}{m}}t\right)$$

The equation of motion shows that the mass will vibrate around the static equilibrium position $\Delta = \frac{mg}{k}$. The following code can help you visualize the effect of changing the mass and the stiffness on the vibration of the system. Copy and paste your code into Mathematica, and press the PLAY button under the time variable t to visualize the vibrations (Figure 7–5). Can you comment on the effect of increasing the stiffness and or the mass on the vibrations?

```
Manipulate[g=1;
Graphics3D[{GrayLevel[0.3],Cuboid[{-1,-1,-5},{0,0,(g*m-
g*m*Cos[Sqrt[k/m]*t])/k}]},Axes->True,Lighting->"Neutral",PlotRange->{{-
1,0},{-1,0},{-5,2*5/1}},BaseStyle->Directive[Bold,12],AxesOrigin-
>{0,0,0},ViewVertical->{0,0,-1}],{t,0,4*Pi*Sqrt[5]},{m,1,5},{k,1,5}]
```

At any point in time, the potential energy and the kinetic energy of the system are equal to:

$$PE = \frac{1}{2}kx^2 - mgx, \quad KE = \frac{1}{2}m\dot{x}^2$$

Notice that the potential energy of the system is not constant, but rather, oscillates and is trans–formed into kinetic energy such that the total sum of both the potential energy, and the kinetic energy remains constant with respect to time:

$$Energy = PE + KE = \frac{1}{2}kx^2 - mgx + \frac{1}{2}m\dot{x}^2$$

$$\frac{dEnergy}{dt} = kx\dot{x} - mg\dot{x} + m\dot{x}\ddot{x} = \dot{x}(kx - mg + m\ddot{x}) = 0$$

A very important difference between this problem and the static problems is that the maximum spring force is higher than the applied weight. The maximum position for the spring is obtained when the cosine function is equal to –1 and thus, $x_{max} = 2mg/k$. The corresponding force in the spring is equal to $F = kx\,|_{max} = k(2mg/k) = 2\,mg$, and thus, the force in the spring is doubled compared to the case of static equilibrium! In the process of designing a structure susceptible to vibrations, a dynamic effect factor is usually considered. This dynamic factor accounts for the increase in the internal forces induced by the vibrations.

In the following section, the concept of "elastic internal energy" in one dimensional springs is extended to the continuum. However, the discussions shown above on the elasto–plastic and

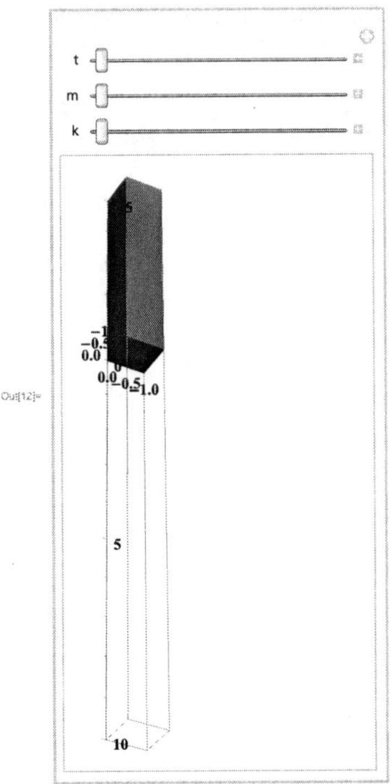

FIGURE 7-5 Vibrations of a mass–spring system.

the dynamic behaviours are limited to one dimensional springs and were presented as alternative analysis tools for cases beyond the scope of this text.

7.2. DEFORMATION ENERGY INSIDE A CONTINUUM

During the time period of application of the external forces on a deforming body, the external forces perform work that is transmitted into internal energy inside the deforming body. To accurately describe this internal energy stored, the power of those external forces (rate of application of work) is used and then integrated over a small period of time to find the increment of the deformation energy stored inside the body. This concept will be used in the following derivation to show that the increment in the deformation energy stored in a continuum arises naturally from the product of the components of the stress tensor and the components of the increments in the small–strain tensor:

$$d\overline{U} = \sum_{i,j=1}^{3} \sigma_{ij} d\varepsilon_{ij}$$

Notice that the overbar is used to differentiate the strain–energy function $\overline{U}$ from U the stretch part in the polar decomposition of F (Section 3.2.3). The above formula has been traditionally derived using two methods. In the first method, the power of the external forces acting on a small infinitesimal cube is analyzed to find the increment in the strain energy. In the second method, the power of the external forces acting on the whole continuum is analyzed, yielding the same results. The two methods are inherently the same, but are both shown here for completeness.

7.2.1. Derivation Using a Differential Volume

Let the current configuration of a material body be represented by the set $\Omega \subset \mathbb{R}^3$ with the orthonormal basis set $B = \{e_1, e_2, e_3\}$ and the coordinates $\{x_1, x_2, x_3\}$. Also, let $\sigma \in \mathbb{M}^3$ be the Cauchy stress tensor at every point and $v \in \mathbb{R}^3$ be a function $v{:}\Omega \rightarrow \mathbb{R}^3$ describing the velocity of each material point inside the deforming body. The body forces vector acting on the continuum body is represented by $b \in \mathbb{R}^3$, and the density is denoted by ρ. The power (rate of change of the internal energy inside a body) of the stresses can be calculated by considering a small cube of material with dimensions dx_1, dx_2, and dx_3. First, we consider the total power TP of the stress component σ_{11} and the horizontal body force b_1 (Figure 7–6a):

$$
\begin{aligned}
TP &= -\sigma_{11} v_1 dx_2 dx_3 + \left(\sigma_{11} + \frac{\partial \sigma_{11}}{\partial x_1} dx_1\right)\left(v_1 + \frac{\partial v_1}{\partial x_1} dx_1\right) dx_2 dx_3 + \rho b_1 v_1 \\
&= \left(\frac{\partial\left(\sigma_{11} v_1\right)}{\partial x_1} + \rho b_1 v_1\right) dx_1 dx_2 dx_3
\end{aligned}
$$

The total power TP due to the stress component σ_{11}, and the horizontal body force b_1 is then equal to:

$$
P = \frac{\partial(\sigma_{11} v_1)}{\partial x_1} + \rho b_1 v_1
$$

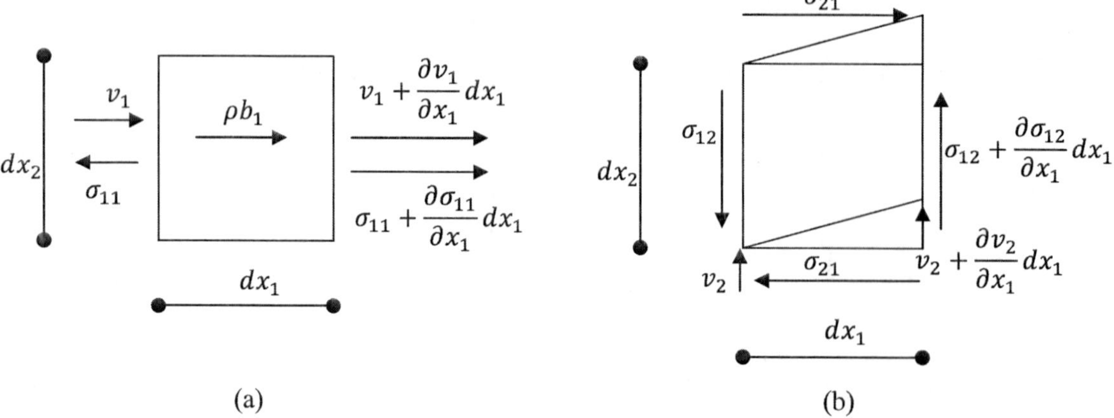

FIGURE 7-6 (a) Horizontal velocity components and horizontal body forces for the calculation of the deformation power (strain energy per unit time) stored in the shown cube due to σ_{11} and ρb_1 during deformation. (b) Vertical velocity components for the calculation of the deformation power (strain energy per unit time) stored in the shown cube due to σ_{12} during deformation.

Similarly, the total power supplied by the stress components σ_{22} and σ_{33} and the body forces b_2 and b_3 have the following forms:

$$TP = \left(\frac{\partial\left(\sigma_{22}v_2\right)}{\partial x_2} + \frac{\partial\left(\sigma_{33}v_3\right)}{\partial x_3} + \rho b_2 v_2 + \rho b_3 v_3\right) dx_1 dx_2 dx_3$$

The total power of the stress components σ_{12} (Figure 7–6b) can be calculated as follows:

$$TP = \sigma_{12}\left(\frac{\partial v_2}{\partial x_1}dx_1\right) dx_2 dx_3 + v_2\left(\frac{\partial \sigma_{12}}{\partial x_1}dx_1\right) dx_2 dx_3 = \frac{\partial(\sigma_{12}v_2)}{\partial x_1} dx_1 dx_2 dx_3$$

Similar results could be obtained for the remaining shear stress components.

Combining all the total power terms due to the different stress and body force components, the following expression for the total power TP is reached:

$$TP = \left(\sum_{i,j=1}^{3} \frac{\partial \sigma_{ij} v_j}{\partial x_i} + \sum_{i=1}^{3} \rho b_i v_i\right) dv = \left(\operatorname{div} \sigma v + \rho b \cdot v\right) dv$$

where $dv = dx_1 dx_2 dx_3$ is the volume of the infinitesimal cube. Using the equilibrium equa– tions derived in section 4.3, then the expression for the total power can be further simplified to (considering the summation convention defined in section 1.2.5.a):

$$TP = \left(\sum_{j=1}^{3} \rho a_j v_j + \sum_{i,j=1}^{3} \sigma_{ij}\frac{\partial v_j}{\partial x_i}\right) dv = \left(\rho a_j v_j + \sigma_{ij}\frac{\partial v_j}{\partial x_i}\right) dv$$

where a_j are the components of the acceleration vector.

▶ **BONUS EXERCISE:** Show the detailed derivation of the above simplified expression.

Thus, the total power of the external forces acting on the differential volume is composed of two terms. Using the continuity of mass equation $\left(\frac{d(\rho J)}{dt} = 0\right)$ shown in section 4.2.2.b, the first term $(\rho a_j v_j dv)$ can be shown to represent the change in the kinetic energy $(K.E.)$ of the differential volume, as follows:

$$\begin{aligned} \frac{d\,(K.E.)}{dt} &= \frac{d\left(\frac{1}{2}\rho\,(v\cdot v)\,dv\right)}{dt} = \frac{d\left(\frac{1}{2}\rho\,(v_i v_i)\,dv\right)}{dt} = \rho a_j v_j dv + \frac{1}{2}\,(v_i v_i)\,\frac{d\,(\rho dv)}{dt} \\ &= \rho a_j v_j dv + \frac{1}{2}\,(v_i v_i)\,dV\frac{d\,(\rho J)}{dt} = \rho a_j v_j dv \end{aligned}$$

Thus, the total power of the external forces acting on the infinitesimal cube can be represented as the sum of the following two terms:

$$TP = \frac{d\,(K.E.)}{dt} + \left(\sigma_{ij}\frac{\partial v_j}{\partial x_i}\right) dv$$

The second term in the total power expression is called the **stress power**. The **stress power** per unit volume is thus equal to:

$$P = \sigma_{ij}\frac{\partial v_j}{\partial x_i}$$

The arbitrariness of i and j in the above sums can be utilized to rewrite the stress power in the following form:

$$\sigma_{ij}\frac{\partial v_j}{\partial x_i} = \sigma_{ji}\frac{\partial v_i}{\partial x_j} \Longrightarrow P = \frac{1}{2}\left(\sigma_{ij}\frac{\partial v_j}{\partial x_i} + \sigma_{ji}\frac{\partial v_i}{\partial x_j}\right)$$

The symmetry of the stress components can be utilized to rewrite the stress power expression, as follows:

$$\begin{aligned} P &= \frac{1}{2}\sigma_{ij}\left(\frac{\partial v_i}{\partial x_j} + \frac{\partial v_j}{\partial x_i}\right) \\ &= \sigma_{11}\frac{\partial v_1}{\partial x_1} + \sigma_{22}\frac{\partial v_2}{\partial x_2} + \sigma_{33}\frac{\partial v_3}{\partial x_3} + \frac{1}{2}\sigma_{12}\left(\frac{\partial v_1}{\partial x_2} + \frac{\partial v_2}{\partial x_1}\right) + \frac{1}{2}\sigma_{21}\left(\frac{\partial v_2}{\partial x_1} + \frac{\partial v_1}{\partial x_2}\right) + \frac{1}{2}\sigma_{23}\left(\frac{\partial v_2}{\partial x_3} + \frac{\partial v_3}{\partial x_2}\right) \\ &+ \frac{1}{2}\sigma_{32}\left(\frac{\partial v_3}{\partial x_2} + \frac{\partial v_2}{\partial x_3}\right) + \frac{1}{2}\sigma_{13}\left(\frac{\partial v_1}{\partial x_3} + \frac{\partial v_3}{\partial x_1}\right) + \frac{1}{2}\sigma_{31}\left(\frac{\partial v_3}{\partial x_1} + \frac{\partial v_1}{\partial x_3}\right) \end{aligned}$$

Note that the matrix $\frac{1}{2}\left(\frac{\partial v_i}{\partial x_j} + \frac{\partial v_j}{\partial x_i}\right)$, which is often denoted by D, is the symmetric part of the veloc– ity gradient matrix $L = \frac{\partial v_i}{\partial x_j}$ introduced in section 3.4.1. When the matrix D is multiplied by an infinitesimal time increment dt, the increment in the infinitesimal strain tensor with respect to the current coordinates $d\varepsilon = Ddt$ arises naturally:

$$du = \begin{pmatrix} du_1 \\ du_2 \\ du_3 \end{pmatrix} = vdt = \begin{pmatrix} v_1dt \\ v_2dt \\ v_3dt \end{pmatrix}, \quad d\varepsilon = Ddt = \begin{pmatrix} \frac{\partial du_1}{\partial x_1} & \frac{1}{2}\left(\frac{\partial du_1}{\partial x_2} + \frac{\partial du_2}{\partial x_1}\right) & \frac{1}{2}\left(\frac{\partial du_1}{\partial x_3} + \frac{\partial du_3}{\partial x_1}\right) \\ & \frac{\partial du_2}{\partial x_2} & \frac{1}{2}\left(\frac{\partial du_2}{\partial x_3} + \frac{\partial du_3}{\partial x_2}\right) \\ symm & & \frac{\partial du_3}{\partial x_3} \end{pmatrix}$$

An increment in the strain energy per unit current volume $d\bar{U}$ is thus equal to:

$$d\bar{U} = Pdt = \sum_{i,j=1}^{3}\sigma_{ij}D_{ij}dt = \sum_{i,j=1}^{3}\sigma_{ij}d\varepsilon_{ij}$$

The strain energy stored in the whole body when the stress changes from stress state represented by the Cauchy stress $\sigma_1 \in \mathbb{M}^3$ to another represented by $\sigma_2 \in \mathbb{M}^3$ is thus:

$$Strain\ Energy = \int\int_{\bar{U}_1}^{\bar{U}_2} d\bar{U}\,dV = \int\int_{\sigma_1}^{\sigma_2}\sum_{i,j=1}^{3}\sigma_{ij}d\varepsilon_{ij}\,dV = \int\int_{\sigma_1}^{\sigma_2}\sigma_{ij}d\varepsilon_{ij}\,dV$$

7.2.2. Derivation Using Integration over the Continuum Body

Let the current configuration of a material body be represented by the set $\Omega \subset \mathbb{R}^3$ with the orthonormal basis set $B = \{e_1, e_2, e_3\}$ and the coordinates $\{x_1, x_2, x_3\}$. Also, let $\sigma \in \mathbb{M}^3$ be the Cauchy stress tensor at every point and $v \in \mathbb{R}^3$ be a function $v{:}\Omega \to \mathbb{R}^3$ describing the velocity of each material point inside the deforming body. If the body forces vector acting on the continuum body is represented by $b \in \mathbb{R}^3$, and if the density is denoted by ρ, then the total power of the external forces acting on Ω is given by:

$$\textit{External Power} = \int_{\Omega} \rho b \cdot v dv + \int_{\partial\Omega} t_n \cdot v dS$$

where $t_n = \sigma^T n$ is the distribution of the traction vector on the surface of Ω denoted by $\partial\Omega$ with n being the unit normal vector to the surface. The external power supplied by the boundary traction vectors can be expanded using the divergence theorem (section 1.3.2), as follows:

$$\int_{\partial\Omega} t_n \cdot v dS = \int_{\partial\Omega} \sigma_{ji} n_j v_i dS = \int_{\Omega} \frac{\partial \sigma_{ji} v_i}{\partial x_j} dv = \int_{\Omega} \frac{\partial v_i}{\partial x_j} \sigma_{ji} + v_i \frac{\partial \sigma_{ji}}{\partial x_j} dv = \int_{\Omega} \text{trace } (L\sigma) + \text{div}\sigma \cdot v dv$$

Substituting the above expression in the external power equation leads to the following expression:

$$\textit{External Power} = \int_{\Omega} (\rho b + \text{div}\sigma) \cdot v dv + \int_{\Omega} \text{trace } (L\sigma)\, dv$$

The external power expression can be further manipulated using the equilibrium equation derived in section 4.3 to be composed of the following two terms:

$$\textit{External Power} = \int_{\Omega} (\rho \dot{v}) \cdot v dv + \int_{\Omega} \text{trace } (L\sigma)\, dv$$

Notice that the first term on the right–hand side is identically the rate of change of the kinetic energy $K.E.$ of the system:

$$\dot{\overline{(K.E.)}} = \frac{d}{dt} \int_{\Omega} \frac{1}{2} \rho\, (v \cdot v)\, dv = \frac{d}{dt} \int_{\Omega_0} \frac{1}{2} \rho J\, (v \cdot v)\, dV = \int_{\Omega_0} \frac{1}{2} \frac{d\rho J}{dt} (v \cdot v)\, dV + \int_{\Omega} (\rho \dot{v}) \cdot v dv = \int_{\Omega} (\rho \dot{v}) \cdot v dv$$

The physical laws of energy conservation predict that the external power supplied to a system will be equal to the rate of change of the kinetic energy plus the rate of change of the internal energy ($I.E.$) or strain energy of the system. Thus, the rate of change of the internal energy of the system is given by:

$$\dot{\overline{(I.E.)}} = \int_{\Omega} \text{trace } (L\sigma)\, dv$$

The internal power per unit volume is thus given by:

$$P = \text{trace}\,(L\sigma) = \sum_{i,j=1}^{3} \sigma_{ij}\frac{\partial v_j}{\partial x_i} = \sigma_{ij}\frac{\partial v_j}{\partial x_i}$$

The symmetry of the stress tensor can be used to replace L with D in the above expression, and thus, the rate of change of the internal energy per unit volume of the system is equal to:

$$\dot{\bar{U}} = \text{trace}\,(\sigma L) = \text{trace}\,(\sigma D)$$

▶ **BONUS EXERCISE:** Show that trace (σL) = trace $(L\sigma)$ = trace (σD)

An increment in the internal energy per unit volume of the system is thus given by:

$$d\bar{U} = \sigma_{ij}D_{ij}\,dt = \sigma_{ij}d\varepsilon_{ij}$$

The increment in the internal energy per unit volume during deformation of a continuum can also be written in terms of other measures of stress and deformation. For example, if the Cauchy stress tensor is replaced with the first Piola Kirchhoff stress P presented in section 2.5.2, then the increment in the strain energy per unit volume of the deformed configuration can be written in terms of P, the deformation gradient F, and its determinant J, as follows:

$$\dot{\bar{U}} = \text{trace}\,(\sigma D) = \text{trace}\,(L\sigma) = \frac{1}{J}\text{trace}\left(LFP^T\right) = \frac{1}{J}\text{trace}\left(\dot{F}P^T\right) = \frac{1}{J}\sum_{i,j=1}^{3} P_{ij}\dot{F}_{ij}$$

Since $\det F$ (denoted by J) represents the local ratio of volume in the deformed configuration to the volume in the reference configuration, then the rate of change of the strain–energy function per unit volume of the undeformed (reference) configuration can be written as:

$$\dot{W} = J\dot{\bar{U}} = \sum_{i,j=1}^{3} P_{ij}\dot{F}_{ij}$$

When comparing the strain–energy increment per unit volume of the deformed and the undeformed configurations, it can be observed that the strain–energy increment per unite volume stored inside a continuum is given as a product of a stress measure and a strain measure. These two corresponding measures are often called the **energy conjugates** of each other. The stretch D can thus be considered as the energy conjugate of the Cauchy stress σ, while the deformation gradient F is considered to be the energy conjugate of the first Piola Kirchhoff stress P.

▶ **BONUS EXERCISE:** Show that the second Piola Kirchhoff stress S is the energy conjugate of F^TF. In particular, show that $\dot{\bar{U}} = \frac{1}{(2J)}\text{trace}\,(S\dot{C})$ where $C = F^TF$.

7.3. EXPRESSIONS FOR THE ELASTIC STRAIN-ENERGY FUNCTION

7.3.1. Strain-Energy Density Function in Linear-Elastic Materials in Small Deformations

Consider a small cube of material with dimensions dX_1, dX_2, and dX_3 under a general state of stress. If the strains and the deformations are small enough (the differences between the undeformed and the current configurations and their respective volumes are negligible), then the strain–energy density (potential energy per unit volume) $\overline{U}$ stored inside this cube can be calculated using the components of the stress tensor and the small infinitesimal strain tensor:

$$\overline{U} = \int \sigma_{11} d\varepsilon_{11} + \sigma_{22} d\varepsilon_{22} + \sigma_{33} d\varepsilon_{33} + \sigma_{12} d\varepsilon_{12} + \sigma_{21} d\varepsilon_{21} + \sigma_{13} d\varepsilon_{13} + \sigma_{31} d\varepsilon_{31} + \sigma_{23} d\varepsilon_{23} + \sigma_{32} d\varepsilon_{32}$$

which, when considering the summation convention defined in section 1.2.5.a, can be written as:

$$\overline{U} = \int \sigma_{ij} d\varepsilon_{ij}$$

Notice that when the material is considered to be elastic, the deformation energy is independent of the path leading to the deformation, the integral above has to be an "exact integral," and thus, the stresses are the derivative of the potential energy (strain–energy) function $\overline{U}$:

$$\sigma_{ij} = \frac{\partial \overline{U}}{\partial \varepsilon_{ij}}$$

For linear–elastic small deformations materials, since a linear relationship exists between the stress and the strain entries, the strain energy per unit volume stored inside a deformed continuum is given by:

$$\overline{U} = \int_0^{\varepsilon_{ij}} \sigma_{ij} d\varepsilon_{ij} = \frac{1}{2} \sigma_{ij} \varepsilon_{ij}$$

The factor $\frac{1}{2}$ was introduced in the final expressions for $\overline{U}$ because the strain energy is assumed to be due to a gradual increase in stresses accompanied by a gradual increase in the strains (Figure 7–7).

If a linear–elastic isotropic small deformations material is considered, then the constitutive law described in section 5.2.5 can be used to write the strain–energy density function in terms of the stresses and the material constants E and ν:

$$\overline{U} = \frac{1}{2} \sigma_{ij} \varepsilon_{ij} = \frac{(1+\nu)}{2E} \left(\sigma_{11}^2 + \sigma_{22}^2 + \sigma_{33}^2 + 2\sigma_{12}^2 + 2\sigma_{23}^2 + 2\sigma_{13}^2 \right) - \frac{\nu}{2E} (\sigma_{11} + \sigma_{22} + \sigma_{33})^2$$

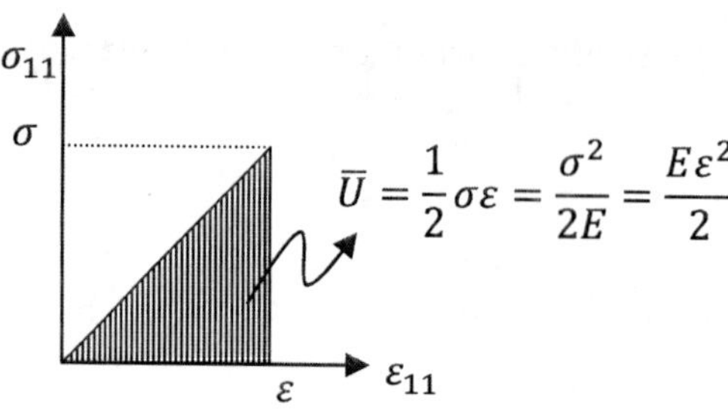

FIGURE 7-7 Relationship between σ_{11} and ε_{11} in a uniaxial state of stress.

If the stresses are replaced with expressions of the strains, the strain–energy density function can also be given in terms of the strains and the Lamé's material constants λ and μ:

$$\bar{U} = \mu\left(\varepsilon_{11}^2 + \varepsilon_{22}^2 + \varepsilon_{33}^2 + \frac{1}{2}\gamma_{12}^2 + \frac{1}{2}\gamma_{13}^2 + \frac{1}{2}\gamma_{23}^2\right) + \frac{\lambda}{2}(\varepsilon_{11} + \varepsilon_{22} + \varepsilon_{33})^2$$

▸ **BONUS EXERCISE:** Show the details of the derivation of the above formulae.

7.3.2. Volumetric and Deviatoric Strain-Energy Functions for Linear-Elastic Isotropic Materials

The strain energy of a linear–elastic isotropic material per unit volume can be divided into two additive components: a volumetric strain–energy function that accompanies volumetric changes inside the material and a deviatoric strain–energy function that accompanies shape changes at a constant volume. Such decomposition is valid for small strains when the volumetric strain can be measured as the trace of the infinitesimal strain matrix (see section 3.3.3.a). In this case, the relationship between the volumetric strain and the applied stresses using the linear–elastic constitutive laws described in section 5.2.5 is:

$$\text{trace}(\varepsilon) = \varepsilon_{11} + \varepsilon_{22} + \varepsilon_{33} = \frac{(\sigma_{11} + \sigma_{22} + \sigma_{33})(1 - 2\nu)}{E} = \frac{\text{trace}(\sigma)(1 - 2\nu)}{E}$$

The above relationship shows that nonzero volumetric strains exist *only* when the hydrostatic stress p defined in section 2.2.1 is nonzero. To separate the volumetric from the deviatoric strain–energy functions, the deviatoric stress defined in section 2.2.2 as $S = \sigma - pI$ will be used. The strain–energy density function can be written as follows:

$$\bar{U} = \frac{1}{2}\sigma_{ij}\varepsilon_{ij} = \frac{1}{2}\left((S_{11} + p)\varepsilon_{11} + (S_{22} + p)\varepsilon_{22} + (S_{33} + p)\varepsilon_{33} + S_{12}2\varepsilon_{12} + S_{13}2\varepsilon_{13} + S_{23}2\varepsilon_{23}\right)$$
$$= \frac{1}{2}(S_{11}\varepsilon_{11} + S_{22}\varepsilon_{22} + S_{33}\varepsilon_{33} + S_{12}2\varepsilon_{12} + S_{13}2\varepsilon_{13} + S_{23}2\varepsilon_{23}) + \frac{1}{2}p(\varepsilon_{11} + \varepsilon_{22} + \varepsilon_{33})$$

The first term is thus due to the deviatoric stress components S_{ij} while the second term is due to the hydrostatic stress p. By replacing the strains with expressions in the stress using the linear–elastic isotropic constitutive law (section 5.2.5), the strain energy per unit volume can be written as:

$$\begin{aligned}\overline{U} &= \frac{1}{2}\left(\frac{(\sigma_{11}-\sigma_{22})^2+(\sigma_{22}-\sigma_{33})^2+(\sigma_{11}-\sigma_{33})^2}{6G}+\frac{\sigma_{12}^2+\sigma_{13}^2+\sigma_{23}^2}{G}\right)\\ &\quad+\frac{(1-2\nu)}{6E}(\sigma_{11}+\sigma_{22}+\sigma_{33})^2\\ &=\frac{1}{12G}\left((\sigma_{11}-\sigma_{22})^2+(\sigma_{22}-\sigma_{33})^2+(\sigma_{11}-\sigma_{33})^2+6\left(\sigma_{12}^2+\sigma_{13}^2+\sigma_{23}^2\right)\right)\\ &\quad+\frac{(1-2\nu)}{6E}(\sigma_{11}+\sigma_{22}+\sigma_{33})^2\end{aligned}$$

▶ **BONUS EXERCISE:** Show the missing details used to derive the above relationship.

Thus, the strain–energy density function can be decomposed into:

$$\overline{U}=\overline{U}_{deviatoric}+\overline{U}_{volumetric}$$

Notice that the strain–energy density function is independent of the coordinate system since it is a function of the invariants of the stress tensor:

$$\begin{aligned}\overline{U}_{volumetric} &= \frac{(1-2\nu)}{6E}(\sigma_{11}+\sigma_{22}+\sigma_{33})^2=\frac{(1-2\nu)}{6E}(\text{trace}(\sigma))^2\\ \overline{U}_{deviatoric} &= \frac{1}{12G}\left((\sigma_{11}-\sigma_{22})^2+(\sigma_{22}-\sigma_{33})^2+(\sigma_{11}-\sigma_{33})^2+6\left(\sigma_{12}^2+\sigma_{13}^2+\sigma_{23}^2\right)\right)\\ &= \frac{(von\ Mises\ stress)^2}{6G}\end{aligned}$$

Problem 1: If the state of stress in a linear–elastic isotropic material is given by the stress matrix:

$$\sigma=\begin{pmatrix}20 & 30 & 0\\ 30 & -10 & 0\\ 0 & 0 & 25\end{pmatrix} MPa$$

Find the strain–energy density and its deviatoric and volumetric components stored inside the material if $E = 10$ GPa and $\nu = 0.1$.

Solution: Using the expressions for the strain–energy function and its deviatoric and volumetric components, the elastic–strain energy and its deviatoric and volumetric components stored in the material under the effect of the given stress are:

$$\begin{aligned}
\overline{U} &= \frac{(1+\nu)}{2E}\left(\sigma_{11}^2+\sigma_{22}^2+\sigma_{33}^2+2\sigma_{12}^2+2\sigma_{23}^2+2\sigma_{13}^2\right)-\frac{\nu}{2E}(\sigma_{11}+\sigma_{22}+\sigma_{33})^2\\
&= 0.155\,\text{MN.m/m}^3\\
\overline{U}_{deviatoric} &= \frac{1}{12G}\left((\sigma_{11}-\sigma_{22})^2+(\sigma_{22}-\sigma_{33})^2+(\sigma_{11}-\sigma_{33})^2+6\left(\sigma_{12}^2+\sigma_{13}^2+\sigma_{23}^2\right)\right)\\
&= 0.138\,\text{MN.m/m}^3\\
\overline{U}_{volumetric} &= \frac{(1-2\nu)}{6E}(\sigma_{11}+\sigma_{22}+\sigma_{33})^2 = 0.016\,\text{MN.m/m}^3
\end{aligned}$$

The following is the Mathematica code utilized:

```
s={{20,30,0},{30,-10,0},{0,0,25}};
Nu=0.1;
Ee=10000;
G=Ee/2/(1+Nu);
U=(1+Nu)/2/Ee*(Sum[s[[i,j]]^2,{i,1,3},{j,1,3}])-
Nu/2/Ee*(Sum[s[[i,i]],{i,1,3}])^2
Udeviatoric=1/12/G*((s[[1,1]]-s[[2,2]])^2+(s[[3,3]]-s[[2,2]])^2+(s[[1,1]]-
s[[3,3]])^2+6*(s[[1,2]]^2+s[[1,3]]^2+s[[2,3]]^2))
Uvolumetric=(1-2Nu)/6/Ee*(s[[1,1]]+s[[2,2]]+s[[3,3]])^2
```

Problem 2: If the strains in a linear–elastic orthotropic material are given by the strain matrix:

$$\varepsilon = \begin{pmatrix} 0.01 & 0.015 & 0.001\\ 0.015 & -0.02 & 0.002\\ 0.001 & 0.002 & 0.001 \end{pmatrix}$$

If the material axes of orthotropy are aligned with the orthonormal basis used, then find the stresses and the strain energy per unit volume stored inside the material if $E_{11} = E_{22} = 100\,\text{MPa}$, $E_3 = 200\,\text{MPa}$. $\nu_{12} = \nu_{13} = \nu_{23} = 0$ and $G_{12} = G_{13} = G_{23} = 35\,\text{MPa}$.

Solution: The stress–strain relationships for orthotropic material shown in section 5.2.3 can be utilized to find the stresses, and since Poisson's ratio for all the different directions is equal to zero, then the stresses are simply:

$$\sigma_{11} = E_{11}\varepsilon_{11},\quad \sigma_{22} = E_{22}\varepsilon_{22},\quad \sigma_{33} = E_{33}\varepsilon_{33},\quad \sigma_{12} = 2G_{12}\varepsilon_{12},\quad \sigma_{13} = 2G_{13}\varepsilon_{13},\quad \sigma_{23} = 2G_{23}\varepsilon_{23}$$

Thus, the stress matrix is:

$$\sigma = \begin{pmatrix} 1 & 1.05 & 0.07\\ 1.05 & -2 & 0.14\\ 0.07 & 0.14 & 0.2 \end{pmatrix}$$

The strain energy per unit volume for a general linear–elastic material is given by the expression:

$$\overline{U} = \frac{1}{2}\sigma_{ij}\varepsilon_{ij} = 0.0412\,\text{MN.m/m}^3$$

The Mathematica code utilized is:

```
E11=100;
E22=100;
E33=200;
G12=G13=G23=35;
eps={{0.01,0.015,0.001},{0.015,-0.02,0.002},{0.001,0.002,0.001}};
stress=Table[0,{i,1,3},{j,1,3}];
stress[[1,1]]=eps[[1,1]]*E11;
stress[[2,2]]=eps[[2,2]]*E22;
stress[[3,3]]=eps[[3,3]]*E33;
stress[[1,2]]=stress[[2,1]]=2eps[[1,2]]*G12;
stress[[1,3]]=stress[[3,1]]=2eps[[1,3]]*G13;
stress[[2,3]]=stress[[3,2]]=2eps[[2,3]]*G23;
stress//MatrixForm
eps//MatrixForm
Energy=Sum[stress[[i,j]]*eps[[i,j]]/2,{i,1,3},{j,1,3}]
```

Problem 3: The position function in the dimensions of m of a plate follows the following form:

$$\begin{aligned} x_1 &= X_1 + 0.001X_2 + 0.0002X_1^2 \\ x_2 &= 1.001X_2 - 0.0002X_2^2 \\ x_3 &= X_3 \end{aligned}$$

The plate has dimensions 2 m in the direction of e_1 and 1 m in the direction of e_2 with the bottom left corner coinciding with the origin of the coordinate system. Assuming that the material is linear–elastic isotropic with Young's modulus of 210 GPa and Poisson's ration of 0.3, then:

- Find an expression for the strain–energy density as a function of the position inside the plate, and draw the contour plot of the strain–energy density function on the plate.
- If the thickness of the plate is 10 mm, find the total strain energy stored inside the plate.
- Find an expression for the deviatoric and volumetric strain–energy density functions, and draw the contour plot of each on the plate.

Solution: The displacement function is:

$$u = x - X = \begin{pmatrix} 0.0002X_1^2 + 0.001X_2 \\ 0.001X_2 - 0.0002X_2^2 \\ 0 \end{pmatrix}$$

The gradient of the displacement tensor is:

$$M = \text{Grad}\, u = \begin{pmatrix} 0.0004X_1 & 0.001 & 0 \\ 0 & 0.001 - 0.0004X_2 & 0 \\ 0 & 0 & 0 \end{pmatrix}$$

The engineering small strain matrix is given by:

$$\varepsilon_{eng} = \frac{1}{2}\left(M + M^T\right) = \begin{pmatrix} 0.0004X_1 & 0.0005 & 0 \\ 0.0005 & 0.001 - 0.0004X_2 & 0 \\ 0 & 0 & 0 \end{pmatrix}$$

The strain vector required for the constitutive law is:

$$\varepsilon = \begin{pmatrix} \varepsilon_{11} \\ \varepsilon_{22} \\ \varepsilon_{33} \\ 2\varepsilon_{12} \\ 2\varepsilon_{13} \\ 2\varepsilon_{23} \end{pmatrix} = \begin{pmatrix} 0.0004X_1 \\ 0.001 - 0.0004X_2 \\ 0 \\ 0.001 \\ 0 \\ 0 \end{pmatrix}$$

Using the constitutive equation for the linear–elastic isotropic material, the stress vector is:

$$\sigma = \begin{pmatrix} 121.154 + 113.077X_1 - 48.4615X_2 \\ 282.692 + 48.4615X_1 - 113.077X_2 \\ 121.154 + 48.4615X_1 + 48.4615X_2 \\ 80.7692 \\ 0 \\ 0 \end{pmatrix} MPa$$

The strain–energy density function is given by:

$$\begin{aligned} \overline{U} &= \frac{1}{2}\sigma_{ij}\varepsilon_{ij} \\ &= 0.181731 + 0.022615X_1 (2.14286 + X_1) - 0.1131X_2 - 0.0194X_1X_2 + 0.02262X_2^2 \left(\text{MN} \cdot \frac{\text{m}}{m^3}\right) \end{aligned}$$

The total energy is given by:

$$Total\ Energy = \int \overline{U} dV = \int_0^{0.01}\int_0^{1}\int_0^{2} \overline{U} dX_1 dX_2 dX_3 = 0.004 \text{ MN.m}.$$

The deviatoric and volumetric strain–energy density functions are given by:

$$\begin{aligned}\overline{U}_{Deviatoric} &= \frac{1}{12G}\left((\sigma_{11}-\sigma_{22})^2 + (\sigma_{22}-\sigma_{33})^2 + (\sigma_{11}-\sigma_{33})^2 + 6\left(\sigma_{12}^2+\sigma_{13}^2+\sigma_{23}^2\right)\right)\\ &= 0.0942 + 0.00862\,(-2.5 + X_1)\,X_1 - 0.04308X_2 + 0.0086X_1X_2 + 0.0086X_2^2\ \left(\text{MN.}\frac{\text{m}}{m^3}\right)\\ \overline{U}_{volumetric} &= \frac{(1-2\nu)}{6E}(\sigma_{11}+\sigma_{22}+\sigma_{33})^2 = 3.1746\times 10^{-7}\,(525 + 210X_1 - 210X_2)^2\ \left(\text{MN.}\frac{\text{m}}{m^3}\right)\end{aligned}$$

The required contour plots are:

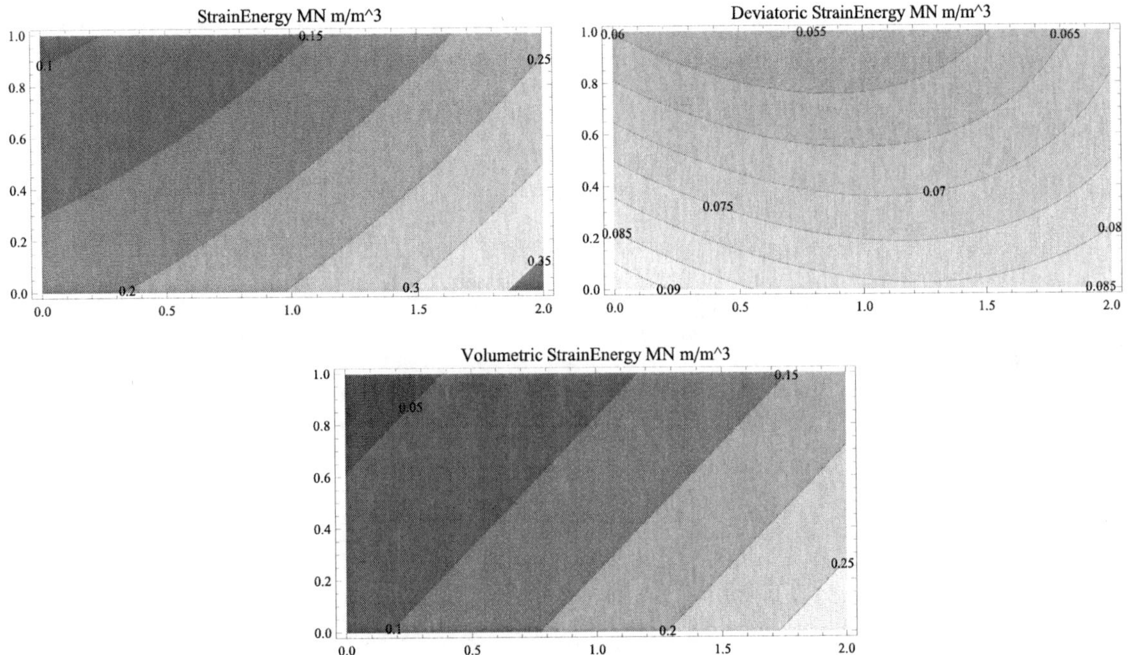

The Mathematica code utilized is:

```
Clear[X1,X2,X3,x1,x2,x3,Ee,Nu]
X={X1,X2,X3};
x={x1,x2,x3};
x1=X1+0.001X2+0.0002X1^2;
x2=1.001X2-0.0002X2^2;
```

```
x3=X3;
u=x-X;
gradu=Table[D[u[[i]],X[[j]]],{i,1,3},{j,1,3}];
esmall=1/2(gradu+Transpose[gradu]);
esmall//MatrixForm
Strainvector={esmall[[1,1]],esmall[[2,2]],esmall[[3,3]],2*esmall[[1,2]],2*es
mall[[1,3]],2*esmall[[2,3]]};
Strainvector//MatrixForm
Ee=210000;
Nu=0.3;
G=Ee/2/(1+Nu);
Cc={{1/Ee,-Nu/Ee,-Nu/Ee,0,0,0},{-Nu/Ee,1/Ee,-Nu/Ee,0,0,0},{-Nu/Ee,-
Nu/Ee,1/Ee,0,0,0},{0,0,0,1/G,0,0},{0,0,0,0,1/G,0},{0,0,0,0,0,1/G}};
Dd=FullSimplify[Inverse[Cc]];
stressvector=FullSimplify[Chop[Dd.Strainvector]];
stressvector//MatrixForm
S={{stressvector[[1]],stressvector[[4]],stressvector[[5]]},{stressvector[[4]
],stressvector[[2]],stressvector[[6]]},{stressvector[[5]],stressvector[[6]],
stressvector[[3]]}};
S=Chop[FullSimplify[S]];
S//MatrixForm
StrainEnergy=FullSimplify[Sum[1/2*S[[i,j]]*esmall[[i,j]],{i,1,3},{j,1,3}]]
TotalEnergy=Integrate[StrainEnergy,{X1,0,2},{X2,0,1},{X3,0,0.01}]
Udeviatoric=FullSimplify[1/12/G*((S[[1,1]]-S[[2,2]])^2+(S[[3,3]]-
S[[2,2]])^2+(S[[1,1]]-S[[3,3]])^2+6*(S[[1,2]]^2+S[[1,3]]^2+S[[2,3]]^2))]
Uvolumetric=FullSimplify[(1-2Nu)/6/Ee*(S[[1,1]]+S[[2,2]]+S[[3,3]])^2]
(*Check that the above expressions for the energy add up*)
Chop[FullSimplify[StrainEnergy-Udeviatoric-Uvolumetric]]
ContourPlot[StrainEnergy,{X1,0,2},{X2,0,1},AspectRatio-
>Automatic,ContourLabels->All,PlotLabel->"StrainEnergy MN
m/m^3",ColorFunction->"GrayTones"]
ContourPlot[Udeviatoric,{X1,0,2},{X2,0,1},AspectRatio-
>Automatic,ContourLabels->All,PlotLabel->"Deviatoric StrainEnergy MN
m/m^3",ColorFunction->"GrayTones"]
ContourPlot[Uvolumetric,{X1,0,2},{X2,0,1},AspectRatio-
>Automatic,ContourLabels->All,PlotLabel->"Volumetric StrainEnergy MN
m/m^3",ColorFunction->"GrayTones"]
```

7.3.3. Linear-Elastic Strain-Energy in Euler Bernoulli Beams with Small Deformations

Let B represent a linear–elastic Euler Bernoulli beam with a constant cross–sectional area and with a length L with its neutral axis aligned with the X_1 axis. Then, using the expressions for the stresses and strains derived in section 6.1.1, the total strain energy stored in the beam can be evaluated using one of the forms of the following integral:

$$Total\ Strain\ Energy = \int \bar{U}dV = \iint \int \sigma_{ij}d\varepsilon_{ij}dAdX_1 = \iint \int \sigma_{11}d\varepsilon_{11}dAdX_1 = \iint \frac{\sigma_{11}\varepsilon_{11}}{2}dAdX_1$$

$$= \frac{1}{2}\int_0^L M\frac{d^2y}{dX_1^2}dX_1 = \frac{1}{2EI}\int_0^L M^2dX_1 = \frac{EI}{2}\int_0^L \left(\frac{d^2y}{dX_1^2}\right)^2 dX_1$$

Problem: Consider a linear–elastic small deformations cantilever beam. If a load of value P is applied at the free end, find an expression for the strain energy per unit volume as a function of the position, P, and the beam length L. Also find the total strain energy stored. Assume Young's modulus to be E and that the beam has a rectangular cross section with a moment of inertia I.

Solution: The bending moment in a cantilever beam loaded by a concentrated load at the free end is:

$$M = -P(L - X_1)$$

The stress component σ_{11} and the strain component ε_{11} are given as:

$$\sigma_{11} = -\frac{MX_2}{I}, \quad \varepsilon_{11} = \frac{\sigma_{11}}{E} = -\frac{MX_2}{EI}$$

Therefore, the strain energy per unit volume at any point is given by:

$$\bar{U} = \frac{\sigma_{11}\varepsilon_{11}}{2} = \frac{M^2X_2^2}{2EI^2} = \frac{P^2(L - X_1)^2X_2^2}{2EI^2}$$

The total strain energy stored inside the beam is given by:

$$Total\ Strain\ Energy = \int \bar{U}dV = \frac{P^2}{2EI}\int_0^L (L - X_1)^2 dX_1 = \frac{P^2L^3}{6EI}$$

The displacement at the free end can be shown to be equal to (**How?**):

$$\Delta = \frac{PL^3}{3EI}$$

Notice that the worked done by the external force P during the gradual application of the displacement is equal to the stored total strain energy:

$$W = \frac{P\Delta}{2} = \frac{P^2L^3}{6EI}$$

7.3.4. Linear-Elastic Strain Energy in Timoshenko Beams with Small Deformations

Let B represent a linear–elastic Timoshenko beam with a constant cross–sectional area and with a length L with its neutral axis aligned with the X_1 axis. Then, using the nonzero expressions for the stresses and the strains derived in section 6.2.1, the total strain energy stored in the beam can be evaluated using the following integral:

$$\text{Total Strain Energy} = \int \bar{U} dV = \iint \int \sigma_{ij} d\varepsilon_{ij} dA dX_1 = \iint \int \sigma_{11} d\varepsilon_{11} dA dX_1 + \iint \int \sigma_{12} d\gamma_{12} dA dX_1$$

$$= \frac{EI}{2} \int_0^L \left(\frac{d\psi}{dX_1} \right)^2 dX_1 + \frac{kGA}{2} \int_0^L \gamma^2 dX_1 = \frac{1}{2EI} \int_0^L M^2 dX_1 + \frac{1}{2kGA} \int_0^L V^2 dX_1$$

7.3.5. Elastic Large (Hyper-Elastic) Deformations Materials

"Hyper–Elastic Materials" or, alternatively, "Green–Elastic Materials" are ideally elastic materi–als for which the stress–strain relationship derives from a strain–energy potential function. For general elastic materials with large deformations accompanied by large volumetric changes, it is difficult to write a closed–form expression for the strain–energy function in terms of the Cauchy stress σ since its energy conjugate is the stretching part D of the velocity gradient L, which is a rather instantaneous measure of deformation. Additionally, the small infinitesimal strain measure is not adequate when a continuum body exhibits large rotations. In addition, many elastic materi–als exhibiting large deformations possess a nonlinear relationship between the stress and the strain measures used. In such materials, the strain–energy density function is generally written in terms of the principal stretches or the singular values λ_1, λ_2, and λ_3 of the deformation gradient. (See section 3.2.3.b.)

7.3.5.a. Frame Invariant Isotropic Hyper-Elastic Potential Energy Functions

As per section 7.2.2, the increment in the elastic potential–energy function inside a continuum per unit undeformed volume is given by the expression.

$$\dot{W} = J\dot{\bar{U}} = \sum_{i,j=1}^{3} P_{ij} \dot{F}_{ij} \Longrightarrow dW = \sum_{i,j=1}^{3} P_{ij} dF_{ij}$$

If the material is elastic and W is independent of the path, then the First Piola Kirchhoff stress is derived from the strain–energy function W, as follows:

$$dW = \sum_{i,j=1}^{3} P_{ij} dF_{ij} = \sum_{i,j=1}^{3} \frac{\partial W}{\partial F_{ij}} dF_{ij}$$

i.e., the components of the first Piola Kirchhoff stress can be obtained by differentiating the strain–energy density function W, as follows:

$$P_{ij} = \frac{\partial W}{\partial F_{ij}} \Longrightarrow P = \frac{\partial W}{\partial F}$$

The above expression indicates that the hyper–elastic strain–energy function per unit undeformed volume could be written as a function of the deformation gradient entries:

$$W = W(F)$$

Frame Invariance[1]**:** The term "Frame Invariance" is used to indicate that the strain energy stored inside the material should be independent of the coordinate system used to describe the deformed configuration. If a rotation $Q \in \mathbb{M}^3$ is applied to the deformed configuration, then the strain–energy function before applying the rotation should be equal to that after applying rotation:

$$\forall Q \in \mathbb{M}^3,\ Q \text{ is a rotation: } W(F) = W(QF)$$

This automatically leads to the result that W is a function of the right–stretch tensor of the polar decomposition of the deformation gradient, as follows: Let $F \in \mathbb{M}^3$ with $\det F > 0$, and then:

$$W(F) = W(RU) = W\left(R^T RU\right) = W(U)$$

Isotropy[1]**:** If a material is isotropic, then the strain–energy function is independent of the orientation of the material vectors. In other words, if a rotation $Q \in \mathbb{M}^3$ is applied to the undeformed configuration and then the material is deformed such that $F \in \mathbb{M}^3$ describes the new deformation, then the strain–energy function of an isotropic material should be equal irrespective of the value of the applied rotation $Q \in \mathbb{M}^3$:

$$\forall Q \in \mathbb{M}^3,\ Q \text{ is a rotation: } W(F) = W(FQ)$$

Using the singular–value decomposition of the deformation gradient (See section 3.2.3.b.) $F = PDQ^T$, isotropy and frame invariance automatically lead to the following result:

$$W(F) = W\left(PDQ^T\right) = W\left(P^T PDQ^T Q\right) = W(D) = W(\lambda_1, \lambda_2, \lambda_3)$$

Thus, the strain–energy function of an isotropic hyper–elastic material can be described as a function of the principal stretches (the singular values of F). Moreover, W should be isotropic in its variables $\lambda_1, \lambda_2, \lambda_3$ **(Why?)**.

---

[1]For a rigorous proof of the assertions in this section, see: Ciarlet, P. (2004). Mathematical Elasticity, Volume: 1. Three Dimensional Elasticity. *Elsevier*.

Physical Restrictions on W: There are a few restrictions on the possible forms of W derived from physical reasoning on the possible deformations of elastic materials. The first is that W has to be minimum when the stretching part of the deformation gradient is the identity matrix. In other words:

$$W(Q) = W(I) = W_{\lambda_1=\lambda_2=\lambda_3=1} = minimum$$

The second restriction is that as the material is compressed such that the volume approaches zero or the material is stretched such that the volume approaches infinity, then the strain energy is expected to approach infinity:

$$\det(F) = \lambda_1\lambda_2\lambda_3 \to 0^+ \Longrightarrow W(\lambda_1,\lambda_2,\lambda_3) \to \infty$$
$$\det(F) = \lambda_1\lambda_2\lambda_3 \to \infty \Longrightarrow W(\lambda_1,\lambda_2,\lambda_3) \to \infty$$

Additionally, if the stretch in any direction approaches infinity, then the strain energy is expected to approach infinity as well:

$$\forall i{:}\lambda_i \to \infty \Longrightarrow W(\lambda_1, \lambda_2, \lambda_3) \to \infty$$

Normally, W, the elastic potential–energy density function of an isotropic material is written as a function of the quantities: $(\lambda_1 + \lambda_2 + \lambda_3)$, $(\lambda_1^2 + \lambda_2^2 + \lambda_3^2)$, $(\lambda_1\lambda_2 + \lambda_1\lambda_3 + \lambda_2\lambda_3)$ and $\lambda_1\lambda_2\lambda_3$ with the following physical interpretation of each quantity:

$$\lambda_i = \text{Principal stretch}$$
$$\lambda_1\lambda_2\lambda_3 = \det F = \frac{\text{Deformed volume}}{\text{Original volume}}$$
$$\lambda_1\lambda_2 = \text{Change in the areas of a face perpendicular to the direction of } \lambda_3$$

7.3.5.b. Examples of Isotropic Hyper-Elastic Potential Energy Functions

Hyper–elastic potential–energy functions are often developed by proposing certain forms and then calibrating material coefficients, according to experimental results. The proposed forms naturally have to abide by frame invariance and other physical restrictions, such as isotropy. There are many examples of hyper–elastic potential energy functions. An interested reader should consult a more detailed reference [for example, Holzapfel, G. (2000)[2]]. In this section, some brief examples of hyper–elastic potential energy functions will be introduced. However, first it will be illustrated that a linear–elastic material is in fact a special example of a hyper–elastic material:

---

[2]Holzapfel, G. (2000). Nonlinear Solid Mechanics, A Continuum Approach for Engineering, *John Wiley & Sons, Ltd.*

Linear–Elastic Materials as a Special Example of Hyper–Elastic Materials:

If a deformation is characterized by stretching without any rotation or shear deformations, the deformation gradient has the following form:

$$F = \begin{pmatrix} \lambda_1 & 0 & 0 \\ 0 & \lambda_2 & 0 \\ 0 & 0 & \lambda_3 \end{pmatrix}$$

Where λ_1, λ_2 and λ_3 are the principal stretches. In this case, the small strain matrix is given by:

$$\varepsilon_{small} = \frac{1}{2}\left(\text{Grad}\, u + \text{Grad}\, u^T\right) = \frac{1}{2}\left(F + F^T - 2I\right) = \begin{pmatrix} \lambda_1 - 1 & 0 & 0 \\ 0 & \lambda_2 - 1 & 0 \\ 0 & 0 & \lambda_3 - 1 \end{pmatrix}$$

Thus, in this case, the relationship between the principal stretches and the normal (axial) strains in the directions of the coordinate system chosen is:

$$\lambda_1 = 1 + \varepsilon_{11} = \frac{l_1}{l_{01}}, \quad \lambda_2 = 1 + \varepsilon_{22} = \frac{l_2}{l_{02}}, \quad \lambda_3 = 1 + \varepsilon_{33} = \frac{l_3}{l_{03}}$$

where, l_{01}, l_{02} and l_{03} are the original lengths of a cuboid and l_1, l_2, and l_3 are its final lengths.

If a coordinate system is adopted such that the shear strains are equal to zero and if the deformation is small such that $W \simeq \overline{U}$ (i.e., $J \simeq 1$), then the strain–energy density function of a linear–elastic material in terms of the small strains and Lamé's constants in a coordinate system with no shear strains can be written as (See section 7.3.1.):

$$W \simeq \overline{U} = \mu\left(\varepsilon_{11}^2 + \varepsilon_{22}^2 + \varepsilon_{33}^2\right) + \frac{\lambda}{2}\left(\varepsilon_{11} + \varepsilon_{22} + \varepsilon_{33}\right)^2$$

If the stretches are substituted instead of the strains, then $\overline{U}$ has the following form:

$$W \simeq \overline{U} = \mu\left((1-\lambda_1)^2 + (1-\lambda_2)^2 + (1-\lambda_3)^2\right) + \frac{\lambda}{2}\left(3 - \lambda_1 - \lambda_2 - \lambda_3\right)^2$$

This indeed shows that the strain–energy function of a linear–elastic material is a special example of frame–invariant isotropic hyper–elastic materials.

Incompressible Hyper–Elastic Isotropic Strain–Energy Potential Functions in Terms of $U^2 = F^T F$

Many of the elastic rubber materials can be considered incompressible. Incompressibility implies that the deformation is zero for any given hydrostatic state of the stress. For this type of materials, the strain–energy potential functions are usually given in terms of the invariants of the square of the stretch part $U^2 = F^T F$ of the deformation gradient. The invariants are related to the eigenvalues of U^2, as follows (See section 3.2.3 for the details of the polar decomposition of F.):

$$I_1(U^2) = \lambda_1^2 + \lambda_2^2 + \lambda_3^2, \quad I_2(U^2) = \lambda_1^2\lambda_2^2 + \lambda_1^2\lambda_3^2 + \lambda_2^2\lambda_3^2, \quad I_3(U^2) = 1$$

▶ **BONUS EXERCISES:** Verify the relationships for the invariants of U^2.

Notice that, for incompressible materials, the third invariant of U^2 is always equal to 1. The following are different forms of strain–energy density functions written in terms of the first two invariants of U^2:

- Ogden material model for rubber like materials:

$$W(U^2) = \sum_{p=1}^{N} \frac{\mu_p}{\alpha_p}\left(\lambda_1^{\alpha_p} + \lambda_2^{\alpha_p} + \lambda_3^{\alpha_p} - 3\right)$$

- Mooney–Rivlin material model:

$$W(U^2) = \frac{\mu_1}{2}(I_1(U^2) - 3) + \frac{\mu_2}{2}(I_2(U^2) - 3)$$

- Neo–Hookean material model:

$$W(U^2) = 2\mu(I_1(U^2) - 3)$$

where the constants: μ, μ_i, α_p are to be calibrated from experiments.

▶ **BONUS EXERCISE:** Show that, for the three incompressible hyper–elastic material models described above, the minimum value attained by $W\left(U^2\right)$ is equal to 0 and that this value is attained when F is a rotation matrix.

Stress Tensor of Incompressible Hyper–Elastic Isotropic Materials:

Since the material is incompressible, the hydrostatic stress does not cause any deformation and thus is not accompanied by strain (deformation) energy. For this reason, the first Piola Kirchhoff stress and the Cauchy stress tensors are written as:

$$\sigma = pI + \frac{1}{J}\frac{\partial W\left(U^2\right)}{\partial F}F^T, \quad P = JpF^{-T} + \frac{\partial W\left(U^2\right)}{\partial F}$$

where p is an undetermined hydrostatic pressure and can be obtained from the boundary conditions of the problem.

▶ BONUS EXERCISE: Assume that the energy function of incompressible materials is given as the additive form:

$$W_{incomp} = W(U^2) + p(J - 1)$$

where p is an undetermined hydrostatic pressure. Notice that the second term is always equal to zero since the material is incompressible. Show that the relationships of the first Piola Kirchhoff stress and the Cauchy stress tensors given above can be directly obtained, as follows:

$$P = \frac{\partial W_{incomp}}{\partial F}, \quad \sigma = \frac{1}{J} P F^T$$

Problem 1: Assume that a unit length cube of material deforms such that the length of the sides parallel to e_1, e_2, and e_3 become 0.8, 0.625 and l units, respectively, without any rotations. If the material follows a Neo–Hookean material model with $\mu = 1$ units, find W, $\overline{U}$, l, the Cauchy stress tensor and the first Piola Kirchhoff stress tensor after deformation in terms of the unknown hydrostatic pressure p.

Solution: The deformation gradient of the above deformation is:

$$F = \begin{pmatrix} 0.8 & 0 & 0 \\ 0 & 0.625 & 0 \\ 0 & 0 & l \end{pmatrix}$$

Since the material is incompressible, $\det F = 1 \Longrightarrow l = 2$ units.

Since F is diagonal, then the right–polar decomposition is equal to the left–polar decomposition:

$$U = V = F$$

The first invariant of U^2 is:

$$I_1\left(U^2\right) = \lambda_1^2 + \lambda_2^2 + \lambda_3^2 = 0.8^2 + 0.625^2 + 2^2 = 5.03$$

The strain energy per unit volume of the undeformed configuration is:

$$W = 2\mu(I_1(U^2) - 3) = 2\,(2.03) = 4.06 \text{ units of energy per volume}$$

since $J = 1 \Longrightarrow \overline{U} = W = 4.06$ units of energy per volume.

The relationship between the components of F and the singular values λ_1, λ_2, and λ_3 is as follows (no summation):

$$F_{ij} = \lambda_i \delta_{ij}$$

The transpose of the inverse of F is:

$$F^{-T} = \begin{pmatrix} 1.25 & 0 & 0 \\ 0 & 1.6 & 0 \\ 0 & 0 & 0.5 \end{pmatrix}$$

The components of the first Piola Kirchhoff stress can be found by taking the derivatives of W, with respect to the components of F, as follows:

$$P_{ij} = p(F^{-T})_{ij} + \frac{\partial W}{\partial F_{ij}} = p(F^{-T})_{ij} + \frac{\partial W}{\partial \lambda_1}\frac{\partial \lambda_1}{\partial F_{ij}} + \frac{\partial W}{\partial \lambda_2}\frac{\partial \lambda_2}{\partial F_{ij}} + \frac{\partial W}{\partial \lambda_3}\frac{\partial \lambda_3}{\partial F_{ij}}$$

Notice that the off–diagonal components of the first Piola Kirchhoff stress tensor are zero:

$$P = \begin{pmatrix} 3.2 + 1.25p & 0 & 0 \\ 0 & 2.5 + 1.6p & 0 \\ 0 & 0 & 8 + \frac{p}{2} \end{pmatrix}$$

The components of the Cauchy stress tensor are:

$$\sigma = \frac{1}{J}PF^T = \begin{pmatrix} 2.56 + p & 0 & 0 \\ 0 & 1.5625 + p & 0 \\ 0 & 0 & 16 + p \end{pmatrix}$$

Notice that the unknown hydrostatic pressure p can only be obtained from a boundary condition for the stress.

The following is the Mathematica code utilized:

```
mu=1;
F={{lambda1,0,0},{0,lambda2,0},{0,0,lambda3}};
U=F;
I1=Sum[U[[i,i]]^2,{i,1,3}]
W=2*mu*(I1-3)
DWD11=D[W,lambda1]
DWD12=D[W,lambda2]
DWD13=D[W,lambda3]
FnT=Transpose[Inverse[F]];
DiagonalP=Table[0,{i,1,3},{j,1,3}]
P=p*FnT+{{DWD11,0,0},{0,DWD12,0},{0,0,DWD13}};
P//MatrixForm
(P/.{lambda1->0.8,lambda2->0.625,lambda3->2})//MatrixForm
```

```
(FnT/.{lambda1->0.8,lambda2->0.625,lambda3->2})//MatrixForm;
sigma=FullSimplify[1/Det[F]*P.Transpose[F]];
sigma//MatrixForm
(FullSimplify[sigma/.{lambda1->0.8,lambda2->0.625,lambda3->2}])//MatrixForm
```

Problem 2: Assume that a unit length cube of material deforms such that the deformation gradient is:

$$F = \begin{pmatrix} \frac{1}{4} & -\frac{1}{2\sqrt{2}} & \frac{1}{4} \\ \frac{1}{4} & \frac{1}{2\sqrt{2}} & \frac{1}{4} \\ -2\sqrt{2} & 0 & 2\sqrt{2} \end{pmatrix}$$

If the material follows a Neo–Hookean material model with $\mu = 1$ units, find $W, \overline{U}$, the Cauchy stress tensor and the first Piola Kirchhoff stress tensor after deformation in terms of the unknown hydrostatic pressure p.

Solution: The determinant of F is equal to 1, and thus, the material indeed behaves in an incom–pressible manner. The singular–value decomposition of the deformation gradient has to be applied to find the principal stretches (see section 3.2.3.c):

$$F = PD_\lambda Q^T = \begin{pmatrix} 0 & \frac{1}{\sqrt{2}} & -\frac{1}{\sqrt{2}} \\ 0 & \frac{1}{\sqrt{2}} & \frac{1}{\sqrt{2}} \\ 1 & 0 & 0 \end{pmatrix} \begin{pmatrix} 4 & 0 & 0 \\ 0 & \frac{1}{2} & 0 \\ 0 & 0 & \frac{1}{2} \end{pmatrix} \begin{pmatrix} -\frac{1}{\sqrt{2}} & 0 & \frac{1}{\sqrt{2}} \\ \frac{1}{\sqrt{2}} & 0 & \frac{1}{\sqrt{2}} \\ 0 & 1 & 0 \end{pmatrix}$$

where D_λ, is a diagonal matrix with the principal stretches of F. The principal stretches λ_1, λ_2 and λ_3 are equal to 4, $\frac{1}{2}$, and $\frac{1}{2}$, respectively. The relationship between the principal stretches and the components of F is:

$$D_\lambda = P^T F Q$$

Therefore, the following relationships can be obtained:

$$\frac{\partial \lambda_1}{\partial F} = \begin{pmatrix} 0 & 0 & 0 \\ 0 & 0 & 0 \\ -\frac{1}{\sqrt{2}} & 0 & \frac{1}{\sqrt{2}} \end{pmatrix}, \quad \frac{\partial \lambda_2}{\partial F} = \begin{pmatrix} \frac{1}{2} & 0 & \frac{1}{2} \\ \frac{1}{2} & 0 & \frac{1}{2} \\ 0 & 0 & 0 \end{pmatrix}, \quad \frac{\partial \lambda_3}{\partial F} = \begin{pmatrix} 0 & -\frac{1}{\sqrt{2}} & 0 \\ 0 & \frac{1}{\sqrt{2}} & 0 \\ 0 & 0 & 0 \end{pmatrix}$$

The strain energy per unit volume of the undeformed configuration is:

$$W = 2\mu\left(I_1\left(U^2\right) - 3\right) = 2\left(\frac{33}{2} - 3\right) = 27 \text{ units of energy per volume}$$

since $J = 1 \Longrightarrow \overline{U} = W = 27$ units of energy per volume.

The components of the first Piola Kirchhoff stress can be found by taking the derivatives of W with respect to the components of F.

$$P_{ij} = p\left(F^{-T}\right)_{ij} + \frac{\partial W}{\partial F_{ij}} = p\left(F^{-T}\right)_{ij} + \frac{\partial W}{\partial \lambda_1}\frac{\partial \lambda_1}{\partial F_{ij}} + \frac{\partial W}{\partial \lambda_2}\frac{\partial \lambda_2}{\partial F_{ij}} + \frac{\partial W}{\partial \lambda_3}\frac{\partial \lambda_3}{\partial F_{ij}}$$

$$P = \begin{pmatrix} 1+p & -\sqrt{2}-\sqrt{2}p & 1+p \\ 1+p & \sqrt{2}+\sqrt{2}p & 1+p \\ -8\sqrt{2}-\dfrac{p}{4\sqrt{2}} & 0 & 8\sqrt{2}+\dfrac{p}{4\sqrt{2}} \end{pmatrix}$$

The components of the Cauchy stress tensor are:

$$\sigma = \frac{1}{J}PF^T = \begin{pmatrix} 1+p & 0 & 0 \\ 0 & 1+p & 0 \\ 0 & 0 & 64+p \end{pmatrix}$$

Notice that the unknown hydrostatic pressure p can only be obtained from a boundary condition for the stress.

The following is the Mathematica code utilized:

```
F={{1/4, -(1/(2*Sqrt[2])), 1/4}, {1/4, 1/(2*Sqrt[2]), 1/4},{-
2*Sqrt[2],0,2*Sqrt[2]}};
Det[F]
{P,Dd,Q}=SingularValueDecomposition[F];
P//MatrixForm
Dd//MatrixForm
Qt=Transpose[Q];
Qt//MatrixForm
Fmat=Table[Row[{f,i,j}],{i,1,3},{j,1,3}];
Fmat//MatrixForm
Dl=FullSimplify[Transpose[P].Fmat.Q];
Dl//MatrixForm
Dl1DF=Table[D[Dl[[1,1]],Fmat[[i,j]]],{i,1,3},{j,1,3}];
Dl2DF=Table[D[Dl[[2,2]],Fmat[[i,j]]],{i,1,3},{j,1,3}];
Dl3DF=Table[D[Dl[[3,3]],Fmat[[i,j]]],{i,1,3},{j,1,3}];
Dl1DF//MatrixForm
Dl2DF//MatrixForm
Dl3DF//MatrixForm
W=2*mu*(Sum[Dd[[i,i]]^2,{i,1,3}]-3)
W=2*mu*(lambda1^2+lambda2^2+lambda3^2-3);
FnT=Transpose[Inverse[F]];
FnT//MatrixForm
P=p*FnT+D[W,lambda1]*Dl1DF+D[W,lambda2]*Dl2DF+D[W,lambda3]*Dl3DF;
(P/.{lambda1->Dd[[1,1]],lambda2->Dd[[2,2]],lambda3->Dd[[3,3]]})//MatrixForm
```

```
s=1/Det[F]*P.Transpose[F];
FullSimplify[s]//MatrixForm
FullSimplify[(s/.{lambda1->Dd[[1,1]],lambda2->Dd[[2,2]],lambda3-
>Dd[[3,3]]})]//MatrixForm
```

Compressible Hyper–Elastic Isotropic Strain–Energy Potential Functions with Additive Volumetric and Deviatoric Components:

For compressible hyper–elastic isotropic materials, some authors propose[3] to decompose the strain–energy potential function into two additive components, a volumetric and a deviatoric component. For such decomposition, the deformation gradient is first decomposed into a vol–umetric component $J^{1/3}I$ and a deviatoric component $\overline{F}$, as follows:

$$F = \overline{F}\begin{pmatrix} J^{1/3} & 0 & 0 \\ 0 & J^{1/3} & 0 \\ 0 & 0 & J^{1/3} \end{pmatrix}, \quad \overline{F} = \frac{F}{J^{\frac{1}{3}}} \Longrightarrow \det\overline{F} = 1$$

The singular–value decomposition of the deviatoric component $\overline{F}$ can be written as:

$$\overline{F} = P\begin{pmatrix} \overline{\lambda_1} & 0 & 0 \\ 0 & \overline{\lambda_2} & 0 \\ 0 & 0 & \overline{\lambda_3} \end{pmatrix}Q^T$$

where $\overline{\lambda_i} = \frac{\lambda_i}{J^{1/3}}$. The invariants of $\overline{F}^T\overline{F}$ can be written as:

$$\overline{I_1}\left(\overline{F}^T\overline{F}\right) = \overline{\lambda_1}^2 + \overline{\lambda_2}^2 + \overline{\lambda_3}^2, \quad \overline{I_2}\left(\overline{F}^T\overline{F}\right) = \overline{\lambda_1}^2\overline{\lambda_2}^2 + \overline{\lambda_1}^2\overline{\lambda_3}^2 + \overline{\lambda_2}^2\overline{\lambda_3}^2 = \frac{1}{\overline{\lambda_3}^2} + \frac{1}{\overline{\lambda_2}^2} + \frac{1}{\overline{\lambda_1}^2},$$

$$\overline{I_3}\left(\overline{F}^T\overline{F}\right) = 1$$

The following are different forms of strain–energy density functions written in terms of the above invariants[4]:

- Mooney–Rivlin Material Model:

  $$W = C_{10}\left(\overline{I_1}-3\right) + C_{01}\left(\overline{I_2}-3\right) + \frac{1}{D_1}(J-1)^2$$

  When the deformations are small, this material behaves like a linear–elastic material with a bulk modulus $K = \frac{2}{D_1}$ and a shear modulus $G = 2\left(C_{10} + C_{01}\right)$.

---

[3] Holzapfel, G. (2000). Nonlinear Solid Mechanics: A Continuum Approach for Engineering. *John Wiley & Sons Ltd.*

[4] ABAQUS 6.9 Finite Element Analysis Software. SIMULIA.

- Neo–Hookean material model:

$$W = C_{10}\left(\overline{I_1} - 3\right) + \frac{1}{D_1}(J-1)^2$$

When the deformations are small, this material behaves like a linear–elastic material with a bulk modulus $K = \frac{2}{D_1}$ and a shear modulus $G = 2\,(C_{10})$.

- Ogden material model:

$$W = \sum_{i=1}^{N} \frac{2\mu_i}{\alpha_i^2}\left(\overline{\lambda_1^{\alpha_i}} + \overline{\lambda_2^{\alpha_i}} + \overline{\lambda_3^{\alpha_i}} \quad - 3\right) + \sum_{i=1}^{N} \frac{1}{D_i}(J-1)^{2i}$$

When the deformations are small, this material behaves like a linear–elastic material with a bulk modulus $K = \frac{2}{D_1}$ and a shear modulus $G = \sum_{i=1}^{N} \mu_i$.

- Polynomial form material model:

$$W = \sum_{i+j=1}^{N} C_{ij}\left(\overline{I_1} - 3\right)^i \left(\overline{I_2} - 3\right)^j + \sum_{i=1}^{N} \frac{1}{D_i}(J-1)^{2i}$$

When the deformations are small, this material behaves like a linear–elastic material with a bulk modulus $K = \frac{2}{D_1}$ and a shear modulus $G = 2\,(C_{10} + C_{01})$.

Problem: Draw the relationship between the component P_{11} of the first Piola Kirchhoff stress and λ_1 in a uniaxial state of stress using the three material models (Linear Elastic, compressible Neo–Hookean, and compressible Mooney–Rivlin material models). Assume that the material is isotropic and no rotations occur during deformation. Assume also that in small deformations the equivalent Young's modulus and Poisson's ratio are: $E = 20{,}000$ MPa and $\nu = 0.49$ respectively. Recall: $K = \frac{E}{3(1-2\upsilon)}$, $G = \frac{E}{2(1+\upsilon)}$.

Solution: The material parameters of the hyper–elastic material models can be calibrated as follows:

Neo–Hookean material model: $D_1 = \frac{2}{K} = 6 \times 10^{-6}\text{MPa}^{-1}$, $C_{10} = \frac{G}{2} = 3355.7\text{MPa}$

Mooney–Rivlin Material model: $D_1 = \frac{2}{K} = 6 \times 10^{-6}\,\text{MPa}^{-1}$, $C_{10} = C_{01} = \frac{G}{4} = 1667.85\,\text{MPa}$

The first Piola Kirchhoff stress in a uniaxial state of stress is:

$$P = \begin{pmatrix} P_{11} & 0 & 0 \\ 0 & 0 & 0 \\ 0 & 0 & 0 \end{pmatrix}$$

To draw the relationship between P_{11} and λ_1, a value for λ_1 is first assumed and then the values of λ_2 and λ_3 are determined given that the stresses P_{22} and P_{33} are both zero:

$$P_{22} = P_{33} = \frac{\partial W}{\partial \lambda_2} = \frac{\partial W}{\partial \lambda_3} = 0$$

The two equations $\frac{\partial W}{\partial \lambda_2} = 0$ and $\frac{\partial W}{\partial \lambda_3} = 0$ are nonlinear in the unknowns λ_2, and λ_3, and so the Newton Raphson method is used to solve for λ_2 and λ_3. The plot between P_{11} and λ_1 shows that the different materials exhibit the same force–displacement relationship at small deformations (Figure 7–8). At higher strains ($\lambda \gg 1$ or $\lambda \ll 1$), the different material models exhibit different behavior.

The following Mathematica code is utilized to draw the required graph:

```
J=l1*l2*l3;
l1d=l1/J^(1/3);
l2d=l2/J^(1/3);
l3d=l3/J^(1/3);
I1=l1d^2+l2d^2+l3d^2;
I2=l1d^-2+l2d^-2+l3d^-2;
Ee=20000;
Nu=49/100;
K=Ee/3/(1-2*Nu);

G=Ee/2/(1+Nu);
D1=2/K;
C1=N[G/4]
C2=N[G/4]
Wmooney=C1*(I1-3)+C2 (I2-3)+1/D1*(J-1)^2;
p1=D[Wmooney,l1]/.l2→l3;
p2=D[Wmooney,l2]/.l2→l3;
p3=D[Wmooney,l3]/.l2→l3;
stress=Table[0,{i,150}];
lambda=Table[0,{i,150}];
Do[lambda[[i]]=0.4+i/50;
  s=l1→lambda[[i]];
  s2=FindRoot[(p2/.s)==0,{l3,0.99}];
  stress[[i]]=(p1/.s2)/.s,{i,1,150}];
Rel1=Table[{lambda[[i]],stress[[i]]},{i,1,50}];

C1=N[G/2]
D1=2/K;
Wneo=C1*(I1-3)+1/D1*(J-1)^2;
p1=D[Wneo,l1]/.l2→l3;
p2=D[Wneo,l2]/.l2→l3;
p3=D[Wneo,l3]/.l2→l3;
stress=Table[0,{i,150}];
lambda=Table[0,{i,150}];
Do[lambda[[i]]=0.4+i/50;
  s=l1→lambda[[i]];
```

```
  s2=FindRoot[(p2/.s)==0,{13,0.99}];
  stress[[i]]=(p1/.s2)/.s,{i,1,150}];
Rel2=Table[{lambda[[i]],stress[[i]]},{i,1,50}];

lameconstant=K-2G/3;
Welastic=G*((1-11)^2+(1-12)^2+(1-13)^2)+lameconstant*(3-11-12-13)^2
p1=D[Welastic,11]/.12→13;
p2=D[Welastic,12]/.12→13;
p3=D[Welastic,13]/.12→13;
stress=Table[0,{i,150}];
lambda=Table[0,{i,150}];
Do[lambda[[i]]=0.4+i/50;
  s=11→lambda[[i]];
  s2=FindRoot[(p2/.s)==0,{13,0.99}];
  stress[[i]]=(p1/.s2)/.s,{i,1,150}];
Rel3=Table[{lambda[[i]],stress[[i]]},{i,1,50}];

Needs["PlotLegends'"];
ListPlot[{Rel1,Rel2,Rel3},BaseStyle→Directive[Bold,12],AxesLabel→{"11","P11
(stress
units)"},PlotStyle→Black,Joined→{False,False,True},PlotMarkers→Automatic,Pl
otLegend→{Style["Mooney-Rivlin",Bold,12],Style["Neo-
Hookean",Bold,12],Style["Linear Elastic",Bold,12]},LegendPosition→{1.1,-
0.4},LegendSize→{1.5,0.5}]
```

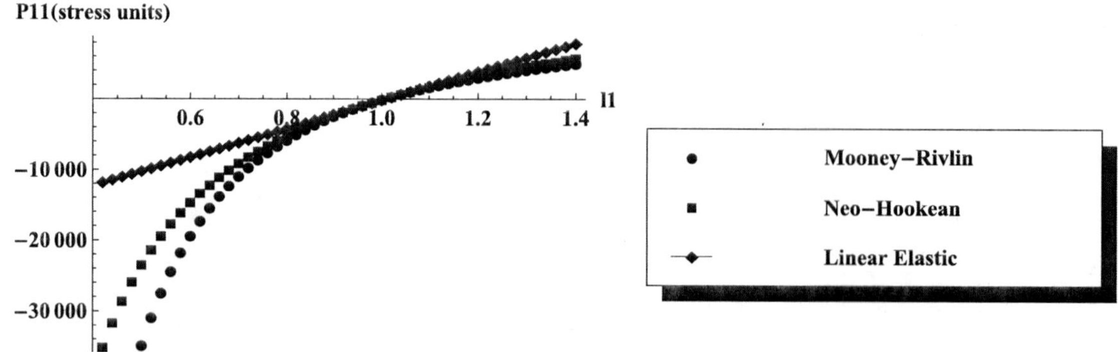

FIGURE 7-8 Relationship between P_{11} and λ_1 in a uniaxial state of stress for different compressible hyper–elastic material models that have the same modulus at small deformations.

7.4. THE PRINCIPLE OF VIRTUAL WORK

The principle of virtual work is widely used to solve a variety of continuum and solid mechanics problems. The statement of the principle of virtual work usually involves the phrase "virtual displacement field," which is designed to engage the intuition by attempting to give a physical explanation to a mathematical statement. However, as will be shown in the following sections, the statement itself can be derived solely from mathematical, rather than physical explanations.

7.4.1. Virtual Work Principle for a Mass Spring System

Consider the mass spring system in section 7.1. At equilibrium, the sum of the vertical forces is equal to zero, and the force F in the spring is equal to the applied external load mg. Let the position of equilibrium be at $x = \Delta$ (Figure 7–9). At this position, the equilibrium equation is:

$$F - mg = 0$$

If the position of the spring is perturbed by an arbitrary small displacement Δ^* (Figure 7–9) and if the equilibrium equation above is multiplied by Δ^*, then:

$$(F - mg)\,\Delta^* = 0 \Rightarrow F\Delta^* = mg\Delta^*$$

The above equation represents the statement of the principle of virtual work in a single–degree–of–freedom system. From an equilibrium position, the external work done by the external forces mg during the application of a small virtual displacement is equal to the internal work done by the spring force during the application of that small virtual displacement (Figure 7–9c).

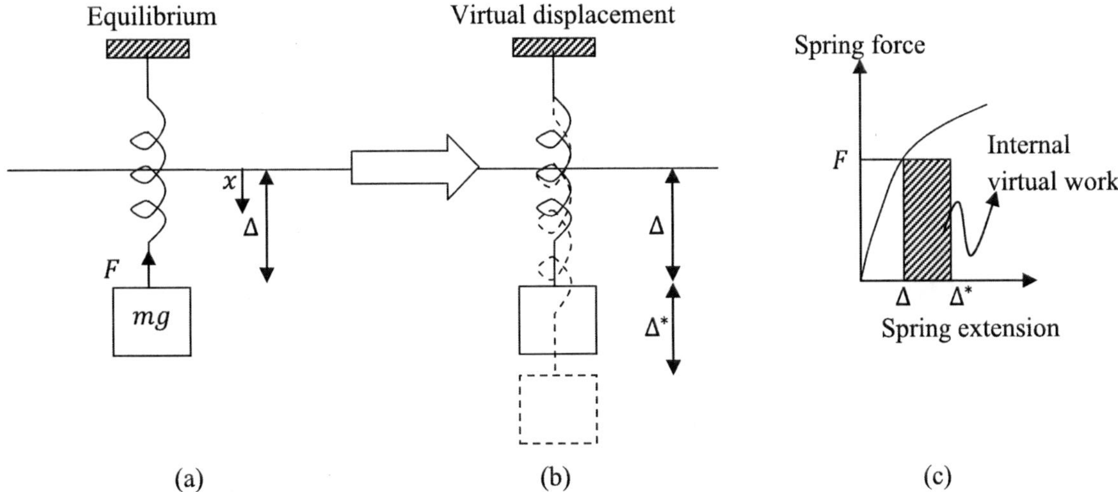

FIGURE 7-9 The principle of virtual work in a mass–spring system. (a) Equilibrium position with the external force. (b) Application of an arbitrary differentiable (small) virtual displacement. (c) Internal virtual work during the application of the virtual displacement.

7.4.2. Virtual Work Principle for a General Continuum

We first start by recalling the equations of static equilibrium of a continuum:

Let $\Omega_0 \in \mathbb{R}^3$ be a set representing a body in its reference configuration and $\Omega \in \mathbb{R}^3$ be the set representing the body in its current configuration with the basis set $\{e_1, e_2, e_3\}$ with the coordinates $\{x_1, x_2, x_3\}$. Given a constant (or variable) body force vector $b \in \mathbb{R}^3$, then the stresses at any point inside the body at static equilibrium satisfy the following equations:

$$\frac{\partial \sigma_{11}}{\partial x_1} + \frac{\partial \sigma_{21}}{\partial x_2} + \frac{\partial \sigma_{31}}{\partial x_3} + \rho b_1 = 0$$
$$\frac{\partial \sigma_{12}}{\partial x_1} + \frac{\partial \sigma_{22}}{\partial x_2} + \frac{\partial \sigma_{32}}{\partial x_3} + \rho b_2 = 0$$
$$\frac{\partial \sigma_{13}}{\partial x_1} + \frac{\partial \sigma_{23}}{\partial x_2} + \frac{\partial \sigma_{33}}{\partial x_3} + \rho b_3 = 0$$

Let $u^* \in \mathbb{R}^3$ be an ARBITRARY smooth function that could be viewed as a virtual displacement defined on Ω (Figure 7–10). When each of the equilibrium equations is multiplied by the corresponding component of the vector u^*, and then the three equations are added together, the following equation is obtained:

$$\left(\frac{\partial \sigma_{11}}{\partial x_1} + \frac{\partial \sigma_{21}}{\partial x_2} + \frac{\partial \sigma_{31}}{\partial x_3} + \rho b_1\right) u_1^* + \left(\frac{\partial \sigma_{12}}{\partial x_1} + \frac{\partial \sigma_{22}}{\partial x_2} + \frac{\partial \sigma_{32}}{\partial x_3} + \rho b_2\right) u_2^*$$
$$+ \left(\frac{\partial \sigma_{13}}{\partial x_1} + \frac{\partial \sigma_{23}}{\partial x_2} + \frac{\partial \sigma_{33}}{\partial x_3} + \rho b_3\right) u_3^* = 0$$

The following equality will be used to manipulate the above expression (using summation convention):

$$\frac{\partial \sigma_{ji}}{\partial x_j} u_i^* = \frac{\partial \sigma_{ji} u_i^*}{\partial x_j} - \sigma_{ji} \frac{\partial u_i^*}{\partial x_j}$$

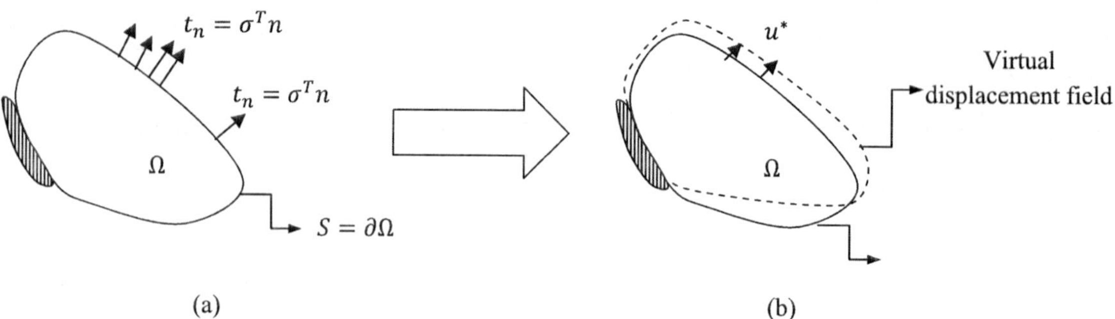

FIGURE 7-10 The principle of virtual work in a continuum. (a) Equilibrium position with external forces. (b) Application of an arbitrary differentiable (small) virtual displacement field.

In addition, the "virtual" strain field ε^* will be defined as follows:

$$\varepsilon_{ij}^* = \frac{1}{2}\left(\frac{\partial u_i^*}{\partial x_j} + \frac{\partial u_j^*}{\partial x_i}\right)$$

After using the above manipulation and introducing the virtual strain field, the three equations of equilibrium can now be written in the combined form:

$$\operatorname{div} \sigma u^* + \rho b \cdot u^* = \sum_{i,j=1}^{3} \sigma_{ij}\varepsilon_{ij}^*$$

When the above equation is integrated over the whole volume of the set Ω, the following integral form is obtained:

$$\int \operatorname{div} \sigma u^* dv + \int \rho b \cdot u^* dv = \int \sum_{i,j=1}^{3} \sigma_{ij}\varepsilon_{ij}^* dv$$

Finally, the divergence theorem (see section 1.3.1.c) can be utilized to replace the first volume integral $\int \operatorname{div} \sigma u^* dv$ with the surface integral $\int \sigma u^* \cdot n dS = \int u^* \cdot \sigma^T n dS = \int u^* \cdot t_n dS$, resulting in the mathematical expression for the principle of virtual work:

$$\int u^* \cdot t_n dS + \int \rho b \cdot u^* dv = \int \sum_{i,j=1}^{3} \sigma_{ij}\varepsilon_{ij}^* dv$$

The left–hand side represents the external work done by the external traction vectors (external forces) $t_n = \sigma^T n$ on the surfaces and the external body forces ρb during the application of an arbi–trary small and *differentiable* displacement field. The right–hand side represents the increment in the strain energy or the internal work done by the internal forces during the application of the arbitrary differentiable displacement field. Thus, from an equilibrium position (since the starting point was the equilibrium equations), the external work done during the application of an arbi–trary small and differentiable "virtual" displacement field is equal to the internal work done by the internal forces during the application of the virtual displacement field.

Illustrative Example: The Cauchy stress in units of kN/m^2 inside the shown plate has the following form:

$$\sigma = \begin{pmatrix} x_2 & 5 & 0 \\ 5 & x_1 & 0 \\ 0 & 0 & 0 \end{pmatrix}$$

where x_1 and x_2 are the coordinates of the position inside the plate with units of m. Verify that this stress is in equilibrium with a zero–body forces vector. Find the traction forces on the boundary edges A, B, C and D of the plate. Verify the principal of virtual work in the following two cases:

- A virtual horizontal displacement field $u_1^* = ax_1$, $a > 0$ is applied to the plate
- A virtual vertical displacement field $u_2^* = bx_2$, $b > 0$ is applied to the plate

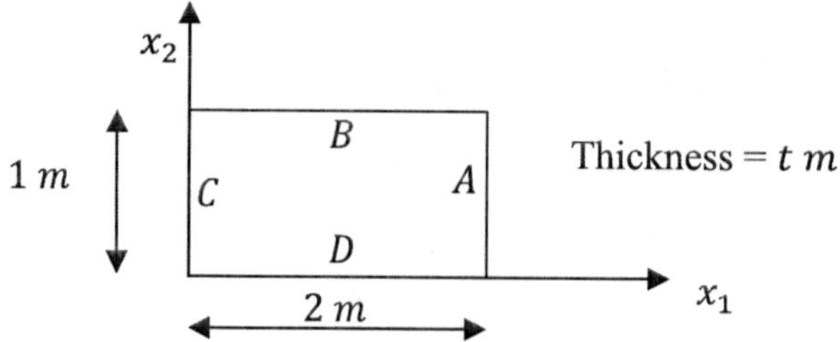

Solution: The equilibrium equations are indeed satisfied with a zero–body forces vector for the plate, as follows:

$$\frac{\partial\sigma_{11}}{\partial x_1}+\frac{\partial\sigma_{21}}{\partial x_2}+\frac{\partial\sigma_{31}}{\partial x_3}=\frac{\partial\sigma_{12}}{\partial x_1}+\frac{\partial\sigma_{22}}{\partial x_2}+\frac{\partial\sigma_{32}}{\partial x_3}=\frac{\partial\sigma_{13}}{\partial x_1}+\frac{\partial\sigma_{23}}{\partial x_2}+\frac{\partial\sigma_{33}}{\partial x_3}=0$$

Since this is a state of plane stress, the problem can be reduced to $\mathbb{R}^2$ and the area vectors on the boundary edges A, B, C, and D in this case, are, respectively:

$$n_A=\begin{pmatrix}1\\0\end{pmatrix},\quad n_B=\begin{pmatrix}0\\1\end{pmatrix},\quad n_C=\begin{pmatrix}-1\\0\end{pmatrix},\quad n_D=\begin{pmatrix}0\\-1\end{pmatrix}$$

The tractions in units of kN/m^2 on the boundary edges A, B, C, and D are, respectively:

$$t_A=\sigma^T\begin{pmatrix}1\\0\end{pmatrix}=\begin{pmatrix}x_2\\5\end{pmatrix},\quad t_B=\begin{pmatrix}5\\x_1\end{pmatrix},\quad t_C=\begin{pmatrix}-x_2\\-5\end{pmatrix},\quad t_D=\begin{pmatrix}-5\\-x_1\end{pmatrix}$$

The following figure illustrates the traction forces per unit area in units of kN/m^2 on the edges of the plate:

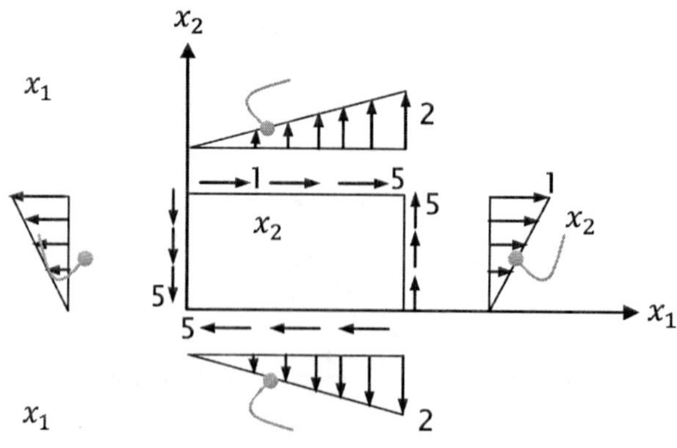

If a virtual horizontal displacement of $u_1^* = ax_1$ is applied to the plate, then the displacement vec– tor is:

$$u^* = \begin{pmatrix} ax_1 \\ 0 \end{pmatrix}$$

The virtual displacement vectors on the edges *A*, *B*, *C* and *D* are, respectively:

$$u_A^* = \begin{pmatrix} 2a \\ 0 \end{pmatrix}, \quad u_B^* = \begin{pmatrix} ax_1 \\ 0 \end{pmatrix}, \quad u_C^* = \begin{pmatrix} 0 \\ 0 \end{pmatrix}, \quad u_D^* = \begin{pmatrix} ax_1 \\ 0 \end{pmatrix}$$

The virtual strain field associated with u^* is:

$$\varepsilon^* = \begin{pmatrix} a & 0 & 0 \\ 0 & 0 & 0 \\ 0 & 0 & 0 \end{pmatrix}$$

The external virtual work supplied by the forces on the edges *A*, *B*, *C*, and *D* is:

$$EVW_A = \int_0^t \int_0^1 t_A \cdot u_A^* dx_2 dx_3 = \int_0^t \int_0^1 x_2 2a dx_2 dx_3 = at$$

$$EVW_B = \int_0^t \int_0^2 t_B \cdot u_B^* dx_1 dx_3 = \int_0^t \int_0^2 5ax_1 dx_1 dx_3 = 10at$$

$$EVW_C = \int_0^t \int_0^1 t_C \cdot u_C^* dx_2 dx_3 = 0$$

$$EVW_D = \int_0^t \int_0^2 t_D \cdot u_D^* dx_1 dx_3 = \int_0^t \int_0^2 -5ax_1 dx_1 dx_3 = -10at$$

The total external virtual work is:

$$EVW = EVW_A + EVW_B + EVW_C + EVW_D = at$$

The only nonzero virtual strain component is ε_{11}^*; therefore, the internal virtual work is:

$$IVW = \int_0^t \int_0^1 \int_0^2 \sigma_{11}\varepsilon_{11}^* dx_1 dx_2 dx_3 = \int_0^t \int_0^1 \int_0^2 ax_2 dx_1 dx_2 dx_3 = \int_0^t \int_0^1 2ax_2 dx_2 dx_3 = at$$

Thus, when virtual horizontal displacement of $u_1^* = ax_1$ is applied to the plate, the external virtual work and the internal virtual work are equal:

$$EVW = IVW = at$$

If a virtual vertical displacement of $u_2^* = bx_2$ is applied to the plate, then the displacement vector is:

$$u^* = \begin{pmatrix} 0 \\ bx_2 \end{pmatrix}$$

The virtual displacement vectors on the edges A, B, C and D are, respectively:

$$u_A^* = \begin{pmatrix} 0 \\ bx_2 \end{pmatrix}, \quad u_B^* = \begin{pmatrix} 0 \\ b \end{pmatrix}, \quad u_C^* = \begin{pmatrix} 0 \\ bx_2 \end{pmatrix}, \quad u_D^* = \begin{pmatrix} 0 \\ 0 \end{pmatrix}$$

The virtual strain field associated with u^* is:

$$\varepsilon^* = \begin{pmatrix} 0 & 0 & 0 \\ 0 & b & 0 \\ 0 & 0 & 0 \end{pmatrix}$$

The external virtual work supplied by the forces on the edges A, B, C, and D is:

$$EVW_A = \int_0^t \int_0^1 t_A \cdot u_A^* dx_2 dx_3 = \int_0^t \int_0^1 5bx_2 dx_2 dx_3 = \frac{5bt}{2}$$

$$EVW_B = \int_0^t \int_0^2 t_B \cdot u_B^* dx_1 dx_3 = \int_0^t \int_0^2 bx_1 dx_1 dx_3 = 2bt$$

$$EVW_C = \int_0^t \int_0^1 t_C \cdot u_C^* dx_2 dx_3 = \int_0^t \int_0^1 5bx_2 dx_2 dx_3 = -\frac{5bt}{2}$$

$$EVW_D = \int_0^t \int_0^2 t_D \cdot u_D^* dx_1 dx_3 = 0$$

The total external virtual work is:

$$EVW = EVW_A + EVW_B + EVW_C + EVW_D = 2bt$$

The only nonzero virtual strain component is ε^*_{22}; therefore, the internal virtual work is:

$$IVW = \int_0^t\int_0^1\int_0^2 \sigma_{22}\varepsilon^*_{22}dx_1dx_2dx_3 = \int_0^t\int_0^1\int_0^2 bx_1dx_1dx_2dx_3 = 2bt$$

Thus, when a virtual vertical displacement of $u^*_2 = bx_2$ is applied to the plate the external virtual work and the internal virtual work are equal:

$$EVW = IVW = 2bt$$

The Mathematica code used is:

```
(*Firstvitualdisplacementu1star=a*x1*)
ustar={a*x1,0};
x={x1,x2};
ustara=ustar/.x1->2;
ustarb=ustar/.x2->1;
ustarc=ustar/.x1->0;
ustard=ustar/.x2->0;
Gradustar=Table[D[ustar[[i]],x[[j]]],{i,1,2},{j,1,2}];
estar=1/2*(Gradustar+Transpose[Gradustar]);
s={{x2,5},{5,x1}};
na={1,0};
nb={0,1};
nc={-1,0};
nd={0,-1};
ta=s.na
tb=s.nb
tc=s.nc
td=s.nd
EVWa=Integrate[(ta.ustara)/.x1->2,{x2,0,1},{x3,0,t}]
EVWb=Integrate[(tb.ustarb)/.x2->1,{x1,0,2},{x3,0,t}]
EVWc=Integrate[(tc.ustarc)/.x1->0,{x2,0,1},{x3,0,t}]
EVWd=Integrate[(td.ustard)/.x2->0,{x1,0,2},{x3,0,t}]
EVW=EVWa+EVWb+EVWc+EVWd
ststrain=Sum[s[[i,j]]*estar[[i,j]],{i,1,2},{j,1,2}]
IVW=Integrate[ststrain,{x1,0,2},{x2,0,1},{x3,0,t}]
(*Secondvitualdisplacementu2star=b*x2*)
ustar={0,b*x2};
x={x1,x2};
ustara=ustar/.x1->2;
ustarb=ustar/.x2->1;
ustarc=ustar/.x1->0;
ustard=ustar/.x2->0;
Gradustar=Table[D[ustar[[i]],x[[j]]],{i,1,2},{j,1,2}];
estar=1/2*(Gradustar+Transpose[Gradustar]);
s={{x2,5},{5,x1}};
na={1,0};
```

```
nb={0,1};
nc={-1,0};
nd={0,-1};
ta=s.na
tb=s.nb
tc=s.nc
td=s.nd
EVWa=Integrate[(ta.ustara)/.x1->2,{x2,0,1},{x3,0,t}]
EVWb=Integrate[(tb.ustarb)/.x2->1,{x1,0,2},{x3,0,t}]
EVWc=Integrate[(tc.ustarc)/.x1->0,{x2,0,1},{x3,0,t}]
EVWd=Integrate[(td.ustard)/.x2->0,{x1,0,2},{x3,0,t}]
EVW=EVWa+EVWb+EVWc+EVWd
ststrain=Sum[s[[i,j]]*estar[[i,j]],{i,1,2},{j,1,2}]
IVW=Integrate[ststrain,{x1,0,2},{x2,0,1},{x3,0,t}]
```

7.4.3. Virtual Work Principle for an Euler Bernoulli Beam

The Euler Bernoulli Beam is a special case of the general continuum. The solution of the equations of equilibrium is the function y that describes the vertical motion of the neutral axis (section 6.1.1 Figure 7–10). The statement of the principle of virtual work applied to an Euler Bernoulli beam can be derived from the general statement given in the previous section. However, the procedure described in section 7.4.1 will be repeated again. Using the assumptions behind an Euler Bernoulli beam (section 6.1.1), we first start by recalling the equation of static equilibrium of an Euler Bernoulli Beam and the moment, shear displacement relationships:

$$EI\frac{d^4y}{dX_1^4} = q, \quad M = EI\frac{d^2y}{dX_1^2} = EI\frac{d\theta}{dX_1}, \quad V = EI\frac{d^3y}{dX_1^3}$$

Let V_1, V_2, M_1, and M_2 be the boundary conditions for the shear and moment, as shown in Figure 7–11. Then, similar to the previous section, a virtual arbitrary differentiable displacement field y^* is assumed and multiplied by the equilibrium equation and then integrated across the beam length:

$$\int_0^L EI\frac{d^4y}{dX_1^4}y^*\,dX_1 = \int_0^L qy^*\,dX_1$$

Using integration by parts, the integral on the left–hand side can be manipulated as follows:

$$EI\frac{d^3y}{dX_1^3}y^*\Bigg|_0^L - \int_0^L EI\frac{d^3y}{dX_1^3}\frac{dy^*}{dX_1}dX_1 = \int_0^L qy^*\,dX_1$$

The integral on the left–hand side can be further manipulated by repeating the integration by parts:

$$EI\frac{d^3y}{dX_1^3}y^*\Bigg|_0^L - EI\frac{d^2y}{dX_1^2}\frac{dy^*}{dX_1}\Bigg|_0^L + \int_0^L EI\frac{d^2y}{dX_1^2}\frac{d^2y^*}{dX_1^2}dX_1 = \int_0^L qy^*\,dX_1$$

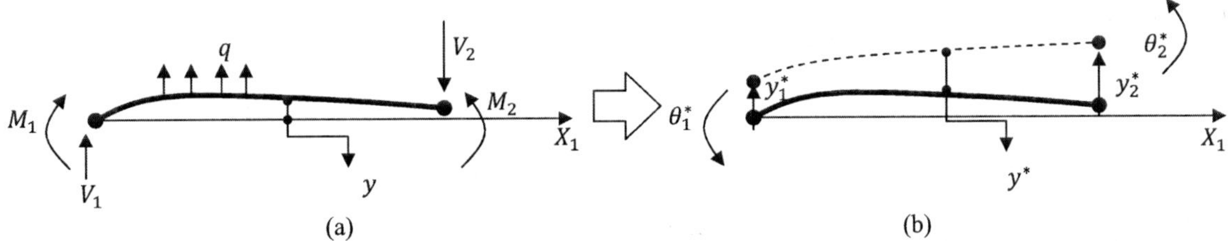

FIGURE 7-11 The principle of virtual work in an Euler Bernoulli Beam. (a) Equilibrium position with external forces. (b) Application of an arbitrary differentiable (small) virtual displacement field.

Finally, the virtual work expression for an Euler Bernoulli beam is:

$$\int_0^L EI\frac{d^2y}{dX_1^2}\frac{d^2y^*}{dX_1^2}dX_1 = \int_0^L qy^*dX_1 - V_2y_2^* + V_1y_1^* + M_2\theta_2^* - M_1\theta_1^*$$

The right–hand side represents the work done by the external forces applied to the beam during the application of the virtual small differentiable displacement field y^* while the left–hand side represents the internal work done by the internal forces $M = EI\frac{d^2y}{dX_1^2}$ during the application of the virtual displacement.

Illustrative Example: A simply supported Euler Bernoulli beam is subjected to a distributed load $q = -5X_1\ kN/m$. Assuming that Young's modulus, the length, and the moment of inertia for the beam are E, L, and I, respectively, verify that the principle of virtual work applies when a virtual cubic displacement $y^* = aX_1^2(X_1 - L)$, $a > 0$ is applied to the beam.

Solution: The bending moments and the shearing force are obtained by solving the beam differ–ential equation of equilibrium:

$$EI\frac{d^4y}{dX_1^4} = q = -5X_1 \Longrightarrow EIy = -\frac{X_1^5}{24} + \frac{C_1X_1^3}{6} + \frac{C_2X_1^2}{2} + C_3X_1 + C_4$$

The constants C_1, C_2, C_3, and C_4 are integration constants and can be obtained from the following boundary conditions:

$$@X_1 = 0 \Longrightarrow y = 0$$

$$@X_1 = 0 \Longrightarrow M = EI\frac{d^2y}{dX_1^2} = 0$$

$$@X_1 = L \Longrightarrow y = 0$$

$$@X_1 = L \Longrightarrow M = EI\frac{d^2y}{dX_1^2} = 0$$

Thus, the displacement function y and the bending moment M functions are:

$$y = \frac{-7L^4X_1 + 10L^2X_1^3 - 3X_1^5}{72EI}, \qquad M = \frac{5}{6}X_1(L - X_1)(L + X_1)$$

When a virtual displacement $y^* = aX_1^2(X_1 - L)$ is applied, the external virtual work is:

$$EVW = \int_0^L qy^* dX_1 = \int_0^L -5X_1 aX_1^2(X_1 - L)\,dX_1 = \frac{aL^5}{4}$$

Notice that the virtual displacement applied is equal to zero at the supports; therefore, no external virtual work is applied by the reactions.

The internal virtual work is:

$$IVW = \int_0^L M\frac{d^2y^*}{dX_1^2}dX_1 = \int_0^L \frac{5}{6}X_1(L - X_1)(L + X_1)(6aX_1 - 2aL)dX_1 = \frac{aL^5}{4}$$

Thus, when a virtual cubic displacement $y^* = aX_1^2(X_1 - L)$ is applied to the beam, the external virtual work is equal to the internal virtual work:

$$EVW = IVW = \frac{aL^5}{4}$$

The following is the Mathematica code that was utilized for the calculations:

```
Clear[M,y,EI,x,s]
q=-5x
s=DSolve[{EI*y''''[x]==q,y[0]==0,EI*y''[0]==0,y[L]==0,EI*y''[L]==0},y[x],x]
y=y[x]/.s[[1]]
M=FullSimplify[EI*D[y,{x,2}]]
ystar=a*x^2(x-L);
IVW=Integrate[M*D[ystar,{x,2}],{x,0,L}]
EVW=Integrate[q*ystar,{x,0,L}]
```

7.4.4. Virtual Work Principle for a Timoshenko Beam

The Timoshenko Beam is a special case of the general continuum. The solutions of the equations of equilibrium are the functions y and ψ that describe the vertical motion and the rotation of the neutral axis, respectively. (See section 6.2.1.) Similar to the derivation for the Euler Bernoulli beam in section 7.4.3, the statement of virtual work for a Timoshenko beam starts by assuming an arbitrary differentiable smooth "virtual" function y^*. In addition to that, an arbitrary smooth function γ^* is assumed such that the total virtual cross–sectional rotation is the sum of the two virtual components:

$$\psi^* = \frac{dy^*}{dX_1} + \gamma^*$$

Then, the equilibrium equation is multiplied by y^* and integrated over the whole domain:

$$\int_0^L EI\frac{d^3\psi}{dX_1^3}y^*\,dX_1 = \int_0^L qy^*\,dX_1$$

Using integration by parts:

$$EI\frac{d^2\psi}{dX_1^2}y^*\Big|_0^L - \int_0^L EI\frac{d^2\psi}{dX_1^2}\frac{dy^*}{dX_1}dX_1 = \int_0^L qy^*\,dX_1$$

$$EI\frac{d^2\psi}{dX_1^2}y^*\Big|_0^L - \int_0^L EI\frac{d^2\psi}{dX_1^2}\left(\psi^* - \gamma^*\right)dX_1 = \int_0^L qy^*\,dX_1$$

$$EI\frac{d^2\psi}{dX_1^2}y^*\Big|_0^L - EI\frac{d\psi}{dX_1}\psi^*\Big|_0^L + \int_0^L EI\frac{d\psi}{dX_1}\frac{d\psi^*}{dX_1}dX_1 + \int_0^L EI\frac{d^2\psi}{dX_1^2}\gamma^*\,dX_1 = \int_0^L qy^*\,dX_1$$

Finally, the statement of the virtual work principle has the form:

$$\int_0^L V\gamma^*\,dX_1 + \int_0^L EI\frac{d\psi}{dX_1}\frac{d\psi^*}{dX_1}dX_1 = \int_0^L qy^*\,dX_1 - V_2y_2^* + V_1y_1^* + M_2\psi_2^* - M_1\psi_1^*$$

7.5. SOME APPLICATIONS OF THE PRINCIPLE OF VIRTUAL WORK

7.5.1. Solving for Reactions in Statically Determinate Beams

A statically determinate beam is a beam whose external reactions can be obtained by solving the equilibrium equations without the need for using the constitutive equations. In case of a plane beam, the equilibrium equations: the sums of the horizontal and vertical forces, separately, are equal to zero, and the sum of moments is equal to zero. For such beams, it will be shown that the principle of virtual work can be used to generate the same equations by simply applying a virtual displacement field that produces no internal forces. Such a displacement field that produces no internal forces is termed: "**Rigid Body Displacement**." Recalling the principle of virtual work derived in section 7.4.3, if the arbitrary smooth displacement field y^* is such that $\frac{d^2y^*}{dX_1^2}$ is equal to zero, then the internal work done is equal to zero and the statement of the principle of virtual work is reduced to: The work done by the external forces through an arbitrary and smooth displacement field is equal to zero:

$$\int_0^L qy^* dX_1 - V_2 y_2^* + V_1 y_1^* + M_2\theta_2^* - M_1\theta_1^* = 0$$

The following examples illustrate the application of such principle to obtain the equations of equi–librium.

7.5.1.a. Example 1: Euler Bernoulli Beam with a Vertical Force

The beam shown in Figure 7–12 has its neutral axis aligned with the X_1 axis. Find the reactions R_1, R_2, and R_3 using the equilibrium equations and then using the principle of virtual work.

Solution: The equilibrium equations are:

$$\sum F_{X_1} = -R_3 = 0, \quad \sum F_{X_2} = R_1 - P + R_2 = 0, \quad \sum M_{@X_1=0} = -Pa + R_2 b = 0$$

Thus, the reactions are:

$$R_1 = \frac{Pb}{L}, \quad R_2 = \frac{Pa}{L}, \quad R_3 = 0$$

To apply the principle of virtual work, an arbitrary virtual smooth rigid–body displacement is applied to the whole beam (Figure 7–12b). The equation for the virtual displacement field is

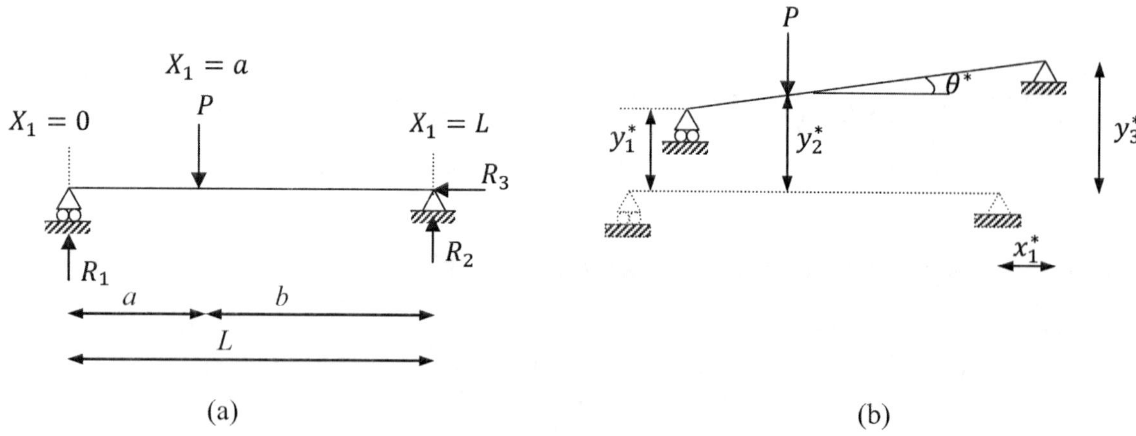

FIGURE 7-12 Virtual Work Principle in an Euler Bernoulli Beam. (a) Geometry and Loading, (b) Rigid displacement field.

$y^* = y_1^* + X_1 \sin\theta^*$. A horizontal arbitrary displacement x_1^* is also applied to the whole system. The statement of virtual work of the system is:

$$R_1 y_1^* - P y_2^* + R_2 y_3^* - R_3 x_1^* = R_1 y_1^* - P(y_1^* + a\sin\theta^*) + R_2(y_1^* + L\sin\theta^*) - R_3 x_1^* = 0$$

The above equation can be rearranged to have the following form:

$$(R_1 - P + R_2)y_1^* + (-Pa + R_2 L)\sin\theta^* - R_3 x_1^* = 0$$

Since the displacement field y^* is arbitrary, we can choose any values for y_1^*, $\sin\theta^*$, and x_1^*. Setting each two of the former arbitrary displacement values to zero, we obtain the equations of equilib–rium:

$$(R_1 - P + R_2) = 0, \quad (-Pa + R_2 L) = 0, \quad R_3 = 0$$

Notice that the arbitrary displacement was chosen such that $\theta^* =$ constant, so that no internal forces would be generated in the beam.

7.5.1.b. Example 2: Euler Bernoulli Cantilever Beam

The beam shown in Figure 7–13 has its neutral axis aligned with the X_1 axis. Find the reactions R_1, R_2, and M_3 using the equilibrium equations and then using the principle of virtual work.

Solution: The equilibrium equations are:

$$\sum F_{X_1} = R_2 = 0, \quad \sum F_{X_2} = R_1 - qL = 0, \quad \sum M_{@X_1=0} = \frac{qL^2}{2} - M_3 = 0$$

Thus, the reactions are:

$$R_1 = qL, \quad R_2 = 0, \quad M_3 = \frac{qL^2}{2}$$

To apply the principle of virtual work, an arbitrary smooth virtual rigid–body displacement is applied to the whole beam (Figure 7–13b). The equation for the virtual displacement field is

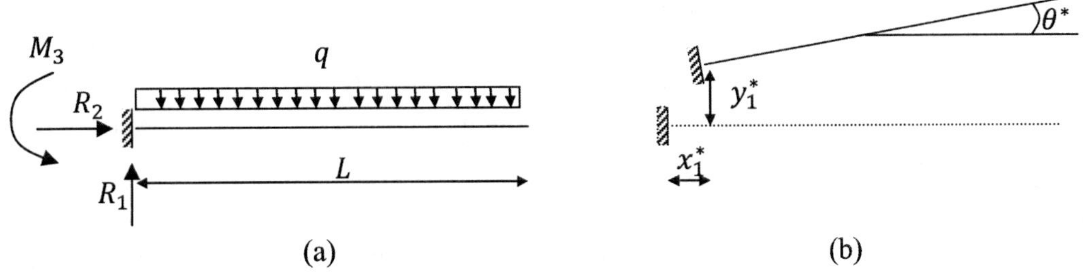

FIGURE 7-13 Virtual Work Principle in an Euler Bernoulli Cantilever Beam. (a) Geometry and Loading, (b) Rigid displacement field.

$y^* = y_1^* + X_1 \sin\theta^*$. A horizontal arbitrary displacement x_1^* is also applied to the whole system. The statement of the virtual work of the system is:

$$R_1 y_1^* + R_2 x_1^* + M_3\theta^* - \int_0^L q y^* dX_1 = R_1 y_1^* + R_2 x_1^* + M_3\theta^* - \int_0^L q(y_1^* + X_1 \sin\theta^*)dX_1 = 0$$

Rearranging the above equation yields the following:

$$R_1 y_1^* + R_2 x_1^* + M_3\theta^* - q y_1^* L - \frac{qL^2}{2}\sin\theta^* = 0$$

Notice that the displacement field is chosen small enough such that $\theta^* = \sin\theta^* = \tan\theta^*$. The above equations can then be rearranged to have the following form:

$$(R_1 - qL)y_1^* + \left(M_3 - \frac{qL^2}{2}\right)\theta^* - R_2 x_1^* = 0$$

Similar to the previous example, setting each two of the arbitrary displacement values to zero, we obtain the equations of equilibrium:

$$(R_1 - qL) = 0, \quad \left(M_3 - \frac{qL^2}{2}\right) = 0, \quad R_2 = 0$$

Notice that the arbitrary displacement was chosen such that $\theta^* =$ constant so that no internal forces would be generated in the beam.

7.5.2. Finding Displacements at Specific Points for Linear-Elastic Small Deformations Beams

Structural engineers often use the method of virtual work to find the displacement at specific points in statically determinate structures when the bending moment, the shearing force, and the normal force diagrams can be computed for the structure under consideration. In this procedure, the internal forces diagram for the structure is solved for twice, once with the original set of external forces and once with a unit load applied at the point of interest. Then, the displacement field from the original set of external forces is considered to be analogous to the virtual displacement field y^* in the derivation of the equations of the Virtual Work principle in section 7.4.3. The displacement field obtained for the structure with the applied unit load is considered to be analogous to the displacement field y. The equation of the principle of virtual work then becomes:

$$\int EI \frac{d^2y}{dX_1^2}\frac{d^2y^*}{dX_1^2} dX_1 = P \times \Delta^*$$

Where:

y is the displacement field obtained through applying a unit load to the point of interest.

y^* is the displacement field obtained from the original structure.

P is a unit load applied at the point of interest.

Δ^* is the unknown displacement.

However, since the bending moment diagram for both structures is readily available, the equation becomes:

$$\int \frac{M_0 m_1}{EI} dX_1 = P \times \Delta^*$$

If the shear and normal forces deformations are to be considered as well, the equation becomes:

$$\int \frac{V_0 v_1}{kGA} dX_1 + \int \frac{N_0 n_1}{EA} dX_1 + \int \frac{M_0 m_1}{EI} dX_1 = P \times \Delta^*$$

Where:

V_0, N_0 and M_0 are the shearing force, normal force, and bending moment equations for the original set of forces on the structure,

v_1, n_1 and m_1 are the shearing force, normal force, and bending moment equations for the beam structure with a unit load applied at the point of interest,

kGA, EA and EI are the shear stiffness, normal–force stiffness, and bending stiffness for the individual beam members.

Problem: Find the vertical displacement at the cantilever end due to the loads shown in Figure 7–14, given that EI is the bending stiffness of the beam. Ignore the shear and normal force deformations.

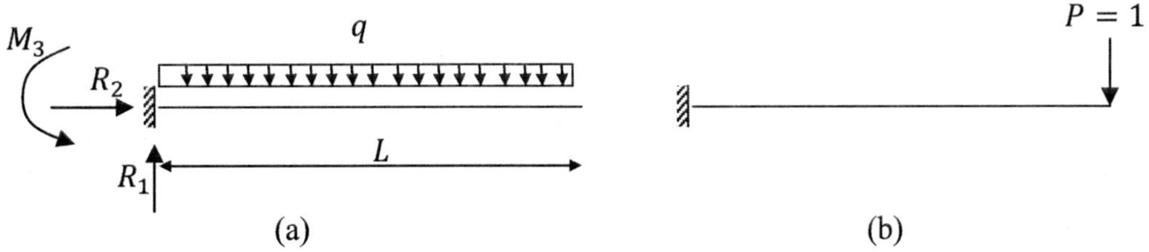

FIGURE 7-14 **Virtual Work Principle in an Euler Bernoulli Cantilever Beam. (a) Geometry and Loading, (b) Unit load.**

Solution: The bending moment equation for the structure with the original set of loads is given by the equation:

$$M_0 = -\frac{qL^2}{2} + qLX_1 - \frac{qX_1^2}{2}$$

After removing the loads from the structure and applying a unit load at the point of interest (Figure 7–14b), the bending moment equation for the structure with the unit load is given by the equation:

$$m_1 = -PL + PX_1 = X_1 - L$$

Applying the statement of the virtual work as described above gives the sought displacement:

$$\int_0^L \frac{M_0 m_1}{EI} dX_1 = \int_0^L \left(-\frac{qL^2}{2} + qLX_1 - \frac{qX_1^2}{2}\right)\frac{(X_1 - L)}{EI} dX_1 = 1 \times \Delta^* \Longrightarrow \Delta^* = \frac{qL^4}{8EI}$$

7.5.3. Finding Approximate Solutions

In sections 7.5.1 and 7.5.2, the principle of virtual work was used to find the reactions of statically determinate beam structures and to also find the displacements at certain points in statically determinate beam structures. However, such applications are probably only useful for illustrations, since structural–analysis computer software based on the stiffness method are now capable of solving most structures in a much simpler fashion. However, the principle of virtual work plays a significant role in the development of approximate solutions to continuum mechanics problems, in particular the finite element analysis method. The principle of virtual work is applied by first approximating the unknown displacement field with a shape or a form with unknown parameters. Then, the virtual displacement field is applied by varying the unknown parameters. This method results in a set of equations that are sufficient to find the unknown parameters. The method is akin to the Rayleigh Ritz method that will be presented in the next chapter.

An Illustrative Example: Use a cubic polynomial approximation and the virtual work method to find a displacement shape for the shown beam (Figure 7–15) with Young's modulus E and moment of inertia I and ignoring Poisson's ratio. Compare with the exact solution for $P/EI = 1$ unit and $L = 1$ unit.

Solution: Exact Displacement Shape: The exact displacement shape of the shown beam can be found by solving the differential equation of equilibrium with the symmetry boundary conditions shown in Figure 7–15.

$$EI\frac{d^4y}{dX_1^4} = q = 0 \Longrightarrow EIy = \frac{C_1X_1^3}{6} + \frac{C_2X_1^2}{2} + C_3X_1 + C_4$$

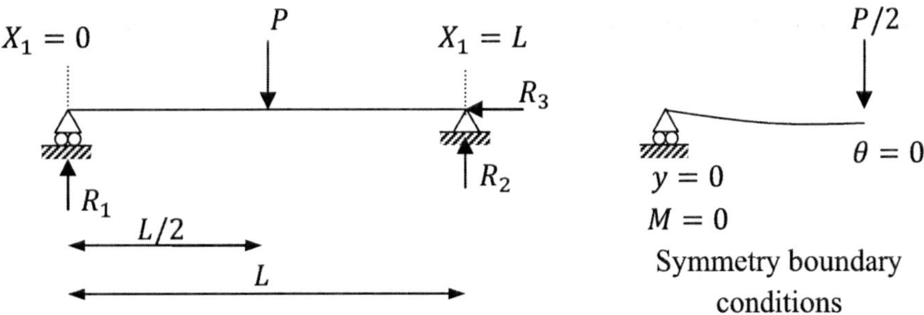

FIGURE 7-15 Geometry, loading and symmetry boundary conditions for an Euler Bernoulli beam.

The constants C_1, C_2, C_3 and C_4 are integration constants and can be obtained from the following boundary conditions:

$$@X_1 = 0 \Longrightarrow y = 0$$

$$@X_1 = 0 \Longrightarrow M = EI\frac{d^2y}{dX_1^2} = 0$$

$$@X_1 = L/2 \Longrightarrow V = EI\frac{d^3y}{dX_1^3} = P/2$$

$$@X_1 = L/2 \Longrightarrow \theta = \frac{dy}{dX_1} = 0$$

Thus, the displacement function y for $0 \le X_1 \le L/2$ is:

$$y = \frac{PX_1}{48EI}(-3L^2 + 4X_1^2)$$

The displacement function for $L/2 \le X_1 \le L$ can be obtained by replacing X_1 with $L - X_1$ in y, and the final displacement shape has the form:

$$y = \begin{cases} \dfrac{PX_1}{48EI}\left(-3L^2 + 4X_1^2\right) & 0 \le X_1 \le L/2 \\ \dfrac{P(L - X_1)}{48EI}\left(L^2 - 8X_1L + 4X_1^2\right) & L/2 \le X_1 \le L \end{cases}$$

Approximate Displacement Shape: To find the required approximate solution, we first start by assuming a cubic displacement function:

$$y = a_0 + a_1X_1 + a_2X_1^2 + a_3X_1^3$$

The displacement function has to satisfy the boundary conditions of zero displacements on both ends of the beams and therefore y takes the following form:

$$@X_1 = 0, \quad y = 0 \Longrightarrow a_0 = 0$$

$$@X_1 = L, \quad y = 0 \Longrightarrow a_1 = -a_2L - a_3L^2$$

Thus, the approximate displacement shape has the form:

$$y = a_2X_1(X_1 - L) + a_3X_1(X_1^2 - L^2)$$

The bending moment diagram in this case has the following form:

$$M = EI\frac{d^2y}{dX_1^2} = EI(2a_2 + 6a_3X_1)$$

A small perturbation in the displacement can be assumed as:

$$y^* = a_2^*X_1(X_1 - L) + a_3^*X_1(X_1^2 - L^2), \quad \frac{d^2y^*}{dX_1^2} = (2a_2^* + 6a_3^*X_1)$$

The internal virtual work and the external virtual work can be evaluated as follows:

$$IVW = \int_0^L EI\frac{d^2y}{dX_1^2}\frac{d^2y^*}{dX_1^2}dX_1 = \int_0^L EI(2a_2 + 6a_3X_1)(2a_2^* + 6a_3^*X_1)dX_1$$

$$= EI(4a_2La_2^* + 6a_3L^2a_2^* + 6a_2L^2a_3^* + 12a_3L^3a_3^*)$$

$$= EI(4a_2L + 6a_3L^2)a_2^* + EI(6a_2L^2 + 12a_3L^3)a_3^*$$

$$EVW = -Py^*|_{X_1=L/2} = -P\left(-\frac{a_2^*L^2}{4} - \frac{3a_3^*L^3}{8}\right)$$

Since the small perturbation is arbitrary, the multipliers of a_2^* and a_3^* on both sides of the virtual work equation have to be equal, leading to the following set of two equations:

$$EI(4a_2L + 6a_3L^2) = \frac{PL^2}{4}, \quad EI(6a_2L^2 + 12a_3L^3) = \frac{3PL^3}{8}$$

Solving the above equations yields the following values for a_2 and a_3:

$$a_2 = \frac{PL}{16EI}, \quad a_3 = 0$$

Therefore, the approximate solution for the displacement shape is:

$$y = \frac{PL}{16EI} X_1(X_1 - L)$$

This approximate solution can be compared with the exact solution when $P/EI = 1$ unit and $L = 1$ unit. The plot of y versus X_1 shows that the approximate solution under–predicts the displacement of the structure as shown in the following illustration:

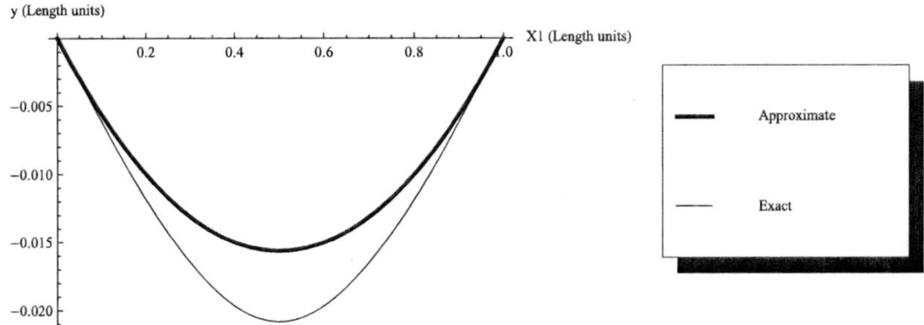

Note that the approximate solution gives a structure that has less displacement than the exact solution. As will be shown later, approximate solutions of structures normally produce a "stiffer" structure. In addition, the approximate solution is in fact continuous and differentiable in com–parison with the exact solution, which has a point of discontinuity at mid–span. The Mathematica code utilized is:

```
Clear[y,P,X1,EI,L,yexact]
(*Exact solution*)
s=DSolve[{y''''[X1]==0,y[0]==0,y''[0]==0,y'[L/2]==0,y'''[L]==P/2/EI},y[X1],X1]
y1=FullSimplify[y[X1]/.s[[1]]]
y2=FullSimplify[y1/.X1->L-X1]
yexact=Piecewise[{{y1,0<=X1<L/2},{y2,L/2<=X1<=L}}]
(*Approximate solution*)
y=a2*X1*(X1-L)+a3*X1*(X1^2-L^2);
ystar=y/.{a2->a2s,a3->a3s};
EVW=-P*ystar/.X1->L/2
IVW=Integrate[EI*D[y,{X1,2}]*D[ystar,{X1,2}],{X1,0,L}]
Eq1=Coefficient[IVW,a2s]-Coefficient[EVW,a2s]
Eq2=Coefficient[IVW,a3s]-Coefficient[EVW,a3s]
s=Solve[{Eq1==0,Eq2==0}{a2,a3}]
y=y/.s[[1]]
yp=y/.{P->1,EI->1,L->1};
yexactp=yexact/.{P->1,EI->1,L->1};
Needs["PlotLegends'"];
Plot[{yp,yexactp},{X1 ,0,1},PlotLegend->{"Approximate","Exact"},AxesLabel-
>{"X1 (Length units)","y (Length units)"},LegendPosition->{1.1,-
0.4},PlotStyle->{Directive[Black,Thick],Directive[Black,Thin]}]
```

7.6. THE PRINCIPLE OF MINIMUM POTENTIAL ENERGY FOR CONSERVATIVE SYSTEMS

A **conservative system** is defined as a system whose energy function is independent of the path between different deformation configurations, while a **conservative force** is defined as a force that exerts the same work to move a particle between two fixed points independent of the path taken. A conservative system would generally be composed of an elastic body acted upon by conservative forces. Forces arising due to gravity are in general conservative forces. On the other hand, non-conservative systems are those that usually contain frictional forces or nonelastic materials. For such systems, energy is lost during loading cycles that return the material back to its original position. For systems composed of elastic bodies and external forces that are conservative in nature, it will be shown that, provided a certain restriction on the constitutive equations for the elastic body, the state of static equilibrium corresponds to the state of minimum potential energy for the system.

7.6.1. Stable, Neutral and Unstable Equilibrium

For the purpose of illustration, the differences between stable, neutral, and unstable equilibrium states are often depicted though the use of schematics, such as Figure 7–16. In the three cases shown in the figure, the equations of equilibrium (sum of vertical forces is equal to zero) are achieved when the small cylinder is at position *A*. In Figure 7–16a, any perturbation (small change) in the position of the cylinder from point *A* will not upset equilibrium because the cylinder will roll back to its equilibrium position. Clearly this is the position of "Stable Equilibrium," and in fact, this is the position of the minimum potential energy of the system (**Why?**). In Figure 7–16b, the small cylinder is in equilibrium at any point on the flat surface since all those points correspond to the same potential energy; such equilibrium is termed "Neutral Equilibrium". In Figure 7–16c, while

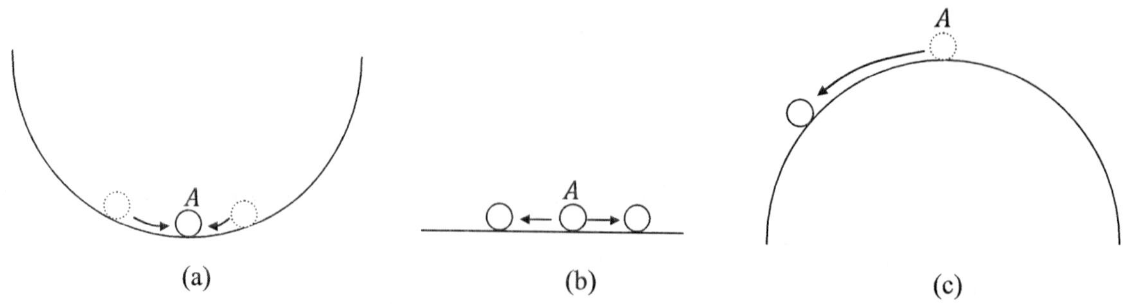

FIGURE 7-16 Types of equilibrium (a) Stable, (b) Neutral, and (c) Unstable equilibrium.

the small cylinder is in static equilibrium on top of the larger cylinder, any small perturbation in the position of the smaller cylinder will upset equilibrium, and the small cylinder will roll away from the equilibrium position. It is clear that the equilibrium position in Figure 7–16c corresponds to the maximum potential energy of the system, and the system cannot sustain this position. Such type of equilibrium is termed "Unstable Equilibrium."

7.6.2. Principle of Stationary Potential Energy at Equilibrium

Let's assume that a system composed of an **elastic** body and a set of conservative external and body forces is in equilibrium with the equilibrium displacement field $u: \Omega \to \mathbb{R}^3$. Then, the total strain energy stored inside the body is the integral of the strain–energy density function $\overline{U}$ associated with the displacement field u and integrated over the whole volume V. Assume also that the external tractions and body forces are not functions of the displacement. The total potential energy function Π of the system is defined as:

$$\Pi = \int \overline{U} dV - W$$

where W is the lost potential of the external tractions and body forces during the application of the displacement field u:

$$W = \int u \cdot t_n dS + \int u \cdot \rho b dV$$

Thus, the total potential energy of the system is:

$$\Pi = \int \overline{U} dV - \int u \cdot t_n dS - \int u \cdot \rho b dV$$

The purpose of the next step is to investigate the variation of Π denoted by $\delta\Pi$ when the displacement field u is perturbed by a small change denoted δu such that the body forces and the surface tractions are not altered. In addition, the perturbation δu is assumed to be smooth such that it is associated with a small perturbed strain tensor $\delta\varepsilon$; then:

$$\delta\Pi = \int \sum_{i,j=1}^{3} \frac{\partial U}{\partial \varepsilon_{ij}} \delta\varepsilon_{ij} dV - \int \delta u \cdot t_n dS - \int \delta u \cdot \rho b dV$$

However, since the material is assumed to be elastic, then the Cauchy stress is derived from the strain–energy function and thus:

$$\delta\Pi = \int \sum_{i,j=1}^{3} \sigma_{ij} \delta\varepsilon_{ij} dV - \int \delta u \cdot t_n dS - \int \delta u \cdot \rho b dV$$

The first term in $\delta\Pi$ is equal to the work done by the internal forces, while the second and third terms represent minus the work done by the external forces when the equilibrium position is perturbed by a differentiable displacement field δu. Using the principle of virtual work, at equilibrium, the terms in $\delta\Pi$ equate to zero, and thus, the potential energy defined above for a system composed of an elastic body and with conservative forces that are not function in the displacement is stationary at equilibrium (has a slope of zero with respect to the displacements).

7.6.3. Principle of Minimum Potential Energy at Equilibrium

In section 7.6.2, it was shown that the potential energy of a conservative system composed of an elastic body and external forces that are independent of the displacement is stationary (has a slope of zero with respect to the displacement variables) at equilibrium. Therefore, the potential energy at the equilibrium position is minimum, maximum, an inflection point or the potential energy function is in fact a constant value independent of the position. If the potential energy is minimum, then this is analogous to the stable equilibrium shown in Figure 7–16a. Otherwise, the equilibrium position is either an unstable or a neutral position. For the potential energy to be minimum, the second variation in the potential energy has to be positive:

$$\delta^2\Pi = \frac{1}{2}\int \sum_{i,j=1}^{3} \frac{\partial^2 \overline{U}}{\partial\varepsilon_{ij}\partial\varepsilon_{kl}}\delta\varepsilon_{ij}\delta\varepsilon_{kl}dV - \frac{1}{2}\sum_{i,j=1}^{3}\frac{\partial^2 W}{\partial u_i \partial u_j}\delta u_i \delta u_j > 0$$

However, since by assumption, W is a linear function in the displacements (the forces are independent of the displacements), then the second variation in the potential energy of the system is due solely to the elastic strain–energy function term. Then, for the system to have a stable equilibrium, the following second variation of the elastic potential energy function has to be positive:

$$\delta^2\Pi = \frac{1}{2}\int \sum_{i,j=1}^{3} \frac{\partial^2 \overline{U}}{\partial\varepsilon_{ij}\partial\varepsilon_{kl}}\delta\varepsilon_{ij}\delta\varepsilon_{kl}dV > 0$$

The above equation imposes a severe mathematical restriction on the strain–energy function $\overline{U}$ such that a stable equilibrium is always achieved; this mathematical restriction is called positive definiteness. Fortunately, for small deformations linear–elastic materials, such condition is readily satisfied when the material constants follow the restrictions imposed in section 5.2.6. Recall that for the linear isotropic material law to be positive definite, Young's modulus E and Poisson's ratio ν have the following restrictions: $E > 0$ and $-1 < \nu < 0.5$.

7.6.3.a. Principle of Minimum Potential Energy Functions for Linear-Elastic Materials

For linear–elastic small deformations materials when the infinitesimal strain matrix is used as the strain measure, the positive definiteness of the internal energy is satisfied. The strain–energy function $\overline{U}$ is quadratic, and thus, it is indeed convex in the variables ε_{ij} (**Bonus exercise:** What are convex functions?). For such materials, rigid body displacements (but not necessarily rotations!) result in no strain components ε_{ij} and thus are not accompanied by a change in the internal energy. For linear–elastic small deformations, the potential energy is indeed minimum at equilibrium when the infinitesimal strain matrix is used. It should also be noted that in the derivation in the previous section, the components ε_{ij} were taken as gradients with respect to the deformed configuration coordinate system. However, the infinitesimal strain tensor is in fact defined as gradients with respect to the undeformed configuration coordinate system. For small deformations, the following approximation is traditionally used:

$$\varepsilon_{ij} = \frac{1}{2}\left(\frac{\partial u_i}{\partial X_j} + \frac{\partial u_j}{\partial X_i}\right) \approx \frac{1}{2}\left(\frac{\partial u_i}{\partial x_j} + \frac{\partial u_j}{\partial x_i}\right)$$

Such approximation is valid only if the deformation gradient is small enough such that it can be approximated by the identity matrix I plus a matrix of small numbers $\delta \in \mathbb{M}^3$:

$$\text{Grad}\, u = \frac{\partial u}{\partial X} = \frac{\partial u}{\partial x}\frac{\partial x}{\partial X} = \text{grad}\, u F = \text{grad}\, u\,(I + \delta) = \text{grad}\, u + \text{grad}\, u\,\delta \approx \text{grad}\, u$$

The principle of minimum potential energy for linear–elastic small deformation materials is very successful in solving most solid mechanics engineering applications. However, the principle of minimum potential energy when the infinitesimal strain tensor is used can fail in predicting phenomena like buckling, since buckling involves large rotations. In this case, the use of the infinitesimal strain matrix is erroneous since it predicts nonzero values for the strain components when large rotations are encountered. These erroneous nonzero strain values can inhibit predicting certain types of mechanical behaviour accompanied by large rotations.

7.6.3.b. Principle of Minimum Potential Energy Functions for Elastic Materials with Large Deformations

In large deformations, the expressions for the virtual work and the second derivative of the potential energy of the system are derived by defining ε^*_{ij} using the deformed coordinates:

$$\varepsilon^*_{ij} = \frac{1}{2}\left(\frac{\partial u^*_i}{\partial x_j} + \frac{\partial u^*_j}{\partial x_i}\right)$$

Phenomenon like buckling, which are accompanied by large rotations without any change in the internal energy, can be modelled using the principle of virtual work since, during rotations, the terms $\delta\varepsilon^*_{ij}$ are all equal to zero (For pure rotation, the stretch rate D is equal to zero and, as shown in section 7.2, $\dot{\varepsilon}_{ij} = D_{ij}$.). Constitutive models for large deformation materials are either given in terms of the Green strain $\varepsilon_{Green} = \frac{1}{2}\left(F^TF - I\right)$ or in terms of the principal stretches λ_1, λ_2, and λ_3, which are the positive square roots of the eigenvalues of F^TF. Such materials have a strain–energy potential that is a function in the principal stretches λ_1, λ_2, and λ_3, and thus, large rotations are accompanied by zero strain energy.

▶ EXERCISES:

(Answer key is given in round brackets for some problems):

1. A spring manufacturer supplies three different elastic springs with the following three force–versus–displacement relationships:

$$f_1 = 100x, \quad f_2 = 100x + 100x^2, \quad f_3 = 100x + 1000x^3$$

with the units of f and x being N and mm, respectively. The manufacturer warns that the above relationships are only valid when the spring is stretched or contracted with a distance of 0.5 mm.

   - Draw the relationship between the force and the displacement in the specified range.
   - Find the strain energy stored in each spring at full extension and at full contraction (**Answer:** 12.5, 16.67, 28.125, 12.5, 8.33, 28.125 N.mm).
   - If a spring is required to carry a force of 45N (tension), find the displacement exhibited by each spring, and comment on which spring you would choose for this application.
   - If a spring is required to carry a force of –45N (compression), find the displacement exhibited by each spring, and comment on which spring you would choose for this application.
   - Comment on the stability of spring number 2 at full contraction.

2. A linear–elastic isotropic material with a shear modulus $G = 15$ MPa and a bulk modulus $K = 10$ MPa is stretched such that the strain matrix is given by the following:

$$\varepsilon = \begin{pmatrix} 0.02 & 0.0015 & 0 \\ 0.0015 & -0.001 & 0 \\ 0 & 0 & 0.01 \end{pmatrix}$$

Find the strain–energy density stored during the application of the deformation, and find its deviatoric and volumetric components.(**Answer:** 7.58, 3.38, 4.21 kN.m/m^3)

3. Two states of stress are given by the stress matrices:

$$\sigma = pI, \quad \sigma = \begin{pmatrix} p & 0 & 0 \\ 0 & p & 0 \\ 0 & 0 & 0 \end{pmatrix}$$

where $p \in \mathbb{R}$, $p > 0$ and $I \in \mathbb{M}^3$ is the identity matrix. Find the deviatoric and the volumetric strain–energy density functions associated with both matrices if the material is linear–elastic isotropic with Young's modulus E and Poisson's ratio ν

4. Compare the total strain energy stored in a simple versus fixed ends beam of length L, Young's modulus E, moment of inertia I loaded with a constant distributed load $q = -125\,\text{kN/m}$. (**Answer:** $\frac{3125L^5}{48EI}$, $\frac{3125L^5}{288EI}$, Simple beam has 6 times more energy stored for the same loading!)

5. For a linear isotropic elastic material with a Young's modulus E and Poisson's ratio ν, express the strain–energy density function under the conditions of plane stress in terms of:

   a. σ_{11}, σ_{22} and σ_{12}

   b. $\varepsilon_{11}, \varepsilon_{22}$ and ε_{12}

   c. $\sigma_{11}, \sigma_{22}, \sigma_{12}, \varepsilon_{11}, \varepsilon_{22}$ and ε_{12}

6. Assume that a material that follows the incompressible Neo–Hookean hyper–elastic material model exhibits the following two different cases of deformation:

   a. Pure shear deformation, as described in section 3.1.1.e

   b. A biaxial extension with $\lambda_1 = \lambda_2 = 1.25$

   Find W, $\overline{U}$, σ and P in terms of the undetermined hydrostatic stress p for each case.

7. Assume that a compressible isotropic hyper–elastic material model exhibits the following two different cases of deformation:

   a. Pure shear deformation, as described in section 3.1.1.e

   b. A biaxial extension with $\lambda_1 = \lambda_2 = 1.25$ and $\lambda_3 = 1$ with zero rotations

   If the material behaves like a linear–elastic material with Young's modulus $E = 1$unit and Poisson's ratio $= 0.45$ when the deformations are small. Find W, $\overline{U}$, σ and P in each of the above two cases of deformation and for each of the following material models.

   - The Neo–Hookean material model
   - The Mooney–Rivlin material model

8. Use the virtual work principle to find the reactions for the following statically determinate structures.

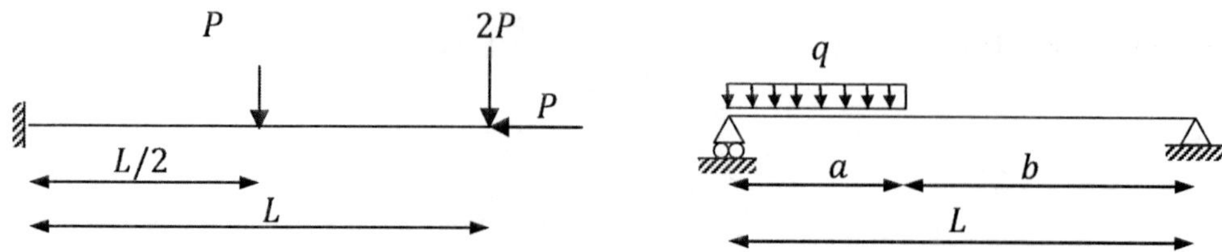

9. The shown beams have Young's modulus and moment of inertia E and I, respectively. Verify that the virtual work principle applies:
   - Assuming a virtual displacement field $y^* = aX_1(X_1 - L)$
   - Assuming a virtual displacement field $y^* = aX_1^2$

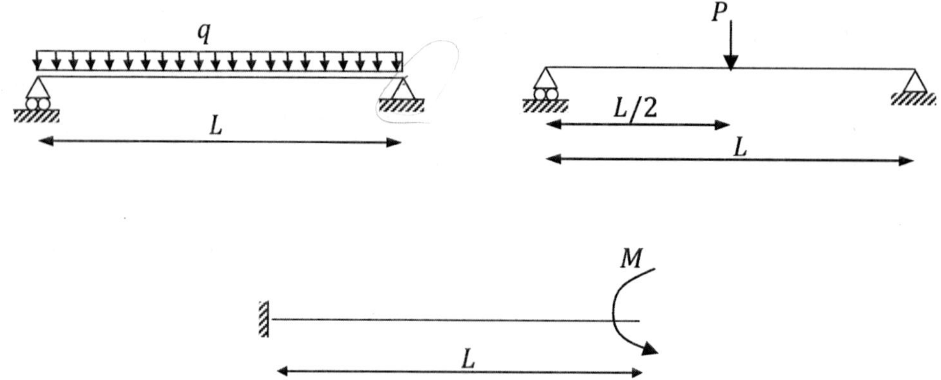

10. The Cauchy stress in units of kN/m^2 inside the shown plate has the following form:

$$\sigma = \begin{pmatrix} x_1 & 3 & 0 \\ 3 & 0 & 0 \\ 0 & 0 & 0 \end{pmatrix}$$

where x_1 and x_2 are the coordinates inside the plate with units of m. Find the equilibrium body forces vector applied to the plate. Find the traction forces on the boundary edges A, B, C, and D of the plate. Verify the principle of virtual work in the following two cases:

   - A virtual horizontal displacement field $u_1^* = ax_1,\ a > 0$ is applied to the plate
   - A virtual vertical displacement field $u_2^* = bx_1,\ b > 0$ is applied to the plate

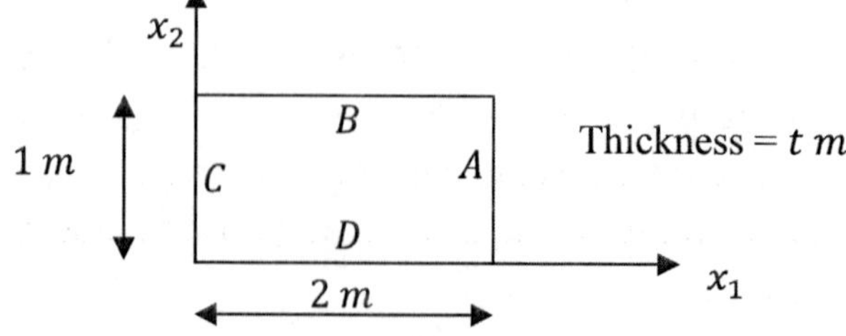

CHAPTER 8
Approximate Methods

The solution to most solid mechanics problems require finding the displacement vector, strain tensor, and stress tensor fields that would satisfy the equations of equilibrium. As shown in Chapter 4, finding a closed–form solution to the equations might not be feasible for most applications, and hence, approximate methods are needed to find solutions that "almost" satisfy the equations of equilibrium. In these approximate methods, a form or a shape of the distribution of the solution variables is first assumed. The solution variable that best satisfies the equilibrium equations while having the assumed shape or distribution is calculated based on some optimization criteria. Many approximate methods are available; however, in this chapter, the methods that serve as the precursors to the finite element analysis method are introduced, in particular the Rayleigh Ritz and the Galerkin methods. The Rayleigh Ritz method is based on the principle of minimum potential energy described in the previous chapter, while the Galerkin method is a more general method that can be applied to situations where the minimum potential energy principle does not apply. A few one–dimensional examples are introduced to illustrate the applications of these methods. In most of the examples presented in this chapter, a closed–form exact solution is available, and thus, it is possible to compare the approximate solution to the exact solution to understand the strength and limitations of the approximate method.

8.1. THE RAYLEIGH RITZ METHOD

The Rayleigh Ritz method is a classical method to find the displacement function, which could be an approximate or exact solution to the equilibrium equations. It is regarded as an ancestor of the widely used approximate method, the Finite Element Method (FEM), to find an approximate solution to the equilibrium equations. The Rayleigh Ritz method relies on the principle of minimum potential energy for conservative systems. The method involves *assuming* a form or a shape for the unknown displacement functions, and thus, the displacement functions would have a few unknown parameters. These assumed shape functions are termed **Trial Functions**. Afterwards, the potential–energy function of the system is written in terms of those few parameters, and the values of those parameters that would minimize the potential energy function of the system are calculated. Mathematically speaking, we assume that the unknown–displacement function u is a member of a certain space of functions; for example, we can assume that $u \in V$ where V is all the possible vector–valued linear functions defined on the body represented by the set $\Omega_0 \subset \mathbb{R}^3$: $V = \{u:\Omega_0 \to \mathbb{R}^3 | x \in \Omega_0, u = Bx, B \in \mathbb{M}^3\}$. With this assumption, we restrict the possible solutions to those in the form $u = Bx$, and thus, the unknown parameters are restricted to the matrix B, which in this particular example, has nine components. Finally, the nine components that would minimize the potential–energy function are obtained. The method will be illustrated using various one–dimensional examples.

8.1.1. Bars under Axial Loads

The Rayleigh Ritz method starts by choosing a form for the displacement function. This form has to satisfy the "Essential" boundary conditions, which for this particular type of problems, are those boundary conditions imposed on the primary variable (the displacement).

8.1.1.a. Example 1: A Bar under Axial Forces

Let a bar with a length L and a constant cross–sectional area A be aligned with the coordinate axis X_1. Assume that the bar is subjected to a horizontal body force (in the direction of the axis X_1) $q = cX_1$ units of force/unit length where c is a known constant. Also, assume that the bar has a constant force of value P applied at the end $X_1 = L$. If the bar is fixed at the end $X_1 = 0$, find the displacement of the bar by directly solving the differential equation of equilibrium. Also, find the displacement function using the Rayleigh Ritz method and the virtual work principle assuming a polynomial function for the displacement of the degrees (0, 1, 2, 3, and 4) (Figure 8–1). Assume that the bar is linear elastic with Young's modulus E and that the small strain tensor is the appropriate measure of strain. Ignore Poisson's ratio effect.

Solution:

Exact solution: The first part involves finding the exact solution that would satisfy the equations of equilibrium. To do so, the equation of equilibrium of this bar is derived, as shown in Figure 8–1:

$$-\sigma_{11}A + cX_1 dX_1 + \sigma_{11}A + \frac{\partial \sigma_{11}}{\partial X_1} dX_1 A = 0$$

And thus, the equilibrium equation is:

$$\frac{\partial \sigma_{11}}{\partial X_1} = -\frac{cX_1}{A}$$

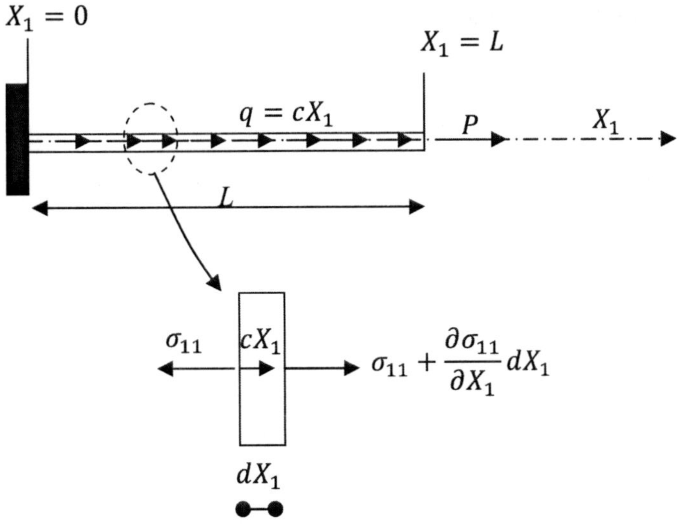

FIGURE 8-1 A bar under axial body forces and an axial load.

Using the constitutive equation of linear–elastic materials while ignoring Piosson's ratio effects leads to:

$$\sigma_{11} = E\varepsilon_{11} = E\frac{\partial u_1}{\partial X_1}$$

The equilibrium equation can thus be written in terms of the displacement function u_1, as follows (Notice that the operator $\frac{\partial}{\partial X_1}$ was replaced with $\frac{d}{dX_1}$ since X_1 is the only independent variable.):

$$\frac{d^2 u_1}{dX_1^2} = -\frac{cX_1}{EA}$$

The solution to the above differential equation is:

$$u_1 = -\frac{cX_1^3}{6EA} + C_1 X_1 + C_2$$

where C_1 and C_2 are constants that can be obtained using the two boundary conditions:

$$X_1 = 0 \Longrightarrow u_1 = 0 \Rightarrow C_2 = 0$$

$$X_1 = L \Longrightarrow \sigma_{11} = E\frac{du_1}{dX_1} = \frac{P}{A} \Rightarrow C_1 = \frac{P}{AE} + \frac{cL^2}{2EA}$$

Thus, the exact solution for the displacement function is:

$$u_1 = -\frac{cX_1^3}{6EA} + \left(\frac{P}{AE} + \frac{cL^2}{2EA}\right) X_1$$

The solution for the stresses distribution along the bar is:

$$\sigma_{11} = -\frac{cX_1^2}{2A} + \left(\frac{P}{A} + \frac{cL^2}{2A}\right)$$

The following Mathematica Code can also be used to solve the differential equation of equilibrium in one step:

```
Clear[u,x]
DSolve[{u''[x]==-c*x/EA,u'[L]==P/EA,u[0]==0},u,x]
```

The Rayleigh Ritz method:

The Rayleigh Ritz method starts by writing the potential energy equation of the system:

$$\Pi = \int \bar{U}dV - W = \int \frac{\sigma_{11}\varepsilon_{11}}{2} AdX_1 - Pu_1|_{X_1=L} - \int qu_1 dX_1$$

Notice that the potential energy lost by the action of the end force P is equal to the product of P, and the displacement u_1 evaluated at $X_1 = L$ while the potential energy lost by the action of the distributed body forces is an integral since it acts on each point along the beam length.

Further, we will use the constitutive equation to rewrite the potential energy function in terms of the function u_1:

$$\Pi = \int \bar{U}dV - W = \int \frac{EA}{2}\left(\frac{du_1}{dX_1}\right)^2 dX_1 - Pu_1|_{X_1=L} - \int cX_1 u_1 dX_1$$

The solution to our problem is to find a function u_1 that would minimize Π. To find an approximate solution, an assumption for the shape or the form of u_1 has to be introduced.

Polynomial of the zero degree:

Assume that u_1 is a constant function (a polynomial of the zero degree):

$$u_1 = a_0$$

However, to satisfy the "essential" boundary condition that $u_1 = 0$ when $X_1 = 0$, the constant a_0 has to equal to zero; this leads to the trivial solution $u_1 = 0$, which will automatically be rejected.

Polynomial of the first degree:

Assume that u_1 is a linear function (a polynomial of the first degree):

$$u_1 = a_1X_1 + a_0$$

However, to satisfy the essential boundary condition that $u_1 = 0$ when $X_1 = 0$, the coefficient a_0 has to be set to zero, then:

$$u_1 = a_1 X_1, \quad \frac{du_1}{dX_1} = a_1$$

Substituting into the equation for the potential–energy function of the system, the following equation is obtained:

$$\Pi = \int \frac{EA}{2} a_1^2 dX_1 - Pa_1 L - \int ca_1 X_1^2 dX_1 = \frac{EA}{2} a_1^2 L - Pa_1 L - \frac{ca_1 L^3}{3}$$

The only variable that can be controlled in the above expression is the coefficient a_1, and thus, the minimizer of the potential energy function can be obtained, as follows:

$$\frac{\partial \Pi}{\partial a_1} = EAa_1 L - PL - \frac{cL^3}{3} = 0 \Longrightarrow a_1 = \frac{P + \frac{cL^2}{3}}{EA}$$

Thus, the best linear function that could be a solution to the problem is:

$$u_1 = a_1 X_1 = \left(\frac{3P + cL^2}{3EA}\right) X_1$$

When the displacement is a linear function, the stress along the length of the bar is a constant function (**Why?**):

$$\sigma_{11} = E\frac{du_1}{dX_1} = \left(\frac{3P + cL^2}{3A}\right)$$

Unfortunately, this solution is far from being accurate, since it does not satisfy the differential equation of equilibrium (**Why?**). However, within the only possible or allowed linear shapes, the obtained solution is the minimizer of the potential energy.

Polynomial of the second degree:

$$u_1 = a_2 X_1^2 + a_1 X_1 + a_0, \quad \frac{du_1}{dX_1} = 2a_2 X_1 + a_1$$

Once more, the essential boundary conditions need to be satisfied; therefore, $a_0 = 0$.

The next step is to write the potential energy of the system:

$$\Pi = \int_0^L \frac{EA}{2} (2a_2 X_1 + a_1)^2 dX_1 - P\left(a_2 L^2 + a_1 L\right) - \int_0^L c\left(a_2 X_1^3 + a_1 X_1^2\right) dX_1$$

$$\Pi = \frac{2EA}{3} a_2^2 L^3 + EAa_2 a_1 L^2 + \frac{EA}{2} a_1^2 L - P\left(a_2 L^2 + a_1 L\right) - \frac{ca_2 L^4}{4} - \frac{ca_1 L^3}{3}$$

The only variables that can be controlled are the coefficients a_2 and a_1, and these can be found by minimizing Π:

$$\frac{\partial \Pi}{\partial a_2} = \frac{4EA}{3} a_2 L^3 + EAa_1 L^2 - PL^2 - \frac{cL^4}{4} = 0$$

$$\frac{\partial \Pi}{\partial a_1} = EAa_2 L^2 + EAL - PL - \frac{cL^3}{3} = 0$$

Solving the above set of two equations in the two unknowns a_1 and a_2 yields:

$$a_1 = \frac{(7cL^2 + 12P)}{12EA}, \quad a_2 = -\frac{cL}{4EA}$$

Therefore, the "best" solution for the displacement function in the space of second degree poly–nomials is:

$$u_1 = -\frac{cL}{4EA} X_1^2 + \frac{(7cL^2 + 12P)}{12EA} X_1$$

and the stresses are:

$$\sigma_{11} = -\frac{cL}{2A} X_1 + \frac{(7cL^2 + 12P)}{12A}$$

The Mathematica code that is used for the above calculations is:

```
Clear[u,x,a2,a1,Ee,A,c,P,L]
u=a2*x^2+a1*x;
PE=Integrate[(1/2)*Ee*A*D[u,x]^2,{x,0,L}]-P*(u/.x->L)-Integrate[c*x*u,{x,0,L}];
Eq1=D[PE,a2]
Eq2=D[PE,a1]
s=Solve[{Eq1==0,Eq2==0},{a2,a1}]
u=u/.s[[1]]
stress=FullSimplify[Ee*D[u,x]]
```

Polynomial of the third degree:

$$u_1 = a_3 X_1^3 + a_2 X_1^2 + a_1 X_1 + a_0, \quad \frac{du_1}{dX_1} = 3a_3 X_1^2 + 2a_2 X_1 + a_1$$

Similar to the previous solutions, the essential boundary conditions need to be satisfied; therefore, $a_0 = 0$.

The next step is to write the potential energy of the system:

$$\Pi = \int_0^L \frac{EA}{2}\left(3a_3 X_1^2 + 2a_2 X_1 + a_1\right)^2 dX_1 - P\left(a_3 L^3 + a_2 L^2 + a_1 L\right) - \int_0^L c\left(a_3 X_1^4 + a_2 X_1^3 + a_1 X_1^2\right) dX_1$$

It is clear that while the method is simple, it generates very large equations, and thus, Mathematica will be used to find the coefficients a_3, a_2 and a_1 that would minimize Π:

```
Clear[u,x,a2,a1,a3,Ee,A,c,P,L]
u=a3*x^3+a2*x^2+a1*x;
PE=Integrate[(1/2)*Ee*A*D[u,x]^2,{x,0,L}]-P*(u/.x->L)-Integrate[c*x*u,{x,0,L}];
Eq1=D[PE,a2]
Eq2=D[PE,a1]
Eq3=D[PE,a3]
s=Solve[{Eq1==0,Eq2==0,Eq3==0},{a2,a1,a3}]
u=u/.s[[1]]
stress=FullSimplify[Ee*D[u,x]]
```

Since the assumed trial function is a polynomial of the third degree, which is similar to the exact solution, the Rayleigh Ritz method was able to produce the exact solution to the differential equation:

$$u_1 = -\frac{cX_1^3}{6EA} + \left(\frac{P}{AE} + \frac{cL^2}{2EA}\right)X_1$$

$$\sigma_{11} = -\frac{cX_1^2}{2A} + \left(\frac{P}{A} + \frac{cL^2}{2A}\right)$$

Note that the exact solution was obtained because the assumed shape of the trial function was a cubic polynomial. In other words, when the Rayleigh Ritz method is used and when the assumed displacement function already has the shape or the form of the exact solution, the solution obtained is, in fact, the exact solution that satisfies the equilibrium equation.

The Principle of Virtual Work:

The principle of virtual work can be used in a manner similar to the Rayleigh Ritz method to obtain the same approximate solutions. Similar to the Rayleigh Ritz method, the approximate solution has to satisfy the essential boundary conditions, and thus, polynomials of the zero degree cannot be used. The virtual work method will be shown using a polynomial of the first and the second degrees.

Polynomial of the first degree:

Using a polynomial of the first degree, the displacement function has the form:

$$u_1 = a_1X_1 + a_0$$

However, to satisfy the essential boundary condition that $u_1 = 0$ when $X_1 = 0$, the coefficient a_0 has to be set to zero. Then:

$$u_1 = a_1X_1, \quad \varepsilon_{11} = \frac{du_1}{dX_1} = a_1$$

Then, an arbitrary smooth virtual displacement field is assumed to have the form:

$$u_1^* = a_1^*X_1, \quad \varepsilon_{11}^* = \frac{du_1^*}{dX_1} = a_1^*$$

The principle of virtual work can now be applied; the work done by the external forces during the arbitrary smooth virtual displacement is:

$$Pu_1^*|_{X_1=L} + \int_0^L cX_1 u_1^* dX_1 = Pa_1^* L + \frac{ca_1^* L^3}{3}$$

The internal virtual work is:

$$\int \sum_{i,j=1}^{3} \sigma_{ij}\varepsilon_{ij}^* dV = \int_0^L \sigma_{11}\varepsilon_{11}^* A dX_1 = \int_0^L Ea_1 a_1^* A dX_1 = Ea_1 a_1^* AL$$

Equating the internal with the external virtual work:

$$Ea_1 a_1^* AL = Pa_1^* L + \frac{ca_1^* L^3}{3} \Longrightarrow a_1 = \frac{P}{EA} + \frac{cL^2}{3EA}$$

which is exactly the same solution using the Rayleigh Ritz method using a linear–trial function.

Polynomial of the second degree:

Using a polynomial of the second degree, the displacement function has the form:

$$u_1 = a_2 X_1^2 + a_1 X_1 + a_0$$

However, to satisfy the essential boundary condition that $u_1 = 0$ when $X_1 = 0$, the coefficient a_0 has to be set to zero. Then:

$$u_1 = a_2 X_1^2 + a_1 X_1, \quad \varepsilon_{11} = \frac{du_1}{dX_1} = 2a_2 X_1 + a_1$$

Then, an arbitrary smooth virtual displacement field is assumed to have the form:

$$u_1^* = a_2^* X_1^2 + a_1^* X_1, \quad \varepsilon_{11}^* = \frac{du_1^*}{dX_1} = 2a_2^* X_1 + a_1^*$$

The principle of virtual work can now be applied; the work done by the external forces during the arbitrary smooth virtual displacement is:

$$Pu_1^*|_{X_1=L} + \int_0^L cX_1 u_1^* dX_1 = \frac{a_1^* cL^3}{3} + \frac{a_2^* cL^4}{4} + P\left(a_1^* L + a_2^* L^2\right)$$

The internal virtual work is:

$$\int \sum_{i,j=1}^{3} \sigma_{ij}\varepsilon_{ij}^* dV = \int_0^L \sigma_{11}\varepsilon_{11}^* A dX_1 = \int_0^L E\varepsilon_{11}\varepsilon_{11}^* A dX_1$$

$$= a_1 a_1^* EAL + a_2 a_1^* EAL^2 + a_1 a_2^* EAL^2 + \frac{4}{3} a_2 a_2^* EAL^3$$

Equating the internal with the external virtual work and rearranging:

$$\left(\frac{cL^3}{3} + PL - a_1EAL - a_2EAL^2\right)a_1^* + \left(\frac{cL^4}{4} + PL^2 - a_1EAL^2 - \frac{4}{3}a_2EAL^3\right)a_2^* = 0$$

Since the virtual displacement is arbitrary, we can set $a_1^* = 0, a_2^* \neq 0$ or we can set $a_2^* = 0, a_1^* \neq 0$. Thus, each of the coefficients of a_1^* and a_2^* has to be equal to zero, which gives rise to the following set of linear equations in two unknowns a_1 and a_2:

$$\left(\frac{cL^3}{3} + PL - a_1EAL - a_2EAL^2\right) = 0$$

$$\left(\frac{cL^4}{4} + PL^2 - a_1EAL^2 - \frac{4}{3}a_2EAL^3\right) = 0$$

These are exactly the same equations obtained using the Rayleigh Ritz method, and thus, the same solution is obtained for the coefficients a_1 and a_2:

$$a_1 = \frac{(7cL^2 + 12P)}{12EA}, \quad a_2 = -\frac{cL}{4EA}$$

The following Mathematica code is used to solve the above equations. Note the use of the function Coefficient to find the coefficients of a_1^* and a_2^* in the virtual work expressions.

```
u=a2*x^2+a1*x;
eps=D[u,x];
ust=a2st*x^2+a1st*x;
epsst=D[ust,x];
EVW=P*(ust/.x->L)+Integrate[c*x*ust,{x,0,L}];
IVW=Integrate[EA*eps*epsst,{x,0,L}];
VW=EVW-IVW
Eq1=Coefficient[VW,a1st];
Eq2=Coefficient[VW,a2st];
Solve[{Eq1==0,Eq2==0},{a1,a2}]
```

8.1.1.b. Example 2: A Bar under Axial Forces with Varying Cross-Sectional Area

(Compare this example with the example in section 6.4.3.c.) Let a bar with a length 2 m be aligned with the coordinate axis X_1, and let the width of the bar be equal to 250 mm while the height varies linearly such that the height is equal to 500 mm when $X_1 = 0$ and is equal to 250 mm when $X_1 = 2$ m. Assume that the bar also has a constant force of value 200 N applied at the end $X_1 = 2$ m. Assume that the bar is fixed at the end $X_1 = 0$. Find the displacement of the bar by directly solving the differential equation of equilibrium and then using the Rayleigh Ritz method assuming a polynomial of the degrees (0, 1, 2 and 3) (Figure 8–2). Assume the bar to be linear elastic with Young's

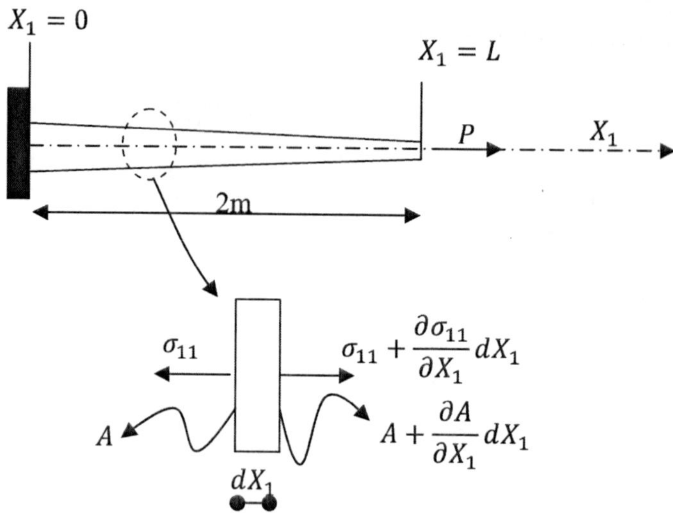

FIGURE 8-2 A bar with a varying cross–sectional area under an axial load.

modulus $E = 100{,}000$ Pa and assume that the small–strain tensor is the appropriate measure of strain. Ignore Poisson's ratio effect. Compare the solution obtained using the Rayleigh Ritz method with the exact solution.

Solution:

Exact solution: The first part involves finding the exact solution that would satisfy the equations of equilibrium. To do so, the equation of equilibrium of this bar is derived, as shown in Figure 8–2:

$$-\sigma_{11}A + +\sigma_{11}A + \sigma_{11}\frac{\partial A_1}{\partial X_1}dX_1 + \frac{\partial \sigma_{11}}{\partial X_1}dX_1 A + \frac{\partial A_1}{\partial X_1}\frac{\partial \sigma_{11}}{\partial X_1}(dX_1)^2 = 0$$

And thus, the equilibrium equation is:

$$\frac{d\sigma_{11}A}{dX_1} = 0$$

The constitutive equation is:

$$\sigma_{11} = E\varepsilon_{11} = E\frac{du_1}{dX_1}$$

Thus, the equilibrium equation in terms of the displacement function is:

$$EA\frac{d^2u_1}{dX_1^2} + E\frac{du_1}{dX_1}\frac{dA}{dX_1} = 0$$

The area varies linearly with X_1 according to the equation:

$$A = 0.25\left(0.5 - \frac{0.25}{2}X_1\right)$$

The two boundary conditions are:

$$X_1 = 0 \Longrightarrow u_1 = 0$$

$$X_1 = L \Longrightarrow \sigma_{11} = E\frac{du_1}{dX_1} = \frac{P}{A|_{X_1=L}}$$

Thus, the exact solution using Mathematica is:

$$u_1 = \frac{8}{125}(\ln 4 - \ln(4 - X_1))$$

$$\sigma_{11} = \frac{6400}{4 - X_1}$$

The Rayleigh Ritz method:

As in the previous example, we will choose a displacement function that would satisfy that the displacement is equal to zero when $X_1 = 0$, and thus, a polynomial of the zero degree cannot be used. Using a polynomial of the first degree, the trial function has the following form:

$$u_1 = a_1 X_1, \quad \frac{du_1}{dX_1} = a_1$$

The potential energy of the system is:

$$\Pi = \int \bar{U} dV - W = \int_0^L \frac{EA}{2}\left(\frac{du_1}{dX_1}\right)^2 dX_1 - Pu_1|_{X_1=L} = \int_0^L \frac{E0.25(0.5 - \frac{0.25}{2}X_1)}{2} a_1^2 dX_1 - Pa_1 L$$

The coefficient a_1 can be obtained by setting:

$$\frac{\partial \Pi}{\partial a_1} = 0$$

Similar to the previous example, the same procedure should be followed for higher degree poly–nomials. The following Mathematica code is used for all the calculations using polynomials of the different degrees for the displacement function:

```
Clear[u, u1, u2, u3, u4, stress, stress1, stress2, stress3, x]
(*Exact Solution *)
Ax=25/100(5/10-25/200*x);
Ee=100000;
L=2;
P=200;
s=DSolve[{Ee*u''[x]*Ax+Ee*u'[x]*D[Ax,x]==0,u'[L]==200/Ee/(Ax/.x-
>L),u[0]==0},u,x]
```

```
u=u/.s[[1]];
stress=Ee*D[u[x],x]

(*First Degree *)
u1=a1*x;
PE=Integrate[1/2*Ee*D[u1,x]^2*Ax,{x,0,L}]-P*u1/.x->L;
Eq1=D[PE,a1];
s=Solve[Eq1==0,a1];
u1=u1/.s[[1]];
stress1=Ee*D[u1,x];

(*Second Degree *)
u2=a1*x+a2*x^2;
PE=Integrate[1/2*Ee*D[u2,x]^2*Ax,{x,0,L}]-P*u2/.x-> L;
Eq1=D[PE,a1];
Eq2=D[PE,a2];
s=Solve[{Eq1==0,Eq2==0},{a1,a2}];
u2=u2/.s[[1]];
stress2=Ee*D[u2,x];

(*Third Degree *)
u3=a1*x+a2*x^2+a3*x^3;
PE=Integrate[1/2*Ee*D[u3,x]^2*Ax,{x,0,L}]-P*u3/.x->L;
Eq1=D[PE,a1];
Eq2=D[PE,a2];
Eq3=D[PE,a3];
s=Solve[{Eq1==0,Eq2==0,Eq3==0},{a1,a2,a3}];
u3=u3/.s[[1]];
stress3=Ee*D[u3,x];
(*comparison*)
Needs["PlotLegends'"];
Plot[{u[x],u1,u2,u3},{x,0,L},PlotLegend->{Style["Exact",Bold,12],Style["u1",
Bold,12],Style["u2",Bold,12],Style["u3",Bold,12]},AxesLabel-> {"X1 m","u(m)"},
BaseStyle->Directive[Bold,12],LegendPosition->{1.1,-0.4},PlotStyle->
{Directive[Black,Dashing[0.005],Thickness[0.005]],Directive[Black,Dashing[0.01],
Thickness[0.01]],Directive[Black,Dashing[0.015],Thickness[0.015]],Directive
[Black,Dashing[0.1],Thickness[0.02]]}]
Plot[{stress,stress1,stress2,stress3},{x,0,L},AxesLabel-> {"X1
m","stress(Pa)"},BaseStyle->Directive[Bold,12],PlotLegend-
>{Style["Exact",Bold,12],Style["u1",Bold,12],Style["u2",Bold,12],Style["u3",
Bold,12]},LegendPosition-> {1.1,-0.4},PlotStyle->{Directive[Black,
Dashing[0.005],Thickness[0.005]],Directive[Black,Dashing[0.01],Thickness[0.01]],
Directive[Black,Dashing[0.015],Thickness[0.015]],Directive[Black,Dashing[0.1],
Thickness[0.02]]}]
```

The exact solution obtained is a logarithmic function, while the approximations used for the Rayleigh Ritz method are polynomial functions. The Taylor series of a logarithmic function can be represented by a linear combination of polynomial functions, and so, the higher the degree of approximation, the closer the approximation to the exact solution. As seen in Figure 8–3, good

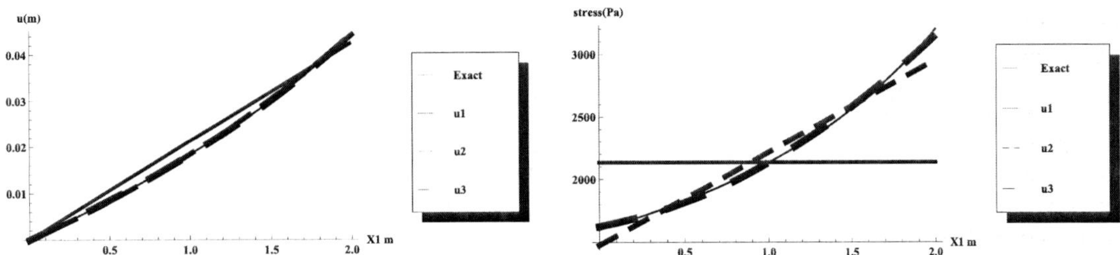

FIGURE 8-3 Comparison between the exact solution and the different approximations using the Rayleigh Ritz method for the (a) displacements and (b) stresses.

accuracy is obtained for a polynomial of the second degree for the displacement function; however, the stresses in that case have higher errors. A better approximation for the stresses is obtained using a polynomial of the third degree. In general, the error in the approximate solution is better than the error in its derivatives.

The principle of Virtual Work:

Follow the same procedure as in the previous example (section 8.1.1.a) to verify that the same approximate solution obtained using the Rayleigh Ritz method can be obtained using the principle of virtual work when the same displacement approximation function is used.

8.1.1.c. Example 3: A Bar under Varying Body Force with Fixed Ends

Let a bar with a length 2 m be aligned with the coordinate axis X_1, and let the cross–sectional area A of the bar be equal to $250 \times 250\ \text{mm}^2$. Assume that the bar is fixed at both ends $X_1 = 0$ and $X_1 = 2$ m (Figure 8–4). Assume that the bar is subjected to a varying body force that is equal to $q = 5X_1^2$ N/m in the direction of the coordinate axis X_1. Find the displacement of the bar by directly solving the differential equation of equilibrium and then using the Rayleigh Ritz method assuming a polynomial of the degrees (0, 1, 2, 3, and 4). Assume that the bar material is linear elastic with Young's modulus $E = 100,000$ Pa and that the small strain tensor is the appropriate measure of strain. Ignore Poisson's ratio effect. Compare the solution obtained using the Rayleigh Ritz method using three– and four–degree polynomials as approximations with the exact solution.

Solution:

Exact solution: Similar to section 8.1.1.a, the equilibrium equation is:

$$\frac{d\sigma_{11}}{dX_1} = -\frac{q}{A} = -\frac{5X_1^2}{0.25^2}$$

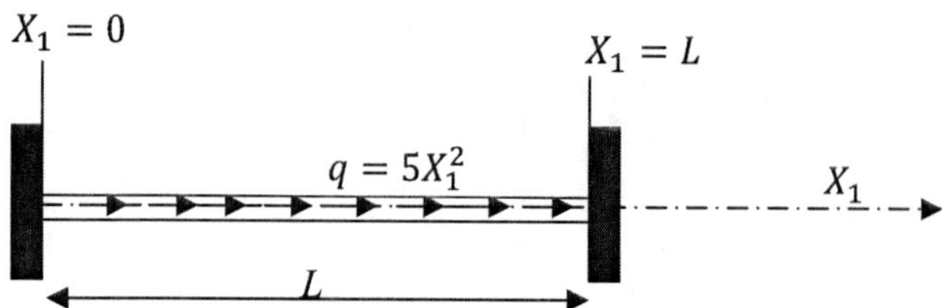

FIGURE 8-4 Bar under a varying body axial load with both ends fixed.

Using the constitutive relationship: $\sigma_{11} = E\varepsilon_{11} = E\frac{du_1}{dX_1}$ the ordinary differential equation to be solved is:

$$E\frac{d^2u_1}{dX_1^2} = -\frac{5X_1^2}{0.25^2}$$

Then the solution for u_1 is:

$$u_1 = -\frac{5X_1^4}{12E(0.25)^2} + C_1X_1 + C_2$$

The boundary conditions for this example are different from the previous two examples in that the displacement is held at zero at both ends:

$$X_1 = 0 \Longrightarrow u_1 = 0 \Rightarrow C_2 = 0$$

$$X_1 = L \Longrightarrow u_1 = 0 \Rightarrow C_1 = \frac{5L^3}{12E(0.25)^2}$$

Thus, the exact solutions for the displacement and the stress functions are:

$$u_1 = -\frac{5X_1^4}{12E(0.25)^2} + \frac{5L^3}{12E(0.25)^2}X_1 = \frac{8X_1 - X_1^4}{15000}\,\text{m}$$

$$\sigma_{11} = \frac{20}{3}\left(8 - 4X_1^3\right) \qquad \text{N/m}^2$$

The following Mathematica code can also be used to solve the differential equation of equilibrium (Notice that, for exact results, the use of decimal points is avoided in the expression for the area.):

```
(*Exact Solution*)
Clear[u,x]
A=(25/100)^2;
Ee=100000;
L=2;
q=5x^2;
s=DSolve[{Ee*u''[x]==-q/A,u[L]==0,u[0]==0},u,x]
u=u/.s[[1]];
stress=Ee*D[u[x],x]
```

The Rayleigh Ritz method

Polynomial of the zero degree:

Let the trial function for u_1 be a constant function (a polynomial of the zero degree):

$$u_1 = a_0$$

The "essential" boundary condition that $u_1 = 0$ when $X_1 = 0$ leads to automatically rejecting this trial function.

Polynomial of the first degree:

Let the trial function for u_1 be a linear function (a polynomial of the first degree):

$$u_1 = a_1 X_1 + a_0$$

The essential boundary condition that $u_1 = 0$ at $X_1 = 0$ leads to the coefficient a_0 being equal to zero. In addition, to satisfy the essential boundary condition that $u_1 = 0$ when $X_1 = 2\,\text{m}$, then a_1 has to be set to zero as well. Therefore, this solution is automatically rejected.

Polynomial of the second degree:

$$u_1 = a_2 X_1^2 + a_1 X_1 + a_0, \quad \frac{du_1}{dX_1} = 2a_2 X_1 + a_1$$

Once again, the essential boundary conditions need to be satisfied; therefore:

$$X_1 = 0 \Rightarrow u_1 = 0 \Rightarrow a_0 = 0$$

$$X_1 = L \Rightarrow u_1 = 0 \Rightarrow a_2 = -\frac{a_1}{L}$$

Thus, the displacement function has only one parameter a_1 that can be controlled:

$$u_1 = a_1 X_1 \left(1 - \frac{X_1}{L}\right), \quad \frac{du_1}{dX_1} = a_1 \left(1 - \frac{2X_1}{L}\right)$$

The next step is to write the potential energy of the system:

$$\Pi = \int_0^L \frac{EA}{2} a_1^2 \left(1 - \frac{2X_1}{L}\right)^2 dX_1 - \int_0^L 5a_1 X_1^3 \left(1 - \frac{X_1}{L}\right) dX_1 = \frac{6250a_1^2}{3} - 4a_1$$

The coefficient a_1 can be obtained by minimizing Π:

$$\frac{\partial \Pi}{\partial a_1} = \frac{12500}{3} a_1 - 4 = 0 \Rightarrow a_1 = \frac{3}{3125}$$

Therefore, the "best" solutions for the displacement and stress functions in the space of second–degree polynomials are:

$$u_1 = \frac{3X_1}{3125}\left(1 - \frac{X_1}{2}\right)$$

$$\sigma_{11} = 96\,(1 - X_1)$$

The Mathematica code that is used for the above calculations is:

```
(*Second Degree *)
A=(25/100)^2;
Ee=100000;
u2=a1*x(1-x/L);
q=5x^2;
PE=Integrate[1/2*Ee*D[u2,x]^2*A,{x,0,L}]-Integrate[q*u2,{x,0,L}]
Eq1=D[PE,a1]
s=Solve[Eq1==0,a1]
u2=FullSimplify[u2/.s[[1]]]
stress2=FullSimplify[Ee*D[u2,x]]
```

Polynomials of the third and fourth degree:

The results from the following Mathematica code show that the third–degree approximation is very close to the actual solution for both displacements and stresses. The polynomial of the fourth–degree approximation is in fact exact since the exact solution is a polynomial of the fourth degree. Study the Mathematica code carefully to understand how the second boundary condition of displacement is incorporated ($u_1 = 0$ at $X_1 = L$). The graph comparing the results is shown in Figure 8–5.

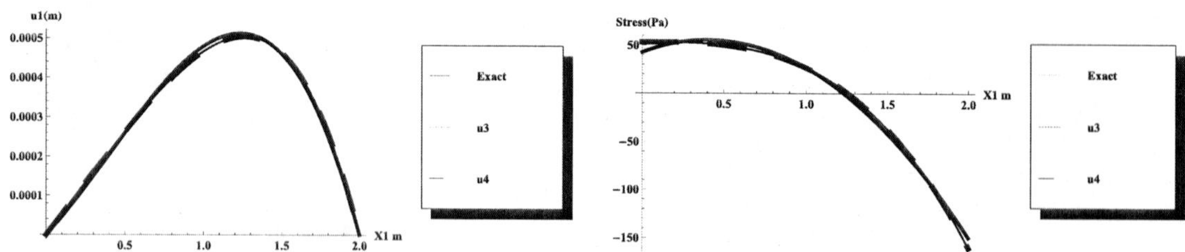

FIGURE 8-5 Comparison between the exact solution and the different approximations using the Rayleigh Ritz method for the (a) displacements and (b) stresses for problem in section 8.1.1.c.

```
(*Exact Solution*)
Clear[u,x,a1,a2,a3,a4,a5,a0,Ee,A,L,q,u3,u4]
q=5x^2;
s=DSolve[{Ee*u''[x]==-q/A,u[L]==0,u[0]==0},u,x];
u=u/.s[[1]];
stress=Ee*D[u[x],x];

(*Third Degree *)
u3=a3*x^3+a2*x^2+a1*x;
sol=Solve[(u3/.x->L)==0,a3];
u3=u3/.sol[[1]];
PE=1/2*Integrate[Ee*A*D[u3,x]^2,{x,0,L}]-Integrate[q*u3,{x,0,L}];
Eq1=D[PE,a1];
Eq2=D[PE,a2];
sol1=Solve[{Eq1==0,Eq2==0},{a2,a1}];
u3=FullSimplify[u3/.sol1[[1]]];
stress3=FullSimplify[Ee*D[u3,x]];

(*Fourth Degree *)
u4=a4*x^4+a3*x^3+a2*x^2+a1*x;
sol=Solve[(u4/.x->L)==0,a4]
u4=u4/.sol[[1]];
PE=1/2*Integrate[Ee*A*D[u4,x]^2,{x,0,L}]-Integrate[q*u4,{x,0,L}];
Eq1=D[PE,a1];
Eq2=D[PE,a2];
Eq3=D[PE,a3];
sol1=Solve[{Eq1==0,Eq2==0,Eq3==0},{a1,a2,a3}];
u4=FullSimplify[u4/.sol1[[1]]];
stress4=FullSimplify[Ee*D[u4,x]];
A=(25/100)^2;
Ee=100000;
L=2;

(*comparison*)
Needs["PlotLegends'"];
Plot[{u[x],u3,u4},{x,0,L},PlotLegend-
>{Style["Exact",Bold,12],Style["u3",Bold,12],Style["u4",Bold,12]},AxesLabel-
>{"X1 m","u1(m)"},PlotLegend->{"Exact","u3","u4"},BaseStyle->Directive[Bold,12],
LegendPosition->{1.1,-0.4},PlotStyle->{Directive[Black,Dashing[0.005],
Thickness[0.005]],Directive[Black,Dashing[0.01],Thickness[0.01]],Directi
ve[Black,Dashing[0.1],Thickness[0.015]]}]
Plot[{stress,stress3,stress4},{x,0,L},PlotLegend-
>{Style["Exact",Bold,12],Style["u3",Bold,12],Style["u4",Bold,12]},AxesLabel-
>{"X1 m","Stress(Pa)"},PlotLegend->{"Exact","u3","u4"},LegendPosition->{1.1,
-0.4},BaseStyle->Directive[Bold,12],PlotStyle->{Directive[Black,Dashing[0.005],
Thickness[0.005]],Directive[Black,Dashing[0.01],Thickness[0.01]],Directi
ve[Black,Dashing[0.1],Thickness[0.015]]}]
```

The Principle of Virtual Work:

It will be shown here that the principle of virtual work using a polynomial of the second degree for the displacement function will yield the same solution as the Rayleigh Ritz method using a polynomial of the second degree:

$$u_1 = a_2X_1^2 + a_1X_1 + a_0, \quad \frac{du_1}{dX_1} = 2a_2X_1 + a_1$$

The chosen displacement function has to satisfy the essential boundary conditions; therefore:

$$X_1 = 0 \Rightarrow u_1 = 0 \Rightarrow a_0 = 0$$

$$X_1 = L \Rightarrow u_1 = 0 \Rightarrow a_2 = -\frac{a_1}{L}$$

Therefore, the displacement function has only one parameter a_1 that can be controlled:

$$u_1 = a_1X_1\left(1 - \frac{X_1}{L}\right), \quad \varepsilon_{11} = \frac{du_1}{dX_1} = a_1\left(1 - \frac{2X_1}{L}\right)$$

The next step is to assume an arbitrary, smooth, and small virtual displacement function:

$$u_1^* = a_1^*X_1\left(1 - \frac{X_1}{L}\right), \quad \varepsilon_{11}^* = \frac{du_1^*}{dX_1} = a_1^*\left(1 - \frac{2X_1}{L}\right)$$

The principle of virtual work can now be applied; the work done by the external forces during the arbitrary smooth virtual displacement is:

$$\int_0^L 5X_1^2u_1^*dX_1 = 5a_1^*\left(\frac{L^4}{4} - \frac{L^4}{5}\right) = \frac{a_1^*L^4}{4}$$

The internal virtual work is:

$$\int \sum_{i,j=1}^{3} \sigma_{ij}\varepsilon_{ij}^*dV = \int_0^L \sigma_{11}\varepsilon_{11}^*AdX_1 = \int_0^L E\varepsilon_{11}\varepsilon_{11}^*AdX_1 = \frac{a_1a_1^*EAL^3}{3}$$

Equating the internal with the external virtual work and rearranging:

$$\left(\frac{L^4}{4} - \frac{a_1EAL^3}{3}\right)a_1^* = 0$$

Since the virtual displacement is arbitrary, we can set $a_1^* \neq 0$. Thus, the coefficient of a_1^* has to be equal to zero, which will give the following solution to the coefficient a_1:

$$a_1 = \frac{3L}{4EA} = \frac{3}{3125}$$

This is exactly the same solution obtained using the Rayleigh Ritz method. The following Math–ematica code is used to solve the above equations. Note the use of the function `Coefficient` to find the coefficient of a_1^* in the virtual work expressions. Verify that similar results will be obtained for higher–order polynomial approximation functions.

```
Clear[L,EA,a1,u]
u=a1*x*(1-x/L);
eps=D[u,x];
ust=a1st*x*(1-x/L);
epsst=D[ust,x];
EVW=Integrate[5*x^2*ust,{x,0,L}];
IVW=Integrate[EA*eps*epsst,{x,0,L}];
VW=EVW-IVW;
Eq1=Coefficient[VW,a1st];
s=Solve[{Eq1==0},{a1}]
a1=a1/.s[[1]]
a1/.{L-> 2,EA->100000*(25/100)^2}
```

8.1.2. Euler Bernoulli Beams under Lateral Loading

A few examples to solve for the displacements, rotations, shear forces, and moments for an Euler Bernoulli beam under lateral loading are presented in this section. Similar to the previous section, the Rayleigh Ritz method starts by choosing a form for the displacement function. This form has to satisfy the "Essential" boundary conditions, which for this particular type of problems are those boundary conditions imposed on the primary variables (the displacements and the rotations). Similar to the previous section, verify that the virtual work principle can also be used to generate the same approximate solutions.

8.1.2.a. Example 1: Cantilever Beam

Let a beam with a length L and a constant cross–sectional parameters area A and moment of inertia I be aligned with the coordinate axis X_1. Assume that the beam is subjected to a constant force of value P applied at the end $X_1 = L$ in the direction shown in Figure 8–6. Assume that the bar is fixed (rotation and displacement are equal to zero) at the end $X_1 = 0$. Find the displacement, rotation, bending moment, and shearing forces on the beam by directly solving the differential equation of equilibrium and then using the Rayleigh Ritz method assuming a polynomial of the degrees (0, 1, 2, 3, and 4). Assume the beam to be linear elastic with Young's modulus E and that the Euler Bernoulli Beam is an appropriate beam approximation.

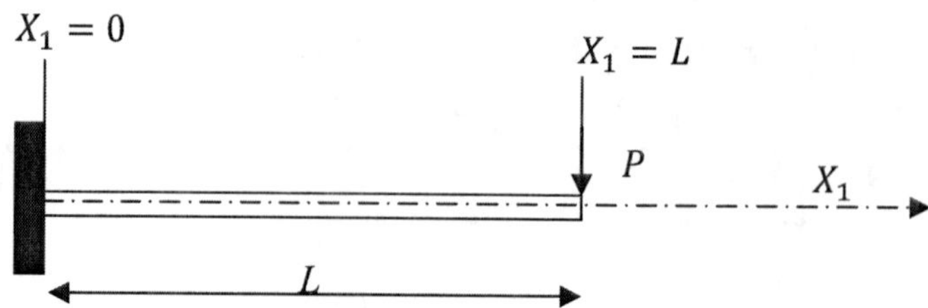

FIGURE 8-6 Cantilever beam with a concentrated load at the tip.

Solution:

Exact solution: The differential equation of equilibrium for the Euler Bernoulli beam is:

$$EI\frac{d^4y}{dX_1^4} = q$$

For this particular problem, $q = 0$; therefore, the solution to this differential equation is:

$$y = \frac{C_1X_1^3}{6} + \frac{C_2X_1^2}{2} + C_3X_1 + C_4$$

The following four boundary conditions: zero displacement and zero rotation at $X_1 = 0$, zero moment at $X_1 = L$ and a shear of value P at the end $X_1 = L$ can be used to find the values of the unknown coefficients C_1, C_2, C_3 and C_4:

$$X_1 = 0 \Longrightarrow y = 0 \Longrightarrow \frac{C_1(0)^3}{6} + \frac{C_2(0)^2}{2} + C_3(0)^3 + C_4 = 0$$

$$X_1 = 0 \Longrightarrow \frac{dy}{dX_1} = 0 \Longrightarrow \frac{C_1(0)^2}{2} + C_2(0) + C_3 = 0$$

$$X_1 = L \Longrightarrow M = 0 \Longrightarrow EI\frac{d^2y}{dX_1^2} = 0 \Longrightarrow C_1(L) + C_2 = 0$$

$$X_1 = L \Longrightarrow V = P \Longrightarrow EI\frac{d^3y}{dX_1^3} = P \Longrightarrow C_1 = \frac{P}{EI}$$

The values of the four coefficients can be found from the above equations, as follows:

$$C_4 = 0, \quad C_3 = 0, \quad C_2 = -\frac{PL}{EI}, \quad C_1 = \frac{P}{EI}$$

Therefore, the equations for the displacement, rotation, moment, and shear are:

$$y = \frac{PX_1^3}{6EI} - \frac{PLX_1^2}{2EI}$$

$$\frac{dy}{dX_1} = \frac{PX_1^2}{2EI} - \frac{PLX_1}{EI}$$

$$M = EI\frac{d^2y}{dX_1^2} = PX_1 - PL$$

$$V = EI\frac{d^3y}{dX_1^3} = P$$

The following Mathematica code was used to solve the above equations:

```
Clear[a,x,V,M,y,L,EI]
M=EI*D[y[x],{x,2}];
V=EI*D[y[x],{x,3}];
th=D[y[x],x];
q=0;
M1=M/.x->0;
M2=M/.x->L;
V1=V/.x->0;
V2=V/.x->L;
y1=y[x]/.x->0;
y2=y[x]/.x->L;
th1=th/.x->0;
th2=th/.x->L;
s=DSolve[{EI*y''''[x]==q,y1==0,th1==0,V2==P,M2==0},y,x];
y=y/.s[[1]]
th
M
V
```

The Rayleigh Ritz method

Polynomial of the zero degree:

Let y be a constant function (a polynomial of the zero degree):

$$y = a_0$$

This assumption however leads to the trivial solution $y = 0$, and thus, is automatically rejected.

Polynomial of the first degree:

Let y be a linear function (a polynomial of the first degree):

$$y = a_1X_1 + a_0$$

However, to satisfy the essential boundary condition that $\frac{dy}{dX_1} = 0$ when $X_1 = 0$, the coefficient a_1 has to be set to zero. In addition, to satisfy the essential boundary condition that $y = 0$ when $X_1 = 0$ m, then a_0 has to be set to zero as well. Therefore, this solution is automatically rejected.

Polynomial of the second degree:

If the trial function for y is assumed to be a cubic polynomial:

$$y = a_2X_1^2 + a_1X_1 + a_0, \quad \frac{dy}{dX_1} = 2a_2X_1 + a_1, \quad M = EI\frac{d^2y}{dX_1^2} = 2EIa_2, \quad V = 0$$

then to satisfy the essential boundary conditions, a_0 and a_1 have to be set to equal to zero:

$$X_1 = 0 \Rightarrow y = 0 \Rightarrow a_0 = 0$$

$$X_1 = 0 \Rightarrow \frac{dy}{dX_1} = 0 \Rightarrow a_1 = 0$$

Therefore, the displacement function has only one parameter a_2 that can be controlled:

$$y = a_2X_1^2, \quad \frac{dy}{dX_1} = 2a_2X_1, \quad M = EI\frac{d^2y}{dX_1^2} = 2EIa_2, \quad V = 0$$

The next step is to write the potential energy of the system:

$$\Pi = \frac{EI}{2}\int_0^L \left(\frac{d^2y}{dX_1^2}\right)^2 dX_1 - (-P)\,y|_{X_1=L} = \frac{EI}{2}\int_0^L (2a_2)^2\,dX_1 + Pa_2L^2 = 2a_2^2EIL + Pa_2L^2$$

The coefficient a_2 can be obtained by minimizing Π:

$$\frac{\partial\Pi}{\partial a_2} = 4a_2EIL + PL^2 = 0 \Rightarrow a_2 = \frac{-PL}{4EI}$$

Therefore, the "best" solutions for the displacement, rotation, moment, and shear are:

$$y = -\frac{PL}{4EI}X_1^2, \quad \frac{dy}{dX_1} = -\frac{PL}{2EI}X_1, \quad M = EI\frac{d^2y}{dX_1^2} = -\frac{PL}{2}, \quad V = 0$$

Polynomials of the third degree:

Assuming that the displacement is a polynomial of the third degree:

$$y = a_3X_1^3 + a_2X_1^2 + a_1X_1 + a_0, \quad \frac{dy}{dX_1} = 3a_3X_1^2 + 2a_2X_1 + a_1,$$

$$M = EI\frac{d^2y}{dX_1^2} = 6EIa_3X_1 + 2EIa_2, \quad V = 6EIa_3$$

then to satisfy the essential boundary conditions, the coefficients a_0 and a_1 are equal to zero:

$$X_1 = 0 \Rightarrow y = 0 \Rightarrow a_0 = 0$$

$$X_1 = 0 \Rightarrow \frac{dy}{dX_1} = 0 \Rightarrow a_1 = 0$$

Therefore, the displacement function has only two parameters a_2 and a_3 that can be controlled:

$$y = a_3X_1^3 + a_2X_1^2, \quad \frac{dy}{dX_1} = 3a_3X_1^2 + 2a_2X_1, \quad M = EI\frac{d^2y}{dX_1^2} = 6EIa_3X_1 + 2EIa_2, \quad V = 6EIa_3$$

The next step is to write the potential energy of the system:

$$\Pi = \frac{EI}{2}\int_0^L \left(\frac{d^2y}{dX_1^2}\right)^2 dX_1 - (-P)\,y|_{X_1=L} = \frac{EI}{2}\int_0^L (6a_3X_1 + 2a_2)^2\, dX_1 + P\left(a_3L^3 + a_2L^2\right)$$

The coefficients a_2 and a_3 can be obtained by minimizing Π:

$$\frac{\partial \Pi}{\partial a_2} = \frac{EI}{2}\left(8a_2L + 12a_3L^2\right) + PL^2 = 0$$

$$\frac{\partial \Pi}{\partial a_3} = \frac{EI}{2}\left(12a_2L^3 + 24a_3L^3\right) + PL^3 = 0$$

Solving the above two equations for the parameters a_2 and a_3 leads to:

$$a_2 = -\frac{PL}{2EI}, \quad a_3 = \frac{P}{6EI}$$

Therefore, the "best" solutions for the displacement, rotation, moment, and shear are:

$$y = \frac{PX_1^3}{6EI} - \frac{PLX_1^2}{2EI}, \quad \frac{dy}{dX_1} = \frac{PX_1^2}{2EI} - \frac{PLX_1}{EI}, \quad M = EI\frac{d^2y}{dX_1^2} = PX_1 - PL, \quad V = EI\frac{d^3y}{dX_1^3} = P$$

The exact solution for the displacement function is in fact a polynomial of the third degree. There–fore, the two approximations (polynomials of the third and fourth degrees) are expected to give the exact solution.

Polynomials of the fourth degree:

Assuming that the displacement is a polynomial of the fourth degree:

$$y = a_4X_1^4 + a_3X_1^3 + a_2X_1^2 + a_1X_1 + a_0, \quad \frac{dy}{dX_1} = 4a_4X_1^3 + 3a_3X_1^2 + 2a_2X_1 + a_1,$$

$$M = EI\frac{d^2y}{dX_1^2} = 12a_4X_1^2 + 6EIa_3X_1 + 2EIa_2, \quad V = 24a_4X_1 + 6EIa_3,$$

then the essential boundary conditions need to be satisfied, leads to:

$$X_1 = 0 \Rightarrow y = 0 \Rightarrow a_0 = 0 \quad X_1 = 0 \Rightarrow \frac{dy}{dX_1} = 0 \Rightarrow a_1 = 0$$

Therefore, the displacement function has the parameters a_2, a_3 and a_4 that can be controlled:

$$y = a_4X_1^4 + a_3X_1^3 + a_2X_1^2, \quad \frac{dy}{dX_1} = 4a_4X_1^3 + 3a_3X_1^2 + 2a_2X_1,$$

$$M = EI\frac{d^2y}{dX_1^2} = 12a_4X_1^2 + 6EIa_3X_1 + 2EIa_2, \quad V = 24a_4X_1 + 6EIa_3$$

The next step is to write the potential energy of the system:

$$\Pi = \frac{EI}{2}\int_0^L \left(\frac{d^2y}{dX_1^2}\right)^2 dX_1 - (-P)\,y|_{X_1=L}$$

The coefficients a_2 and a_3 can be obtained by minimizing Π:

$$\frac{\partial\Pi}{\partial a_2} = 0, \quad \frac{\partial\Pi}{\partial a_3} = 0, \quad \frac{\partial\Pi}{\partial a_4} = 0$$

Solving the above three equations for the parameters a_2, a_3 and a_4: $a_2 = -\frac{PL}{2EI}$, $a_3 = \frac{P}{6EI}$, $a_4 = 0$.

This is exactly the same solution obtained using a third degree polynomial ($a_4 = 0$)!

The Mathematica code for the above calculations is:

```
(*Third Degree*)
Clear[y,x,M,V,th,EI,PE]
y=a3*x^3+a2*x^2
th=D[y,x];
M=EI*D[y,{x,2}];
V=EI*D[y,{x,3}];
PE=EI/2*Integrate[D[y,{x,2}]^2,{x,0,L}]+P*(y/.x->L);
Eq1=D[PE,a3];
Eq2=D[PE,a2];
Sol=Solve[{Eq1==0,Eq2==0},{a3,a2}]
y=y/.Sol[[1]]
th/.Sol[[1]]
FullSimplify[M/.Sol[[1]]]
FullSimplify[V/.Sol[[1]]]

(*Fourt Degree*)
Clear[y,x,M,V,th,EI,PE]
y=a4*x^4+a3*x^3+a2*x^2
th=D[y,x];
M=EI*D[y,{x,2}];
V=EI*D[y,{x,3}];
PE=EI/2*Integrate[D[y,{x,2}]^2,{x,0,L}]+P*(y/.x->L);
Eq1=D[PE,a3];
```

```
Eq2=D[PE,a2];
Eq3=D[PE,a4];
Sol=Solve[{Eq1==0,Eq2==0,Eq3==0},{a3,a2,a4}]
y=y/.Sol[[1]]
th/.Sol[[1]]
FullSimplify[M/.Sol[[1]]]
FullSimplify[V/.Sol[[1]]]
```

8.1.2.b. Example 2: Simple Beam

Let a beam with a length $L = 10$ m and a constant moment of inertial $I = 4 \times 10^8$ mm^4 be aligned with the coordinate axis X_1. Assume that the beam is subjected tó a distributed load $q = -25$ kN/m applied in the direction shown in Figure 8–7 (Notice that the negative sign was added since q is positive when it is in the direction of X_2). Assume that the beam is hinged at the ends $X_1 = 0$ and $X_1 = L$. Find the displacement, rotation, bending moment, and shearing forces on the beam by directly solving the differential equation of equilibrium and then using the Rayleigh Ritz method assuming a polynomial of the degrees (0, 1, 2, 3 and 4). Assume the beam to be linear elastic with Young's modulus $E = 200$ GPa and that the Euler Bernoulli Beam is an appropriate beam approx–imation. Compare the Rayleigh Ritz method with the exact solution.

Solution:

Exact solution: The differential equation of equilibrium for the Euler Bernoulli beam is:

$$EI\frac{d^4y}{dX_1^4} = q$$

For this particular problem, $q = -25$ kN/m; therefore, the solution to this differential equation is:

$$EIy = \frac{qX_1^4}{24} + \frac{C_1X_1^3}{6} + \frac{C_2X_1^2}{2} + C_3X_1 + C_4$$

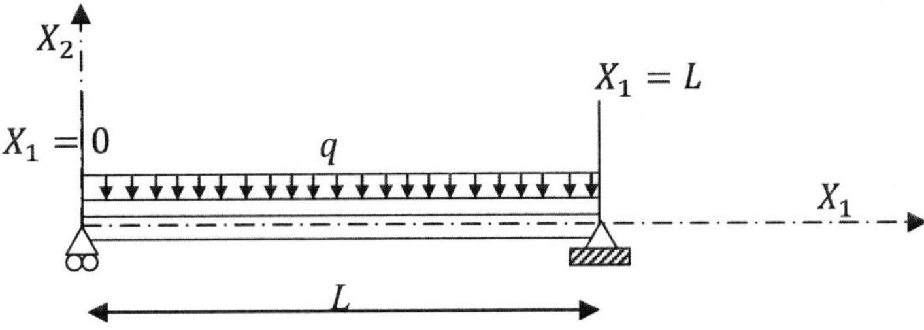

FIGURE 8-7 Simple beam with downward distributed load.

The four boundary conditions: zero displacement at $X_1 = 0$ and $X_1 = L$, zero moment at $X_1 = 0$ and $X_1 = L$ can be used to find the values of the unknown coefficients C_1, C_2, C_3 and C_4:

$$X_1 = 0 \Longrightarrow y = 0 \Longrightarrow \frac{q(0)^4}{24} + \frac{C_1(0)^3}{6} + \frac{C_2(0)^2}{2} + C_3(0)^3 + C_4 = 0$$

$$X_1 = 0 \Longrightarrow EI\frac{d^2y}{dX_1^2} = 0 \Longrightarrow \frac{q(0)^2}{2} + C_1(0) + C_2 = 0$$

$$X_1 = L \Longrightarrow y = 0 \Longrightarrow EI\frac{d^2y}{dX_1^2} = 0 \Longrightarrow \frac{q(L)^2}{2} + C_1(L) + C_2 = 0$$

$$X_1 = L \Longrightarrow y = 0 \Longrightarrow \frac{q(L)^4}{24} + \frac{C_1(L)^3}{6} + \frac{C_2(L)^2}{2} + C_3(L)^3 + C_4 = 0$$

Solving the above four equations for the four coefficients leads to:

$$C_4 = 0, \quad C_3 = \frac{qL^3}{24}, \quad C_2 = 0, \quad C_1 = -\frac{qL}{2}$$

Therefore, the equations for the displacement, rotation, moment, and shear are:

$$y = \frac{qX_1^4}{24EI} - \frac{qLX_1^3}{12EI} + \frac{qL^3X_1}{24EI}, \quad \frac{dy}{dX_1} = \frac{qX_1^3}{6EI} - \frac{qLX_1^2}{4EI} + \frac{qL^3}{24EI}, \quad M = EI\frac{d^2y}{dX_1^2} = \frac{qX_1^2}{2} - \frac{qLX_1}{2}$$

$$V = EI\frac{d^3y}{dX_1^3} = qX_1 - \frac{qL}{2}$$

The Rayleigh Ritz method

Polynomial of the zero and of the first degree:

Similar to the previous example, this solution is automatically rejected since there are two "essen–tial" boundary conditions of zero displacement at both ends of the beam that have to be satisfied.

Polynomials of higher degrees:

The same procedure will be followed according to the following steps:

- Assume a polynomial function for the displacement function y.
- Find values for the coefficients by satisfying the essential boundary conditions of displacement and rotation, which in this case are: $X_1 = 0 \Rightarrow y = 0$ and $X_1 = L \Rightarrow y = 0$.
- The next step is to write the potential energy of the system in terms of the unknown coefficients of the polynomial:

$$\Pi = \frac{EI}{2}\int_0^L \left(\frac{d^2y}{dX_1^2}\right)^2 dX_1 - \int_0^L qy dX_1$$

- The coefficients are obtained by minimizing Π: (by solving the set of equations $\frac{\partial \Pi}{\partial a_i} = 0$).

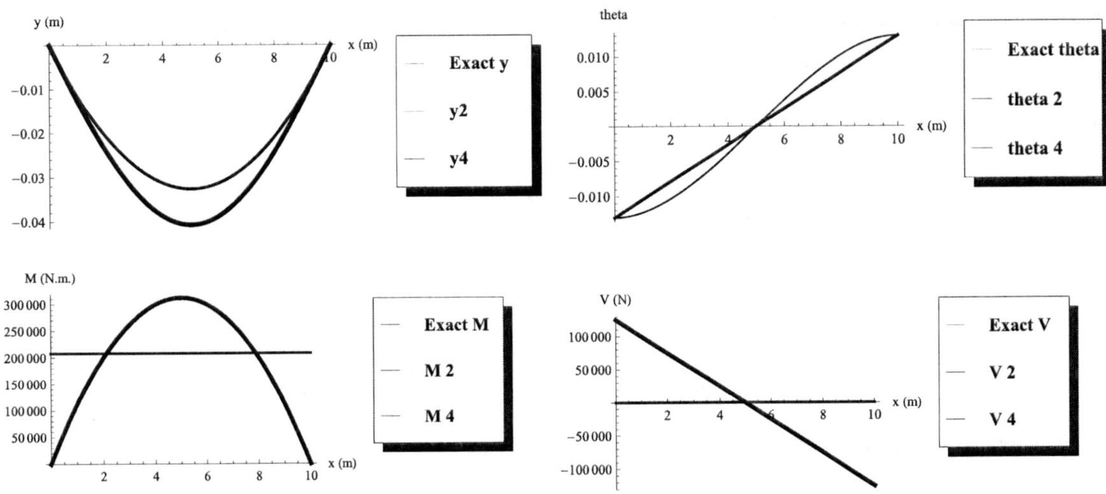

FIGURE 8-8 Comparison between the solution using the Rayleigh Ritz method and the exact solution for a simple beam. (The units used are metre for distances and Newton for forces)

Note that the second– and third–degree trial functions give the same result, while the fourth– and fifth–degree trial functions give the same result. The solution is shown to give the accurate result when a polynomial of the fourth or fifth degree is used. Figure 8–8 shows the comparison between the exact solution and the polynomial of second–degree trial function. The Mathematica code for comparison is:

```
(*Exact Solution*)
Clear[q,EI,y,x,c1,c2,c3,c4,L]
y=(q*x^4/24+c1*x^3/6+c2*x^2/2+c3*x+c4)/EI;
Eq1=y/.x-> 0;
Eq2=y/.x-> L;
Eq3=EI*D[y,{x,2}]/.x-> 0;
Eq4=EI*D[y,{x,2}]/.x-> L;
a=Solve[{Eq1==0,Eq2==0,Eq3==0,Eq4==0},{c1,c2,c3,c4}];
y=y/.a[[1]];
theta=D[y,x]/.a[[1]];
M=FullSimplify[EI*D[y,{x,2}]/.a[[1]]];
V=FullSimplify[EI*D[y,{x,3}]/.a[[1]]];

(*Second Degree*)
y2=a2*x^2+a1*x+a0;
s=Solve[{(y2/.x-> 0)==0,(y2/.x->L)==0},{a0,a1}];
y2=y2/.s[[1]];
theta2=D[y2,x];
M2=EI*D[y2,{x,2}];
V2=EI*D[y2,{x,3}];
PE=EI/2*Integrate[(D[y2,{x,2}])^2,{x,0,L}]-Integrate[q*y2,{x,0,L}];
Eq1=D[PE,a2];
```

```
a=Solve[{Eq1==0},{a2}];
y2=y2/.a[[1]];
theta2=D[y2,x]/.a[[1]];
M2=FullSimplify[EI*D[y2,{x,2}]/.a[[1]]];
V2=FullSimplify[EI*D[y2,{x,3}]/.a[[1]]];

(*Third Degree*)
y3=a3*x^3+a2*x^2+a1*x+a0;
s=Solve[{(y3/.x-> 0)==0,(y3/.x->)==0},{a0,a1}];
y3=y3/.s[[1]];
theta3=D[y3,x];
M3=EI*D[y3,{x,2}];
V3=EI*D[y3,{x,3}];
PE=EI/2*Integrate[(D[y3,{x,2}])^2,{x,0,L}]-Integrate[q*y3,{x,0,L}];
Eq1=D[PE,a2];
Eq2=D[PE,a3];
a=Solve[{Eq1==0,Eq2==0},{a2,a3}];
y3=y3/.a[[1]];
theta3=D[y3,x]/.a[[1]];
M3=FullSimplify[EI*D[y3,{x,2}]/.a[[1]]];
V3=FullSimplify[EI*D[y3,{x,3}]/.a[[1]]];

(*fourth Degree*)
y4=a4*x^4+a3*x^3+a2*x^2+a1*x+a0;
s=Solve[{(y4/.x-> 0)==0,(y4/.x->L)==0},{a0,a1}];
y4=y4/.s[[1]];
theta4=D[y4,x];
M4=EI*D[y4,{x,2}];
V4=EI*D[y4,{x,3}];
PE=EI/2*Integrate[(D[y4,{x,2}])^2,{x,0,L}]-Integrate[q*y4,{x,0,L}];
Eq1=D[PE,a2];
Eq2=D[PE,a3];
Eq3=D[PE,a4];
a=Solve[{Eq1==0,Eq2==0,Eq3==0},{a2,a3,a4}];
y4=y4/.a[[1]];
theta4=D[y4,x]/.a[[1]];
M4=FullSimplify[EI*D[y4,{x,2}]/.a[[1]]];
V4=FullSimplify[EI*D[y4,{x,3}]/.a[[1]]];

(*fifth Degree*)
y5=a5*x^5+a4*x^4+a3*x^3+a2*x^2+a1*x+a0;
s=Solve[{(y5/.x-> 0)==0,(y5/.x->L)==0},{a0,a1}];
y5=y5/.s[[1]];
theta4=D[y5,x];
M5=EI*D[y5,{x,2}];
V5=EI*D[y5,{x,3}];
PE=EI/2*Integrate[(D[y5,{x,2}])^2,{x,0,L}]-Integrate[q*y5,{x,0,L}];
Eq1=D[PE,a2];
Eq2=D[PE,a3];
Eq3=D[PE,a4];
```

```
Eq4=D[PE,a5];
a=Solve[{Eq1==0,Eq2==0,Eq3==0,Eq4==0},{a2,a3,a4,a5}];
y5=y5/.a[[1]];
theta5=D[y5,x]/.a[[1]];
M5=FullSimplify[EI*D[y5,{x,2}]/.a[[1]]];
V5=FullSimplify[EI*D[y5,{x,3}]/.a[[1]]];

EI = 200000*1000000*4*100000000/1000000000000;
q = -25*1000;
L = 10;
Needs["PlotLegends'"];
Plot[{y,y2,y4},{x,0,L},AxesLabel-> {"x (m)","y (m)"},PlotLegend-
>{Style["Exact
y",Bold,12],Style["y2",Bold,12],Style["y4",Bold,12]},LegendPosition-> {1.1,
-0.4},PlotStyle-
>{Directive[Black,Thickness[0.005]],Directive[Black,Thickness[0.01]],Directi
ve[Black,Thickness[0.015]]}]
Plot[{theta,theta2,theta4},{x,0,L},AxesLabel-> {"x
(m)","theta"},LegendPosition-> {1.1,-0.4},PlotLegend-> {Style["Exact
theta",Bold,12],Style["theta 2",Bold,12],Style["theta
4",Bold,12]},PlotStyle-
>{Directive[Black,Thickness[0.005]],Directive[Black,Thickness[0.01]],Directi
ve[Black,Thickness[0.015]]}]
Plot[{M,M2,M4},{x,0,L},AxesLabel-> {"x (m)","M (N.m.)"},LegendPosition-
>{1.1,-0.4},PlotLegend-> {Style["Exact M",Bold,12],Style["M
2",Bold,12],Style["M 4",Bold,12]},PlotStyle-
>{Directive[Black,Thickness[0.005]],Directive[Black,Thickness[0.01]],Directi
ve[Black,Thickness[0.015]]}]
Plot[{V,V2,V4},{x,0,L},AxesLabel-> {"x (m)","V (N)"},LegendPosition-> {1.1,-
0.4},PlotLegend->{Style["Exact V",Bold,12],Style["V 2",Bold,12],Style["V
4",Bold,12]},PlotStyle-
>{Directive[Black,Thickness[0.005]],Directive[Black,Thickness[0.01]],Directi
ve[Black,Thickness[0.015]]}]
```

8.2. THE GALERKIN AND OTHER APPROXIMATE METHODS

As mentioned earlier, the approximate methods to solve the differential equation of equilibrium involve assuming a trial function for the unknown displacements and restricting the unknown dis–placement to a few parameters. Then, by solving a set of equations that would minimize a certain quantity, the unknown parameters can be obtained. The quantity minimized in the Rayleigh Ritz method was the potential energy of the system. For other methods, the quantity to be minimized does not necessarily have a physical interpretation, but rather is a measure of the error due to using the approximate solution. For example, let's assume that the differential equation of equilibrium has the following form:

$$Du - f = 0$$

Subject to the boundary conditions:

$$D_s u - g = 0$$

where $u: \Omega_0 \to \mathbb{R}^3$ is a smooth unknown displacement function. D and D_s are differential oper– ators. f and g are smooth functions of Ω_0 and $\partial\Omega_0$. Then, a trial displacement function can be assumed which has the form:

$$\tilde{u} = \sum_{i=1}^{n} A_i \phi_i(x)$$

where, A_i are n unknown constants and $\phi_i(x)$ are n assumed shape functions that describe the shape of the displacement over the domain Ω_0 at every element $x \in \Omega_0$. Naturally, if $\tilde{u}$ is chosen arbitrarily, it will not satisfy the differential equation of equilibrium, nor the boundary conditions. The errors produced when $\tilde{u}$ is substituted in the equations of equilibrium and in the equations of the boundary conditions are termed the residuals R and R_s, respectively:

$$R = D\tilde{u} - f$$

$$R_s = D_s\tilde{u} - f$$

Approximate methods would aim at finding the n unknown parmeters A_i by minimizing R and R_s. Examples of methods that operate on R and R_s are the **point collocation method** and the **Galerkin method**. In the following sections, the point collocation method and the Galerkin method are introduced through two examples, and a few other methods are also listed.

8.2.1. Point-Collocation Method

The point–collocation method seeks to find the n parameters A_i such that the differential equation or the boundary conditions are satisfied at n chosen points.

Problem: Using a polynomial trial function of the third degree, find the displacement function of the shown bar (Figure 8–9) by satisfying the essential boundary conditions, the differential equa– tion of equilibrium at $X_1 = \frac{L}{2}, L$, and the nonessential boundary condition at $X_1 = L$. Assume that the bar is linear elastic with Young's modulus E and cross–sectional area A and that the small– strain tensor is the appropriate measure of strain. Ignore Poisson's ratio effect.

Solution: The assumed trial function that satisfies the *essential boundary conditions* is:

$$\tilde{u} = A_1X_1 + A_2X_1^2 + A_3X_1^3$$

$$\frac{d\tilde{u}}{dX_1} = A_1 + 2A_2X_1 + 3A_3X_1^2$$

$$\frac{d^2\tilde{u}}{dX_1^2} = 2A_2 + 6A_3X_1$$

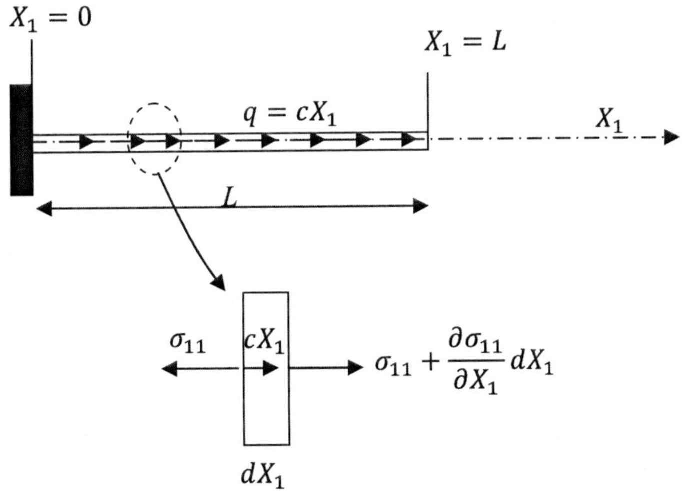

FIGURE 8-9 Bar under axial body force using point collocation method.

The differential equation of equilibrium is:

$$\frac{d\sigma_{11}}{dX_1} = \frac{-cX_1}{A}$$

Using the constitutive law for linear elasticity the differential equation of equilibrium has the form:

$$\frac{d^2u_1}{dX_1^2} + \frac{cX_1}{EA} = 0$$

The function $\tilde{u}$ is then substituted in the above differential equation at $X_1 = \frac{L}{2}, L$, to obtain the following two equations:

$$2A_2 + \frac{6A_3L}{2} + \frac{cL}{2EA} = 0, \quad 2A_2 + \frac{6A_3L}{4} + \frac{cL}{4EA} = 0$$

Finally, the equation for the nonessential boundary condition at $X_1 = L$ is:

$$\sigma_{11} = 0 = A_1 + 2A_2L + 3A_3L^2 = 0$$

The above three equations can be solved to find the approximate solution:

$$\tilde{u} = \frac{cL^2}{2EA}X_1 - \frac{c}{6EA}X_1^3$$

Notice that since the exact solution is contained in the space of functions of the trial solution, the solution obtained is indeed the exact solution. The following is the Mathematica code utilized:

```
u=a1*x1+a2*x1^2+a3*x1^3;
ux=D[u,x1]
uxx=D[ux,x1]
Eq1=(uxx+c*x1/EA)/.x1-> L/2;
Eq2=(uxx+c*x1/EA)/.x1-> L;
Eq3=(ux)/.x1-> L;
s=Solve[{Eq1==0,Eq2==0,Eq3==0},{a1,a2,a3}]
u/.s[[1]]
```

8.2.2. Other Methods

Examples of other methods involve:

- Finding the n unknown parameters A_i by equating the integral of the the residuals R and R_s over n chosen sub–domains ω_i to zero:

$$I_i = \int_{\omega_i} R dV = 0, \quad or, \quad I_i = \int_{\partial\omega_i} R_s dA = 0$$

- Finding the n unknown parameters A_i by minimizing the integral of the squares of R and R_s

$$I = \int_{\Omega_0} R^2 dV + \int_{\partial\Omega_0} R_s^2 dA = min$$

8.2.3. The Method of Weighted Residuals, the Galerkin Method

The weighted residuals method involves finding the n unknown parameters A_i such that the weighted integrals of R over the volume are equal to zero:

$$I = \int_{\Omega_0} R W_i dV$$

The **Galerkin method** is a form of the weighted residuals method such that the weight functions W_i are chosen as the functions $\phi_i(x)$. When the Galerkin method is employed, the equations developed for structural mechanics problems are exactly the virtual work equations. For conservative systems when the Rayleigh Ritz method is applicable, the equations developed by the three methods: the Galerkin, the Rayleigh Ritz, and the virtual work methods are identical.

Problem: Using a polynomial trial function of the second degree, find the displacement function of the shown bar (Figure 8–9) using the Galerkin method. Assume that the bar is linear elastic with Young's modulus E and cross–sectional area A and that the small strain tensor is the appropriate measure of strain. Ignore Poisson's ratio effect.

Solution: The differential equation of equilibrium is:

$$\frac{d\sigma_{11}}{dX_1} = \frac{-cX_1}{A}$$

Using the constitutive law for linear elasticity, the differential equation of equilibrium has the form:

$$\frac{d^2u_1}{dX_1^2} + \frac{cX_1}{EA} = 0$$

The assumed trial function that *satisfies the essential boundary conditions* is:

$$\tilde{u} = A_1X_1 + A_2X_1^2 + A_3X_1^3$$
$$\frac{d\tilde{u}}{dX_1} = A_1 + 2A_2X_1 + 3A_3X_1^2$$
$$\frac{d^2\tilde{u}}{dX_1^2} = 2A_2 + 6A_3X_1$$

The residuals are obtained by substituting $\tilde{u}$ in the differential equation of equilibrium:

$$R = \frac{d^2\tilde{u}}{dX_1^2} + \frac{cX_1}{EA}$$

The weight functions for the Galerkin Method are:

$$W_1 = \phi_1(X_1) = X_1, \quad W_2 = \phi_2(X_1) = X_1^2, \quad W_3 = \phi_3(X_1) = X_1^3$$

The integral of the weighted residuals is:

$$I = \int_0^L W_iRdX_1 = \int_0^L W_i\left(\frac{d^2\tilde{u}}{dX_1^2} + \frac{cX_1}{EA}\right)dX_1 = 0$$

$$\int_0^L W_i\left(\frac{d^2\tilde{u}}{dX_1^2}\right)dX_1 = -\int_0^L W_i\left(\frac{cX_1}{EA}\right)dX_1$$

Using integration by parts, the left–hand side of the above equation is manipulated to include the nonessential boundary conditions:

$$\int_0^L W_i\left(\frac{d^2\tilde{u}}{dX_1^2}\right)dX_1 = W_i\frac{d\tilde{u}}{dX_1}\bigg|_0^L - \int_0^L \frac{dW_i}{dX_1}\left(\frac{d\tilde{u}}{dX_1}\right)dX_1 = 0 - \int_0^L \frac{dW_i}{dX_1}\left(\frac{d\tilde{u}}{dX_1}\right)dX_1$$

Notice that the stress at the far right end of the bar is equal to zero and thus, $\left.\frac{d\tilde{u}}{dX_1}\right|_{X_1=L} = 0$.

The following equality was utilized:

$$\int_0^L df_1 f_2 = f_1 f_2\big|_0^L = \int_0^L f_1 df_2 + \int_0^L f_2 df_1$$

With $f_1 = W_i$ and $f_2 = \frac{d\tilde{u}}{dX_1}$.

Thus, the three equations ($i = 1, 2$ and 3) obtained are:

$$\int_0^L \frac{dW_i}{dX_1}\left(\frac{d\tilde{u}}{dX_1}\right) dX_1 = \int_0^L W_i \left(\frac{cX_1}{EA}\right) dX_1$$

By solving the three equations, the following approximate solution is obtained:

$$\tilde{u} = \frac{cL^2}{2EA} X_1 - \frac{c}{6EA} X_1^3$$

This solution is equal to the exact solution since the exact solution is an element of the space of the trial functions. The following Mathematica code was utilized:

```
w1=x1;
w2=x1^2;
w3=x1^3;
u=a1*w1+a2*w2+a3*w3;
Eq1=Integrate[D[w1,x1]*D[u,x1],{x1,0,L}]-Integrate[w1*c*x1/EA,{x1,0,L}];
Eq2=Integrate[D[w2,x1]*D[u,x1],{x1,0,L}]-Integrate[w2*c*x1/EA,{x1,0,L}];
Eq3=Integrate[D[w3,x1]*D[u,x1],{x1,0,L}]-Integrate[w3*c*x1/EA,{x1,0,L}];
s=Solve[{Eq1==0,Eq2==0,Eq3==}{a1,a2,a3}]
u/.s[[1]]
```

8.3. THE STRONG FORM VERSUS THE WEAK FORM

The virtual work method, the Rayleigh Ritz method, the weighted residuals, and the Galerkin methods are designed to find approximate solutions using assumed trial functions for the displacement field sought. The terminology "weak formulation" is often used to describe the equations developed in those methods in comparison to the "strong formulation," which is used to describe the original differential equation. The displacement function that satisfies the differential equation at every point is called the "strong form solution," while the approximate displacement function that satisfies the weak formulation is termed the "weak form solution." The term "strong" refers to the strict requirement that the solution is exact, while the term "weak" refers to the fact that the obtained solution satisfies fewer requirements than the "strong form solution." Another major difference between the "strong form" and the "weak form" of the problem is that the strong form often involves a differential equation, which is harder to deal with at locations of discontinuities. When dealing with a strong form, the equations often involve dividing the domain into different locations at points of discontinuities. When dealing with a weak form, however, the equations involve an integral, which can sustain discontinuities, and thus, while the exact solution (strong form) is discontinuous, the weak form obtained can still be continuous.

General Notes on the Weak Form of Structural Mechanics Problems:

- The trial functions assumed should be as close as possible to the exact solution.
- Since in the weak form the system is constrained to a specified displacement shape, the displacements are under–predicted and the total energy stored in the system is also under–predicted. On the other hand, if the boundary conditions involve prescribed displacements, then the loads producing such displacements are over–predicted using the weak formulation.
- As with any approximate solution, the accuracy of the principal unknown is higher than the accuracy of its derivatives. For the problems described in this text, the displacement is the primary unknown function. Thus, the accuracy of the approximate displacement function is better than the accuracy of its derivatives (the strain components), and thus, is better than the accuracy of the stress components (which are functions of the strain components).
- The term "essential boundary conditions" in solid mechanics problems is used to refer to the boundary conditions of the displacements in solids and rotations in beams and shells. The term "non–essential boundary conditions" is used to refer to the boundary conditions of the stress in solids and of bending moments and shearing forces in beams and shells.

▶ EXERCISES:

(Answer key is given in round brackets for some problems):

1. The exact displacement in meters of the shown Euler Bernoulli beam follows the function:

$$y = \begin{cases} 0.0003375X_1^2(3X_1 - 20) & 0 \leq X_1 \leq 5 \\ -0.0003125\,(X_1 - 8)^2\,(7X_1 - 20) & 5 < X_1 \leq 8 \end{cases}$$

The beam's Young's modulus and moment of inertia are $E = 20000$ MPa and $I = 0.00260417\ \mathrm{m}^4$

   a. Find the strain energy stored in the beam (**Answer:** 7.776 N.m).

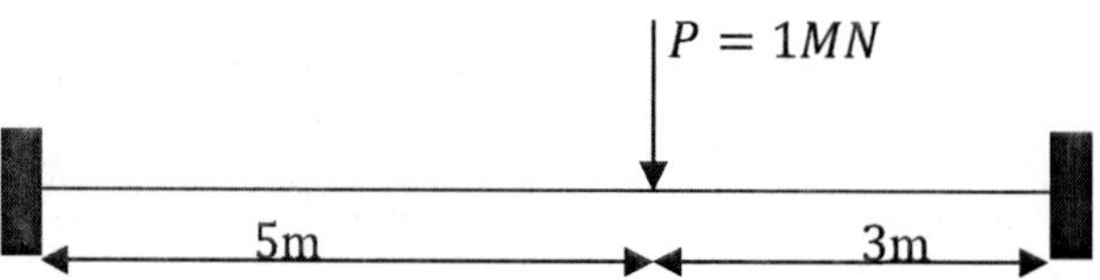

b. Use the Rayleigh Ritz method to find approximate solutions for the displacement across the whole beam. Assume the approximations to be: polynomials of the zero, first, second, third, fourth, fifth and sixth degrees.

c. Plot the exact solution versus the approximate solutions, and comment on your results.

2. Solve the following two linear–elastic isotropic small deformations beam structures:

   a. Using the strong form differential equation

   b. Using the Rayleigh Ritz method with polynomial trial functions satisfying the essential boundary conditions. Assume the trial functions to be of zero, first, second, third and fourth degrees. Comment on the solution in comparison to the exact solution obtained.

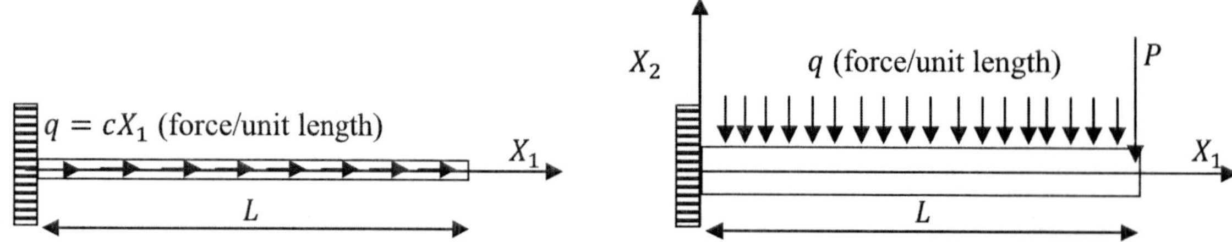

3. A beam of length L with both ends fixed is loaded by a concentrated force P at the center of the beam. Assume that the beam is made out of a linear–elastic isotropic material and that the deformations are small. The material's Young's modulus is E, and moment of inertia is I. Find:

   a. The strain energy in terms of the displacement function y.

   b. Expressions for the potential energy Π of the system.

   c. Show that the following series satisfies the essential boundary conditions (displacements and rotations):

$$w = \sum_{n=2}^{\infty} A_n \left(\cos \frac{n\pi x}{L} - 1\right) \quad n = 2, 4, 6 \ldots$$

d. Take one term and two terms from the series above, and determine the constants by the Rayleigh–Ritz method.

e. Find the exact solution by solving the differential equation of the Euler Bernoulli beam, and compare with the approximate solution obtained from the Rayleigh–Ritz method.

4. Find the displacement of the shown two beam structures

   a. Using the strong form differential equation.

   b. Using the Rayleigh Ritz method with polynomial trial functions of the zero, one, two, three, and four degrees along the whole domain.

   c. Using the virtual work method with a third– and fourth–degree polynomial functions defined over the whole domain.

   d. Using the Galerkin method with a third– and fourth–degree polynomial functions defined over the whole domain.

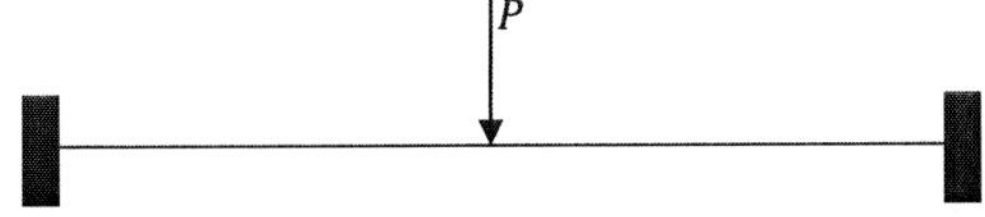

Euler–Bernoulli Beam whose Young's modulus, moment of inertia, and length are E, I, L, respectively. P is applied mid–span.

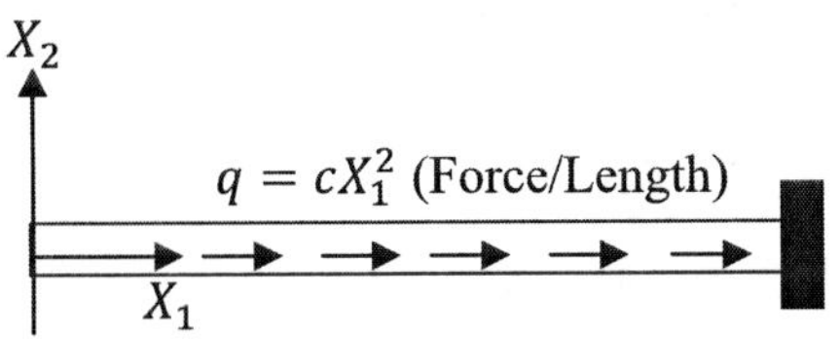

Bar under axial load whose Young's modulus, area, and length are E, A, L, respectively.

CHAPTER 9

Introduction to Finite Element Analysis

Finite element analysis is an approximate method to solve the differential equation of equilibrium by means of a weak formulation. The domain is divided into sub–domains (called elements) on which the displacement functions are assumed to be linear or nonlinear. The displacement is assumed to be continuous across the boundaries of the sub–domains, but the gradients are not necessarily continuous. In other words, strictly speaking, the strains and the stresses across boundaries are not well defined. It can be shown using Mathematical arguments that the higher the number of sub–domains, the closer the approximate solution is to the exact solution. This process of refining the approximation (refining the mesh) until achieving the desired accuracy is termed mesh convergence. In this chapter, the basic theory of finite element analysis applied to problems in small–deformation linear elasticity will be presented along with some of the most common elements used.

9.1. FINITE ELEMENT ANALYSIS APPROXIMATION IN ONE DIMENSION

As shown in the previous chapter, the approximate methods involve assuming trial functions for the unknown variables. The trial functions that were used previously were continuous and differentiable along the whole domain (Figure 9–1a). The finite element analysis, however, involves using piecewise linear or nonlinear functions for the approximate solution (Figure 9–1b). The basic theory behind finding an approximate solution using the weak formulation has been developed in Chapters 7 and 8. In the following, a one–dimensional illustrative example will be used to showcase the piecewise approximation for linear–elastic small deformations structures.

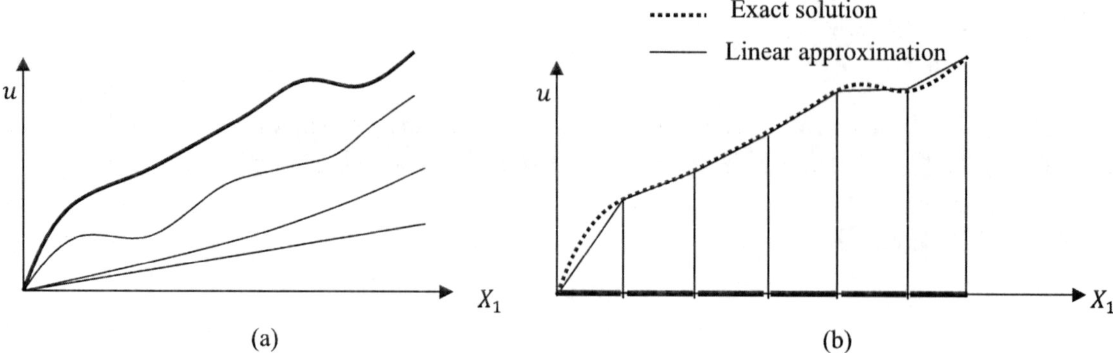

FIGURE 9-1 The final displacement function as (a) the sum of continuously differentiable group of functions versus (b) the sum of a group of piecewise linear functions.

9.1.1. Illustrative Example, One-Dimensional Linear Element Case

Using a four–piecewise linear trial function, find the approximate displacement function of the shown bar (Figure 9–2). Assume that the bar is linear elastic with Young's modulus E and cross–sectional area A and that the small strain tensor is the appropriate measure of strain. Ignore Poisson's ratio effect.

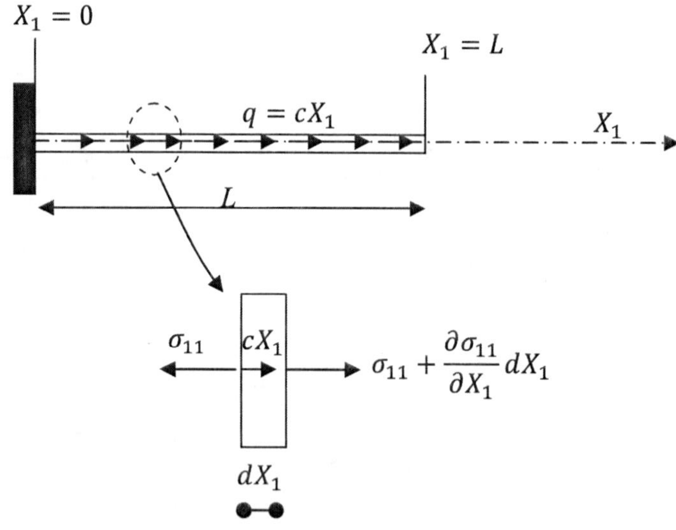

FIGURE 9-2 Bar under axial load.

9.1.1.a. Trial Functions

In traditional finite element analysis, the values of the displacements at specific points (**nodes**) across the domain are the main unknowns to be calculated using the method. The displacements at intermediate locations between the nodes are interpolated according to a chosen interpolation function. Figure 9–3 shows three nodes (node 1, node 2 and node 3) and three different interpolation functions for the displacement at intermediate locations between the nodes. An interpolation function that is discontinuous across nodes is termed C–1. A C*n* interpolation function is an interpolation function that can be differentiated *n* times. Figure 9–3 illustrates discontinuous C–1, continuous C0, and once differentiable C1 interpolation functions.

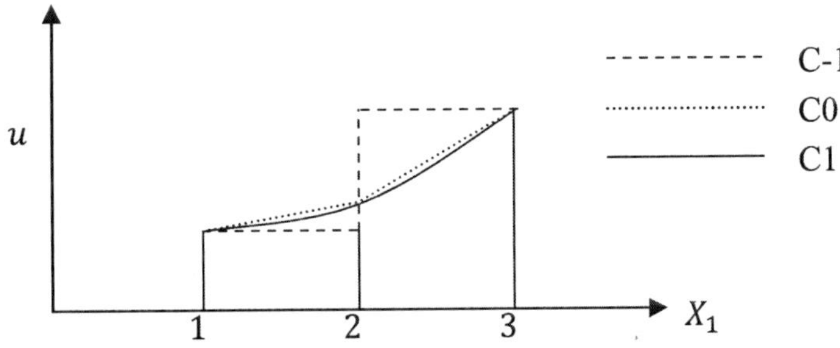

FIGURE 9-3 Examples of order of interpolation across nodes.

In this example, a piecewise C0 linear trial function (Figure 9–4) for the displacement, interpolated between four nodes, is chosen with the following form:

$$\tilde{u} = u_1N_1 + u_2N_2 + u_3N_3 + u_4N_4 = \sum_{i=1}^{4} u_iN_i$$

where u_i are multipliers that happen to be the displacements at the nodes, while N_i are the interpolation functions that are traditionally termed the "hat" functions, for obvious reasons.

The derivatives of the trial function are readily available as:

$$\frac{d\tilde{u}}{dX_1} = \sum_{i=1}^{4} u_i \frac{dN_i}{dX_1}$$

9.1.1.b. Solution Using the Galerkin Method

The Galerkin method, as shown in Chapter 8, involves equating the integral of the weighted residuals to zero. The shape functions N_i are used as the weight functions, and thus, the Galerkin

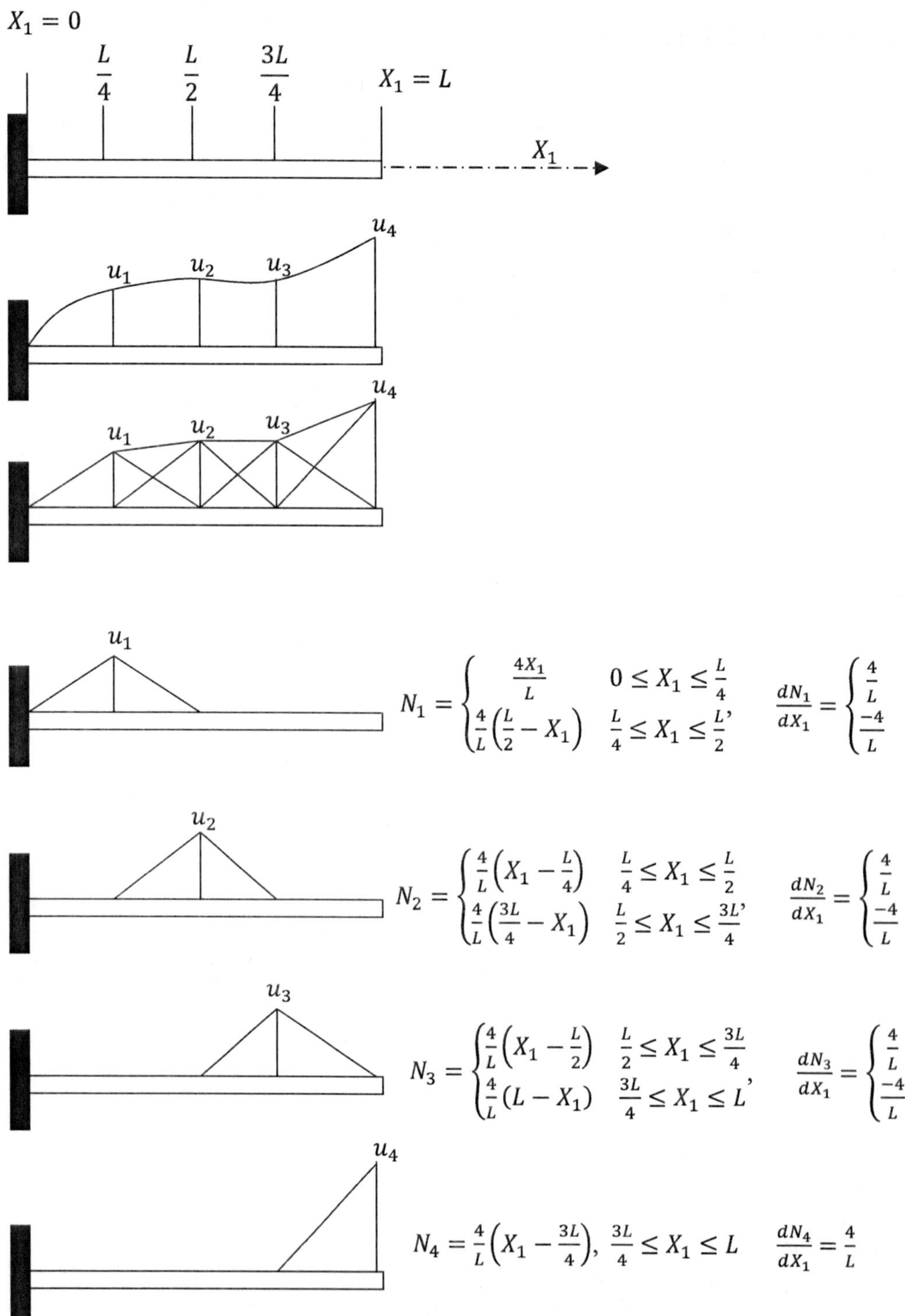

$X_1 = 0$
$\frac{L}{4}$
$\frac{L}{2}$
$\frac{3L}{4}$
$X_1 = L$
X_1
u_1
u_2
u_3
u_4
$N_1 = \begin{cases} \frac{4X_1}{L} & 0 \le X_1 \le \frac{L}{4} \\ \frac{4}{L}\left(\frac{L}{2} - X_1\right) & \frac{L}{4} \le X_1 \le \frac{L}{2} \end{cases}, \quad \frac{dN_1}{dX_1} = \begin{cases} \frac{4}{L} \\ \frac{-4}{L} \end{cases}$
$N_2 = \begin{cases} \frac{4}{L}\left(X_1 - \frac{L}{4}\right) & \frac{L}{4} \le X_1 \le \frac{L}{2} \\ \frac{4}{L}\left(\frac{3L}{4} - X_1\right) & \frac{L}{2} \le X_1 \le \frac{3L}{4} \end{cases}, \quad \frac{dN_2}{dX_1} = \begin{cases} \frac{4}{L} \\ \frac{-4}{L} \end{cases}$
$N_3 = \begin{cases} \frac{4}{L}\left(X_1 - \frac{L}{2}\right) & \frac{L}{2} \le X_1 \le \frac{3L}{4} \\ \frac{4}{L}(L - X_1) & \frac{3L}{4} \le X_1 \le L \end{cases}, \quad \frac{dN_3}{dX_1} = \begin{cases} \frac{4}{L} \\ \frac{-4}{L} \end{cases}$
$N_4 = \frac{4}{L}\left(X_1 - \frac{3L}{4}\right), \ \frac{3L}{4} \le X_1 \le L \quad \frac{dN_4}{dX_1} = \frac{4}{L}$

FIGURE 9-4 Piecewise linear interpolation functions.

method produces four equations of the form:

$$\int_0^L \frac{dW_i}{dX_1}\left(\frac{d\tilde{u}}{dX_1}\right) dX_1 = \int_0^L W_i\left(\frac{cX_1}{EA}\right) dX_1$$

where, $W_i = N_i$. There are four trial or shape function N_i. Each appears only within a small interval and is equal to zero at the remaining part of the domain. Thus, the integration for each equation is only carried on the part where N_i does not equal to zero.

Equation 1 $\left(X_1 = 0 \to \frac{L}{2}\right)$**:**

$$W_1 = N_1, \quad \frac{dW_1}{dX_1} = \frac{dN_1}{dX_1}, \quad \tilde{u} = u_1 N_1 + u_2 N_2, \quad \frac{d\tilde{u}}{dX_1} = u_1 \frac{dN_1}{dX_1} + u_2 \frac{dN_2}{dX_1}$$

$$u_1 \int_0^{\frac{L}{2}} \left(\frac{dN_1}{dX_1}\right)^2 dX_1 + u_2 \int_0^{\frac{L}{2}} \frac{dN_1}{dX_1}\frac{dN_2}{dX_1} dX_1 = \int_0^{\frac{L}{2}} \frac{cX_1}{EA} N_1 dX_1$$

Equation 2 $\left(X_1 = \frac{L}{4} \to \frac{3L}{4}\right)$**:**

$$W_2 = N_2, \quad \frac{dW_2}{dX_1} = \frac{dN_2}{dX_1}, \quad \tilde{u} = u_1 N_1 + u_2 N_2 + u_3 N_3, \quad \frac{d\tilde{u}}{dX_1} = u_1 \frac{dN_1}{dX_1} + u_2 \frac{dN_2}{dX_1} + u_3 \frac{dN_3}{dX_1}$$

$$u_1 \int_{L/4}^{3L/4} \frac{dN_1}{dX_1}\frac{dN_2}{dX_1} dX_1 + u_2 \int_{L/4}^{3L/4} \left(\frac{dN_2}{dX_1}\right)^2 dX_1 + u_3 \int_{L/4}^{3L/4} \frac{dN_2}{dX_1}\frac{dN_3}{dX_1} dX_1 = \int_{L/4}^{3L/4} \frac{cX_1}{EA} N_2 dX_1$$

Equation 3 $\left(X_1 = \frac{L}{2} \to L\right)$**:**

$$W_3 = N_3, \quad \frac{dW_3}{dX_1} = \frac{dN_3}{dX_1}, \quad \tilde{u} = u_2 N_2 + u_3 N_3 + u_4 N_4, \quad \frac{d\tilde{u}}{dX_1} = u_2 \frac{dN_2}{dX_1} + u_3 \frac{dN_3}{dX_1} + u_4 \frac{dN_4}{dX_1}$$

$$u_2 \int_{L/2}^{L} \frac{dN_2}{dX_1}\frac{dN_3}{dX_1} dX_1 + u_3 \int_{L/2}^{L} \left(\frac{dN_3}{dX_1}\right)^2 dX_1 + u_4 \int_{L/2}^{L} \frac{dN_4}{dX_1}\frac{dN_3}{dX_1} dX_1 = \int_{L/2}^{L} \frac{cX_1}{EA} N_3 dX_1$$

Equation 4 $\left(X_1 = \frac{3L}{4} \to L\right)$**:**

$$W_4 = N_4, \quad \frac{dW_4}{dX_1} = \frac{dN_4}{dX_1}, \quad \tilde{u} = u_3 N_3 + u_4 N_4, \quad \frac{d\tilde{u}}{dX_1} = u_3 \frac{dN_3}{dX_1} + u_4 \frac{dN_4}{dX_1}$$

$$u_3 \int_{3L/4}^{L} \frac{dN_3}{dX_1}\frac{dN_4}{dX_1} dX_1 + u_4 \int_{3L/4}^{L} \left(\frac{dN_4}{dX_1}\right)^2 dX_1 = \int_{3L/4}^{L} \frac{cX_1}{EA} N_4 dX_1$$

The above four equations can be written in matrix form as follows:

$$\begin{pmatrix} \int_0^{\frac{L}{2}} \left(\frac{dN_1}{dX_1}\right)^2 dX_1 & \int_0^{\frac{L}{2}} \frac{dN_1}{dX_1}\frac{dN_2}{dX_1} dX_1 & 0 & 0 \\ \int_{L/4}^{3L/4} \frac{dN_2}{dX_1}\frac{dN_1}{dX_1} dX_1 & \int_{L/4}^{3L/4} \left(\frac{dN_2}{dX_1}\right)^2 dX_1 & \int_{L/4}^{3L/4} \frac{dN_2}{dX_1}\frac{dN_3}{dX_1} dX_1 & 0 \\ 0 & \int_{L/2}^{L} \frac{dN_3}{dX_1}\frac{dN_2}{dX_1} dX_1 & \int_{L/2}^{L} \left(\frac{dN_3}{dX_1}\right)^2 dX_1 & \int_{L/2}^{L} \frac{dN_3}{dX_1}\frac{dN_4}{dX_1} dX_1 \\ 0 & 0 & \int_{3L/4}^{L} \frac{dN_4}{dX_1}\frac{dN_3}{dX_1} dX_1 & \int_{3L/4}^{L} \left(\frac{dN_4}{dX_1}\right)^2 dX_1 \end{pmatrix} \begin{Bmatrix} u_1 \\ u_2 \\ u_3 \\ u_4 \end{Bmatrix} = \begin{Bmatrix} f_1 \\ f_2 \\ f_3 \\ f_4 \end{Bmatrix}$$

where $f_i = \int_0^L \frac{cX_1}{EA} N_i dX_1$. The integrations on both sides are very easy to do. In fact, most of them are equivalent to each other and produce the same results. The final equations have the following form:

$$\begin{pmatrix} 8/L & -4/L & 0 & 0 \\ -4/L & 8/L & -4/L & 0 \\ 0 & -4/L & 8/L & -4/L \\ 0 & 0 & -4/L & 4/L \end{pmatrix} \begin{Bmatrix} u_1 \\ u_2 \\ u_3 \\ u_4 \end{Bmatrix} = \frac{cL^2}{EA} \begin{Bmatrix} 1/16 \\ 1/8 \\ 3/16 \\ 11/96 \end{Bmatrix}$$

The solution is:

$$\begin{Bmatrix} u_1 \\ u_2 \\ u_3 \\ u_4 \end{Bmatrix} = \frac{cL^3}{EA} \begin{Bmatrix} 47/384 \\ 11/48 \\ 39/128 \\ 1/3 \end{Bmatrix}$$

Notice that the equations produced by the Galerkin method have the following form:

$$Ku = F$$

where

$u \in \mathbb{R}^n$ is a vector of the unknown multipliers u_i which are also termed the degrees of freedom,

$F \in \mathbb{R}^n$ is a vector of "nodal forces" with components $F_i = \int \frac{cX_1}{EA} N_i dX_1$ and

$K \in M^n$ is a matrix representing a linear map between the displacement vector and the nodal forces vector with components $K_{ij} = \int \frac{dN_i}{dX_1}\frac{dN_j}{dX_1} dX_1$.

9.1.1.c. Solution Using the Virtual Work Method

As stated previously, the Galerkin method is equivalent to the virtual work method. The principle of virtual work states that, from a state of equilibrium, when a virtual admissible (compatible) displacement is applied, the increment in the work done by the external and body forces during the application of the virtual displacement is equal to the increment in the deformation energy stored.

The stresses due to the assumed–trial displacement function have the following form:

$$\sigma_{11} = E\varepsilon_{11} = E\sum_{i=1}^{4} u_i \frac{dN_i}{dX_1}$$

The virtual displacement function can be assumed to have the same form as the trial displacement function, but with different multipliers:

$$u^* = \sum_{i=1}^{4} u_i^* N_i$$

The virtual strain field associated with the virtual displacement function is:

$$\varepsilon_{11}^* = \sum_{i=1}^{4} u_i^* \frac{dN_i}{dX_1}$$

The equations of the virtual work are:

$$A\int_0^L \sigma_{11}\varepsilon_{11}^* dX_1 = \int_0^L cX_1 u^* dX_1$$

The virtual work equation can be rearranged to have the following form:

$$\{u_1^* \quad u_2^* \quad u_3^* \quad u_4^*\}\begin{pmatrix} \int_0^{\frac{L}{2}}\left(\frac{dN_1}{dX_1}\right)^2 dX_1 & \int_0^{\frac{L}{2}}\frac{dN_1}{dX_1}\frac{dN_2}{dX_1}dX_1 & 0 & 0 \\ \int_{L/4}^{3L/4}\frac{dN_2}{dX_1}\frac{dN_1}{dX_1}dX_1 & \int_{L/4}^{3L/4}\left(\frac{dN_2}{dX_1}\right)^2 dX_1 & \int_{L/4}^{3L/4}\frac{dN_2}{dX_1}\frac{dN_3}{dX_1}dX_1 & 0 \\ 0 & \int_{L/2}^{L}\frac{dN_3}{dX_1}\frac{dN_2}{dX_1}dX_1 & \int_{L/2}^{L}\left(\frac{dN_3}{dX_1}\right)^2 dX_1 & \int_{L/2}^{L}\frac{dN_3}{dX_1}\frac{dN_4}{dX_1}dX_1 \\ 0 & 0 & \int_{3L/4}^{L}\frac{dN_4}{dX_1}\frac{dN_3}{dX_1}dX_1 & \int_{3L/4}^{L}\left(\frac{dN_4}{dX_1}\right)^2 dX_1 \end{pmatrix}\begin{Bmatrix} u_1 \\ u_2 \\ u_3 \\ u_4 \end{Bmatrix}$$

$$= \{u_1^* \quad u_2^* \quad u_3^* \quad u_4^*\}\begin{Bmatrix} f_1 \\ f_2 \\ f_3 \\ f_4 \end{Bmatrix}$$

Since the multipliers $\{u_1^* \quad u_2^* \quad u_3^* \quad u_4^*\}$ are the degrees of freedom of the arbitrary virtual dis–placement field, then the same equations of the Galerkin method are retrieved. (The coefficients of each u_i^* on both sides of the equality are equal.)

9.1.1.d. Comments on the Approximate Solution

The approximate solution involves assuming a trial function for the displacements; it is formed by shape functions, each associated with a multiplier. The multipliers are termed the degrees of free–dom of the system, and in the finite element analysis, they are the displacements of the nodes. The region between the nodes are called **elements**. The right–hand side of the equations is equivalent to a nodal force matrix and formed by lumping the distributed load and forming equivalent con–centrated loads at the nodes. For this particular example the solution is exact at the selected nodes. However, the solution within the elements will be linearly interpolated between the two values on each side and thus is not exact. The value of the gradient of the displacement at each node does not exist, as the function $\tilde{u}$ is not smooth. In the finite element analysis method, the stresses that are related to the strains (the gradients of the displacements) are usually averaged at the nodes, and the degree of variation of the stresses between the different elements could be an indication of the accuracy of the solution. The matrix multiplied by the degrees of freedom is, in conservative systems, symmetric and is termed the **stiffness matrix**.

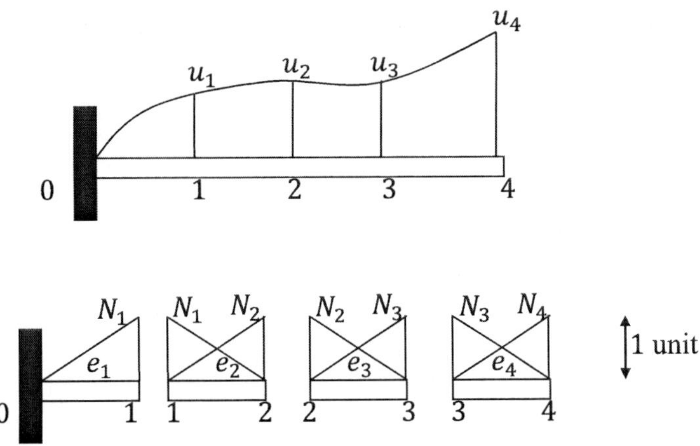

FIGURE 9-5 Dividing the domain into elements and nodes.

9.1.1.e. Nodes, Elements, and Local and Global Stiffness Matrices

One of the major advantages of the finite element analysis method is that the equations developed can easily be implemented in a computer program. This advantage, along with the advance in computational capabilities of modern computers, has led to the widespread use of the method. The choice of the linear interpolation functions between nodes allows for computing the developed integrations on the sub–domains (elements) rather than on the whole domain. As shown in Figure 9–5, the domain in the previous example was divided into four sub–domains, each called an element. Elements are connected to each other, or rather, separated from each other at the "nodes." On each element, the shape functions or interpolation functions are associated with the end nodes. For example, element e_2 connects nodes 1 and 2 with two shape functions, N_1 and N_2. What is remarkable about the method is the similarity between all the elements shown in Figure 9–5. Thus, a local stiffness matrix for a representative element can be developed, and then, the global stiffness matrix can be easily assembled by combining all the local stiffness matrices.

Assuming that the left and right nodes of a general element have the displacements u_{1e} and u_{2e}, respectively, then the assumed–trial function on that general element, the uniaxial strain, and the uniaxial stress have the forms:

$$\tilde{u} = \langle N_1 \quad N_2 \rangle \begin{Bmatrix} u_{1e} \\ u_{2e} \end{Bmatrix}, \quad \varepsilon_{11} = \left\langle \frac{dN_1}{dX_1} \quad \frac{dN_2}{dX_1} \right\rangle \begin{Bmatrix} u_{1e} \\ u_{2e} \end{Bmatrix}, \quad \sigma_{11} = E\varepsilon_{11} = E\left\langle \frac{dN_1}{dX_1} \quad \frac{dN_2}{dX_1} \right\rangle \begin{Bmatrix} u_{1e} \\ u_{2e} \end{Bmatrix}$$

The virtual displacement field on the general element for the virtual work method can be assumed to have the form:

$$u^* = \langle N_1 \quad N_2 \rangle \begin{Bmatrix} u^*_{1e} \\ u^*_{2e} \end{Bmatrix}, \qquad \varepsilon^*_{11} = \left\langle \frac{dN_1}{dX_1} \quad \frac{dN_2}{dX_1} \right\rangle \begin{Bmatrix} u^*_{1e} \\ u^*_{2e} \end{Bmatrix} = \langle u^*_{1e} \quad u^*_{2e} \rangle \begin{Bmatrix} \dfrac{dN_1}{dX_1} \\ \dfrac{dN_2}{dX_1} \end{Bmatrix}$$

If f is the distributed body force, then the equations of virtual work on the general element have the form:

$$A \int_e \varepsilon^*_{11} \sigma_{11} dX_1 = \int_e f u^* dX_1$$

$$EA \langle u^*_{1e} \quad u^*_{2e} \rangle \int_e \begin{Bmatrix} \dfrac{dN_1}{dX_1} \\ \dfrac{dN_2}{dX_1} \end{Bmatrix} \left\langle \frac{dN_1}{dX_1} \quad \frac{dN_2}{dX_1} \right\rangle dX_1 \begin{Bmatrix} u_{1e} \\ u_{2e} \end{Bmatrix} = \langle u^*_{1e} \quad u^*_{2e} \rangle \int_e \begin{Bmatrix} fN_1 \\ fN_2 \end{Bmatrix} dX_1$$

Thus, the local stiffness matrix of the general element has the form:

$$EA \begin{pmatrix} \displaystyle\int_e \left(\frac{dN_1}{dX_1}\right)^2 dX_1 & \displaystyle\int_e \frac{dN_1}{dX_1}\frac{dN_2}{dX_1} dX_1 \\ \displaystyle\int_e \frac{dN_2}{dX_1}\frac{dN_1}{dX_1} dX_1 & \displaystyle\int_e \left(\frac{dN_2}{dX_1}\right)^2 dX_1 \end{pmatrix} \begin{Bmatrix} u_{1e} \\ u_{2e} \end{Bmatrix} = \begin{Bmatrix} \displaystyle\int_e fN_1 dX_1 \\ \displaystyle\int_e fN_2 dX_1 \end{Bmatrix} \Longrightarrow K_e u_e = f_e$$

where, K_e, u_e, f_e are the stiffness matrix, nodal displacements, and nodal forces local to the general element. If the components of K_e and f_e are termed ke_{ij} and fe_i, respectively, then the component form of the above equation can be written as:

$$\begin{pmatrix} ke_{11} & ke_{12} \\ ke_{21} & ke_{22} \end{pmatrix} \begin{Bmatrix} u_{1e} \\ u_{2e} \end{Bmatrix} = \begin{Bmatrix} fe_1 \\ fe_2 \end{Bmatrix}$$

The global equations of the whole domain have the form:

$$Ku = f$$

The global stiffness matrix is formed by placing each component of the local stiffness and local nodal forces in their corresponding locations in the global stiffness matrix and the global nodal forces vector, and adding all the different components from the different elements together. The following example (Figure 9–6) illustrates the procedure of assembly.

Element $e1$ connects nodes 1 and 2, the local stiffness matrix of this element has components $Ke1_{ij}$, and the local–force matrix has components $fe1_i$. This is repeated for elements $e2$, $e3$, and $e4$:

$$\begin{pmatrix} Ke1_{11} & Ke1_{12} \\ Ke1_{21} & Ke1_{22} \end{pmatrix} \begin{Bmatrix} u_1 \\ u_2 \end{Bmatrix} = \begin{Bmatrix} fe1_1 \\ fe1_2 \end{Bmatrix}, \quad \begin{pmatrix} Ke2_{11} & Ke2_{12} \\ Ke2_{21} & Ke2_{22} \end{pmatrix} \begin{Bmatrix} u_2 \\ u_3 \end{Bmatrix} = \begin{Bmatrix} fe2_1 \\ fe2_2 \end{Bmatrix}, \quad \ldots$$

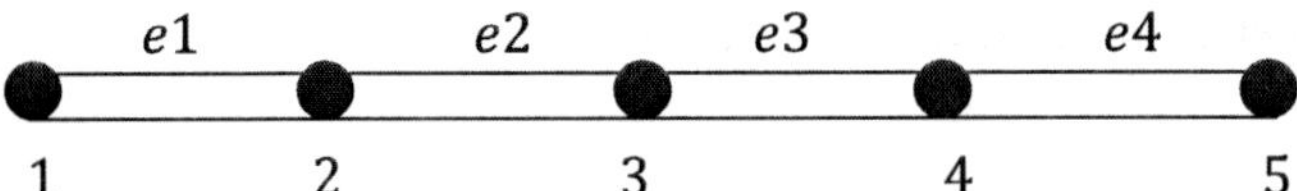

FIGURE 9-6 Example illustrating dividing a domain into elements and then forming the global stiffness matrix from individual local elements.

The global stiffness matrix has the form:

$$\begin{pmatrix} Ke1_{11} & Ke1_{12} & 0 & 0 & 0 \\ Ke1_{21} & Ke1_{22}+Ke2_{11} & Ke2_{12} & 0 & 0 \\ 0 & Ke2_{21} & Ke2_{22}+Ke3_{11} & Ke3_{12} & 0 \\ 0 & 0 & Ke3_{21} & Ke3_{22}+Ke4_{11} & Ke4_{12} \\ 0 & 0 & 0 & Ke4_{21} & Ke4_{22} \end{pmatrix} \begin{Bmatrix} u_1 \\ u_2 \\ u_3 \\ u_4 \\ u_5 \end{Bmatrix} = \begin{Bmatrix} fe1_1 \\ fe1_2+fe2_1 \\ fe2_2+fe3_1 \\ fe3_2+fe4_1 \\ fe4_2 \end{Bmatrix}$$

9.1.1.f. Properties of the Stiffness Matrix of Linear Elastic Structures

The comments in this section apply to the stiffness matrices of problems in one, two, and three dimensions.

- **Sparsity:** The stiffness matrix formed in the traditional finite element analysis is sparse, i.e., contains many zero entries. This property is due to the choice of piecewise–linear interpolation functions, since the shape functions (or the interpolation functions) are "Hat functions" that take values of 1 at each node and decay linearly or in a parabolic fashion across the neighboring elements connected to that node. Nodes that are not connected together with any elements have a corresponding zero entry in the global stiffness matrix.
- **Invertibility (See section 1.2.3.):** The stiffness matrix of a structure is a linear map between $u \in \mathbb{R}^n$ and $f \in \mathbb{R}^n$. Such map is invertible if and only if the map is injective (one to one), which means that the only element in the kernel of the map is zero. Otherwise, if the map is not injective, then there are two nonzero distinct displacement vectors $u, v \in \mathbb{R}^n$ that can be produced by the same force vector $f \in \mathbb{R}^n$ then: $Ku = Kv = f \Rightarrow K(u - v) = 0$, then the nonzero vector displacement $u - v$ can be applied to the structure without any external forces! If no boundary conditions are applied to a structure, then naturally rigid body motions and rigid body rotations can be applied without any external forces. Therefore, a stiffness matrix of a structure that is not restrained is *not* invertible and its determinant is equal to zero. Once enough boundary conditions are applied to the structure, then the only element of the kernel of

the linear map is the zero element, the stiffness matrix is invertible, and its determinant is not equal to zero. Notice, however, that when a structure is not well supported and does not have the correct boundary conditions, then the stiffness matrix is termed "ill conditioned." Some finite–element analysis software will still be able to "invert" the matrix, depending on the solution method. A user should be cautious in such cases since the solution could involve very large numbers for the displacement that would render the solution inaccurate.

- **Positive and Semi–Positive Definiteness (See section 1.2.6.g.):** A stable structure will deform in the direction of increasing external forces. Mathematically speaking, the dot product between the displacement and the force vectors has to be positive. If the structure is well restrained, i.e., K is invertible, then K is positive definite and any displacement field in the structure will produce positive energy. In mathematical terms, this imples: $\forall u \in \mathbb{R}^n : u \cdot f = u \cdot Ku > 0$. However, if the structure is not well restrained, then K is not invertible and $\exists u \in \mathbb{R}^n$ such that $Ku = 0$. If the material of the structure is elastic, then the global stiffness matrix of the structure is semi–positive definite since $\forall u \in \mathbb{R}^n : u \cdot Ku \geq 0$. Most finite–element analysis software will check the invertibility of the stiffness matrix by finding the smallest eigenvalue of the stiffness matrix. If one of the eigenvalues of the stiffness matrix is zero, then the stiffness matrix is not invertible. If one of the eigenvalues is negative, then the stiffness matrix is not positive definite, since positive definiteness ensures positive eigenvalues (section 1.2.6.g). A direct consequence of the positive definiteness is that the diagonal entries of the stiffness matrix have to be positive; since a diagonal element represents the force applied at the corresponding degree of freedom to produce unit displacement at this particular degree of freedom, then the requirement that the energy is positive ensures that the force and the displacement are in the same direction, and thus, the diagonal entry has to be positive.
- Any vertical column of the stiffness matrix is equal to the vector of nodal forces required to move the corresponding degree of freedom by one unit displacement while keeping the remaining degrees of freedom equal to zero.

9.1.2. One-Dimensional 3-Node Quadratic Element

In the previous section, the domain of a one–dimensional problem was discretized using piecewise linear functions. Another approximation is the C0 piecewise quadratic interpolation functions, where on each element, the displacement is quadratic. The displacement function is continuous across nodes. However, the displacement function is not differentiable at the bound–aries between elements. Each linear quadratic element possesses two boundary nodes and an

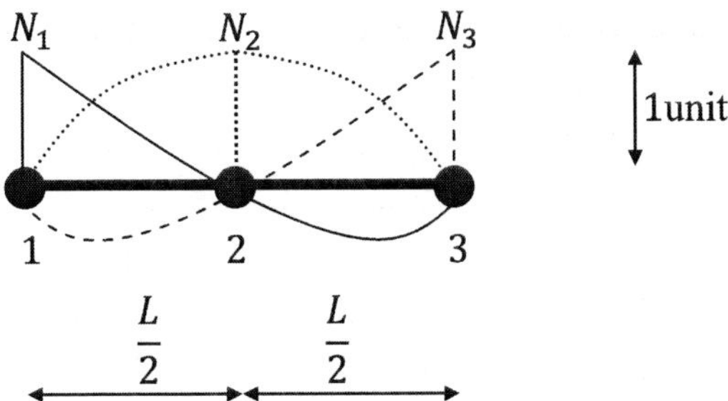

FIGURE 9-7 **Quadratic interpolation functions for a one–dimensional element.**

intermediate node. The trial displacement interpolation function has the form:

$$\tilde{u} = \langle N_1 \quad N_2 \quad N_3 \rangle \begin{Bmatrix} u_{1e} \\ u_{2e} \\ u_{3e} \end{Bmatrix}$$

The shape functions N_i can be obtained intuitively as follows. N_1 is a quadratic shape function that is equal to 1, 0 and 0 at $X_1 = 0, \frac{L}{2}$ and L, respectively. N_2 is a quadratic shape function that is equal to 0, 1 and 0 at $X_1 = 0, \frac{L}{2}$ and L, respectively, and N_3 is a quadratic shape function that is equal to 0, 0 and 1 at $X_1 = 0, \frac{L}{2}$ and L respectively (Figure 9–7). Therefore:

$$N_1 = \frac{2(L - X_1)\left(\frac{L}{2} - X_1\right)}{L^2}, \quad N_2 = \frac{4X_1(L - X_1)}{L^2}, \quad N_3 = -\frac{2X_1\left(\frac{L}{2} - X_1\right)}{L^2}$$

It can be checked that the sum of the shape functions is always equal to 1, which allows for the rigid–body motion or constant displacement of an element (**Why?**). The local stiffness matrix and the local nodal forces vectors can be established similar to the previous section, as follows:

$$\tilde{u} = \langle N_1 \quad N_2 \quad N_3 \rangle \begin{Bmatrix} u_{1e} \\ u_{2e} \\ u_{3e} \end{Bmatrix}, \quad \varepsilon_{11} = \left\langle \frac{dN_1}{dX_1} \quad \frac{dN_2}{dX_1} \quad \frac{dN_3}{dX_1} \right\rangle \begin{Bmatrix} u_{1e} \\ u_{2e} \\ u_{3e} \end{Bmatrix}, \quad \sigma_{11} = E\left\langle \frac{dN_1}{dX_1} \quad \frac{dN_2}{dX_1} \quad \frac{dN_3}{dX_1} \right\rangle \begin{Bmatrix} u_{1e} \\ u_{2e} \\ u_{3e} \end{Bmatrix}$$

The virtual displacement and the associated strain fields are:

$$u^* = \langle N_1 \quad N_2 \quad N_3 \rangle \begin{Bmatrix} u_{1e}^* \\ u_{2e}^* \\ u_{3e}^* \end{Bmatrix}, \quad \varepsilon_{11}^* = \left\langle \frac{dN_1}{dX_1} \quad \frac{dN_2}{dX_1} \quad \frac{dN_3}{dX_1} \right\rangle \begin{Bmatrix} u_{1e}^* \\ u_{2e}^* \\ u_{3e}^* \end{Bmatrix}$$

Denoting the matrix $B = \begin{pmatrix} \frac{dN_1}{dX_1} & \frac{dN_2}{dX_1} & \frac{dN_3}{dX_1} \end{pmatrix}$ and $B^T = \begin{Bmatrix} \frac{dN_1}{dX_1} \\ \frac{dN_2}{dX_1} \\ \frac{dN_3}{dX_1} \end{Bmatrix}$, then the virtual work expression is:

$$A \int_e B^T EBdX_1 \begin{Bmatrix} u_{1e} \\ u_{2e} \\ u_{3e} \end{Bmatrix} = \int_e \begin{Bmatrix} fN_1 \\ fN_2 \\ fN_3 \end{Bmatrix} dX_1$$

The stiffness matrix in this case $K = A \int_e B^T EBdX_1 \in \mathbb{M}^3$ (K has dimensions 3×3).

9.2. FINITE ELEMENT ANALYSIS IN TWO AND THREE DIMENSIONS

The basic types of elements used in two dimensions are either the quadrilateral (four–sided) or the triangular (three–sided) elements. The quadrilateral elements can either have bilinear or quadratic displacement functions. The bilinear displacement function element possesses four nodes at the four corners of the element, while the quadratic displacement function element possesses eight nodes, four at the corners of the element and four at the midpoints of the sides. The triangular elements can have either bilinear or quadratic displacement functions. The bilinear displacement function element possesses three nodes at the corners of the triangle, and the quadratic displacement function element possesses six nodes, three nodes at the corners and three nodes at the midpoints of the sides. For three dimensions, the eight–node trilinear brick element is the three–dimensional extension of the bilinear quadrilateral and the 20–node brick element is the three–dimensional extension of the quadratic quadrilateral. The four–node tetrahedron is the three–dimensional extension of the bilinear triangular element, while the ten–node tetrahedron is the three–dimensional extension of the quadratic displacement function triangular element. The choice of the element type depends on many aspects. For domains with irregular geometry, the triangle or the tetrahedron could be the favorable choice because of the ease of discretizing irregular domains with triangular meshes. However, linear triangular elements (as will be shown later) tend to be stiffer than other types of elements, and thus, finer meshes or the nonlinear triangular elements should be used, which in turn require greater computational resources. The quadrilateral element could provide better accuracy with fewer computational requirements but requires regular geometry to be discretized. In the following section, the general equations for two–dimensional problems in linear elasticity are formulated for general elements. Then, the derivation of the different shape functions for the bilinear and quadratic triangular and quadri–lateral two–dimensional elements is presented.

9.2.1. General Equations for Finite Element Analysis of Problems in Linear Elasticity

In three dimensions, the trial displacement function is a vector–valued function that has three components in the three directions of the coordinates X_1, X_2, and X_3:

$$\tilde{u} = \begin{Bmatrix} \tilde{u}_1 \\ \tilde{u}_2 \\ \tilde{u}_3 \end{Bmatrix} = \begin{Bmatrix} u \\ v \\ w \end{Bmatrix} = \begin{Bmatrix} N_1 u_1 + N_2 u_2 + \cdots \\ N_1 v_1 + N_2 v_2 + \cdots \\ N_1 w_1 + N_2 w_2 + \cdots \end{Bmatrix} = \begin{pmatrix} N_1 & 0 & 0 & N_2 & 0 & 0 & \ldots \\ 0 & N_1 & 0 & 0 & N_2 & 0 & \ldots \\ 0 & 0 & N_1 & 0 & 0 & N_2 & \ldots \end{pmatrix} \begin{Bmatrix} u_1 \\ v_1 \\ w_1 \\ u_2 \\ v_2 \\ w_2 \\ . \\ . \\ . \end{Bmatrix} = N u_e$$

where

u, v and w are used to represent the components of the displacement function $\tilde{u}$ in the direction of the three orthonormal basis vectors.

u_i, v_i and w_i are the three components of the displacement at node number i. These are also called the degrees of freedom of the element since they control the values of the displacements, strains, and stresses across the whole element.

u_e is the vector of the degrees of freedom of the element.

N is the matrix of shape functions.

For simplifying the equations, the vector representation of the stress and strain tensors (See section 5.2.1.) will be used. The small strain associated with the above trial–displacement function has the following form:

$$\varepsilon = \begin{Bmatrix} \varepsilon_{11} \\ \varepsilon_{22} \\ \varepsilon_{33} \\ \gamma_{12} \\ \gamma_{13} \\ \gamma_{23} \end{Bmatrix} = \begin{Bmatrix} \frac{\partial u}{\partial X_1} \\ \frac{\partial v}{\partial X_2} \\ \frac{\partial w}{\partial X_3} \\ \frac{\partial u}{\partial X_2} + \frac{\partial v}{\partial X_1} \\ \frac{\partial u}{\partial X_3} + \frac{\partial w}{\partial X_1} \\ \frac{\partial v}{\partial X_3} + \frac{\partial w}{\partial X_2} \end{Bmatrix} = \begin{pmatrix} \frac{\partial N_1}{\partial X_1} & 0 & 0 & \frac{\partial N_2}{\partial X_1} & 0 & 0 & \cdots \\ 0 & \frac{\partial N_1}{\partial X_2} & 0 & 0 & \frac{\partial N_2}{\partial X_2} & 0 & \cdots \\ 0 & 0 & \frac{\partial N_1}{\partial X_3} & 0 & 0 & \frac{\partial N_2}{\partial X_3} & \cdots \\ \frac{\partial N_1}{\partial X_2} & \frac{\partial N_1}{\partial X_1} & 0 & \frac{\partial N_2}{\partial X_2} & \frac{\partial N_2}{\partial X_1} & 0 & \cdots \\ \frac{\partial N_1}{\partial X_3} & 0 & \frac{\partial N_1}{\partial X_1} & \frac{\partial N_2}{\partial X_3} & 0 & \frac{\partial N_2}{\partial X_1} & \cdots \\ 0 & \frac{\partial N_1}{\partial X_3} & \frac{\partial N_1}{\partial X_2} & 0 & \frac{\partial N_1}{\partial X_3} & \frac{\partial N_1}{\partial X_2} & \cdots \end{pmatrix} \begin{Bmatrix} u_1 \\ v_1 \\ w_1 \\ u_2 \\ v_2 \\ w_2 \\ \cdot \\ \cdot \\ \cdot \end{Bmatrix} = Bu_e$$

For a general linear elastic material, the relationship between the stress and the strain is:

$$\sigma = \begin{Bmatrix} \sigma_{11} \\ \sigma_{22} \\ \sigma_{33} \\ \sigma_{12} \\ \sigma_{13} \\ \sigma_{23} \end{Bmatrix} = C \begin{Bmatrix} \varepsilon_{11} \\ \varepsilon_{22} \\ \varepsilon_{33} \\ \gamma_{12} \\ \gamma_{13} \\ \gamma_{23} \end{Bmatrix}$$

where C is a 6×6 matrix that relates the stresses to the strains (section 5.2.1).

If the principle of virtual work is adopted (section 7.4), the virtual displacement field and the associated virtual strain field can have the following forms with u_e^* being an arbitrary "virtual" vector of degrees of freedom of the system:

$$u^* = Nu_e^*, \quad \varepsilon^* = Bu_e^*$$

The expression for the internal virtual work IVW has the following form:

$$IVW = \int_e \sum_{i,j=1}^{3} \sigma_{ij}\varepsilon_{ij}^* dV = u_e^{*T} \left(\int B^T CBdV \right) u_e$$

The expression for the external virtual work EVW has the following form:

$$EVW = u_e^{*T} \int N^T t_n dS + u_e^{*T} \int N^T \rho b dV$$

Equating the internal virtual work with the external virtual work results in the following scalar–valued equation:

$$u_e^{*T}\left(\int B^T CBdV\right)u_e = u_e^{*T}\int N^T t_n dS + u_e^{*T}\int N^T \rho bdV$$

In the above scalar–valued equation, the arbitrariness of the components of the u_e^{*T} vector allows equating the multipliers of each component of the vector u_e^* from both sides resulting in the n dimensional vector–valued equation:

$$Ku_e = f_e$$

where,

K is the stiffness matrix having the dimensions $n \times n$, with n being the number of degrees of free–dom of the element and $K = \int B^T CBdV$,

$u_e \in \mathbb{R}^n$ is a vector of the nodal displacements of the element.

$f_e = \int N^T t_n dS + \int N^T \rho bdV \in \mathbb{R}^n$ is a vector of the nodal forces of the element.

In the remainder of section 9.2, several different elements with different shape functions will be investigated, and the stiffness matrix of each element will be derived following the procedure out–lined above.

9.2.2. Triangular Elements

9.2.2.a. Linear Triangular Elements (Constant Strain Triangle)

The linear triangular element is obtained by assuming that the displacement within a triangular shape is a linear function of the displacements at the three corner nodes. This linear triangu–lar element is called the constant strain triangle, since as will be shown in the derivation below, the strain across the whole element is always constant (the matrix B has constant values). This is a major disadvantage of such element since it tends to be extremely stiff. Approximating the domain with triangular elements can only be considered within good accuracy if the difference in the strains between neighboring elements is relatively small.

Consider the plane triangle shown in Figure 9–8. For simplicity, two of the triangle sides are assumed to have unit lengths. The trial linear–displacement function across the element has the

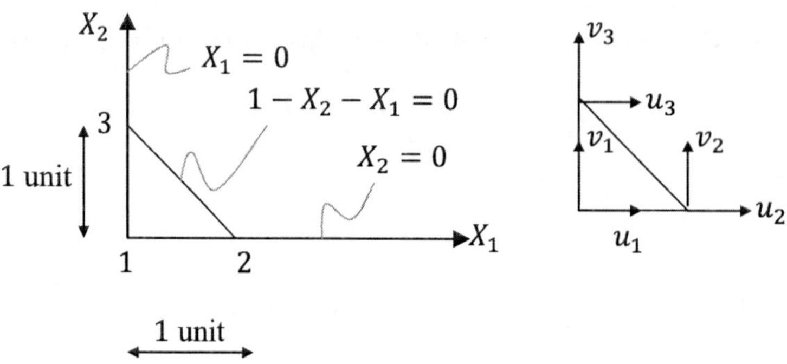

FIGURE 9-8 Constant strain triangle.

following form:

$$\tilde{u} = \begin{Bmatrix} \tilde{u}_1 \\ \tilde{u}_2 \end{Bmatrix} = \begin{Bmatrix} u \\ v \end{Bmatrix} = \begin{Bmatrix} N_1u_1 + N_2u_2 + N_3u_3 \\ N_1v_1 + N_2v_2 + N_3v_3 \end{Bmatrix} = \begin{pmatrix} N_1 & 0 & N_2 & 0 & N_3 & 0 \\ 0 & N_1 & 0 & N_2 & 0 & N_3 \end{pmatrix} \begin{Bmatrix} u_1 \\ v_1 \\ u_2 \\ v_2 \\ u_3 \\ v_3 \end{Bmatrix} = Nu_e$$

The shape functions satisfy the following: $N_1 = 1, 0$ and 0 at nodes 1, 2 and 3, respectively. $N_2 = 0, 1$ and 0 at nodes 1, 2, and 3, respectively, and $N_3 = 0, 0$ and 1 at nodes 1, 2, and 3 respectively. Thus, the shape functions have the following form:

$$N_1 = (1 - X_1 - X_2), \quad N_2 = X_1, \quad N_3 = X_2$$

Notice that it is possible to rewrite the displacement interpolation function using the following form in which the six degrees of freedom u_i and v_i are replaced with six general degrees of freedom a_i and b_i as follows:

$$u = a_1 + a_2X_1 + a_3X_2, \quad v = b_1 + b_2X_1 + b_3X_2$$

▶ **BONUS EXERCISE:** Find the relationship between the degrees of freedom u_i and v_i and the general degrees of freedom a_i and b_i.

The strain associated with the assumed trial–displacement field has the following form:

$$\varepsilon = \begin{Bmatrix} \varepsilon_{11} \\ \varepsilon_{22} \\ \gamma_{12} \end{Bmatrix} = \begin{pmatrix} \frac{\partial N_1}{\partial X_1} & 0 & \frac{\partial N_2}{\partial X_1} & 0 & \frac{\partial N_3}{\partial X_1} & 0 \\ 0 & \frac{\partial N_1}{\partial X_2} & 0 & \frac{\partial N_2}{\partial X_2} & 0 & \frac{\partial N_3}{\partial X_2} \\ \frac{\partial N_1}{\partial X_2} & \frac{\partial N_1}{\partial X_1} & \frac{\partial N_2}{\partial X_2} & \frac{\partial N_2}{\partial X_1} & \frac{\partial N_3}{\partial X_2} & \frac{\partial N_3}{\partial X_1} \end{pmatrix} \begin{Bmatrix} u_1 \\ v_1 \\ u_2 \\ v_2 \\ u_3 \\ v_3 \end{Bmatrix} = \begin{pmatrix} -1 & 0 & 1 & 0 & 0 & 0 \\ 0 & -1 & 0 & 0 & 0 & 1 \\ -1 & -1 & 0 & 1 & 1 & 0 \end{pmatrix} \begin{Bmatrix} u_1 \\ v_1 \\ u_2 \\ v_2 \\ u_3 \\ v_3 \end{Bmatrix}$$

Notice that as mentioned above, the values of the strain components are independent of the location inside the triangle; rather the strain components are constant across the element.

Stiffness Matrix:

The material constitutive relationship C is a 3×3 matrix that depends on whether the material is in a plane strain or a plane stress state (See section 5.2.7.). For a linear–elastic isotropic material in the plane strain state, the material constitutive relationship matrix is:

$$C = \frac{E}{1+\nu} \begin{pmatrix} \frac{1-\nu}{1-2\nu} & \frac{\nu}{1-2\nu} & 0 \\ \frac{\nu}{1-2\nu} & \frac{1-\nu}{1-2\nu} & 0 \\ 0 & 0 & \frac{1}{2} \end{pmatrix}$$

The stiffness matrix of the constant strain triangle can be evaluated by the following integral:

$$K = \int B^T CB dV = \int \begin{pmatrix} -1 & 0 & -1 \\ 0 & -1 & -1 \\ 1 & 0 & 0 \\ 0 & 0 & 1 \\ 0 & 0 & 1 \\ 0 & 1 & 0 \end{pmatrix} C \begin{pmatrix} -1 & 0 & 1 & 0 & 0 & 0 \\ 0 & -1 & 0 & 0 & 0 & 1 \\ -1 & -1 & 0 & 1 & 1 & 0 \end{pmatrix} dV$$

Assuming that the triangle has a thickness t, then the differential volume $dV = tdX_1 dX_2$. In addition, all the components of the matrix $B^T CB$ are constant; therefore, the integral can be evaluated by simply multiplying the matrix $B^T CB$ by $\frac{t}{2}$, which represents the volume of the triangle:

$$K = t\int_0^1\int_0^{1-X_2} B^T CBdX_1dX_2 = \frac{t}{2}B^T CB = \frac{tE}{2(1+\nu)}\begin{pmatrix} \frac{3-4\nu}{2-4\nu} & \frac{1}{2-4\nu} & \frac{1-\nu}{-1+2\nu} & -\frac{1}{2} & -\frac{1}{2} & \frac{\nu}{-1+2\nu} \\ & \frac{3-4\nu}{2-4\nu} & \frac{\nu}{-1+2\nu} & -\frac{1}{2} & -\frac{1}{2} & \frac{1-\nu}{-1+2\nu} \\ & & \frac{1-\nu}{1-2\nu} & 0 & 0 & \frac{\nu}{1-2\nu} \\ & & & \frac{1}{2} & \frac{1}{2} & 0 \\ & symm & & & \frac{1}{2} & 0 \\ & & & & & \frac{1-\nu}{1-2\nu} \end{pmatrix}$$

The following Mathematica code was utilized to obtain the above matrix:

```
B={{-1,0,1,0,0,0},{0,-1,0,0,0,1},{-1,-1,0,1,1,0}};
Cc=Ee/(1+nu)*{{(1-nu)/(1-2nu),nu/(1-2nu),0},{(nu)/(1-2nu),(1-nu)/
(1-2nu),0},{0,0,1/2}};
K=FullSimplify[Transpose[B].Cc.B];
FullSimplify[1/Ee*(1+nu)*K]//MatrixForm
```

Nodal Loads due to Body Forces:

Assuming that the distributed body forces vector per unit mass is $b = \{b_1, b_2\}$, and the density is ρ, then the nodal forces due to the distributed body forces can be obtained using the integral:

$$f_e = \int N^T \rho bdV = \rho t\int_0^1\int_0^{1-X_2}\begin{pmatrix} N_1 & 0 \\ 0 & N_1 \\ N_2 & 0 \\ 0 & N_2 \\ N_3 & 0 \\ 0 & N_3 \end{pmatrix}\begin{pmatrix} b_1 \\ b_2 \end{pmatrix} dX_1dX_2 = \frac{\rho t}{6}\begin{pmatrix} b_1 \\ b_2 \\ b_1 \\ b_2 \\ b_1 \\ b_2 \end{pmatrix}$$

Thus, a *constant* distributed body forces vector can be represented by equal nodal loads applied on the element nodes (Figure 9–9).

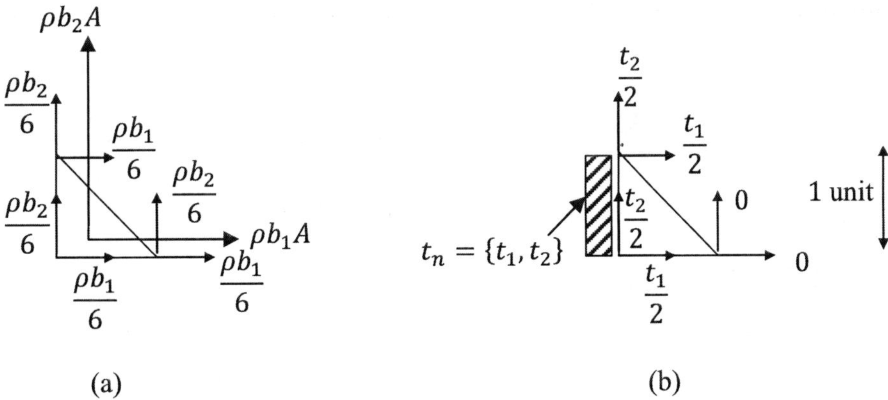

FIGURE 9-9 Discretization of (a) the distributed body forces and (b) the distributed traction vector on the constant strain triangle assuming unit thickness ($A = area = \frac{1}{2}$units, $t = 1$ units).

The following is the Mathematica code used for the above calculations:

```
Shapefun=Table[0,{i,1,3}]
Shapefun[[1]]=1-x1-x2;
Shapefun[[2]]=x1;
Shapefun[[3]]=x2;
Nn=Table[0,{i,1,2},{j,1,6}];
Do[Nn[[1,2i-1]]=Nn[[2,2i]]=Shapefun[[i]],{i,1,3}];
rb={rb1,rb2};
Integrate[Transpose[Nn].rb,{x2,0,1},{x1,0,1-x2}]//MatrixForm
```

Nodal Loads Due to Distributed Traction Forces:

Assuming that a constant distributed pressure per unit area of $t_n = \{t_1, t_2\}$ is acting on the surface joining nodes 1 and 3, then the nodal forces due to the distributed traction vector can be obtained using the following integral evaluated on the left side ($X_1 = 0$):

$$f_e = \int N^T t_n dS = t \int_0^1 \begin{pmatrix} N_1 & 0 \\ 0 & N_1 \\ N_2 & 0 \\ 0 & N_2 \\ N_3 & 0 \\ 0 & N_3 \end{pmatrix} \begin{pmatrix} t_1 \\ t_2 \end{pmatrix} dX_2 = \frac{t}{2} \begin{pmatrix} t_1 \\ t_2 \\ 0 \\ 0 \\ t_1 \\ t_2 \end{pmatrix}$$

Thus, the distributed *constant* traction on one side of the triangle can be discretized into equal nodal loads applied on the nodes of that side (Figure 9–9). The following Mathematica code was used for the above calculations:

```
Shapefun=Table[0,{i,1,3}]
Shapefun[[1]]=1-x1-x2;
Shapefun[[2]]=x1;
Shapefun[[3]]=x2;
Nn=Table[0,{i,1,2},{j,1,6}];
Do[Nn[[1,2i-1]]=Nn[[2,2i]]=Shapefun[[i]],{i,1,3}];
tn={t1,t2};
Integrate[(Transpose[Nn].tn/.x1->0),{x2,0,1}]//MatrixForm
```

9.2.2.b. Quadratic Triangular Element (Linear-Strain Triangle)

The quadratic triangular element offers a better approximation to the displacement field within a triangular element by introducing additional nodes on the straight sides of the triangle. The quadratic triangular element is called the linear–strain triangle since, as will be shown in the derivation below, the matrix B contains linear expressions in the coordinates X_1, and X_2 and there–fore, this element is capable of modeling linear strains (for example, bending). However, the addi–tion of nodes comes with a higher computational price compared to its linear counterpart.

Consider the plane triangle shown in Figure 9–10. For simplicity, two of the triangle sides are assumed to have unit lengths. The trial displacement function can be assumed to have the fol–lowing form:

$$\tilde{u} = \begin{Bmatrix} \tilde{u}_1 \\ \tilde{u}_2 \end{Bmatrix} = \begin{Bmatrix} u \\ v \end{Bmatrix} = \begin{Bmatrix} \sum_{i=1}^{6} N_i u_i \\ \sum_{i=1}^{6} N_i v_i \end{Bmatrix} = \begin{pmatrix} N_1 & 0 & N_2 & 0 & N_3 & 0 & N_4 & 0 & N_5 & 0 & N_6 & 0 \\ 0 & N_1 & 0 & N_2 & 0 & N_3 & 0 & N_4 & 0 & N_5 & 0 & N_6 \end{pmatrix} \begin{Bmatrix} u_1 \\ v_1 \\ u_2 \\ v_2 \\ u_3 \\ v_3 \\ u_4 \\ v_4 \\ u_5 \\ v_5 \\ u_6 \\ v_6 \end{Bmatrix} = N u_e$$

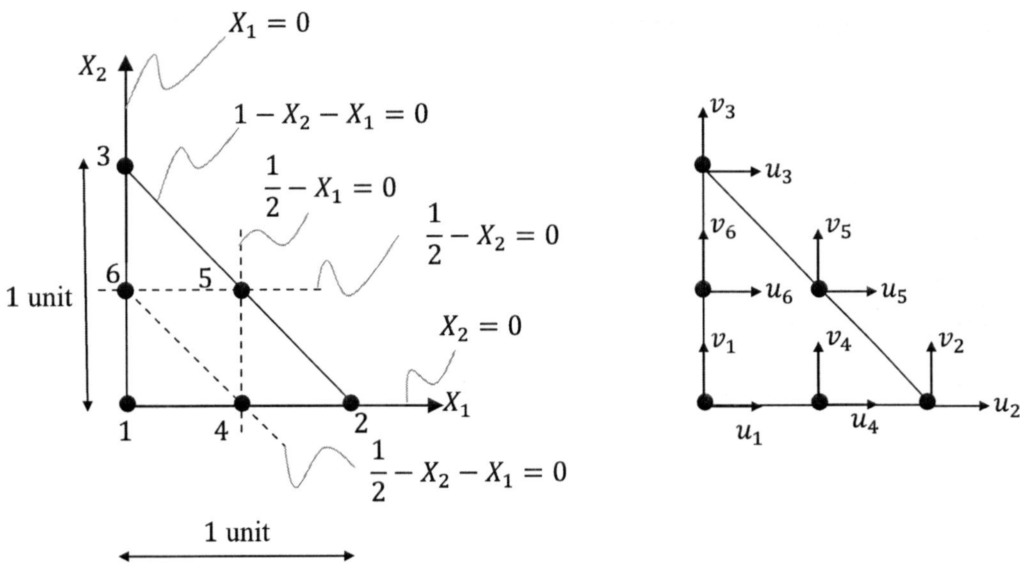

FIGURE 9-10 Quadratic triangle.

The shape functions satisfy the following: $N_i = 1$ at node i and 0 at all the other nodes. Thus, the shape functions have the following form:

$$N_1 = 2(1 - X_1 - X_2)\left(\frac{1}{2} - X_1 - X_2\right), \quad N_2 = 2\left(X_1 - \frac{1}{2}\right)X_1, \quad N_3 = 2\left(X_2 - \frac{1}{2}\right)X_2$$

$$N_4 = 4X_1(1 - X_1 - X_2), \quad N_5 = 4X_1X_2, \quad N_6 = 4X_2(1 - X_1 - X_2)$$

The sum of the shape functions is equal to 1, indicating that a constant displacement can be mod–eled (**Why?**).

Notice that it is possible to rewrite the displacement interpolation function using the following form in which the 12 degrees of freedom u_i and v_i are replaced with 12 general degrees of freedom a_i and b_i as follows:

$$u = a_1 + a_2X_1 + a_3X_2 + a_4X_1^2 + a_5X_1X_2 + a_6X_2^2, \quad v = b_1 + b_2X_1 + b_3X_2 + b_4X_1^2 + b_5X_1X_2 + b_6X_2^2$$

▶ BONUS EXERCISE: Find the relationship between the degrees of freedom u_i and v_i and the general degrees of freedom a_i and b_i.

The strain associated with the assumed–trial displacement field has the following form:

$$\varepsilon = \begin{Bmatrix} \varepsilon_{11} \\ \varepsilon_{22} \\ \gamma_{12} \end{Bmatrix} = Bu_e$$

where B is a 3×12 matrix containing entries in the form of $\frac{dN_i}{dX_j}$:

$$B = \begin{pmatrix} \frac{dN_1}{dX_1} & 0 & \frac{dN_2}{dX_1} & 0 & \frac{dN_3}{dX_1} & 0 & \frac{dN_4}{dX_1} & 0 & \frac{dN_5}{dX_1} & 0 & \frac{dN_6}{dX_1} & 0 \\ 0 & \frac{dN_1}{dX_2} & 0 & \frac{dN_2}{dX_2} & 0 & \frac{dN_3}{dX_2} & 0 & \frac{dN_4}{dX_2} & 0 & \frac{dN_5}{dX_2} & 0 & \frac{dN_6}{dX_2} \\ \frac{dN_1}{dX_2} & \frac{dN_1}{dX_1} & \frac{dN_2}{dX_2} & \frac{dN_2}{dX_1} & \frac{dN_3}{dX_2} & \frac{dN_3}{dX_1} & \frac{dN_4}{dX_2} & \frac{dN_4}{dX_1} & \frac{dN_5}{dX_2} & \frac{dN_5}{dX_1} & \frac{dN_6}{dX_2} & \frac{dN_6}{dX_1} \end{pmatrix}$$

The different entries of the matrix B are indeed linear functions of the coordinates X_1 and X_2 inside the domain. Thus, this element can be used to accurately model a domain with linear strain across the element.

Stiffness Matrix:

Similar to the derivation for the linear triangular element, the material will be assumed to be in the plane strain state (See section 5.2.7.). For a linear–elastic isotropic material in the plane strain state, the material constitutive relationship matrix is:

$$C = \frac{E}{1+\nu} \begin{pmatrix} \frac{1-\nu}{1-2\nu} & \frac{\nu}{1-2\nu} & 0 \\ \frac{\nu}{1-2\nu} & \frac{1-\nu}{1-2\nu} & 0 \\ 0 & 0 & \frac{1}{2} \end{pmatrix}$$

Assuming that t is the thickness of the triangle, then the differential volume $dV = tdX_1dX_2$. The stiffness matrix can be obtained by evaluating the following integral on the triangular area:

$$K = t\int_0^1 \int_0^{1-X_2} B^T CBdX_1dX_2$$

The stiffness matrix has the dimensions of 12×12, and the following Mathematica code can be utilized to view its components:

```
Shapefun=Table[0,{i,1,6}];
Shapefun[[1]]=2(1-x1-x2)(1/2-x1-x2);
Shapefun[[2]]=2x1 (x1-1/2);
Shapefun[[3]]=2x2 (x2-1/2);
Shapefun[[4]]=4x1 (1-x1-x2);
Shapefun[[5]]=4x1*x2;
Shapefun[[6]]=4x2*(1-x1-x2);
B=Table[0,{i,1,3},{j,1,12}];
Do[B[[1,2i-1]]=B[[3,2i]]=D[Shapefun[[i]],x1];B[[2,2i]]=B[[3,2i-
1]]=D[Shapefun[[i]],x2],{i,1,6}];
Cc=Ee/(1+nu)*{{(1-nu)/(1-2nu),nu/(1-2nu),0},{(nu)/(1-2nu),(1-nu)/(1-
2nu),0},{0,0,1/2}};
Kbeforeintegration=t*FullSimplify[Transpose[B].Cc.B];
K=Integrate[Kbeforeintegration,{x2,0,1},{x1,0,1-x2}];
K//MatrixForm
```

Notice that the integration for this element requires high computational time (Try using Math–ematica.). If a structure is composed of hundreds of elements, the construction of the stiffness matrix would require high computational resources.

Nodal Loads Due to Body Forces:

Assuming that the distributed body–forces vector per unit mass is $b = \{b_1, b_2\}$, and the density is ρ, then the nodal forces due to the distributed body forces vector can be obtained using the integral:

$$f_e = \int N^T \rho b dV = \frac{\rho t}{6}\begin{pmatrix} 0 \\ 0 \\ 0 \\ 0 \\ 0 \\ 0 \\ b_1 \\ b_2 \\ b_1 \\ b_2 \\ b_1 \\ b_2 \end{pmatrix}$$

Thus, a *constant* distributed body–forces vector can be discretized into equal loads applied on the mid–side nodes and zero loads applied on the corner nodes (Figure 9–11).

The following is the Mathematica code used for the above calculations:

```
Shapefun=Table[0,{i,1,6}];
Shapefun[[1]]=2(1-x1-x2)(1/2-x1-x2);
Shapefun[[2]]=2x1 (x1-1/2);
Shapefun[[3]]=2x2 (x2-1/2);
Shapefun[[4]]=4x1 (1-x1-x2);
Shapefun[[5]]=4x1*x2;
Shapefun[[6]]=4x2*(1-x1-x2);
Nn=Table[0,{i,1,2},{j,1,12}];
Do[Nn[[1,2i-1]]=Nn[[2,2i]]=Shapefun[[i]],{i,1,6}];
rb={rb1,rb2};
fe=Integrate[Transpose[Nn].rb,{x2,0,1},{x1,0,1-x2}]//MatrixForm
```

Nodal Loads Due to Distributed Traction Forces:

Assuming that a constant distributed pressure per unit area of $t_n = \{t_1, t_2\}$ is acting on the surface joining nodes 1 and 3, then the nodal forces due to the distributed traction vector can be obtained using the following integral evaluated on the left side ($X_1 = 0$):

$$f_e = \int N^T t_n dS = \frac{t}{6}\begin{pmatrix} t_1 \\ t_2 \\ 0 \\ 0 \\ t_1 \\ t_2 \\ 0 \\ 0 \\ 0 \\ 0 \\ 4t_1 \\ 4t_2 \end{pmatrix}$$

The appropriate nodal loads used to discretize a *constant* traction vector on the left side of the quadratic triangular element are shown in Figure 9–11. The following Mathematica code was used for the above calculations:

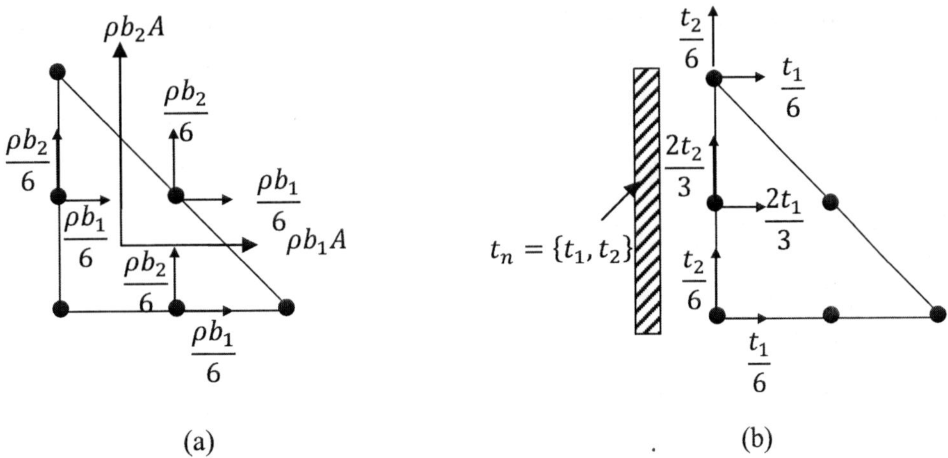

FIGURE 9-11 Discretization of (a) the distributed body forces and (b) the traction vector on the linear strain triangle, assuming unit thickness ($A = area = \frac{1}{2}$units, $t = 1$ units).

```
Shapefun=Table[0,{i,1,3}]
Shapefun[[1]]=1-x1-x2;
Shapefun[[2]]=x1;
Shapefun[[3]]=x2;
Nn=Table[0,{i,1,2},{j,1,6}];
Do[Nn[[1,2i-1]]=Nn[[2,2i]]=Shapefun[[i]],{i,1,3}];
tn={t1,t2};
Integrate[(Transpose[Nn].tn/.x1->0),{x2,0,1}]//MatrixForm
```

9.2.3. Quadrilateral Elements

9.2.3.a. Bilinear Quadrilateral (4-Node Plane Element)

It is usually easier to discretize a domain by using triangular elements; however, when the domain to be discretized has a regular shape, quadrilateral elements can be used to offer a more regular displacement discretization, which, in some cases, might offer a better approximation to the displacement shape. The bilinear quadrilateral element offers a bilinear displacement approximation where the displacement within the element is assumed to vary bilinearly (linear in two directions within the element). This element, however, behaves in a stiff manner when it is used to model linear strains (bending strains) since, as will be shown in the derivation, it is not possible to model pure bending (pure linear normal strain components in one direction) without introducing additional shear strains (usually termed parasitic shear strains).

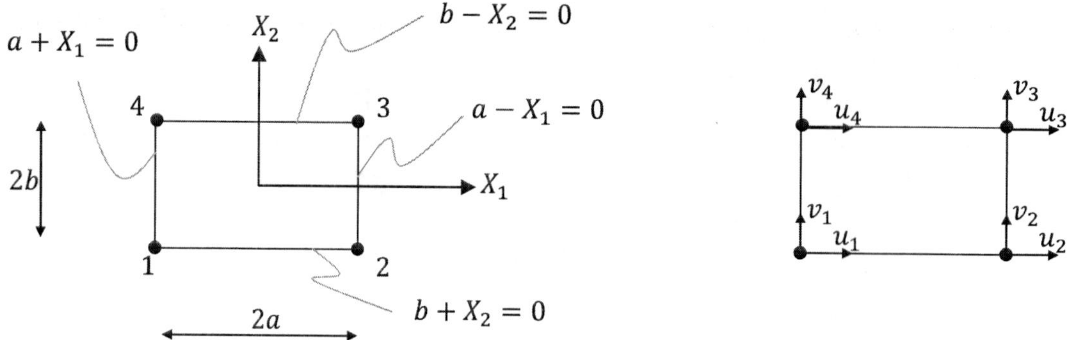

FIGURE 9-12 Bilinear quadrilateral.

Consider the plane quadrilateral shown in Figure 9–12. The trial displacement function has the following form:

$$\tilde{u} = \begin{Bmatrix} \tilde{u}_1 \\ \tilde{u}_2 \end{Bmatrix} = \begin{Bmatrix} u \\ v \end{Bmatrix} = \begin{Bmatrix} \sum_{i=1}^{4} N_i u_i \\ \sum_{i=1}^{4} N_i v_i \end{Bmatrix} = \begin{pmatrix} N_1 & 0 & N_2 & 0 & N_3 & 0 & N_4 & 0 \\ 0 & N_1 & 0 & N_2 & 0 & N_3 & 0 & N_4 \end{pmatrix} \begin{Bmatrix} u_1 \\ v_1 \\ u_2 \\ v_2 \\ u_3 \\ v_3 \\ u_4 \\ v_4 \end{Bmatrix} = N u_e$$

The shape functions satisfy the following: $N_i = 1$ at node i and 0 at all the other nodes. Thus, the shape functions have the following form:

$$N_1 = \frac{(b - X_2)(a - X_1)}{4ab}, \quad N_2 = \frac{(b - X_2)(a + X_1)}{4ab}, \quad N_3 = \frac{(b + X_2)(a + X_1)}{4ab}, \quad N_4 = \frac{(b + X_2)(a - X_1)}{4ab}$$

The sum of the shape functions is equal to 1, indicating that a constant displacement can be mod–eled (**Why?**).

Notice that it is possible to rewrite the displacement interpolation function using the following form in which the eight degrees of freedom u_i and v_i are replaced with eight general degrees of freedom a_i and b_i as follows:

$$u = a_1 + a_2X_1 + a_3X_2 + a_4X_1X_2, \quad v = b_1 + b_2X_1 + b_3X_2 + b_4X_1X_2$$

▶ **BONUS EXERCISE:** Find the relationship between the degrees of freedom u_i and v_i and the general degrees of freedom a_i and b_i.

The strain associated with the assumed trial–displacement field has the following form:

$$\varepsilon = \begin{Bmatrix} \varepsilon_{11} \\ \varepsilon_{22} \\ \gamma_{12} \end{Bmatrix}$$

$$= \begin{pmatrix} -\frac{(b-X_2)}{4ab} & 0 & \frac{(b-X_2)}{4ab} & 0 & \frac{(b+X_2)}{4ab} & 0 & -\frac{(b+X_2)}{4ab} & 0 \\ 0 & -\frac{(a-X_1)}{4ab} & 0 & -\frac{(a+X_1)}{4ab} & 0 & \frac{(a+X_1)}{4ab} & 0 & \frac{(a-X_1)}{4ab} \\ -\frac{(a-X_1)}{4ab} & -\frac{(b-X_2)}{4ab} & -\frac{(a+X_1)}{4ab} & \frac{(b-X_2)}{4ab} & \frac{(a+X_1)}{4ab} & \frac{(b+X_2)}{4ab} & \frac{(a-X_1)}{4ab} & -\frac{(b+X_2)}{4ab} \end{pmatrix} \begin{Bmatrix} u_1 \\ v_1 \\ u_2 \\ v_2 \\ u_3 \\ v_3 \\ u_4 \\ v_4 \end{Bmatrix}$$

Shear Locking (Parasitic Shear Strains):

A quick glance at the matrix B shows that the bilinear quadrilateral can model bending (linear strains) since the strains ε_{11} and ε_{22} contain linear expressions in X_1 and X_2. However, it will be shown here that to model linear strains, the element predicts an associated shear strain ε_{12} as well (termed parasitic shear strains). For simplicity, assume that $a = b = 1$, then the strains have the following forms:

$$\varepsilon_{11} = \frac{(-u_1+u_2+u_3-u_4)}{4} + \frac{(u_1-u_2+u_3-u_4)}{4}X_2$$

$$\varepsilon_{22} = \frac{(-v_1-v_2+v_3+v_4)}{4} + \frac{(v_1-v_2+v_3-v_4)}{4}X_1$$

$$\gamma_{12} = \frac{(-u_1-u_2+u_3+u_4)}{4} + \frac{(-v_1+v_2+v_3-v_4)}{4} + \frac{(u_1-u_2+u_3-u_4)}{4}X_1 + \frac{(v_1-v_2+v_3-v_4)}{4}X_2$$

Careful investigation of the above expressions shows that any bending strain where ε_{11} is a linear function of X_2 produces a "parasitic" shear strain, indicating that pure bending cannot be modeled! This can be noticed by comparing the multipliers of X_1 and X_2 in the expressions of the normal strains and the shear strains. To see this, first introduce the following independent variables:

$$a_0 = \frac{(-u_1+u_2+u_3-u_4)}{4}, \quad a_1 = \frac{(u_1-u_2+u_3-u_4)}{4}, \quad b_0 = \frac{(-v_1-v_2+v_3+v_4)}{4}$$

$$b_1 = \frac{(v_1-v_2+v_3-v_4)}{4}, \quad c_0 = \frac{(-u_1-u_2+u_3+u_4)}{4}, \quad c_1 = \frac{(-v_1+v_2+v_3-v_4)}{4}$$

then the strain expressions have the following forms:

$$\varepsilon_{11} = a_0 + a_1 X_2$$

$$\varepsilon_{22} = b_0 + b_1 X_1$$

$$\gamma_{12} = c_0 + c_1 + a_1 X_1 + b_1 X_2$$

The above expression shows that if a bending state of strain exists such that $a_1 \neq 0$, then the shear strain cannot be equal to zero, indicating that pure bending cannot be modeled with this element.

Stiffness Matrix:

Similar to the derivation for the linear triangular element, the material will be assumed to be in the plane strain state (See section 5.2.7.). For a linear–elastic isotropic material in the plane strain state, the material constitutive relationship matrix is:

$$C = \frac{E}{1+\nu} \begin{pmatrix} \frac{1-\nu}{1-2\nu} & \frac{\nu}{1-2\nu} & 0 \\ \frac{\nu}{1-2\nu} & \frac{1-\nu}{1-2\nu} & 0 \\ 0 & 0 & \frac{1}{2} \end{pmatrix}$$

Assuming that the element has a thickness t, then the differential volume $dV = t dX_1 dX_2$. The stiff–ness matrix can be obtained by evaluating the following integral on the rectangular area:

$$K = t \int_{-b}^{b} \int_{-a}^{a} B^T C B dX_1 dX_2$$

The stiffness matrix has the dimensions of 8×8. The following Mathematica code can be utilized to view its components:

```
Shapefun=Table[0,{i,1,4}];
Shapefun[[1]]=(b-x2)(a-x1)/4/a/b;
Shapefun[[2]]=(b-x2)(a+x1)/4/a/b;
Shapefun[[3]]=(b+x2)(a+x1)/4/a/b;
Shapefun[[4]]=(b+x2)(a-x1)/4/a/b;
B=Table[0,{i,1,3},{j,1,8}];
Do[B[[1,2i-1]]=B[[3,2i]]=D[Shapefun[[i]],x1];B[[2,2i]]=B[[3,2i-
1]]=D[Shapefun[[i]],x2],{i,1,4}];
Cc=Ee/(1+nu)*{{(1-nu)/(1-2nu),nu/(1-2nu),0},{(nu)/(1-2nu),(1-nu)/(1-
2nu),0},{0,0,1/2}};
Kbeforeintegration=t*FullSimplify[Transpose[B].Cc.B];
K=Integrate[Kbeforeintegration,{x2,-b,b},{x1,-a,a}];
K//MatrixForm
```

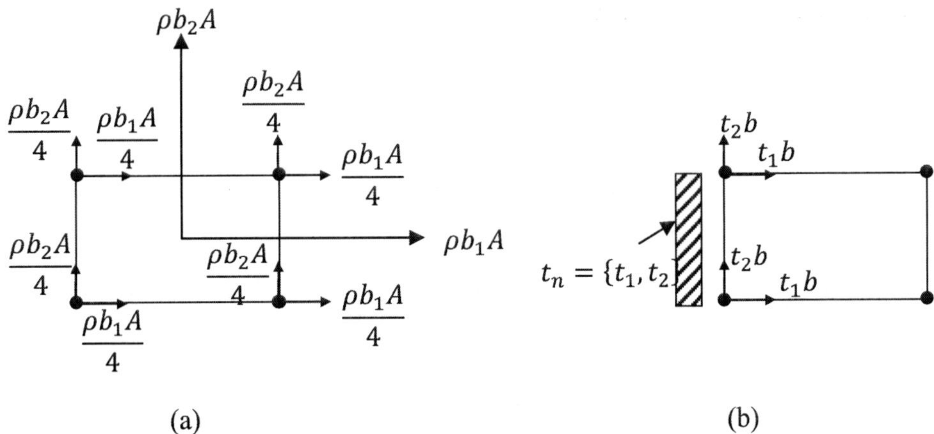

FIGURE 9-13 Discretization of (a) the distributed body forces and (b) the traction vector on the bilinear quadrilateral assuming unit thickness (t = 1units) ($A = 4ab$).

Nodal Loads:

The nodal loads due to distributed body forces and traction forces on the boundaries are equally distributed among the nodes. The same procedure followed for the triangular elements can be repeated to obtain the distribution shown in Figure 9–13.

The following Mathematica code can be utilized for the calculations producing the distribution in Figure 9–13:

```
Shapefun=Table[0,{i,1,4}];
Shapefun[[1]]=(b-x2)(a-x1)/4/a/b;
Shapefun[[2]]=(b-x2)(a+x1)/4/a/b;
Shapefun[[3]]=(b+x2)(a+x1)/4/a/b;
Shapefun[[4]]=(b+x2)(a-x1)/4/a/b;
Nn=Table[0,{i,1,2},{j,1,8}];
Do[Nn[[1,2i-1]]=Nn[[2,2i]]=Shapefun[[i]],{i,1,4}];
tn={t1,t2};
rb={rb1,rb2};
fetraction=Integrate[(Transpose[Nn].tn/.x1->-a),{x2,-b,b}]//MatrixForm
febodyforces=Integrate[(Transpose[Nn].rb),{x2,-b,b},{x1,-a,a}]//MatrixForm
```

9.2.3.b. Quadratic Quadrilateral (8-Node Plane Element)

By introducing four additional nodes in the mid–sides of a quadrilateral element, the displace–ment shape within the element can have a quadratic form, and the parasitic shear stiffness of the

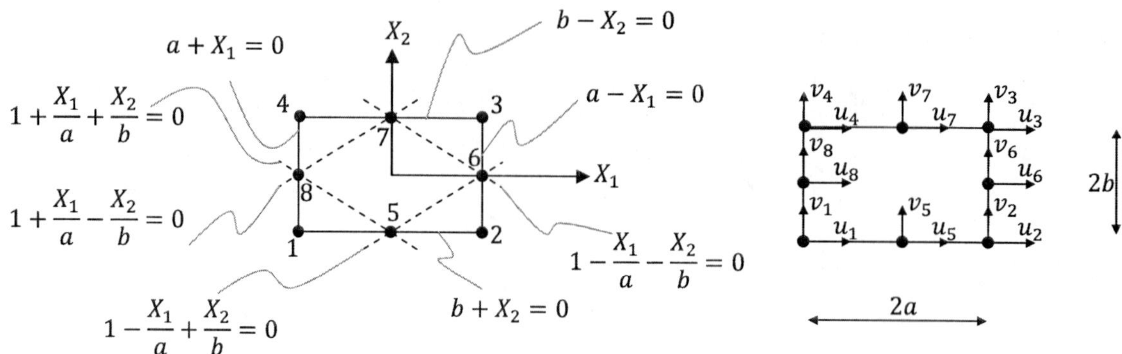

FIGURE 9-14 Quadratic quadrilateral.

bilinear quadrilateral can be avoided. The quadratic quadrilateral offers such advantage with the price of higher computational time due to having additional degrees of freedom.

Consider the plane quadrilateral shown in Figure 9–14. The trial displacement function has the following form:

$$\tilde{u} = \begin{Bmatrix} \tilde{u}_1 \\ \tilde{u}_2 \end{Bmatrix} = \begin{Bmatrix} u \\ v \end{Bmatrix} = \begin{Bmatrix} \sum_{i=1}^{8} N_i u_i \\ \sum_{i=1}^{8} N_i v_i \end{Bmatrix}$$

The shape functions satisfy the following: $N_i = 1$ at node i and 0 at all the other nodes. Thus, the shape functions have the following form:

$$N_1 = \frac{-(b - X_2)(a - X_1)\left(1 + \frac{X_1}{a} + \frac{X_2}{b}\right)}{4ab}, \quad N_2 = \frac{-(b - X_2)(a + X_1)\left(1 - \frac{X_1}{a} + \frac{X_2}{b}\right)}{4ab}$$

$$N_3 = \frac{-(b + X_2)(a + X_1)\left(1 - \frac{X_1}{a} - \frac{X_2}{b}\right)}{4ab}, \quad N_4 = \frac{-(b + X_2)(a - X_1)\left(1 + \frac{X_1}{a} - \frac{X_2}{b}\right)}{4ab}$$

$$N_5 = \frac{(b - X_2)(a - X_1)(a + X_1)}{2a^2 b}, \quad N_6 = \frac{(a + X_1)(b - X_2)(b + X_2)}{2ab^2}$$

$$N_7 = \frac{(b + X_2)(a - X_1)(a + X_1)}{2a^2 b}, \quad N_8 = \frac{(a - X_1)(b - X_2)(b + X_2)}{2ab^2}$$

The sum of the shape functions is equal to 1, indicating that a constant displacement can be mod–eled (**Why?**). Clearly, this element can model bending strains without any difficulty, but requires higher computational resources.

Notice that it is possible to rewrite the displacement interpolation function using the following form in which the eight degrees of freedom u_i and v_i are replaced with eight general degrees of freedom a_i and b_i as follows:

$$u = a_1 + a_2X_1 + a_3X_2 + a_4X_1X_2 + a_5X_1^2 + a_6X_2^2 + a_7X_1X_2^2 + a_8X_1^2X_2$$

$$v = b_1 + b_2X_1 + b_3X_2 + b_4X_1X_2 + b_5X_1^2 + b_6X_2^2 + b_7X_1X_2^2 + b_8X_1^2X_2$$

▶ **BONUS EXERCISE:** Find the relationship between the degrees of freedom u_i and v_i and the general degrees of freedom a_i and b_i.

Stiffness Matrix:

Similar to the derivation for the linear triangular element, the material will be assumed to be in the plane strain state (See section 5.2.7.). For a linear elastic isotropic material in the plane strain state, the material constitutive relationship matrix is:

$$C = \frac{E}{1+\nu}\begin{pmatrix} \frac{1-\nu}{1-2\nu} & \frac{\nu}{1-2\nu} & 0 \\ \frac{\nu}{1-2\nu} & \frac{1-\nu}{1-2\nu} & 0 \\ 0 & 0 & \frac{1}{2} \end{pmatrix}$$

Assuming that the element has a thickness t, then the differential volume $dV = tdX_1dX_2$. The stiff– ness matrix can be obtained by evaluating the following integral on the rectangular area:

$$K = t\int_{-b}^{b}\int_{-a}^{a} B^T CBdX_1dX_2$$

The stiffness matrix has the dimensions of 16×16. The following Mathematica code can be uti– lized to view its components:

```
Shapefun=Table[0,{i,1,8}];
Shapefun[[1]]=(b-x2)(a-x1)/4/a/b*-(1+x1/a+x2/b);
Shapefun[[2]]=(b-x2)(a+x1)/4/a/b*-(1-x1/a+x2/b);
Shapefun[[3]]=(b+x2)(a+x1)/4/a/b*-(1-x1/a-x2/b);
Shapefun[[4]]=(b+x2)(a-x1)/4/a/b*-(1+x1/a-x2/b);
Shapefun[[5]]=(b-x2)(a-x1)(a+x1)/2/a^2/b;
Shapefun[[6]]=(a+x1)(b-x2)(b+x2)/2/a/b^2;
Shapefun[[7]]=(b+x2)(a-x1)(a+x1)/2/a^2/b;
Shapefun[[8]]=(a-x1)(b-x2)(b+x2)/2/a/b^2;
B=Table[0,{i,1,3},{j,1,16}];
Do[B[[1,2i-1]]=B[[3,2i]]=D[Shapefun[[i]],x1];B[[2,2i]]=B[[3,2i-
1]]=D[Shapefun[[i]],x2],{i,1,8}];
B//MatrixForm
Cc=Ee/(1+nu)*{{(1-nu)/(1-2nu),nu/(1-2nu),0},{(nu)/(1-2nu),(1-nu)/(1-
2nu),0},{0,0,1/2}};
```

```
Kbeforeintegration=t*FullSimplify[Transpose[B].Cc.B];
K=Integrate[Kbeforeintegration,{x2,-b,b},{x1,-a,a}];
K//MatrixForm
```

Nodal Loads Due to Body Forces Vectors and Traction Vectors:

Assuming that the distributed body–forces vector per unit mass is $b = \{b_1, b_2\}$ and the density is ρ, and assuming that the traction vector per unit area on the left side is $t_n = \{t_1, t_2\}$, then the nodal forces due to the distributed body forces and the nodal forces due to the traction vector can be obtained as follows:

$$f_e = \int N^T \rho b dV = \frac{\rho t A}{3}\begin{pmatrix} -b_1/4 \\ -b_2/4 \\ -b_1/4 \\ -b_2/4 \\ -b_1/4 \\ -b_2/4 \\ -b_1/4 \\ -b_2/4 \\ b_1 \\ b_2 \\ b_1 \\ b_2 \\ b_1 \\ b_2 \\ b_1 \\ b_2 \end{pmatrix}, \quad f_e = \int N^T t_n dS = \left(\int_{-b}^{b} N^T t_n dX_2\right)_{X_1=-a} = \frac{t}{6}\begin{pmatrix} t_1 \\ t_2 \\ 0 \\ 0 \\ t_1 \\ t_2 \\ 0 \\ 0 \\ 0 \\ 0 \\ 4t_1 \\ 4t_2 \end{pmatrix}$$

The nodal loads due to a *constant* distributed body forces vector and the traction vector on the left side are shown in (Figure 9–15). Notice the surprising result of having small forces applied on the corner nodes opposite in direction to the applied body forces.

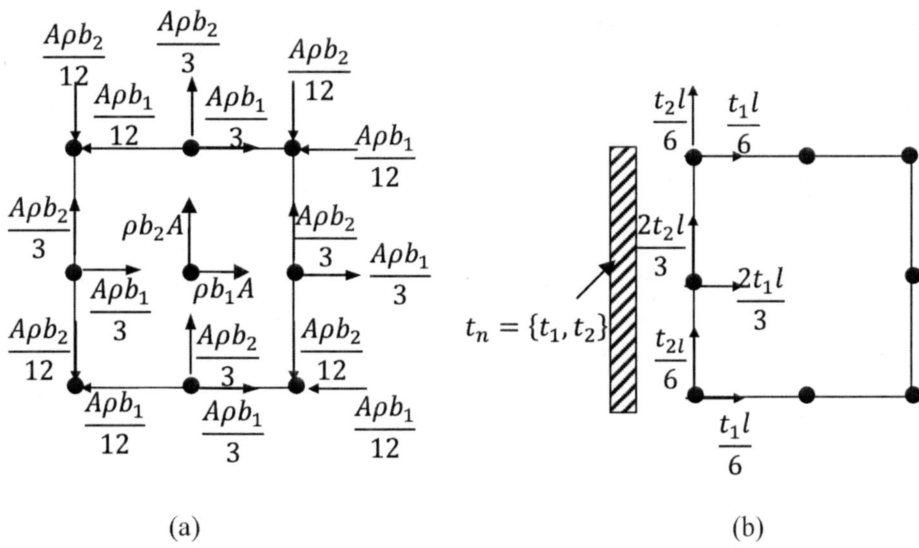

FIGURE 9-15 Discretization of (a) the distributed body forces and (b) the traction vector on the quadratic quadrilateral assuming unit thickness ($t = 1$)($A = 4ab$, $l = 2b$).

The following is the Mathematica code used for the calculations above:

```
Shapefun=Table[0,{i,1,8}];
Shapefun[[1]]=(b-x2)(a-x1)/4/a/b*-(1+x1/a+x2/b);
Shapefun[[2]]=(b-x2)(a+x1)/4/a/b*-(1-x1/a+x2/b);
Shapefun[[3]]=(b+x2)(a+x1)/4/a/b*-(1-x1/a-x2/b);
Shapefun[[4]]=(b+x2)(a-x1)/4/a/b*-(1+x1/a-x2/b);
Shapefun[[5]]=(b-x2)(a-x1)(a+x1)/2/a^2/b;
Shapefun[[6]]=(a+x1)(b-x2)(b+x2)/2/a/b^2;
Shapefun[[7]]=(b+x2)(a-x1)(a+x1)/2/a^2/b;
Shapefun[[8]]=(a-x1)(b-x2)(b+x2)/2/a/b^2;
Nn=Table[0,{i,1,2},{j,1,16}];
Do[Nn[[1,2i-1]]=Nn[[2,2i]]=Shapefun[[i]],{i,1,8}];
tn={t1,t2};
rb={rb1,rb2};
fetraction=Integrate[(Transpose[Nn].tn/.x1->-a),{x2,-b,b}]//MatrixForm
febodyforces=Integrate[(Transpose[Nn].rb),{x2,-b,b},{x1,-a,a}]//MatrixForm
```

9.2.4. Extension to Three-Dimensional Elements

The results of the previous sections can be extended in a straightforward manner to the three–dimensional case. In three dimensions, the 4–node tetrahedron is the three–dimensional version of the linear triangular element, while the 8–node tetrahedron is the nonlinear three–dimensional version corresponding to the quadratic triangular element. The 8–node brick element (trilinear quadrilateral) is the three–dimensional version of the bilinear quadrilateral, while the 20–node quadratic brick element is the nonlinear three–dimensional version corre–sponding to the quadratic quadrilateral. The properties of the two–dimensional elements are inherited by their three–dimensional counterparts.

9.2.5. General Requirements for the Interpolation Functions

Finding an appropriate displacement shape function or interpolation function has been a sub–ject of extensive study since the development of the finite element analysis. While new types of elements are being introduced on a regular basis, some basic requirements characterize a good approximation for the displacement. Among those requirements are *isotropy, ability to model rigid body motions and constant strains* and *element compatibility*. An isotropic interpolation function is a function that would not favor a direction over the other. An interpolation function that has the form $u = a_1 + a_2X_1 + a_3X_2^2$ would clearly produce different results if the structure is rotated such that the coordinates X_1 and X_2 are switched. However, an interpolation function of the form $u = a_1 + a_2X_1 + a_3X_2$ is isotropic. Another basic requirement is that the displacement interpola–tion function should be able to model rigid–body motion. An interpolation function that has the form $u = a_1X_1 + a_2X_2$ cannot model a constant displacement, and thus, cannot be acceptable. Such interpolation function, however, can model constant strain. Another major requirement for the interpolation functions is ensuring the compatibility between elements. All the elements pre–sented in the previous section ensure that if two elements of the same type share the same sides, then the displacement along the boundary (the shared side) between the elements is continuous. This was ensured by choosing interpolation or shape functions that guarantee that the displace–ment on the boundaries (sides) of the element is completely determined by the nodes on that side, and thus, the elements that share that side along with the associated nodes share the same dis–placement across that side. The displacement interpolation functions presented so far are thus truly C0 interpolation functions. In some applications, more (or less) stringent requirements can be considered, and the user should be aware of the capabilities or lack thereof of the elements being used.

9.3. ISOPARAMETRIC ELEMENTS AND NUMERICAL INTEGRATION

In the previous section, the basic elements that are used for discretizing a plane domain were introduced. These elements can be used to mesh material bodies with regular geometries.

For irregular geometries, quadrilateral and triangle elements will not necessarily follow the shapes of the basic elements described (a quadrilateral element will not necessarily be a square or a rectangle, and the triangle element will not necessarily have a right angle, as described before). Isoparametric elements enable meshing an irregular domain with triangular or quadratic elements that do not maintain orthogonality between the sides of the element. The purpose of the isoparametric formulation is to create shape functions that would ensure the compatibility of the displacement between neighboring elements while maintaining the requirements for shape functions mentioned above. When the elements' sides do not necessarily have right angles, the isoparametric formulation is used to map those irregular elements into elements with regular shapes. The mapping functions used are the same shape functions that are used for the displacements, and hence, the name isoparametric (same parameters). After the mapping, integration is performed in the coordinate system of the element with the regular shape. The integration is performed numerically rather than using exact integrals. Numerical integration is preferred over exact integration for two reasons. The first is to reduce the computational resources required, and the second because the irregular geometries render radical (fraction) quantities that could be problematic for exact integrals. Many numerical integration techniques are available but the technique that is widely used for finite element analysis is the Gauss numerical integration technique.

9.3.1. Isoparametric Elements

9.3.1.a. 3-Node Quadratic Isoparametric Element in One Dimension

To introduce the concepts of isoparametric elements, a one–dimensional isoparametric element will first be introduced. Assume that a one–dimensional nonlinear element, as described in section 9.1.2, has nodal coordinates of x_1, x_2 and x_3 as shown in (Figure 9–16). The mapping between the general coordinate system x and the coordinate system ξ shown in Figure 9–16 is defined as follows:

$$x = N_1 x_1 + N_2 x_2 + N_3 x_3$$

where the shape functions are described in the coordinate system of ξ as follows:

$$N_1 = \frac{\xi(1-\xi)}{-2}, \quad N_2 = (1-\xi)(1+\xi), \quad N_3 = \frac{\xi(1+\xi)}{2}$$

To perform the integration in the coordinate system of ξ the gradient of the mapping has to be evaluated:

$$\frac{dx}{d\xi} = \sum_{i=1}^{3} \frac{dN_i}{d\xi} x_i, \quad \frac{d\xi}{dx} = \frac{1}{\frac{dx}{d\xi}} = \frac{1}{\sum_{i=1}^{3} \frac{dN_i}{d\xi} x_i}$$

The displacement function is defined in the ξ coordinate system as follows:

$$u = N_1 u_1 + N_2 u_2 + N_3 u_3$$

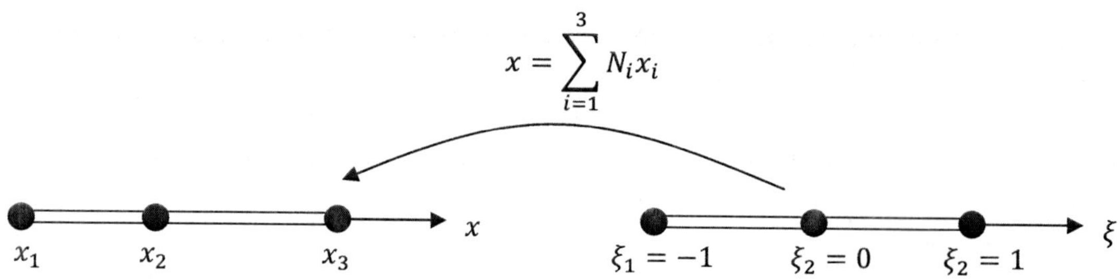

FIGURE 9-16 Isoparametric mapping in the one–dimensional case of a quadratic element.

The normal strain is defined as follows:

$$\varepsilon_{xx} = \frac{du}{dx} = \frac{du}{d\xi}\frac{d\xi}{dx} = \frac{d\xi}{dx}\begin{pmatrix} \frac{dN_1}{d\xi} & \frac{dN_2}{d\xi} & \frac{dN_3}{d\xi} \end{pmatrix}\begin{pmatrix} u_1 \\ u_2 \\ u_3 \end{pmatrix}$$

The virtual displacement u^* and the associated virtual strain field ε^*_{xx} can be chosen as follows:

$$u^* = \sum_{i=1}^{3} N_i u_i^*, \quad \varepsilon^*_{xx} = \frac{du}{dx} = \frac{du}{d\xi}\frac{d\xi}{dx} = \frac{d\xi}{dx}\begin{pmatrix} \frac{dN_1}{d\xi} & \frac{dN_2}{d\xi} & \frac{dN_3}{d\xi} \end{pmatrix}\begin{pmatrix} u_1^* \\ u_2^* \\ u_3^* \end{pmatrix}$$

Assume that f is a body force per unit length of the one–dimensional element shown, then the equation of the virtual work is:

$$A\int_{x_1}^{x_3} \sigma_{xx}\varepsilon^*_{xx}dx = \int_{x_1}^{x_3} fu^* dx$$

The integration is intended to be performed in the coordinate system of ξ so dx is replaced with $\frac{dx}{d\xi}d\xi$. In addition, σ_{xx} is substituted by $E\varepsilon_{xx}$ and the virtual work equation is manipulated to produce:

$$EA\int_{-1}^{1}\left(\frac{d\xi}{dx}\right)^2 \frac{dx}{d\xi}\begin{pmatrix} \frac{dN_1}{d\xi} \\ \frac{dN_2}{d\xi} \\ \frac{dN_3}{d\xi} \end{pmatrix}\begin{pmatrix} \frac{dN_1}{d\xi} & \frac{dN_2}{d\xi} & \frac{dN_3}{d\xi} \end{pmatrix}\begin{pmatrix} u_1 \\ u_2 \\ u_3 \end{pmatrix} d\xi = \int_{-1}^{1} f\begin{pmatrix} N_1 \\ N_2 \\ N_3 \end{pmatrix}\frac{dx}{d\xi}d\xi$$

Problem: Let a nonlinear isoparametric one–dimensional element be placed such that the nodal coordinates x_1, x_2 and x_3 are equal to 0, 4.5, and 10 units, respectively. Assume a body force of f

units per unit area is applied in the direction of x. Find the stiffness matrix and the nodal forces for that element.

Solution: Applying the above equations, the stiffness matrix of the element and the nodal forces has the form:

$$Ku_e = f_e$$

$$EA\begin{pmatrix} 0.2653 & -0.3006 & 0.0353 \\ -0.3006 & 0.5465 & -0.2459 \\ 0.0353 & -0.2459 & 0.21067 \end{pmatrix}\begin{pmatrix} u_1 \\ u_2 \\ u_3 \end{pmatrix} = f\begin{pmatrix} 1.333 \\ 6.667 \\ 2.000 \end{pmatrix}$$

The following Mathematica code was utilized:

```
Clear[x1,x2,x3,x,xi]
N1=-xi(1-xi)/2;
N2=(1-xi)(1+xi);
N3=xi (1+xi)/2;
x=N1*x1+N2*x2+N3*x3;
dxxi=FullSimplify[D[x,xi]];
dxix=1/dxxi;
x1=0;x2=4.5;x3=10;
B=FullSimplify[{{D[N1,xi],D[N2,xi],D[N3,xi]}}]
K1=FullSimplify[EA*dxix*Transpose[B].B];
Kb=Integrate[K1,{xi,-1,1}];
Kb//MatrixForm
a=Integrate[f*N1*dxxi,{xi,-1,1}]
b=Integrate[f*N2*dxxi,{xi,-1.,1}]
c=Integrate[f*N3*dxxi,{xi,-1,1}]
```

9.3.1.b. 4-Node Bilinear Isoparametric Elements in Two Dimensions

Similar to the one–dimensional case presented in the previous example, the irregular 4–node quadrilateral in the coordinate system x and y will have to be mapped to the element in the coordi– nate system of ξ and η as shown in (Figure 9–17). The mapping function between the two elements is performed using the same shape functions for the bilinear quadrilateral, as follows:

$$x = N_1x_1 + N_2x_2 + N_3x_3 + N_4x_4, \quad y = N_1y_1 + N_2y_2 + N_3y_3 + N_4y_4$$

where the shape functions are chosen as:

$$N_1 = \frac{(1-\xi)(1-\eta)}{4}, \quad N_2 = \frac{(1-\eta)(1+\xi)}{4}, \quad N_3 = \frac{(1+\eta)(1+\xi)}{4}, \quad N_4 = \frac{(1+\eta)(1-\xi)}{4}$$

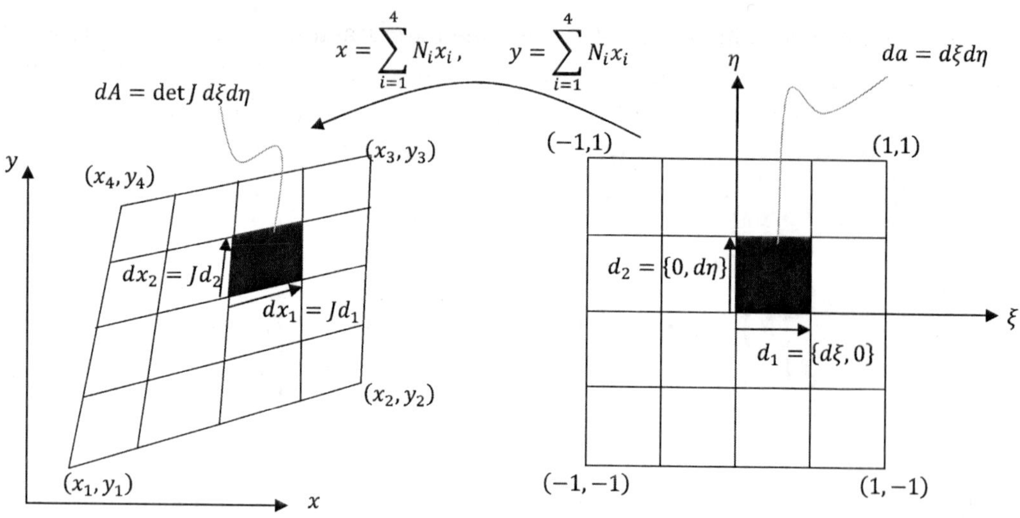

FIGURE 9-17 Isoparametric mapping in the two–dimensional case of a 4–node bilinear element.

Let $u_{\xi\eta} = \{\Delta\xi, \Delta\eta\}$ be a vector inside the square element in the coordinate system of ξ and η. This vector is transformed into the vector $u_{xy} = \{\Delta x, \Delta y\}$ in the general element in the coordinate system of x and y using the following relationship:

$$\begin{pmatrix} \Delta x \\ \Delta y \end{pmatrix} = \begin{pmatrix} \frac{\partial x}{\partial \xi} & \frac{\partial x}{\partial \eta} \\ \frac{\partial y}{\partial \xi} & \frac{\partial y}{\partial \eta} \end{pmatrix} \begin{pmatrix} \Delta \xi \\ \Delta \eta \end{pmatrix}$$

Define the matrices J and its inverse J^{-1} as follows:

$$J = \begin{pmatrix} \frac{\partial x}{\partial \xi} & \frac{\partial x}{\partial \eta} \\ \frac{\partial y}{\partial \xi} & \frac{\partial y}{\partial \eta} \end{pmatrix}, \quad J^{-1} = \begin{pmatrix} \frac{\partial \xi}{\partial x} & \frac{\partial \xi}{\partial y} \\ \frac{\partial \eta}{\partial x} & \frac{\partial \eta}{\partial y} \end{pmatrix}$$

The matrix J is guaranteed to have an inverse since the linear map between vectors in both ele– ments has to invertible (See section 1.2.3) (**Why?**). As shown in (Figure 9–17), let $d_1 = \{d\xi, 0\}$ and $d_2 = \{0, d\eta\}$ represent an infinitesimal area inside the ξ and η element. Let $dx_1 = Jd_1$ and

$dx_2 = Jd_2$ be the corresponding images of d_1 and d_2 in the x and y element. Let dA equal to the area inscribed by the vectors dx_1 and dx_2 Let $da = d\xi d\eta$ be the area inscribed by the vectors d_1 and d_2. Then, according to section 1.2.5.b, the relationship between the corresponding areas is as follows:

$$dA = \det J \, da = \det J \, d\xi d\eta$$

The importance of the above formula is in transforming integrations over the x and y element domain to integrations over the ξ and η element domain. This will allow writing the (integral) equation of the virtual work in the coordinate system of ξ and η. But, first, the displacement function in the ξ and η coordinate system is written as follows:

$$\begin{pmatrix} u \\ v \end{pmatrix} = \begin{pmatrix} N_1 & 0 & N_2 & 0 & N_3 & 0 & N_4 & 0 \\ 0 & N_1 & 0 & N_2 & 0 & N_3 & 0 & N_4 \end{pmatrix} \begin{pmatrix} u_1 \\ v_1 \\ u_2 \\ v_2 \\ u_3 \\ v_3 \\ u_4 \\ v_4 \end{pmatrix}$$

To evaluate the strains, the gradients of the displacement functions in the two different elements need to be evaluated. The gradients of the displacements functions in the ξ and η element have the following form:

$$\begin{pmatrix} \frac{\partial u}{\partial \xi} \\ \frac{\partial u}{\partial \eta} \\ \frac{\partial v}{\partial \xi} \\ \frac{\partial v}{\partial \eta} \end{pmatrix} = \begin{pmatrix} \frac{\partial N_1}{\partial \xi} & 0 & \frac{\partial N_2}{\partial \xi} & 0 & \frac{\partial N_3}{\partial \xi} & 0 & \frac{\partial N_4}{\partial \xi} & 0 \\ \frac{\partial N_1}{\partial \eta} & 0 & \frac{\partial N_2}{\partial \eta} & 0 & \frac{\partial N_3}{\partial \eta} & 0 & \frac{\partial N_4}{\partial \eta} & 0 \\ 0 & \frac{\partial N_1}{\partial \xi} & 0 & \frac{\partial N_2}{\partial \xi} & 0 & \frac{\partial N_3}{\partial \xi} & 0 & \frac{\partial N_4}{\partial \xi} \\ 0 & \frac{\partial N_1}{\partial \eta} & 0 & \frac{\partial N_2}{\partial \eta} & 0 & \frac{\partial N_3}{\partial \eta} & 0 & \frac{\partial N_4}{\partial \eta} \end{pmatrix} \begin{pmatrix} u_1 \\ v_1 \\ u_2 \\ v_2 \\ u_3 \\ v_3 \\ u_4 \\ v_4 \end{pmatrix}$$

The gradients of the displacements functions in the x and y element can be evaluated as functions of the gradients of the displacements in the ξ and η element as follows:

$$\begin{pmatrix} \frac{\partial u}{\partial x} \\ \frac{\partial u}{\partial y} \\ \frac{\partial v}{\partial x} \\ \frac{\partial v}{\partial y} \end{pmatrix} = \begin{pmatrix} \frac{\partial \xi}{\partial x} & \frac{\partial \eta}{\partial x} & 0 & 0 \\ \frac{\partial \xi}{\partial y} & \frac{\partial \eta}{\partial y} & 0 & 0 \\ 0 & 0 & \frac{\partial \xi}{\partial x} & \frac{\partial \eta}{\partial x} \\ 0 & 0 & \frac{\partial \xi}{\partial y} & \frac{\partial \eta}{\partial y} \end{pmatrix} \begin{pmatrix} \frac{\partial u}{\partial \xi} \\ \frac{\partial u}{\partial \eta} \\ \frac{\partial v}{\partial \xi} \\ \frac{\partial v}{\partial \eta} \end{pmatrix}$$

$$= \begin{pmatrix} \frac{\partial \xi}{\partial x} & \frac{\partial \eta}{\partial x} & 0 & 0 \\ \frac{\partial \xi}{\partial y} & \frac{\partial \eta}{\partial y} & 0 & 0 \\ 0 & 0 & \frac{\partial \xi}{\partial x} & \frac{\partial \eta}{\partial x} \\ 0 & 0 & \frac{\partial \xi}{\partial y} & \frac{\partial \eta}{\partial y} \end{pmatrix} \begin{pmatrix} \frac{\partial N_1}{\partial \xi} & 0 & \frac{\partial N_2}{\partial \xi} & 0 & \frac{\partial N_3}{\partial \xi} & 0 & \frac{\partial N_4}{\partial \xi} & 0 \\ \frac{\partial N_1}{\partial \eta} & 0 & \frac{\partial N_2}{\partial \eta} & 0 & \frac{\partial N_3}{\partial \eta} & 0 & \frac{\partial N_4}{\partial \eta} & 0 \\ 0 & \frac{\partial N_1}{\partial \xi} & 0 & \frac{\partial N_2}{\partial \xi} & 0 & \frac{\partial N_3}{\partial \xi} & 0 & \frac{\partial N_4}{\partial \xi} \\ 0 & \frac{\partial N_1}{\partial \eta} & 0 & \frac{\partial N_2}{\partial \eta} & 0 & \frac{\partial N_3}{\partial \eta} & 0 & \frac{\partial N_4}{\partial \eta} \end{pmatrix} \begin{pmatrix} u_1 \\ v_1 \\ u_2 \\ v_2 \\ u_3 \\ v_3 \\ u_4 \\ v_4 \end{pmatrix}$$

The strains are defined in the x and y element as follows:

$$\begin{pmatrix} \varepsilon_{xx} \\ \varepsilon_{yy} \\ \gamma_{xy} \end{pmatrix} = \begin{pmatrix} \frac{\partial u}{\partial x} \\ \frac{\partial v}{\partial y} \\ \frac{\partial u}{\partial y} + \frac{\partial v}{\partial x} \end{pmatrix} = \begin{pmatrix} 1 & 0 & 0 & 0 \\ 0 & 0 & 0 & 1 \\ 0 & 1 & 1 & 0 \end{pmatrix} \begin{pmatrix} \frac{\partial u}{\partial x} \\ \frac{\partial u}{\partial y} \\ \frac{\partial v}{\partial x} \\ \frac{\partial v}{\partial y} \end{pmatrix}$$

Combining the above relationships, the relationship between the strain components and the elements degrees of freedom can be obtained as follows:

$$\begin{pmatrix} \varepsilon_{xx} \\ \varepsilon_{yy} \\ \gamma_{xy} \end{pmatrix} = \begin{pmatrix} 1 & 0 & 0 & 0 \\ 0 & 0 & 0 & 1 \\ 0 & 1 & 1 & 0 \end{pmatrix} \begin{pmatrix} \frac{\partial \xi}{\partial x} & \frac{\partial \eta}{\partial x} & 0 & 0 \\ \frac{\partial \xi}{\partial y} & \frac{\partial \eta}{\partial y} & 0 & 0 \\ 0 & 0 & \frac{\partial \xi}{\partial x} & \frac{\partial \eta}{\partial x} \\ 0 & 0 & \frac{\partial \xi}{\partial y} & \frac{\partial \eta}{\partial y} \end{pmatrix} \begin{pmatrix} \frac{\partial N_1}{\partial \xi} & 0 & \frac{\partial N_2}{\partial \xi} & 0 & \frac{\partial N_3}{\partial \xi} & 0 & \frac{\partial N_4}{\partial \xi} & 0 \\ \frac{\partial N_1}{\partial \eta} & 0 & \frac{\partial N_2}{\partial \eta} & 0 & \frac{\partial N_3}{\partial \eta} & 0 & \frac{\partial N_4}{\partial \eta} & 0 \\ 0 & \frac{\partial N_1}{\partial \xi} & 0 & \frac{\partial N_2}{\partial \xi} & 0 & \frac{\partial N_3}{\partial \xi} & 0 & \frac{\partial N_4}{\partial \xi} \\ 0 & \frac{\partial N_1}{\partial \eta} & 0 & \frac{\partial N_2}{\partial \eta} & 0 & \frac{\partial N_3}{\partial \eta} & 0 & \frac{\partial N_4}{\partial \eta} \end{pmatrix} \begin{pmatrix} u_1 \\ v_1 \\ u_2 \\ v_2 \\ u_3 \\ v_3 \\ u_4 \\ v_4 \end{pmatrix}$$

Denote the matrix B by:

$$B = \begin{pmatrix} 1 & 0 & 0 & 0 \\ 0 & 0 & 0 & 1 \\ 0 & 1 & 1 & 0 \end{pmatrix} \begin{pmatrix} \frac{\partial \xi}{\partial x} & \frac{\partial \eta}{\partial x} & 0 & 0 \\ \frac{\partial \xi}{\partial y} & \frac{\partial \eta}{\partial y} & 0 & 0 \\ 0 & 0 & \frac{\partial \xi}{\partial x} & \frac{\partial \eta}{\partial x} \\ 0 & 0 & \frac{\partial \xi}{\partial y} & \frac{\partial \eta}{\partial y} \end{pmatrix} \begin{pmatrix} \frac{\partial N_1}{\partial \xi} & 0 & \frac{\partial N_2}{\partial \xi} & 0 & \frac{\partial N_3}{\partial \xi} & 0 & \frac{\partial N_4}{\partial \xi} & 0 \\ \frac{\partial N_1}{\partial \eta} & 0 & \frac{\partial N_2}{\partial \eta} & 0 & \frac{\partial N_3}{\partial \eta} & 0 & \frac{\partial N_4}{\partial \eta} & 0 \\ 0 & \frac{\partial N_1}{\partial \xi} & 0 & \frac{\partial N_2}{\partial \xi} & 0 & \frac{\partial N_3}{\partial \xi} & 0 & \frac{\partial N_4}{\partial \xi} \\ 0 & \frac{\partial N_1}{\partial \eta} & 0 & \frac{\partial N_2}{\partial \eta} & 0 & \frac{\partial N_3}{\partial \eta} & 0 & \frac{\partial N_4}{\partial \eta} \end{pmatrix}$$

then the equation of the virtual work can be written as:

$$t \int_{-1}^{1}\int_{-1}^{1} B^T C B \det J \, d\xi \, d\eta u_e = \int_{-1}^{1}\int_{-1}^{1} N^T \rho b \det J \, d\xi \, d\eta + \int_{element\ surface} N^T t_n dS$$

where u_e is the vector of the element degrees of freedom, t is the thickness of the element, C is the material constitutive relationship, ρb is the body forces per unit volume, N^T is the matrix of the shape functions defined in the coordinate system of ξ and η and t_n is the traction vector per unit area on the surface dS. Notice that the surface element dS is given in the x, y coordinate system, and it should be transformed accordingly to the ξ, η coordinate system.

Problem: Find the stiffness matrix and the nodal forces in the 4–node isoparametric plane element whose coordinates are $(0, 0)$, $(1, 0.1)$, $(1.2, 1.2)$, $(0.2, 1)$ with a constant body–force vector $\rho b = \{\rho b1, \rho b2\}$ and a unit thickness. Assume plane stress conditions with $E = 1unit$ and $\nu = 0$.

Solution: The following Mathematica code can be utilized to obtain the stiffness matrix and the nodal forces due to a constant body–force vector per unit volume for any 4–node isoparametric element as a function of the coordinates of the 4 nodes of the element. Study the code carefully. In particular, notice the following: If the nodes of the element are changed such that the they are not defined in a counter–clockwise fashion, then $\det J$ becomes negative and the stiffness matrix has negative diagonal entries (The stiffness matrix is not positive definite in this case.). Most finite–element analysis softwares would give a warning message that such element is not adequately defined since the area of the element is negative. Also, notice that the stiffness matrix is obtained using the built–in numerical integration function `NIntegrate` in Mathematica. Try switching to the built–in exact integration function `Integrate` to notice the difference in the computational resources required. The numerical integration is a much faster technique.

```
coordinates={{0,0},{1,0.1},{1.2,1.2},{0.2,1}};
t=1;
(*Plane Stress *)
Ee=1;
```

```
nu=0;
Cc=Ee/(1+nu)/(1-nu)*{{1,nu,0},{nu,1,0},{0,0,(1-nu)/2}};
a=Polygon[{coordinates[[1]],coordinates[[2]],coordinates[[3]],coordinates[[4
]],coordinates[[1]]}];
Graphics[a]
Shapefun=Table[0,{i,1,4}];
Shapefun[[1]]=(1-xi)(1-eta)/4;
Shapefun[[2]]=(1+xi)(1-eta)/4;
Shapefun[[3]]=(1+xi)(1+eta)/4;
Shapefun[[4]]=(1-xi)(1+eta)/4;
Nn=Table[0,{i,1,2},{j,1,8}];
Do[Nn[[1,2i-1]]=Nn[[2,2i]]=Shapefun[[i]],{i,1,4}];
x=Sum[Shapefun[[i]]*coordinates[[i,1]],{i,1,4}];
y=Sum[Shapefun[[i]]*coordinates[[i,2]],{i,1,4}];
J=Table[0,{i,1,2},{j,1,2}];
J[[1,1]]=D[x,xi];
J[[1,2]]=D[x,eta];
J[[2,1]]=D[y,xi];
J[[2,2]]=D[y,eta];
JinvT=Transpose[Inverse[J]];
mat1={{1,0,0,0},{0,0,0,1},{0,1,1,0}};
mat2={{JinvT[[1,1]],JinvT[[1,2]],0,0},{JinvT[[2,1]],JinvT[[2,2]],0,0},{0,0,J
invT[[1,1]],JinvT[[1,2]]},{0,0,JinvT[[2,1]],JinvT[[2,2]]}};
mat3=Table[0,{i,1,4},{j,1,8}];
Do[mat3[[1,2i-1]]=mat3[[3,2i]]=D[Shapefun[[i]],xi];mat3[[2,2i-
1]]=mat3[[4,2i]]=D[Shapefun[[i]],eta],{i,1,4}];
B=Chop[FullSimplify[mat1.mat2.mat3]];
k1=FullSimplify[Transpose[B].Cc.B];
K=t*NIntegrate[k1*Det[J],{xi,-1,1},{eta,-1,1}];
rb={rb1,rb2};
f3=Integrate[Transpose[Nn].rb*Det[J],{xi,-1,1},{eta,-1,1}];
f3//MatrixForm
K//MatrixForm
```

Note that the results of this section can be extended in a straightforward manner to the three–dimensional 8–node isoparametric brick element.

9.3.1.c. 8-Node Quadratic Isoparametric Elements in Two Dimensions

The 8–node quadratic isoparametric plane element is better suited for curved surfaces since the mapping function between the coordinate system of ξ and η and the coordinate system x and y uses the quadratic interpolation functions (Figure 9–18). The interpolation functions in the ξ and η coordinate system are:

$$N_1 = \frac{-(1-\eta)(1-\xi)(1+\xi+\eta)}{4}, \quad N_2 = \frac{-(1-\eta)(1+\xi)(1-\xi+\eta)}{4}$$

$$N_3 = \frac{-(1+\eta)(1+\xi)(1-\xi-\eta)}{4}, \quad N_4 = \frac{-(1+\eta)(1-\xi)(1+\xi-\eta)}{4}$$

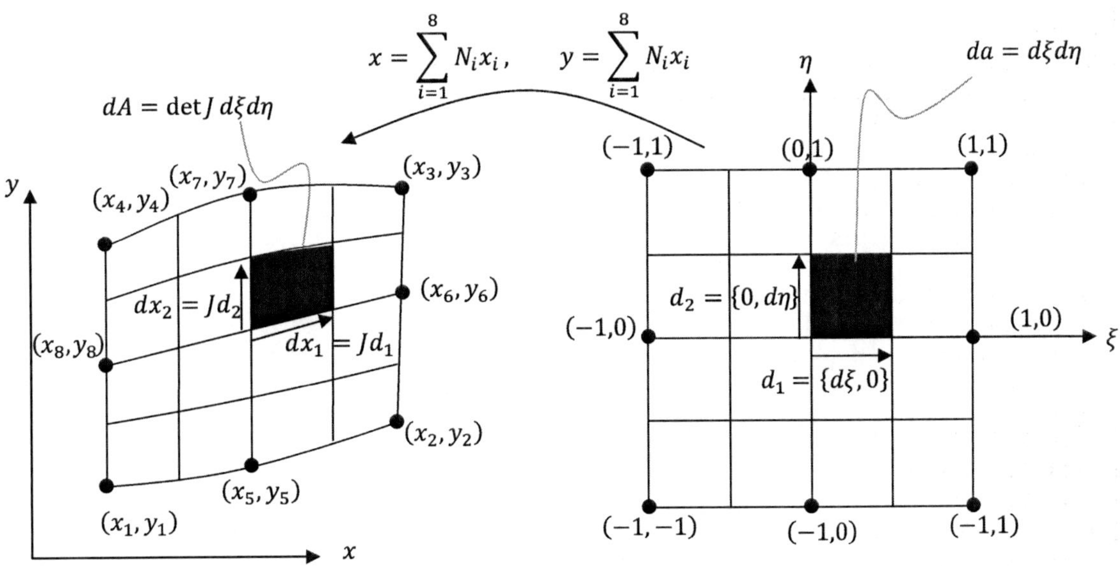

FIGURE 9-18 Isoparametric mapping in the two–dimensional case of an 8–node quadratic element.

$$N_5 = \frac{(1-\eta)(1-\xi)(1+\xi)}{2}, \quad N_6 = \frac{(1+\xi)(1-\eta)(1+\eta)}{2}$$

$$N_7 = \frac{(1+\eta)(1-\xi)(1+\xi)}{2}, \quad N_8 = \frac{(1-\xi)(1-\eta)(1+\eta)}{2}$$

The development of the equations is similar to the previous section and is left to the reader. The only difference is the number of degrees of freedom of the element. The extension to the three–dimensional case (the 20–node brick element) is also straightforward. The following is the Math–ematica code that can be utilized to produce the stiffness matrix and the nodal forces due to a constant body forces vector for the 8–node isoparametric element as a function of the coordi–nates. Notice that the numerical integration built–in function in Mathematica has a higher accu–racy than the traditional Gauss integration scheme (which will be presented later) but requires higher computational resources.

```
coordinates={{0,0},{1,0.1},{1.2,1.2},{0.2,1},{0.5,0},{1.1,0.65},{0.7,1.2},{0
.1,0.5}};
t=1;
Ee=1;
nu=0.3;
(*Plane Stress *)
Cc=Ee/(1+nu)/(1-nu)*{{1,nu,0},{nu,1,0},{0,0,(1-nu)/2}};
a=Polygon[{coordinates[[1]],coordinates[[5]],coordinates[[2]],coordinates[[6
]],coordinates[[3]],coordinates[[7]],coordinates[[4]],coordinates[[8]],coord
```

```
inates[[1]]}];
Graphics[a]
Shapefun=Table[0,{i,1,8}];
Shapefun[[1]]=-(1-eta)(1-xi)(1+xi+eta)/4;
Shapefun[[2]]=-(1-eta)(1+xi)(1-xi+eta)/4;
Shapefun[[3]]=-(1+eta)(1+xi)(1-xi-eta)/4;
Shapefun[[4]]=-(1+eta)(1-xi)(1+xi-eta)/4;
Shapefun[[5]]=(1-eta)(1-xi)(1+xi)/2;
Shapefun[[6]]=(1+xi)(1-eta)(1+eta)/2;
Shapefun[[7]]=(1+eta)(1-xi)(1+xi)/2;
Shapefun[[8]]=(1-xi)(1-eta)(1+eta)/2;
Nn=Table[0,{i,1,2},{j,1,16}];
Do[Nn[[1,2i-1]]=Nn[[2,2i]]=Shapefun[[i]],{i,1,8}];
x=Sum[Shapefun[[i]]*coordinates[[i,1]],{i,1,8}];
y=Sum[Shapefun[[i]]*coordinates[[i,2]],{i,1,8}];
J=Table[0,{i,1,2},{j,1,2}];
J[[1,1]]=D[x,xi];
J[[1,2]]=D[x,eta];
J[[2,1]]=D[y,xi];
J[[2,2]]=D[y,eta];
JinvT=Transpose[Inverse[J]];
mat1={{1,0,0,0},{0,0,0,1},{0,1,1,0}};
mat2={{JinvT[[1,1]],JinvT[[1,2]],0,0},{JinvT[[2,1]],JinvT[[2,2]],0,0},{0,0,J
invT[[1,1]],JinvT[[1,2]]},{0,0,JinvT[[2,1]],JinvT[[2,2]]}};
mat3=Table[0,{i,1,4},{j,1,16}];
Do[mat3[[1,2i-1]]=mat3[[3,2i]]=D[Shapefun[[i]],xi];mat3[[2,2i-
1]]=mat3[[4,2i]]=D[Shapefun[[i]],eta],{i,1,8}];
B=Chop[FullSimplify[mat1.mat2.mat3]];
k1=FullSimplify[Transpose[B].Cc.B];
Ka=t*NIntegrate[k1*Det[J],{xi,-1,1},{eta,-1,1},AccuracyGoal->8];
Ka//MatrixForm
rb={rb1,rb2};
f3=Integrate[Transpose[Nn].rb*Det[J],{xi,-1,1},{eta,-1,1}];
f3//MatrixForm
```

9.3.1.d. 3-Node and 6-Node Triangle Isoparametric Elements in Two Dimensions

The development of the isoparametric triangle elements is also similar to the previous section. The isoparametric mapping between the coordinate system of ξ and η and the coordinate system of x and y is shown in Figure 9–19 for the two types of triangular elements studied. The development of the equations follows the previous section. The shape functions in the ξ and η coordinate system for the triangular elements are similar to the ones obtained in section 9.2.2.

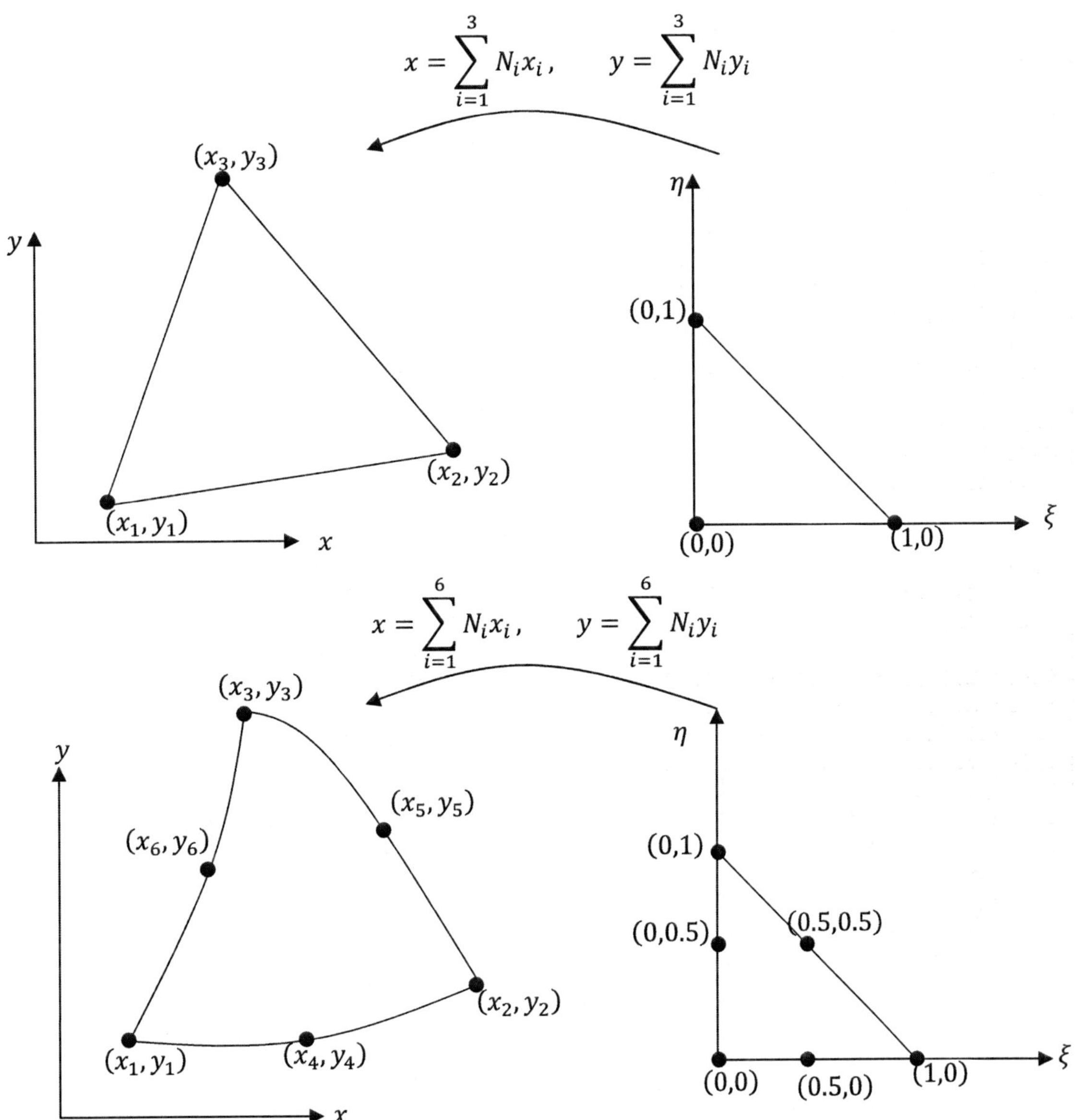

FIGURE 9-19 Isoparametric mapping in the two–dimensional case of the 3–node and the 6–node triangle elements.

9.3.2. Gauss Integration

The Gauss integration scheme is a very efficient method to perform numerical integration over regular domains. In fact, if the function to be integrated is a polynomial of an appropriate degree, then the Gauss integration scheme produces exact results. The Gauss integration scheme has been implemented in almost every finite–element analysis software due to its simplicity and computational efficiency. This section outlines the basic principles behind the Gauss integration scheme, along with its application in the finite element analysis method. In particular, this section will discuss the development of the reduced–integration elements and the benefits and drawbacks of utilizing them in finite element analysis models.

9.3.2.a. Gauss Integration for One-Dimensional Problems

Let $f:[-1,1] \rightarrow \mathbb{R}$. Then, the integral of $\int_{-1}^{1} f dx$ can be approximated by the following formula:

$$\int_{-1}^{1} f dx = \sum_{i=1}^{n} w_i f\left(x_i\right)$$

where $\forall i{:}x_i \in [-1,1]$ and $w_i \in \mathbb{R}$ is a weight factor associated with each x_i. x_i are termed the integration points. The number of integration points required to obtain sufficient accuracy depend on the shape or form of the function f. If f is a linear function, then one integration point is enough to exactly integrate f. If f is a cubic polynomial function, then two integration points are sufficient to obtain the exact solution. Figure 9–20 shows a graphical representation of the Gauss numerical integration scheme for integrating polynomials up to the fifth degree. Following is a table of the number of integration points required to obtain the exact solution for a function f with different polynomial degrees:

Polynomial degree of f	Number of integration points n	Integration points x_i	Associated weight factors w_i
1st degree or lower	1	$x_1 = 0$	$w_1 = 2$
3rd degree or lower	2	$x_1 = 1/\sqrt{3}$ $x_2 = -1/\sqrt{3}$	$w_1 = 1$ $w_2 = 1$
5th degree or lower	3	$x_1 = \sqrt{0.6}$ $x_2 = 0$ $x_3 = -\sqrt{0.6}$	$w_1 = 5/9$ $w_2 = 8/9$ $w_3 = 5/9$

Problem: Verify that the exact integral of a general cubic polynomial on the interval $[-1,1]$ can be obtained by the two integration points $x_1 = 1/\sqrt{3}$ and $x_2 = -1/\sqrt{3}$ with the two weight factors $w_1 = w_2 = 1$.

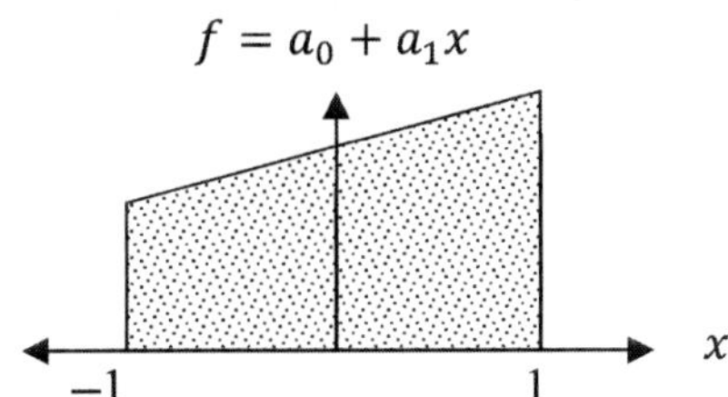

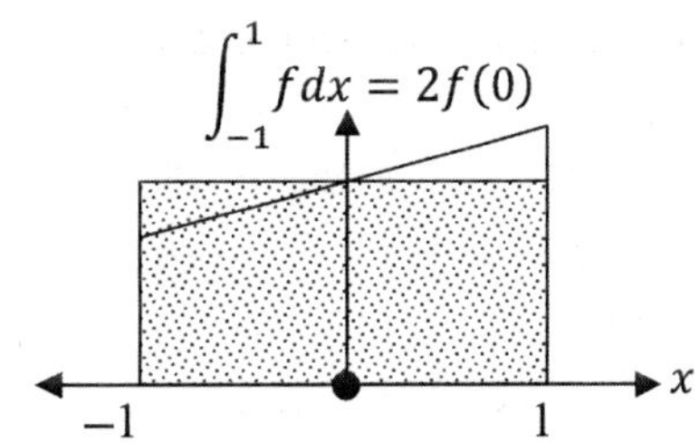

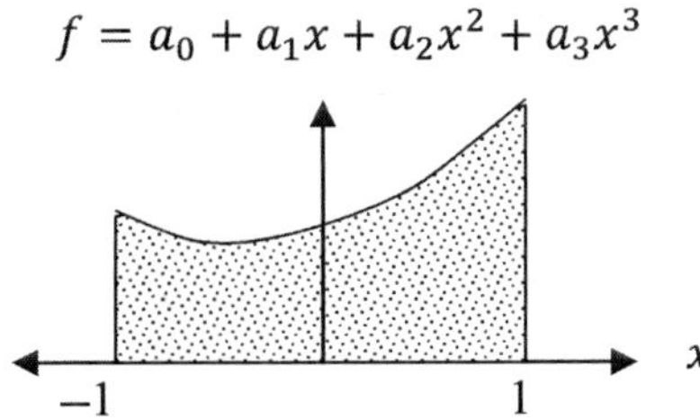

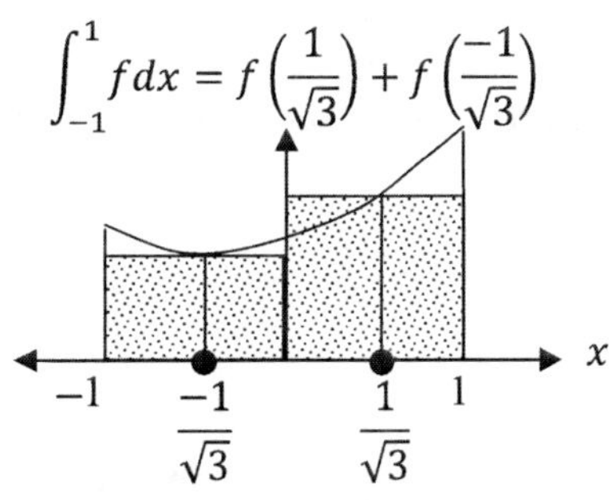

$f = a_0 + a_1 x + a_2 x^2 + a_3 x^3 + a_4 x^4 + a_5 x^5$

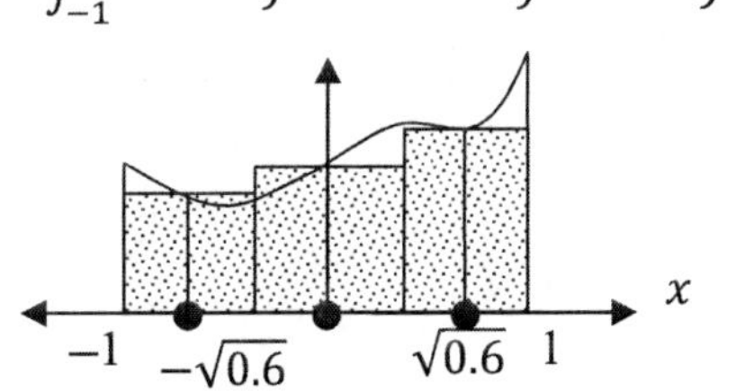

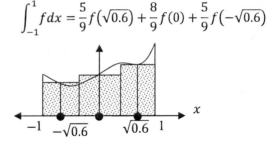

FIGURE 9-20 Gauss integration for polynomials of the 1st, 3rd and 5th degree.

Solution: Let $f = a_0 + a_1 x + a_2 x^2 + a_3 x^3$, then the exact integral is equal to:

$$\int_{-1}^{1} f dx = 2a_0 + 2a_2/3$$

Gauss integration with two integration points is equal to:

$$\int_{-1}^{1} f dx = w_1 \left(a_0 + a_1 x_1 + a_2 x_1^2 + a_3 x_1^3\right) + w_2 \left(a_0 + a_1 x_2 + a_2 x_2^2 + a_3 x_2^3\right)$$

Equating the Gauss integration with the exact solution yields the following four equations:

$$w_1 + w_2 = 2, \quad w_1x_1 + w_2x_2 = 0, \quad w_1x_1^2 + w_2x_2^2 = \frac{2}{3}, \quad w_1x_1^3 + w_2x_2^3 = 0$$

These four equations are obtained by noticing that the exact solution and the Gauss integration solution should be equal given any values for the coefficients a_0, a_1, a_2 and a_3. The solution to the above four equations yields $x_1 = 1/\sqrt{3}$, $x_2 = -1/\sqrt{3}$ and $w_1 = w_2 = 1$.

The Mathematica code to solve the four equations is:

```
Solve[{w1 + w2 == 2, w1*x1^3 + w2*x2^3 == 0, w1*x1^2 + w2*x2^2 == 2/3,
w1*x1 + w2*x2 == 0}, {w1, w2, x1, x2}]
```

▶ **BONUS EXERCISE:** Why can Gaussian integration in one dimension with n points integrate exactly a polynomial of order $2n-1$?

9.3.2.b. Gauss Integration for Two-Dimensional Quadrilaterals

Let $f:[-1,1]\times[-1,1]\to\mathbb{R}$. Assume f is isotropic, then the integral of $\int_{-1}^{1}\int_{-1}^{1} f dx\, dy$ can be approximated by the following formula:

$$\int_{-1}^{1}\int_{-1}^{1} f dx\, dy = \sum_{j=1}^{n}\sum_{i=1}^{n} w_i w_j f\left(x_i, y_j\right)$$

where (x_i, y_j) are $n \times n$ integration points and w_i and w_j are weight factors associated with x_i and y_j respectively. It can be easily seen that the above formula is a direct consequence of the one–dimensional Gauss integration formula as follows:

$$\int_{-1}^{1}\int_{-1}^{1} f\left(x, y\right) dx\, dy = \int_{-1}^{1}\sum_{i=1}^{n} w_i f\left(x_i, y\right) dy = \sum_{j=1}^{n}\sum_{i=1}^{n} w_i w_j f\left(x_i, y_j\right)$$

9.3.2.c. Gauss Integration for Two-Dimensional Quadrilateral Isoparametric Elements

As seen in section 9.2.3.a, the strains of the 4–node elements are linear, and thus, the entries in the matrix $B^T CB$ are quadratic functions of position. Thus, a 2×2 numerical integration scheme is required to produce accurate integrals for the isoparametric 4–node elements (Figure 9–21). While, as seen in section 9.2.3.b, the strains of the 8–node elements are quadratic functions of position, and thus, the entries in the matrix $B^T CB$ are fourth–order polynomials of position. Thus, a 3×3 numerical integration scheme is required to produce accurate integrals for the

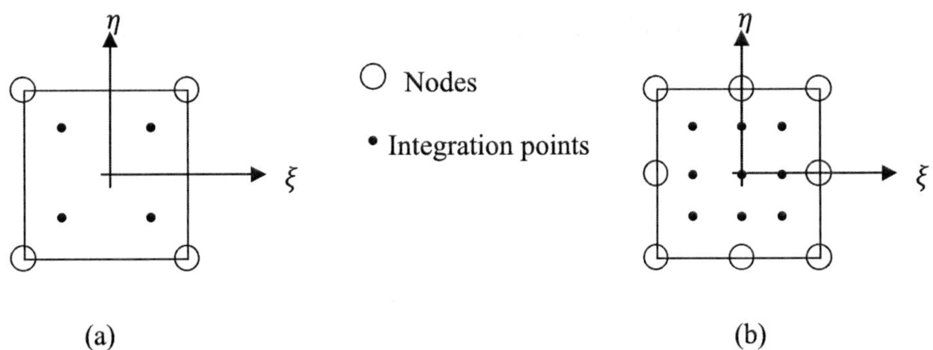

FIGURE 9-21 Gauss integration points for integrating (a) 4–node isoparametric elements $\xi, \eta \in \left\{\frac{1}{\sqrt{3}}, -\frac{1}{\sqrt{3}}\right\}$ with the weight factors $w_i = w_j = 1$. (b) 8–node isoparametric elements $\xi, \eta \in \{-0.6, 0, 0.6\}$ with the weight factors $w_i = \frac{5}{9}, \frac{8}{9}, \frac{5}{9}$ associated with the coordinates $-0.6, 0, 0.6$ respectively.

8–node isoparametric element (Figure 9–21). It should be noted that this can extended the three–dimensional case in a straightforward manner. The 8–node brick element has 8 integration points, while the 20–node brick element has 27 integration points.

9.3.2.d. Reduced Gauss Integration (Under Integration) for Isoparametric Quadrilateral Elements

Elements that are integrated as per the previous section are termed "full integration" elements. For example, to integrate the 8×8 stiffness matrix of the 4–node plane isoparametric quadrilateral element, each entry will be evaluated at the four locations (2×2) of the integration points in the isoparametric element. Thus, the number of computations (excluding the addition) for the integration of the element are $8 \times 8 \times 4 = 256$ computations. The number of computations for the 16×16 stiffness matrix of the 8–node plane isoparametric quadrilateral element is 2,304 computations since it uses a 3×3 Gauss integration points scheme! To save computational resources, a "reduced integration" option for each of those elements is available in finite–element analysis software. The reduced integration version of the 4–node plane isoparametric quadrilateral element uses only one integration point in the center of the element, and the number of computations is reduced from 256 to 64 computations. For the 8–node plane isoparametric quadrilateral, a 2×2 Gauss integration points scheme is used instead of the 3×3 Gauss integration points scheme, which reduces the number of computations from 2,304 to 1,024 computations. The benefit of using reduced integration is not only in saving computational resources but also in balancing out the over–stiffness introduced by assuming a certain deformation field within an element. In fact, reduced integration can lead to better performance in most cases. However, while reduced integration can dramatically reduce the computational time, it can lead to a structure that is too flexible. In fact, the displacements are **over predicted** when a structure is discretized using reduced integration elements. The reason, as will be shown in the following section, is because some information is lost during sampling of the stiffness at the fewer integration points of the reduced integration elements.

9.3.2.e. Spurious Modes Associated with Reduced Integration Elements

Spurious modes, or sometimes called zero energy modes, are combinations of displacements that in general produce strains; however, such strains cannot be detected at the integration or sampling points. So, the internal energy calculated at the integration points is equal to zero for such mode. In other words, a very small amount of load would be enough to produce a very large displacement. If a spurious mode exists for the whole structure, it becomes unstable. Usually spurious modes exist in under–integrated element (reduced integration). This is because the reduced number of integration points cannot detect a spurious strain mode. The two very common spurious modes are those for the bilinear quadrilateral (4–node plane element) and the quadratic displacement quadrilateral (8–node plane element). The spurious modes will be illustrated using the following problem.

Problem 1: Find the forces required to produce the shown displacement and the associated internal energy for the 4–node plane–stress quadrilateral element shown below, assuming a bilinear displacement interpolation function. Use exact integration, full integration (2 × 2 integration points), and reduced integration (1 integration point).

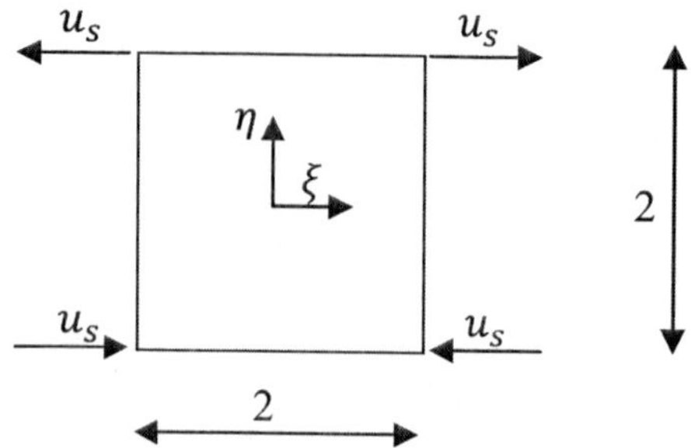

Solution: The shape functions of the shown element are:

$$N_1 = \frac{(1-\xi)(1-\eta)}{4}, \quad N_2 = \frac{(1-\eta)(1+\xi)}{4}, \quad N_3 = \frac{(1+\eta)(1+\xi)}{4}, \quad N_4 = \frac{(1+\eta)(1-\xi)}{4}$$

The displacement functions u and v in the direction of ξ and η, respectively, are:

$$u = u_s N_1 - u_s N_2 + u_s N_3 - u_s N_4$$

$$v = 0$$

The strains are:

$$\begin{Bmatrix} \varepsilon_{\xi\xi} \\ \varepsilon_{\eta\eta} \\ \gamma_{\xi\eta} \end{Bmatrix} = \begin{Bmatrix} u_s\left(\frac{dN_1}{d\xi} - \frac{dN_2}{d\xi} + \frac{dN_3}{d\xi} - \frac{dN_4}{d\xi}\right) \\ 0 \\ u_s\left(\frac{dN_1}{d\eta} - \frac{dN_2}{d\eta} + \frac{dN_3}{d\eta} - \frac{dN_4}{d\eta}\right) \end{Bmatrix} = \begin{Bmatrix} \frac{dN_1}{d\xi} - \frac{dN_2}{d\xi} + \frac{dN_3}{d\xi} - \frac{dN_4}{d\xi} \\ 0 \\ \frac{dN_1}{d\eta} - \frac{dN_2}{d\eta} + \frac{dN_3}{d\eta} - \frac{dN_4}{d\eta} \end{Bmatrix} u_s = Bu_s$$

The stress–strain relationship for plane stress is:

$$\begin{Bmatrix} \sigma_{\xi\xi} \\ \sigma_{\eta\eta} \\ \sigma_{\xi\eta} \end{Bmatrix} = \frac{E}{(1+v)(1-v)} \begin{pmatrix} 1 & v & 0 \\ v & 1 & 0 \\ 0 & 0 & \frac{1-v}{2} \end{pmatrix} \begin{Bmatrix} \varepsilon_{\xi\xi} \\ \varepsilon_{\eta\eta} \\ \gamma_{\xi\eta} \end{Bmatrix} = C \begin{Bmatrix} \varepsilon_{\xi\xi} \\ \varepsilon_{\eta\eta} \\ \gamma_{\xi\eta} \end{Bmatrix}$$

The stiffness matrix K can be evaluated using the following integral (t is the element thickness):

$$K = t \int_{-1}^{1} \int_{-1}^{1} B^T C B \, d\xi \, d\eta = \frac{tE}{2\left(v^2 - 1\right)} \int_{-1}^{1} \int_{-1}^{1} (v-1)\xi^2 - 2\eta^2 d\xi \, d\eta$$

Notice that the system has only one degree of freedom (the unknown displacement variable u_s), and therefore, the stiffness matrix has the dimensions 1×1.

Using Exact Integration:

The stiffness matrix evaluated using exact integration is:

$$K = \frac{2E(v-3)}{3\left(v^2-1\right)}$$

The strain energy associated with this displacement mode is equal to $\frac{1}{2} f \cdot u = \frac{1}{2} K u_s^2 = \frac{2E(v-3)}{3(v^2-1)} u_s^2$.

Using Full Integration:

The stiffness matrix evaluated using full integration is:

$$K = \frac{tE}{2\left(v^2-1\right)} \left(\sum_{j=1}^{2} \sum_{i=1}^{2} w_i w_j \left((v-1)\xi_i^2 - 2\eta_j^2 \right) \right) = \frac{2E(v-3)}{3\left(v^2-1\right)}$$

where $\xi_i, \eta_j \in \left\{\frac{1}{\sqrt{3}}, -\frac{1}{\sqrt{3}}\right\}$ and $w_i = w_j = 2$

The strain energy associated with this displacement mode is equal to $\frac{1}{2}f \cdot u = \frac{1}{2}Ku_s^2 = \frac{2E(\nu-3)}{3(\nu^2-1)}u_s^2$

Thus, the full gauss integration produces the exact results!

Using Reduced Integration:

The stiffness matrix evaluated using reduced integration is:

$$K = \frac{tE}{2(\nu^2-1)}\left(\sum_{j=1}^{1}\sum_{i=1}^{1} w_i w_j \left((\nu-1)\xi_i^2 - 2\eta_j^2\right)\right) = 0$$

where $\xi_i, \eta_j \in \{0\}$ and $w_i = w_j = 2$

The strain energy associated with this displacement mode is equal to $\frac{1}{2}f \cdot u = \frac{1}{2}Ku_s^2 = 0$.

Thus, the reduced integration scheme predicts that the displacement field shown does not require any external forces applied to the element. Thus, this element is unstable if loads produce the displacement mode shown, i.e., bending!

The following is the Mathematica code used for the above calculations:

```
N1=(1-xi)(1-eta)/4;
N2=(1+xi)(1-eta)/4;
N3=(1+xi)(1+eta)/4;
N4=(1-xi)(1+eta)/4;
B={D[N1-N2+N3-N4,xi],0,D[N1-N2+N3-N4,eta]};
Cc=Ee/(1+nu)/(1-nu)*{{1,nu,0},{nu,1,0},{0,0,(1-nu)/2}};
Kbeforeintegration=FullSimplify[B.Cc.B]
(*Using Exact Integration*)
Kexact=Integrate[Kbeforeintegration,{eta,-1,1},{xi,-1,1}]
Energyexact=us*Kexact*us/2
(*Using Full Integration*)
etaset={1/Sqrt[3],-1/Sqrt[3]};
xiset={1/Sqrt[3],-1/Sqrt[3]};
Kfull=FullSimplify[Sum[(Kbeforeintegration/.{xi->xiset[[i]],eta-
>etaset[[j]]}),{i,1,2},{j,1,2}]]
Energyfull=us*Kfull*us/2
(*Using reduced Integration*)
etaset=xiset={0};
Kreduced=Sum[4(Kbeforeintegration/.{xi->xiset[[i]],eta-
>etaset[[j]]}),{i,1,1},{j,1,1}]
Energyreduced=us*Kreduced*us/2
```

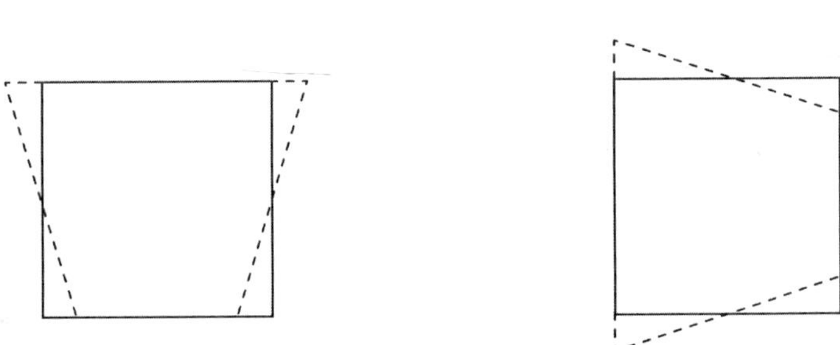

FIGURE 9-22 Spurious modes of the 4–node reduced–integration plane quadrilateral element.

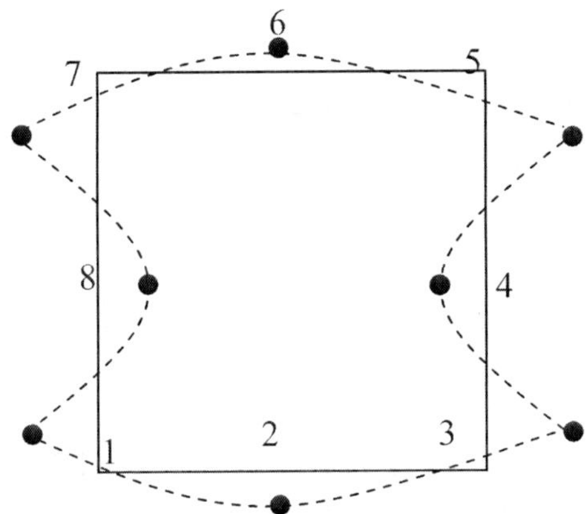

FIGURE 9-23 Spurious mode of an 8–node reduced–integration plane quadrilateral element.

In general, any bending mode is a spurious mode in a 4–node reduced–integration linear element. The strain modes shown cannot be detected at the integration point; the strain at the center of the element (at the location of the single integration point) is equal to zero (Figure 9–22).

The spurious mode of an 8–node reduced–integration quadratic element is called "the hour glass" mode (Figure 9–23) and is produced when:

$$u_1 = u_7 = -1, \quad u_3 = u_5 = 1, \quad v_6 = u_8 = \frac{1}{2},$$

$$v_5 = v_7 = -1, \quad v_1 = v_3 = 1, \quad u_4 = v_2 = -\frac{1}{2}$$

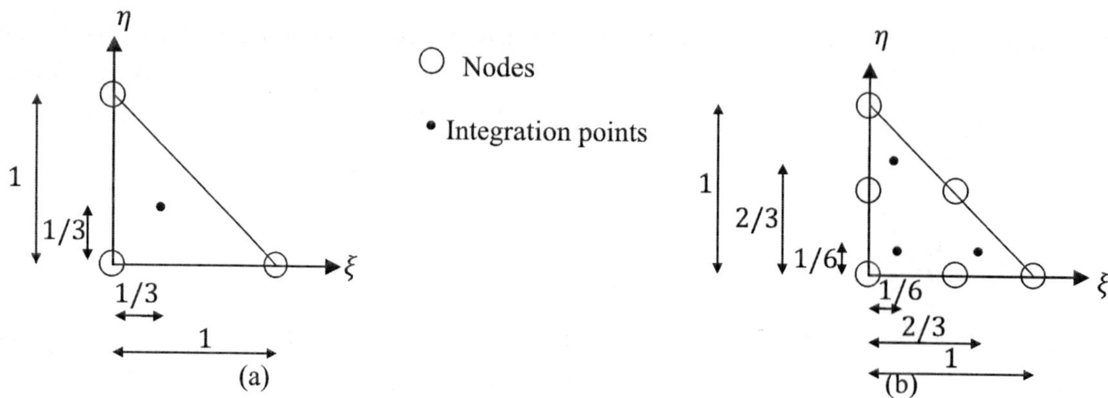

FIGURE 9-24 Gauss integration points for integrating (a) 3 node isoparametric elements: $(\xi, \eta) = \left\{\frac{1}{3}, \frac{1}{3}\right\}$ with the weight factor $w = \frac{1}{2}$. (b) 6 node isoparametric elements $(\xi, \eta) \in \left\{\left\{\frac{1}{6}, \frac{2}{3}\right\}, \left\{\frac{1}{6}, \frac{1}{6}\right\}, \left\{\frac{2}{3}, \frac{1}{6}\right\}\right\}$ with equal weight factors $w = \frac{1}{6}$.

9.3.2.f. Gauss Integration for Two-Dimensional Triangle Isoparametric Elements

As seen in section 9.2.2.a, the strains of the 3–node elements are constant, and thus, the entries in the matrix $B^T CB$ are constant. Thus, a 1–point numerical integration scheme is required to produce accurate integrals for the isoparametric 3–node elements (Figure 9–24). For the 6–node triangular elements, a 3–points numerical integration scheme is required to produce accurate integrals (Figure 9–24). It should be noted that these locations for the integration points can be extended to the three–dimensional case in a straightforward manner.

9.3.3. Final Notes on the Use of Isoparametric Elements and Numerical Integration

The use of isoparametric elements and numerical integration dramatically increases the robustness of the finite–element analysis method. The techniques described in this section allow meshing of irregular domains with *compatible* elements, i.e., elements that ensure the continuity of the displacement field across boundaries (**How?**). However, to ensure the continuity of the displacement field, the elements of the mesh should be of the same type; otherwise, the user has to apply mesh constraints to ensure the compatibility of the displacement between elements. For example, a 4–node bilinear displacement element when connected to an 8–node quadratic displacement element will have incompatible displacements at the side of the connection. This can be overcome by constraining the mid–side node of the quadratic element to always lie on the straight side of the bilinear element. Otherwise, the results at the location of discontinuous displacement fields would be erroneous.

The stiffness matrix and the nodal loads of the structure can be calculated using the Gauss numerical integration scheme, which speeds up the computational time required. However, the price of using the Gauss numerical integration is the loss of accuracy. Gauss numerical integration produces accurate results only when the integrals are exact polynomials. The integrals are exact polynomials only when the elements of the mesh are exactly aligned with the isoparametric mapping into the coordinates of ξ and η. When the shape of the elements in the mesh deviates from the shape of the element in the coordinate system of ξ and η, the functions to be integrated are no longer exact polynomials but rather radicals; thus, the numerical integration scheme is no longer accurate. Most finite element analysis software give warnings when elements are highly distorted since this will lead to inaccurate computations of the stiffness matrix during the numerical integration process. In addition, for nonlinear problems when the stiffness changes with deformation, it is possible that the element will reach a highly distorted shape that the numerical integration scheme is no longer valid.

9.3.4. Example: Obtaining the Stiffness Matrix and the Nodal Loads of a 4-Node Isoparametric Quadrilateral

The stiffness matrix and the nodal loads due to a traction vector and a body forces vector of a plane stress element of a linear–elastic small deformations material whose Young's modulus = 1 unit and Poisson's ratio = 0.3 will be derived using isoparametric formulation with exact integration, full integration, and finally, with reduced integration. The geometry and loading are shown in Figure 9–25. The thickness of the element is assumed to be equal to 1 unit.

Stiffness Matrix:

The isoparametric formulation shown in section 9.3.1.b will be used. The mapping function between the (x, y) coordinate system and the (ξ, η) coordinate system is:

$$x = \sum_{i=1}^{4} N_i x_i = \frac{7 + \eta + 6\xi}{4}$$

$$y = \sum_{i=1}^{4} N_i y_i = \frac{18 + \eta(13 - 2\xi) + 3\xi}{10}$$

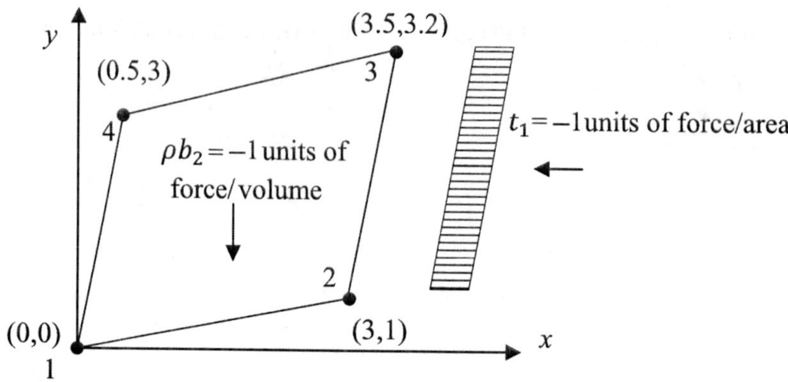

FIGURE 9-25 Geometry and loading for a four noded quadrilateral isoparametric element.

The gradient of the mapping is:

$$J = \begin{pmatrix} \dfrac{\partial x}{\partial \xi} & \dfrac{\partial x}{\partial \eta} \\ \dfrac{\partial y}{\partial \xi} & \dfrac{\partial y}{\partial \eta} \end{pmatrix} = \begin{pmatrix} \dfrac{3}{2} & \dfrac{1}{4} \\ \dfrac{3-2\eta}{10} & \dfrac{13-2\xi}{10} \end{pmatrix}, \quad J^{-1} = \frac{1}{75+2\eta-12\xi}\begin{pmatrix} 52-8\xi & 4(-3+2\eta) \\ -10 & 60 \end{pmatrix}$$

The B matrix is:

$$B = \begin{pmatrix} 1 & 0 & 0 & 0 \\ 0 & 0 & 0 & 1 \\ 0 & 1 & 1 & 0 \end{pmatrix} \begin{pmatrix} \dfrac{\partial \xi}{\partial x} & \dfrac{\partial \eta}{\partial x} & 0 & 0 \\ \dfrac{\partial \xi}{\partial y} & \dfrac{\partial \eta}{\partial y} & 0 & 0 \\ 0 & 0 & \dfrac{\partial \xi}{\partial x} & \dfrac{\partial \eta}{\partial x} \\ 0 & 0 & \dfrac{\partial \xi}{\partial y} & \dfrac{\partial \eta}{\partial y} \end{pmatrix} \begin{pmatrix} \dfrac{\partial N_1}{\partial \xi} & 0 & \dfrac{\partial N_2}{\partial \xi} & 0 & \dfrac{\partial N_3}{\partial \xi} & 0 & \dfrac{\partial N_4}{\partial \xi} & 0 \\ \dfrac{\partial N_1}{\partial \eta} & 0 & \dfrac{\partial N_2}{\partial \eta} & 0 & \dfrac{\partial N_3}{\partial \eta} & 0 & \dfrac{\partial N_4}{\partial \eta} & 0 \\ 0 & \dfrac{\partial N_1}{\partial \xi} & 0 & \dfrac{\partial N_2}{\partial \xi} & 0 & \dfrac{\partial N_3}{\partial \xi} & 0 & \dfrac{\partial N_4}{\partial \xi} \\ 0 & \dfrac{\partial N_1}{\partial \eta} & 0 & \dfrac{\partial N_2}{\partial \eta} & 0 & \dfrac{\partial N_3}{\partial \eta} & 0 & \dfrac{\partial N_4}{\partial \eta} \end{pmatrix}$$

The stiffness matrix is:

$$K = t\int_{-1}^{1}\int_{-1}^{1} B^T CB.\det J d\xi d\eta = t\int_{-1}^{1}\int_{-1}^{1} K_1 \det J.d\xi d\eta$$

where C is the linear elastic isotropic plane stress constitutive relationship matrix and $K_1 = B^T CB$.

Exact integration produces the following stiffness matrix (units of force/length):

$$K = \begin{pmatrix} 0.3228 & 0.1002 & -0.254 & 0.0002 & -0.1009 & -0.1437 & 0.032 & 0.0433 \\ 0.1002 & 0.3896 & 0.0277 & 0.0887 & -0.1437 & -0.1372 & 0.0159 & -0.3411 \\ -0.254 & 0.0277 & 0.7125 & -0.302 & -0.05 & 0.0319 & -0.4085 & 0.2424 \\ 0.0002 & 0.0887 & -0.302 & 0.785 & 0.0594 & -0.4343 & 0.2424 & -0.4395 \\ -0.1009 & -0.1437 & -0.05 & 0.0594 & 0.3911 & 0.0868 & -0.2403 & -0.0025 \\ -0.1437 & -0.1372 & 0.0319 & -0.4343 & 0.0868 & 0.4672 & 0.025 & 0.1043 \\ 0.032 & 0.0159 & -0.4085 & 0.2424 & -0.2403 & 0.025 & 0.6169 & -0.2833 \\ 0.0433 & -0.3411 & 0.2424 & -0.4395 & -0.0025 & 0.1043 & -0.2833 & 0.6763 \end{pmatrix}$$

Full integration produces the following stiffness matrix (units of force/length):

$$K = tK_1 \det J\big|_{\xi=\frac{-1}{\sqrt{3}},\eta=\frac{-1}{\sqrt{3}}} + tK_1 \det J\big|_{\xi=\frac{1}{\sqrt{3}},\eta=\frac{-1}{\sqrt{3}}} + tK_1 \det J\big|_{\xi=\frac{1}{\sqrt{3}},\eta=\frac{1}{\sqrt{3}}} + tK_1 \det J\big|_{\xi=\frac{-1}{\sqrt{3}},\eta=\frac{1}{\sqrt{3}}}$$

$$K = \begin{pmatrix} 0.3226 & 0.1003 & -0.2536 & 0.0001 & -0.1012 & -0.1436 & 0.0322 & 0.0433 \\ 0.1003 & 0.389 & 0.0275 & 0.0896 & -0.1436 & -0.138 & 0.0158 & -0.3405 \\ -0.2536 & 0.0275 & 0.712 & -0.3019 & -0.0495 & 0.0318 & -0.4089 & 0.2425 \\ 0.0001 & 0.0896 & -0.3019 & 0.7839 & 0.0593 & -0.4332 & 0.2425 & -0.4403 \\ -0.1012 & -0.1436 & -0.0495 & 0.0593 & 0.3907 & 0.0869 & -0.24 & -0.0026 \\ -0.1436 & -0.138 & 0.0318 & -0.4332 & 0.0869 & 0.4662 & 0.0249 & 0.105 \\ 0.0322 & 0.0158 & -0.4089 & 0.2425 & -0.24 & 0.0249 & 0.6166 & -0.2832 \\ 0.0433 & -0.3405 & 0.2425 & -0.4403 & -0.0026 & 0.105 & -0.2832 & 0.6758 \end{pmatrix}$$

Reduced integration produces the following stiffness matrix (units of force/length):

$$K = 2 \times 2 \times tK_1 \det J\big|_{\xi=0,\eta=0}$$

$$K = \begin{pmatrix} 0.2266 & 0.119 & -0.1223 & -0.0256 & -0.2266 & -0.119 & 0.1223 & 0.0256 \\ 0.119 & 0.2802 & 0.0018 & 0.2385 & -0.119 & -0.2802 & -0.0018 & -0.2385 \\ -0.1223 & 0.0018 & 0.5321 & -0.2667 & 0.1223 & -0.0018 & -0.5321 & 0.2667 \\ -0.0256 & 0.2385 & -0.2667 & 0.58 & 0.0256 & -0.2385 & 0.2667 & -0.58 \\ -0.2266 & -0.119 & 0.1223 & 0.0256 & 0.2266 & 0.119 & -0.1223 & -0.0256 \\ -0.119 & -0.2802 & -0.0018 & -0.2385 & 0.119 & 0.2802 & 0.0018 & 0.2385 \\ 0.1223 & -0.0018 & -0.5321 & 0.2667 & -0.1223 & 0.0018 & 0.5321 & -0.2667 \\ 0.0256 & -0.2385 & 0.2667 & -0.58 & -0.0256 & 0.2385 & -0.2667 & 0.58 \end{pmatrix}$$

Notice that the exact and full integration produce very similar results. The more distorted the element from a rectangle, the more the full integration technique would deviate from the exact integration. The reduced–integration technique, however, produces numbers that highly deviate from the full integration technique. The structure is expected to be less stiff when the reduced–integration technique is utilized.

Nodal Forces Due to Body Forces:

The body forces vector is:

$$\rho b = \begin{pmatrix} 0 \\ -1 \end{pmatrix} \textit{units of force/volume}$$

The nodal forces due to the body forces vector are:

$$f = t\int_{-1}^{1}\int_{-1}^{1} N^T.\rho b.\det J d\xi d\eta = \int_{-1}^{1}\int_{-1}^{1} f_1 d\xi d\eta$$

Using exact integration, the nodal forces due to the body forces vector are:

$$f = \begin{pmatrix} 0. \\ -1.958 \\ 0. \\ -1.7583 \\ 0. \\ -1.7917 \\ 0. \\ -1.9917 \end{pmatrix} \textit{units of force}$$

Using full integration, the same solution is obtained:

$$f_{full} = f_1\big|_{\xi=\frac{-1}{\sqrt{3}},\eta=\frac{-1}{\sqrt{3}}} + f_1\big|_{\xi=\frac{1}{\sqrt{3}},\eta=\frac{-1}{\sqrt{3}}} + f_1\big|_{\xi=\frac{1}{\sqrt{3}},\eta=\frac{1}{\sqrt{3}}} + f_1\big|_{\xi=\frac{-1}{\sqrt{3}},\eta=\frac{1}{\sqrt{3}}}$$

$$f_{full} = \begin{pmatrix} 0. \\ -1.958 \\ 0. \\ -1.7583 \\ 0. \\ -1.7917 \\ 0. \\ -1.9917 \end{pmatrix} \text{ units of force}$$

Using reduced integration, the solution is:

$$f_{reduced} = 4 \times f_1\big|_{\xi=0,\eta=0}$$

$$f_{reduced} = \begin{pmatrix} 0. \\ -1.875 \\ 0. \\ -1.875 \\ 0. \\ -1.875 \\ 0. \\ -1.875 \end{pmatrix} \text{ units of force}$$

Notice that the reduced–integration technique produces equal numbers on all the nodes. Notice also that the total sum of forces is equal whether the exact, full, or reduced integration is used (**Why?**).

Nodal Forces Due to Traction Vectors:

The traction vector on the side connecting nodes 2 and 3 is:

$$t_n = \begin{pmatrix} -1 \\ 0 \end{pmatrix} \text{ units of force/area}$$

The nodal forces due to the traction forces vector are:

$$f = t\int_{-1}^{1} N^T.t_n.dl = t\int_{-1}^{1} N^T.t_n.\left(\frac{dl}{d\eta}\right)d\eta = \left.\int_{-1}^{1} f_1 d\eta\right|_{\xi=1}$$

The factor that transforms the integration from the spatial coordinate system to the (ξ, η) coordinate system can be obtained as follows:

$$\frac{dl}{d\eta} = \frac{\sqrt{dx^2+dy^2}}{d\eta} = \left.\frac{\sqrt{\left(\frac{\partial x}{\partial \xi}d\xi + \frac{\partial x}{\partial \eta}d\eta\right)^2 + \left(\frac{\partial y}{\partial \xi}d\xi + \frac{\partial y}{\partial \eta}d\eta\right)^2}}{d\eta}\right|_{\xi=1, d\xi=0}$$

$$= \left.\sqrt{\left(\frac{\partial x}{\partial \eta}\right)^2 + \left(\frac{\partial y}{\partial \eta}\right)^2}\right|_{\xi=1} = \frac{\sqrt{509}}{20}$$

Using exact integration, full integration, and reduced integration produces the same result:

$$f = \begin{pmatrix} 0 \\ 0 \\ -1.12805 \\ 0 \\ -1.12805 \\ 0 \\ 0 \\ 0 \end{pmatrix} \textit{units of force}$$

Notice that full integration is obtained by using two integration points, while reduced integration is obtained using one integration point:

$$f_{full} = f_1\big|_{\eta=\frac{1}{\sqrt{3}}} + f_1\big|_{\eta=\frac{-1}{\sqrt{3}}}, \quad f_{reduced} = f_1\big|_{\eta=0}$$

It should be noted that the same results were obtained using the different integration techniques because the traction vector is constant. Different results would be obtained if the traction vector were not constant.

The following is the Mathematica code utilized:

```
(*Using Isoparametric Formulation*)
coordinates={{0,0},{3,1},{35/10,32/10},{5/10,3}};
t=1;
(*Plane Stress*)
Ee=1;
Nu=3/10;
Cc=Ee/(1+Nu)*{{1/(1-Nu),Nu/(1-Nu),0},{Nu/(1-Nu),
1/(1-Nu),0},{0,0,1/2}};
a=Polygon[{coordinates[[1]],coordinates[[2]],coordinates[[3]],
coordinates[[4]],coordinates[[1]]}];
Graphics[a]
Shapefun=Table[0,{i,1,4}];
Shapefun[[1]]=(1-xi)(1-eta)/4;
Shapefun[[2]]=(1+xi)(1-eta)/4;
Shapefun[[3]]=(1+xi)(1+eta)/4;
Shapefun[[4]]=(1-xi)(1+eta)/4;
Nn=Table[0,{i,1,2},{j,1,8}];
Do[Nn[[1,2i-1]]=Nn[[2,2i]]=Shapefun[[i]],{i,1,4}];
x=FullSimplify[Sum[Shapefun[[i]]*coordinates[[i,1]],{i,1,4}]];
y=FullSimplify[Sum[Shapefun[[i]]*coordinates[[i,2]],{i,1,4}]];
J=Table[0,{i,1,2},{j,1,2}];
J[[1,1]]=D[x,xi];
J[[1,2]]=D[x,eta];
J[[2,1]]=D[y,xi];
J[[2,2]]=D[y,eta];
JinvT=Transpose[Inverse[J]];
mat1={{1,0,0,0},{0,0,0,1},{0,1,1,0}};
mat2={{JinvT[[1,1]],JinvT[[1,2]],0,0},{JinvT[[2,1]],
JinvT[[2,2]],0,0},{0,0,JinvT[[1,1]],JinvT[[1,2]]},{0,0,
JinvT[[2,1]],JinvT[[2,2]]}};
mat3=Table[0,{i,1,4},{j,1,8}];
Do[mat3[[1,2i-1]]=mat3[[3,2i]]=D[Shapefun[[i]],xi];
mat3[[2,2i-1]]=mat3[[4,2i]]=D[Shapefun[[i]],eta],{i,1,
4}];
B=Chop[FullSimplify[mat1.mat2.mat3]];
(*Exact integration*)
Kbeforeintegration=FullSimplify[Transpose[B].Cc.B];
K1=t*Integrate[
Kbeforeintegration*Det[J],{xi,-1.,1.},{eta,-1.,1.}];
K1=Chop[K1];
K1=Round[K1,0.0001];
K1//MatrixForm
(*Full integration*)
xiset={-1/Sqrt[3],1/Sqrt[3]};
etaset={-1/Sqrt[3],1/Sqrt[3]};
Kfull=t*
```

```
FullSimplify[
Sum[(Kbeforeintegration*Det[J]/.{xi->iset[[i]],eta-
>etaset[[j]]}),{i,1,2},{j,1,2}]];
Kfull=Round[Kfull,0.0001];
Kfull//MatrixForm
(*Reduced integration*)
xiset={0};
etaset={0};
Kreduced=
4*t*FullSimplify[
Sum[(Kbeforeintegration*Det[J]/.{xi->xiset[[i]],eta-
>etaset[[j]]}),{i,1,1},{j,1,1}]];
Kreduced=Round[Kreduced,0.0001];
Kreduced//MatrixForm

(*Body Forces*)
rb={0,-1};
fbeforeintegration=t*Transpose[Nn].rb*Det[J];
f1=Integrate[fbeforeintegration,{xi,-1,1},{eta,-1,1}];
N[f1]//MatrixForm
xiset={-1/Sqrt[3],1/Sqrt[3]};
etaset={-1/Sqrt[3],1/Sqrt[3]};
ffull=FullSimplify[Sum[(fbeforeintegration/.{xi->xiset[[i]],eta-
>etaset[[j]]}),{i,1,2},{j,1,2}]];
N[ffull]//MatrixForm
xiset={0};
etaset={0};
freduced=4*FullSimplify[Sum[(fbeforeintegration/.{xi->xiset[[i]],eta-
>etaset[[j]]}),{i,1,1},{j,1,1}]];
N[freduced]//MatrixForm

(*Traction Forces*)
dldeta=Sqrt[J[[1,2]]^2+J[[2,2]]^2]/.xi->1
tn={-1,0};
fbeforeintegration=t*Transpose[Nn].tn*dldeta;
f1=Integrate[fbeforeintegration/.xi->1,{eta,-1,1}];
N[f1]//MatrixForm
xiset={-1/Sqrt[3],1/Sqrt[3]};
etaset={-1/Sqrt[3],1/Sqrt[3]};
ffull=FullSimplify[Sum[(fbeforeintegration/.{xi->1,eta-
>etaset[[j]]}),{j,1,2}]];
N[ffull]//MatrixForm
xiset={0};
etaset={0};
freduced=2*FullSimplify[Sum[(fbeforeintegration/.{xi->1,eta-
>etaset[[j]]}),{j,1,1}]];
N[freduced]//MatrixForm
```

9.4. COMPARISON OF DIFFERENT ELEMENTS BEHAVIOR UNDER BENDING

To compare the different elements described in this chapter, the simply supported beam with the distributed load shown in Figure 9–26 was modeled in the finite element analysis software ABAQUS with various different element types. The beam is supporting a distributed load $w = 45\,\text{N/mm}$ and has a Young's modulus $E = 20\,\text{GPa}$ and a length $L = 4000\,\text{mm}$.

The beam was modeled as a 2D plane shell and meshed using 2D plane stress solid elements. Plane stress assumes that the thickness of the beam is small, allowing the material to freely deform in the third direction, thereby resulting in a zero–stress component in the third direction ($\sigma_{13} = \sigma_{23} = \sigma_{33} = 0$). The third direction is the 150 mm dimension in this case. The thickness of the plane stress element was set to 150 mm, while the value of the pressure load applied was set to $45/150 = 0.3\,\text{N/mm}^2$. The imposed boundary conditions are $u_1 = u_2 = 0$ at one end and a roller support $u_2 = 0$ at the other end. Since this is a plane problem, specifying $u_3 = 0$ is redundant. The beam was modeled with various types of elements and mesh sizes. The maximum normal stress components σ_{11} at the top and bottom fibers of the beam at mid–span and the maximum vertical displacement (u_2) were determined in response to the applied distributed load. The following table compares the results for different elements with different mesh sizes measured by the number of elements (layers) in the direction of the second basis vector. Model 11 is a very fine mesh version of model 9 to show the effects of mesh refinement.

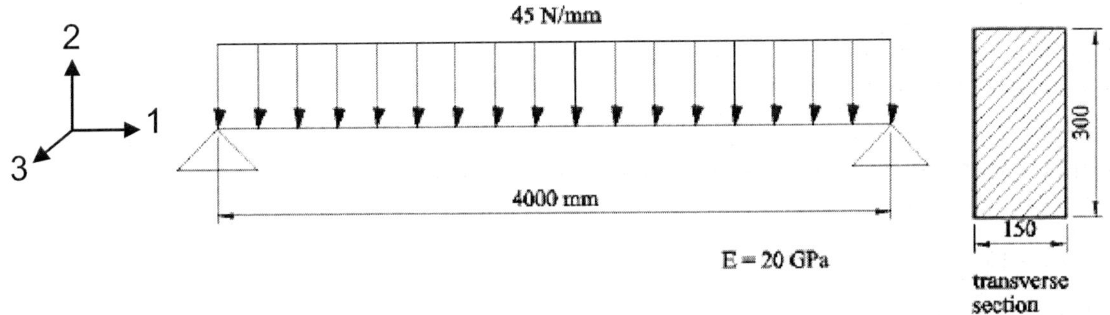

FIGURE 9-26 Geometry and loading for comparing the different elements.

#	Element name	Description	Layers	Integration	σ_{11} Top (Mpa)	σ_{11} Bottom (Mpa)	u_2 (mm)
1	CPS3	3-node Linear triangular	1	F	−9.702	9.702	5.14
2	CPS3	3-node Linear triangular	3	F	−28.782	28.782	16.24
3	CPS4	4-node Quadrilateral	1	F	−29.057	29.057	14.82
4	CPS4	4-node Quadrilateral	3	F	−37.874	37.874	21.32
5	CPS4R	4-node Quadrilateral	1	R	−1.69*E*-10	1.68*E*-10	5680
6	CPS4R	4-node Quadrilateral	3	R	−29.948	29.948	25.57
7	CPS8	8-node Quadrilateral	1	F	−39.92	39.92	22.47
8	CPS8	8-node Quadrilateral	3	F	−40.075	40.075	22.59
9	CPS8R	8-node Quadrilateral	3	R	−40.0537	40.0537	22.611
10	CPS8R	8-node Quadrilateral*	3	R	−40.0537	40.0537	22.611
11	CPS8R	8-node Quadrilateral	60	R	−40.0598	40.0598	22.8036

*Half the beam was modeled with a roller support at one end and imposing symmetry boundary conditions at the other end.

The results according to the Euler–Bernoulli beam theory are as follows. Since no load is applied in the horizontal direction, the reaction in direction 1 should be equal to zero. The vertical reaction force R at each end can be calculated as follows:

$$R = \frac{wL}{2} = \frac{45\,(4000)}{2} = 90\text{ kN}$$

The reaction forces in all of the models matched the ones calculated above.

It is also possible to check that the deformed shape matches the shape that we expect based on the loading environment and boundary conditions. The deformed beam has the shape shown in Figure 9–27 (with a magnified scale factor), which is consistent with the theoretical deformation shape. A similar deformed shape can be seen with all of the tested models.

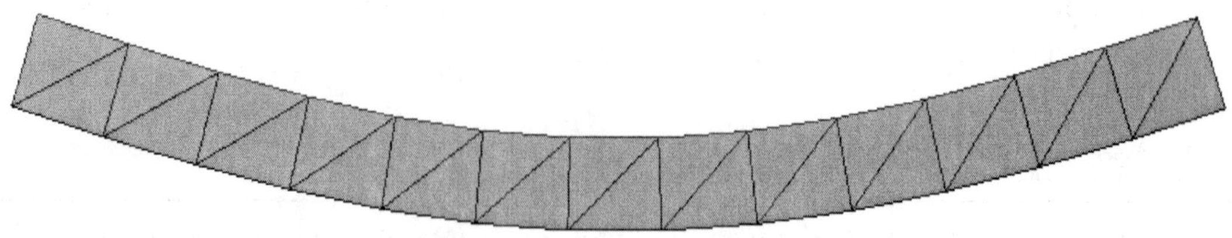

FIGURE 9-27 Deformed shape.

The stress and displacement can be calculated using the Euler–Bernoulli beam theory. For a simply supported beam with a distributed load, the maximum vertical deflection U_{max} is given by the following equation:

$$U_{max} = \frac{5wL^4}{384EI}$$

Where, I is the moment of inertia for the rectangular cross section given by:

$$I = \frac{1}{12}(150)(300)^3 = 3.375 \times 10^8 \text{ mm}^4$$

Thus,

$$U_{max} = \frac{5wL^4}{384EI} = \frac{5(45)(4000)^4}{384(20000)\left(3.375 \times 10^8\right)} = 22.2 \text{ mm}$$

For a simply supported beam with a distributed load, the maximum bending moment is located at the mid–span of the beam and is given by the following equation:

$$M = \frac{wL^2}{8} = \frac{45(4000)^2}{8} = 9 \times 10^7 \text{ N.mm}$$

The normal stress component σ_{11} can be found from the beam equation:

$$\sigma_{11} = \frac{Mc}{I}$$

where c is the distance from the neutral axis to the point of interest. In the modeled beam, the neutral axis is at the center yielding $c = 150$ mm for both the top and bottom fibers. The Euler–Bernoulli beam equation predicts that the stress has a linear profile, with the magnitude of the stress at the top and bottom of the beam being equal. The top of the beam is in compression (negative stress), while the bottom is in tension (positive stress).

The maximum normal stress component σ_{11} according to the Euler–Bernoulli beam theory is located at mid–span and is equal to (positive at the bottom, negative at the top):

$$\sigma_{11} = \frac{Mc}{I} = \frac{9 \times 10^7 \times 150}{3.375 \times 10^8} = 40 \text{ MPa}$$

Notice that these results are not necessarily an exact solution since the Euler Bernoulli beam theory assumes that plane sections perpendicular to the neutral axis before deformation remain plane and perpendicular to the neutral axis after deformation.

9.4.1. Behavior of Linear Triangular Elements

The linear triangular elements dramatically underestimate the stress and the displacement. These are constant strain elements, which is evident when viewing the stress contour plot without averaging across the elements (Figure 28).

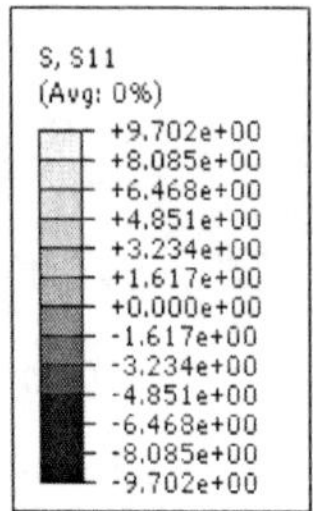

FIGURE 9-28 Highly discontinuous values for the stress component σ_{11} are predicted when a course mesh of linear triangular elements is used.

Linear triangular elements are not appropriate for bending (particularly when only one layer is used) because the strain in bending is not constant, but varies linearly from the top edge to the bottom edge. It is evident from the displacement that these elements produce a very stiff structure when only one layer is used. The result is improved when three layers of elements are used because the strain is forced to be constant over a smaller area, as opposed to constant across the entire cross section of the structure. Using these elements with a very fine mesh (60 layers) comes closer to the beam theory solution with $U_{max} = 22.67$ mm and $\sigma_{11} = 40.02$ MPa.

9.4.2. Behavior of 4-node Quadrilateral Elements

4–node quadrilateral elements offer an improved solution over the linear triangular elements; however, they are still relatively stiff due to shear locking (parasitic shear) described in section 9.2.3.a. These elements allow the stress to vary linearly within the element. In this case, they allow a negative stress at the top and a positive stress at the bottom (Figure 9–29).

Mesh refinement to three layers greatly improves the solution, producing results that approach the Euler–Bernoulli beam solution.

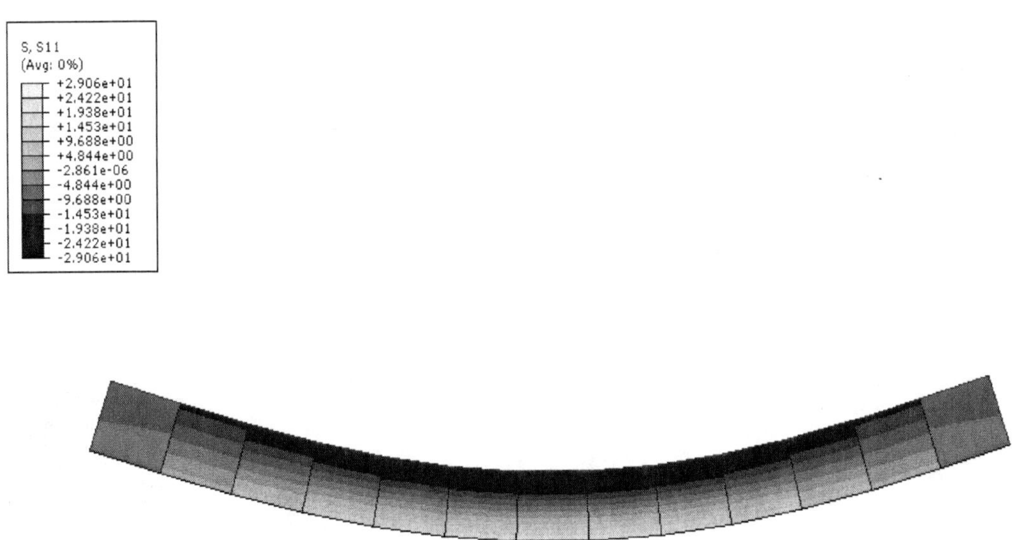

FIGURE 9-29 Stress component σ_{11} when a course mesh of linear quadrilateral elements is used.

9.4.3. Effect of Using Reduced Integration

Using the 4–node quadrilateral elements with reduced integration produces an extremely soft structure with wildly inaccurate results. The 4–node quadrilateral elements have 4 integration points. Using reduced integration, the number of integration points is reduced to one. This is not enough to capture the actual behavior of the element and results in a very soft structure. The integration point is at the center of the element, which is at the neutral axis of the beam when one layer of elements is used. The stresses and strains in the models with a roller support at the right end (engineering beam theory) have zero stress at the neutral axis. So, the results suggest that the elements have zero (or close to zero) stress everywhere and an extremely high displacement. It is clear that a coarse mesh of the 4–node quadrilateral elements with reduced integration cannot be used to model a beam under bending. Refining the mesh to three layers produces much more reasonable results; however, the displacement is still overestimated, meaning that the modeled structure is still softer than the actual solution.

9.4.4. Behavior of 8-node Quadrilateral Elements

8–node quadrilateral elements produce very good results, even with only one layer of elements. The higher number of nodes and integration points allows these elements to model the stress distribution within the beam with only one element in the cross section. Mesh refinement to three layers produces a slightly softer structure, with results very close to the Euler–Bernoulli beam solution.

Using reduced integration with the 8–node quadrilateral elements reduces the number of integration points from 9 to 4 with very little change in the results. Since there is very little change in the results, it would be advantageous to use reduced integration with these elements because the computational time is dramatically reduced. It is evident from the displacement results that the reduced integration 8–node quadrilateral elements have essentially reached a converged solution with a three–layer mesh. When compared to a 60–layer mesh (a huge increase in number of elements), very little change occurs in the results. It can also be noted that the change in the stress is much smaller than the change in the displacement between the two mesh sizes.

9.4.5. Symmetry

In the cases where half the beam was modeled with a symmetry boundary condition imposed on the symmetry plane, the results are exactly the same as in the case with the same mesh and the full beam. This result is to be expected because the beam and the solution are symmetrical. The symmetry boundary condition that was imposed was to constrain the horizontal displacement along the entire symmetry plane ($u_1 = 0$).

9.4.6. Concentrated Loads

For the linear elastic material assumption, the equations of elasticity predict infinite values of the stress at the points where concentrated loads are applied. This indicates that the stress at such locations will never achieve convergence. The boundary conditions used in this example impose a concentrated load at the corners of the beam, causing stress concentrations and a discontinuity in the deformation. The stresses further away from the concentrated load have converged, but since at the tip of the concentrated load, the predicted stresses from the elastic solution are infinite, then the finer the mesh used, the higher the values of the stress at this location. As shown in Figure 9–30, the stresses in the element where the concentrated reaction is applied are over 200 MPa, which is five times the maximum σ_{11} at the mid–span of the beam.

9.4.7. Conclusions

Analyzing this simple beam problem highlights the importance of choosing appropriate elements, integration procedures, and mesh sizes. It was seen that linear–triangular elements are not appropriate in bending unless an extremely fine mesh is used. This is because of their constant strain/stress condition. 4–node quadrilateral elements were seen to behave better than the

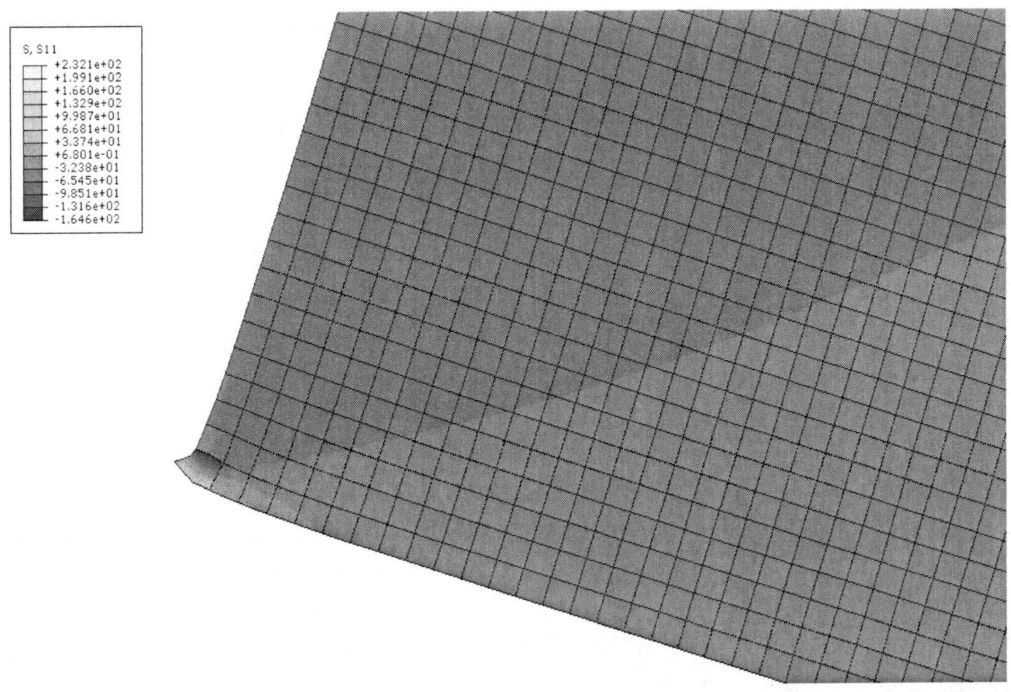

FIGURE 9-30 σ_{11} **in the 60 layers mesh using the 8 nodes reduced integration quadrilateral showing stress concentration due to the concentrated reaction.**

triangular elements, but are still too stiff for this application when a coarse mesh is used. It was determined that the 8–node quadrilateral elements produce very good results for this applica–tion, even when a coarse mesh is used. The different behavior of these elements is a result of their different shape functions. The analysis emphasizes the importance of understanding the shape functions used with each element and understanding how the elements will behave in a given situation.

While reduced integration can save on computational time, it must be applied carefully. With the 4–node quadrilateral elements, reduced integration produced a very soft structure because the one integration point is not enough to capture the behavior of a coarse mesh. It is evident that reduced integration should only be used with these elements if a finer mesh is used.

The 8–node quadrilateral elements with reduced integration seem to be the best option for this application. Using a three–layer mesh, the results are very accurate. Using reduced integration and taking advantage of symmetry saves computational time without compromising the results. In the present example, with the relatively small number of elements, the computational time is not a big factor; however, it would be more important in a larger, more complex model necessitating a finer mesh. In that case, it would be beneficial to use reduced integration and to take advantage of symmetry wherever possible.

▶ EXERCISES

1. A plane element has length of 2 units aligned with the x axis and height of 4 units that is aligned with the y axis. If a pure rigid–body rotation around the z (out of plane) axis of angle θ is applied to the element, find the corresponding variation in the magnitude of the small "apparent" strain components. Then, use a finite–element analysis software to compare your theoretical results with the software results by assigning different values for θ. Comment on the results. Then, if the software used has a geometric non–linearity option, inspect the effect of selecting this option on the magnitudes of the resulting strain components at the different chosen values for θ. (Hint: the rotation of the element can be applied by directly applying the appropriate displacements to the corner nodes.)
2. Using any finite–element analysis software of your choice, find the deflection at point A and the stress components at point B as a function of the number of elements used per the height of the beam. Use 4–node quadrilateral full integration elements. Compare with the Euler–Bernoulli beam solution. Solve twice, once with Poisson's ratio = 0 and another time with Poisson's ratio = 0.3. Consider the thickness to be 1 units of length.

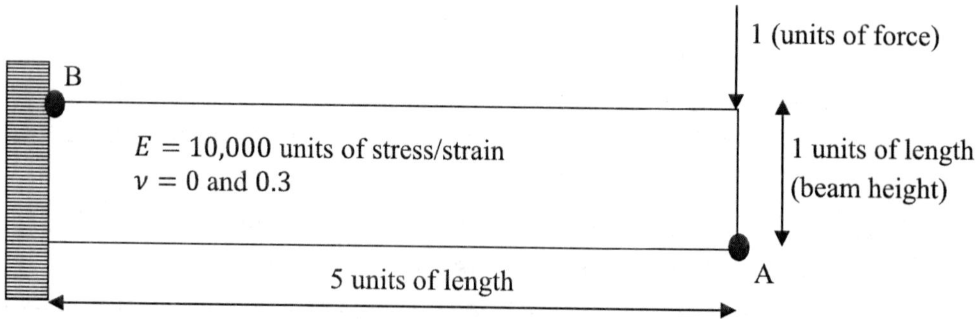

3. Solve problem 2 using the following two types of elements: Reduced–integration 4–node elements, and reduced–integration 8–node elements. Compare with the results obtained in the previous problem. Comment on the **convergence rate** (the rate of change of the results with respect to increasing the number of elements).
4. Find the Jacobian Matrix and its determinant for the following mapping for an isoparametric element

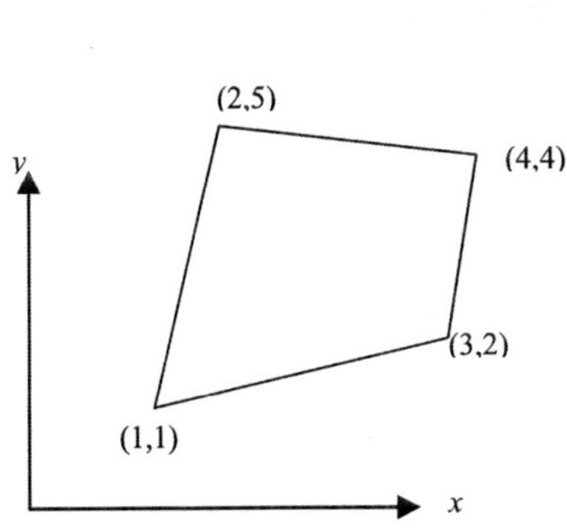

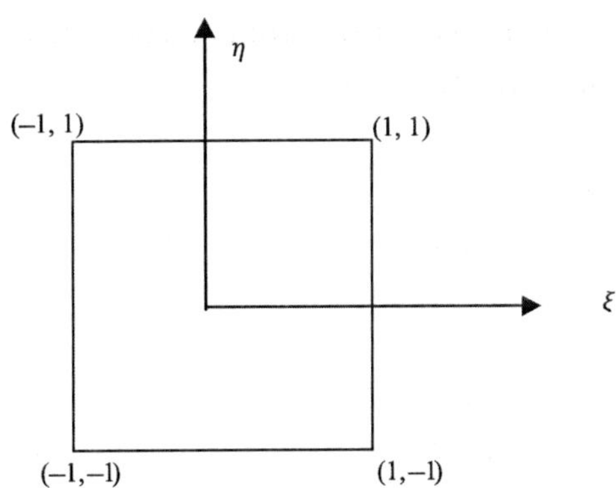

5. Use a suitable quadrature to evaluate the following integrals and compare with the exact solution. Comment on the results in reference to the finite–element analysis method integration scheme.

$$\int_{-1}^{1} 2x^2 + 3dx, \quad \int_{-1}^{1}\int_{-1}^{1} \left(2x^2+3\right)\left(2y^2+3xy\right) dxdy, \quad \int_{-1}^{1}\int_{-1}^{1} \frac{(2x^2+3)\,(2y^2+3xy)}{5+xy} dxdy$$

6. Following the example shown in section 9.3.4 use different element shapes to comment on the following:
   a. The effect of increasing the distortion of the element on the formulae for k_{11} and k_{12} before integration.
   b. The effect of increasing the distortion of the element on the difference between the exactly integrated, fully integrated, and the reduced integrated values of k_{11} and k_{12}.
   c. If the isoparametric element is rectangular in shape but is rotated in space, what is the effect of the angle of rotation on the formulae for k_{11} and k_{12} and the difference between the exactly integrated, fully integrated, and reduced integrated values of k_{11} and k_{12}.
7. The following are two main requirements for the shape functions of a 4–node quadrilateral element that has a general non–rectangular shape:
   a. The sum of all the shape functions has to be equal to unity to ensure that rigid body motion is feasible.
   b. The displacement of the element side is fully determined *only* by the displacement of the nodes to which this side is connected in a manner that ensures element compatibility.

Does the isoparametric element formulation of a general element shape ensure the above requirements?

Show that one or both of those requirements are not met if, for the example shown in section 9.3.4, either of the following two methods was used to find the shape functions:

Method 1: By assuming the following bilinear displacement across the element:

$$u = a_1 + a_2X_1 + a_3X_2 + a_4X_1X_2, \quad v = b_1 + b_2X_1 + b_3X_2 + b_4X_1X_2$$

Method 2: By assuming that the shape function of each node is equal to 1 at the specified node and zero on the two sides not connected to that node.

8. The shown structure has $A_1 = 1\,\text{m}^2, A_2 = \frac{1}{2}\,\text{m}^2, L = 5\text{m}, E = 20\,\text{GPa}, \nu = 0.2$ and is subjected to the shown pressure at the top end. Assume that the structure has a unit thickness of 1m.

    a. Show that the closed form differential equation ignoring Poisson's ratio effect is: $EA\frac{d^2u}{dx^2} + E\frac{du}{dx}\frac{dA}{dx} = 0$.

    b. Find the closed form solution using Mathematica.

    c. Compare your solution to the solution obtained using a finite–element analysis software using $\nu = 0.2$ and $\nu = 0$.

    d. Comment on the results.

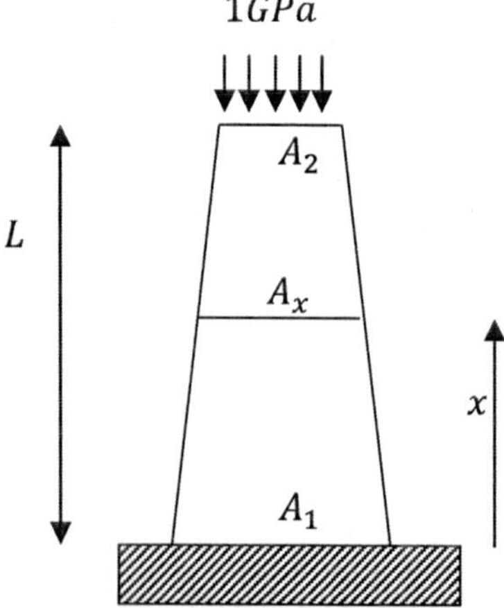